AI-Enhanced Biosurveillance:

Frameworks, Practices, and Futures

Matthew Chidozie Ogwu
https://orcid.org/0000-0001-6054-1667
Appalachian State University, USA

IGI Global
Scientific Publishing
Publishing Tomorrow's Research Today

Vice President of Editorial Melissa Wagner
Director of Acquisitions Mikaela Felty
Director of Book Development Jocelynn Hessler
Production Manager Mike Brehm
Cover Design Jose Rosado

Published in the United States of America by
IGI Global Scientific Publishing
701 East Chocolate Avenue
Hershey, PA, 17033, USA
Tel: 717-533-8845 | Fax: 717-533-7115
Website: https://www.igi-global.com E-mail: cust@igi-global.com

Library of Congress Cataloging-in-Publication Data

Names: Ogwu, Matthew Chidozie editor
Title: AI-enhanced biosurveillance : frameworks, practices, and futures / edited by Matthew Chidozie Ogwu.
Description: Hershey, PA : IGI Global Scientific Publishing, [2027] | Includes bibliographical references and index. | Summary: "The book aims to provide a comprehensive, interdisciplinary reference on the theory, design, implementation, and future trajectories of AI-enhanced biosurveillance systems"-- Provided by publisher.
Identifiers: LCCN 2026009766 (print) | LCCN 2026009767 (ebook) | ISBN 9798337390338 h/c | ISBN 9798337390345 s/c | ISBN 9798337390352 ebook
Subjects: LCSH: Environmental health--Data processing | Public health surveillance--Data processing | Artificial intelligence--Environmental aspects | Bioinformatics
Classification: LCC RA565 .A385 2026 (print) | LCC RA565 (ebook)
LC record available at https://lccn.loc.gov/2026009766
LC ebook record available at https://lccn.loc.gov/2026009767

British Cataloguing in Publication Data
A Cataloguing in Publication record for this book is available from the British Library.

Table of Contents

Detailed Table of Contents

Biosurveillance is becoming a core infrastructure for biological risk governance in an era shaped by accelerating disease emergence, antimicrobial resistance, climate change, ecological disruption, food-system complexity, and expanding human-animal-environment interfaces. This chapter establishes the conceptual and operational foundations of biosurveillance in the age of artificial intelligence by tracing the transition from traditional public health reporting to integrated, anticipatory, AI-enabled intelligence systems. It examines the historical evolution of surveillance, clarifies the distinctions among monitoring, detection, prediction, intelligence, and response, and describes the major domains of modern biosurveillance, including human health, animal health, plant health, food safety, environmental monitoring, wastewater surveillance, genomic surveillance, and biosecurity.

This chapter explores community-centered AI biosurveillance as an ethical and effective approach to modern public health surveillance. It examines how participatory models, trust-building strategies, and public engagement enhance the design, implementation, and governance of AI-driven outbreak detection systems. By integrating citizen-generated data, local knowledge, and transparent governance

standardization of methods, ethical concerns, and integration with conventional surveillance systems. Methodological harmonization and interdisciplinary collaboration are essential to strengthen their role in sustainable public health and environmental monitoring.

Outbreak detection and forecasting are central functions of biosurveillance because public-health response depends on recognizing abnormal disease activity before transmission becomes widespread and estimating how an outbreak may evolve under uncertainty. Traditional surveillance systems remain indispensable, but they are often constrained by reporting delays, incomplete case ascertainment, fragmented data streams, limited spatial resolution, and the difficulty of integrating clinical, laboratory, environmental, behavioral, and digital information in near real time. This chapter examines how machine learning can strengthen outbreak detection and forecasting by transforming heterogeneous data into actionable early-warning intelligence. It reviews core data streams, including notifiable disease reports, syndromic surveillance, electronic health records, laboratory positivity, wastewater signals, mobility data, climate and environmental predictors, genomic information, veterinary and wildlife data.

The growing complexity of biological threats necessitates advanced biosurveillance systems capable of three core functions. Traditional surveillance faces challenges due to fragmented data, delayed information sharing, and weak coordination between public health and environmental sectors. Advances in artificial intelligence, particularly in machine learning, deep learning, and data fusion, enable integration of diverse data sources such as epidemiological, environmental, remote sensing, genomic, and digital health data for active risk assessment. Explainable AI improves transparency and supports informed decision-making, while multi-source analytics strengthen early warning, situational awareness, and risk forecasting for emerging threats linked to zoonoses and climate change. Developing AI-driven biosurveillance systems with ethical, transparent, and resilient One Health approaches is essential to address issues of data quality, bias, privacy, and governance.

Facial expressions are crucial for determining one's emotional state, and they are an important factor for evaluating mental health. Facial Expression Recognition (FER) using deep learning methods, especially CNNs, is one of the fields that these models have brought to satisfactory levels of accuracy; however, their black-box nature makes it hard to rely on them for medical-related tasks. In this chapter, an interpretable deep learning application based on MobileNetV3 with transfer learning and integrated using Gradient-weighted Class Activation Mapping (Grad-CAM) is proposed. MobileNetV3 can be deployed in a lightweight and efficient way with an accuracy of 92.5% on the seven emotion classes using the FER-2013, KDEF, and MUTFER2024 datasets. Grad-CAM offers visual explanations by identifying discriminated facial regions that contribute to model predictions. The technique facilitates reliable mental health observation, therapy help, and remote patient care in AI-augmented bio surveillance systems.

Environmental contamination poses persistent risks to public health and food safety, intensified by complex exposure pathways across air, water, soil, and food systems. This chapter examines artificial intelligence–driven monitoring frameworks for detecting, predicting, and managing environmental and foodborne contaminants. Evidence shows that AI-based approaches outperform conventional methods in real-time surveillance, source attribution, and predictive risk assessment. Integration of machine learning, deep learning, IoT sensors, remote sensing, and big data analytics enables precise tracking of hazards. In food systems, AI enhances pesticide residue detection, microbial risk prediction, and supply chain traceability, reducing foodborne disease burden and strengthening governance. AI models also enable scenario analysis, exposure forecasting, and targeted interventions for vulnerable

Furqan Mazhar, Department of Pharmacy, Faculty of Health and Pharmaceutical Sciences, University of Agriculture, Faisalabad, Pakistan
Eman Fatima, Department of Pharmacy, Faculty of Health and Pharmaceutical Sciences, University of Agriculture, Faisalabad, Pakistan
Abdul Ahad Sajid, Department Chemistry, Quaid e Azam University, Islamabad, Pakistan
Esha Hamdani, Department of Pharmacy, Faculty of Health and Pharmaceutical Sciences, University of Agriculture, Faisalabad, Pakistan
Hassnain Ahmad Naqvi, University of Engineering and Technology, Lahore, Pakistan
Rubab Amjad, Department of Pharmacy, Faculty of Health and Pharmaceutical Sciences, University of Agriculture, Faisalabad, Pakistan
Abu Sulman, University of Agriculture, Faisalabad, Pakistan
Sabira Sultana, Department of Eastern Medicine, Faculty of Medical Sciences, Government College University, Faisalabad, Pakistan
Adnan Afzal, Government College University, Faisalabad, Pakistan
Afsheen Mansoor, Shaheed Zulfiqar Ali Bhutto Medical University, Islamabad, Pakistan

Early bio surveillance is being reformed by the intersection of artificial intelligence (AI), wastewater-based epidemiology (WBE), and distributed environmental sensor networks. Through machine learning, predictive analytics, and anomaly detection based on real-time environmental sampling, AI-enabled systems will provide the opportunity to identify the outbreak of an infectious disease, the trend of antimicrobial resistance, a chemical exposure, and an emerging threat of biological attacks before it manifests itself clinically. This chapter reviews the technological principles of wastewater and environmental monitoring systems that operate on AI, sensor design, edge computing, data fusion, cloud-based analytics, and digital dashboards to support the decision-making process of the general population. It discusses the potential to improve the sensitivity, specificity, and timeliness of outbreak detection using the advanced modeling methods, including deep learning, spatiotemporal forecasting, and interpreting genomic signals.

Preface

INTRODUCTION

The twenty-first century has witnessed an unprecedented convergence of biological threats, environmental change, and rapid technological innovation. Emerging infectious diseases, antimicrobial resistance, climate-driven ecological disruptions, food insecurity, and environmental contamination have exposed the limitations of traditional biosurveillance systems, which often rely on fragmented data streams, delayed reporting, and reactive decision-making. At the same time, advances in artificial intelligence (AI), genomics, remote sensing, sensor technologies, and digital infrastructures are transforming how biological risks are detected, monitored, interpreted, and managed. This convergence has given rise to a new era of biosurveillance—one that is increasingly predictive, integrated, and data-driven.

AI-Enhanced Biosurveillance: Foundations, Technologies, and Intelligent Systems provides a comprehensive exploration of the scientific, technological, and conceptual foundations that underpin this transformation. Rather than viewing biosurveillance solely as disease monitoring, this volume adopts a broader perspective that recognizes biosurveillance as an interdisciplinary enterprise encompassing human, animal, plant, and environmental health within a One Health framework. The chapters collectively examine how artificial intelligence is reshaping biosurveillance by enabling real-time data integration, automated pattern recognition, predictive analytics, and intelligent decision support across increasingly complex biological systems.

The book begins by establishing the theoretical foundations of modern biosurveillance and introducing conceptual frameworks that integrate artificial intelligence into surveillance systems. It then examines governance, intelligent data ecosystems, public trust, and participatory approaches that ensure technological innovation remains transparent, ethical, and socially responsive. Building on these foundations, the volume explores the technological infrastructure supporting next-generation biosurveillance, including genomic and metagenomic sequencing, wastewater-based epidemiology, environmental monitoring, Internet of Things devices, sensor networks, and AI-driven analytical frameworks. The latter chapters demonstrate how these technologies and advanced computational approaches are applied across diverse biosurveillance domains, including outbreak forecasting, environmental monitoring, food systems, climate-linked biological risks, antimicrobial resistance, and agricultural biosecurity, illustrating the transformative potential of AI for predictive, integrated, and evidence-based biosurveillance.

The volume concludes by presenting advanced AI methodologies, including machine learning, deep learning, explainable AI, multimodal data integration, and visualization techniques such as Gradient-weighted Class Activation Mapping (Grad-CAM). Together, these chapters demonstrate how modern computational intelligence is transforming biosurveillance from retrospective observation into proactive prediction and decision support. This volume is intended for researchers, public health professionals, environmental scientists, computer scientists, engineers, policymakers, graduate students, and practitioners working at the interface of artificial intelligence, public health, environmental monitoring, and biosecurity. It provides both conceptual understanding and technical insight into the emerging architectures that will define biosurveillance systems of the future.

CHAPTER OVERVIEW

Chapter 1: Foundations of Biosurveillance in the Age of Artificial Intelligence

This chapter establishes the foundational concepts and evolving landscape of biosurveillance within an increasingly complex global risk environment. It examines how traditional surveillance approaches have transformed into integrated, AI-enabled intelligence systems capable of supporting proactive biological risk governance. The chapter provides an overview of the historical progression of surveillance practices while distinguishing key functions such as monitoring, detection, prediction, intelligence generation, and response. By exploring interconnected biosurveillance domains—including human, animal, and plant health, food safety, environmental monitoring, wastewater analysis, genomic surveillance, and biosecurity—the chapter frames biosurveillance as a multidimensional infrastructure essential for anticipating and managing emerging biological threats.

Chapter 2: Community-Centered AI Biosurveillance: Participatory Models, Trust, and Public Engagement

This chapter explores the critical role of community participation, trust, and public engagement in the development of ethical and effective AI-driven biosurveillance systems. It highlights how participatory frameworks incorporating citizen-generated data, local expertise, and transparent governance mechanisms can improve outbreak detection while addressing limitations associated with traditional surveillance models. The chapter examines key ethical considerations surrounding privacy, consent, accountability, and equity, while emphasizing the importance of emerging approaches such as explainable AI and federated learning. Through a community-centered perspective, this chapter demonstrates how inclusive approaches can strengthen the sustainability, legitimacy, and resilience of future biosurveillance infrastructures.

Chapter 3: Genomic and Metagenomic Approaches to Biosurveillance

This chapter examines the transformative contributions of genomic and metagenomic technologies to modern biosurveillance by enabling biological threats to be identified, characterized, and tracked at unprecedented molecular resolution. It explores the role of whole-genome sequencing in pathogen transmission analysis, variant identification, and outbreak investigation, alongside the ability of metagenomic approaches to detect diverse organisms across clinical, environmental, agricultural, and wastewater contexts. The chapter further considers how AI-enhanced analytical pipelines integrate sequencing data with bioinformatics, machine learning, geospatial intelligence, and visualization tools to support actionable decision-making. By addressing technological foundations, applications, governance considerations, and future opportunities, this chapter highlights the expanding role of molecular intelligence within One Health biosurveillance systems.

Chapter 4: Wastewater-Based Epidemiology and Environmental Genomic Monitoring

This chapter examines wastewater-based epidemiology and environmental genomic monitoring as innovative approaches for population-level health assessment and environmental risk detection. It explores how wastewater and environmental samples can reveal critical indicators, including pathogens, antimicrobial resistance markers, pharmaceuticals, and chemical contaminants, providing timely insights into community health conditions and ecological changes. The chapter discusses the integration of metagenomics, sequencing technologies, and advanced analytical methods to strengthen early warning capabilities while addressing challenges related to standardization, interpretation, ethics, and system integration. By emphasizing interdisciplinary collaboration and methodological advancement, this chapter highlights the importance of environmental surveillance in sustainable public health strategies.

Chapter 5: Machine Learning Approaches for Outbreak Detection and Forecasting

This chapter investigates the application of machine learning techniques in strengthening outbreak detection and forecasting within contemporary biosurveillance systems. It explores how AI-driven models can transform diverse data sources—including disease reports, electronic health records, laboratory data, wastewater indicators, mobility patterns, environmental information, genomic data, and animal health records—into actionable early-warning intelligence. The chapter highlights the limitations of conventional surveillance approaches and demonstrates how machine learning can improve timeliness, predictive accuracy, and situational awareness. Through an examination of data integration strategies, modeling approaches, and implementation considerations, this chapter underscores the potential of AI to enhance preparedness and response to emerging biological threats.

Chapter 6: Explainable and Multi-Source AI Frameworks for Adaptive Biosurveillance in Public Health and Environmental Systems

This chapter explores the development of adaptive biosurveillance frameworks that combine explainable artificial intelligence, multi-source data integration, and One Health approaches to address increasingly complex biological risks. It examines how machine learning, deep learning, and data fusion techniques enable the integration of epidemiological, environmental, genomic, remote sensing, and digital health information for improved risk assessment and forecasting. Particular attention is given to the importance of explainability in supporting transparent decision-making and building confidence in AI-generated insights. The chapter also discusses challenges associated with data quality, bias, privacy, and governance, emphasizing the need for ethical and resilient AI infrastructures capable of supporting public health and environmental protection.

Chapter 7: Visual Interpretation of Deep Learning Decisions in Facial Expression Recognition Using Grad-CAM

This chapter examines the application of interpretable deep learning techniques to facial expression recognition and their potential contributions to AI-enhanced biosurveillance and healthcare monitoring. It explores how convolutional neural networks and lightweight architectures can achieve effective emotion classification while addressing concerns regarding the opacity of AI decision-making. Through the integration of MobileNetV3 and Gradient-weighted Class Activation Mapping (Grad-CAM), the chapter demonstrates approaches for visualizing the facial regions influencing model predictions. By emphasizing transparency and interpretability, this chapter highlights opportunities for responsible AI applications in mental health observation, remote care, and supportive clinical environments.

Chapter 8: AI-Driven Monitoring of Environmental Contaminants: Implications for Public Health and Food Safety

This chapter investigates the role of artificial intelligence in monitoring environmental contaminants and strengthening public health and food safety systems. It explores how machine learning, deep learning, Internet of Things (IoT) technologies, remote sensing, and big data analytics enable real-time detection, prediction, and management of environmental hazards across air, water, soil, and food systems. The chapter examines applications in contaminant tracking, microbial risk prediction, pesticide detection, food safety monitoring, and supply chain governance. By addressing opportunities and challenges related to data quality, interoperability, and implementation, this chapter demonstrates how AI-driven surveillance approaches can support proactive environmental risk assessment and evidence-based interventions.

Chapter 9: Environmental Microbiomics for Public Health Safety: Applications in Food Adulteration and Contamination Detection

This chapter explores the emerging field of environmental microbiomics and its applications in improving food safety, traceability, and contamination detection. It examines how microbial signatures, combined with digital technologies such as artificial intelligence, blockchain, Internet of Things networks, and cloud-based platforms, can enhance the identification of food fraud, adulteration, and contamination events. The chapter highlights the potential of spatial-temporal analytics and advanced data integration to strengthen global food supply chains while addressing challenges related to standardization, scalability, privacy, and accessibility. Through an interdisciplinary perspective, this chapter illustrates how microbiome-based intelligence can contribute to safer and more transparent food systems.

Chapter 10: AI-Enabled Early Warning Systems for Climate-Linked Biological Risks

This chapter examines the growing importance of AI-enabled predictive systems for addressing biological risks associated with climate change. It explores how machine learning, deep learning, Earth observation technologies, and real-time sensing platforms improve the prediction and management of climate-related threats affecting human health, agriculture, wildlife, and ecosystems. The chapter considers applications in vector-borne disease forecasting, zoonotic spillover prediction, agricultural pest management, and ecological risk assessment. By addressing challenges involving data availability, uncertainty, model transferability, and policy integration, this chapter emphasizes the need for scalable and explainable AI frameworks capable of supporting climate-resilient biosurveillance.

Chapter 11: AI-Enabled Wastewater and Environmental Sensor Networks for Early Biosurveillance: Opportunities, Limitations, and Equity Implications

This chapter explores the integration of artificial intelligence, wastewater-based epidemiology, and environmental sensor networks as emerging approaches for early biological threat detection. It examines how AI-enabled systems leverage machine learning, predictive analytics, anomaly detection, edge computing, and data fusion to identify infectious disease trends, antimicrobial resistance patterns, chemical exposures, and potential biological hazards before widespread clinical manifestation. The chapter discusses technological foundations, including sensor design, genomic analysis, cloud-based platforms, and digital dashboards, while considering limitations related to equity, accessibility, governance, and implementation. Through this analysis, the chapter highlights the opportunities and challenges of deploying intelligent environmental surveillance systems for population health protection.

Chapter 12: Biosurveillance of Food Systems: Leveraging Digital Technology for Contamination Detection and Supply Chain Resilience

This chapter examines how digital technologies are transforming food system biosurveillance by improving contamination detection, traceability, and supply chain resilience. It explores the integration of IoT sensors, artificial intelligence analytics, blockchain technologies, and big data modeling to support rapid identification of microbial and chemical hazards throughout farm-to-fork pathways. The chapter highlights how digital platforms enhance transparency, coordination, and predictive capabilities among stakeholders involved in food safety and public health. It also considers ongoing challenges related to infrastructure, data governance, and algorithmic bias, emphasizing the importance of inclusive and trustworthy approaches to modern food surveillance.

Chapter 13: Plant-Based RNA Interference Powered by AI for Agricultural Safety

This chapter explores the intersection of artificial intelligence and plant RNA interference technologies as an emerging strategy for advancing agricultural safety and sustainability. It examines how AI-driven approaches enhance gene target identification, RNA sequence prediction, delivery optimization, and the development of environmentally responsible crop protection solutions. The chapter highlights the potential of AI-supported RNA interference to improve disease resistance, pest management, and crop resilience while reducing reliance on conventional chemical interventions. Through its focus on biotechnology innovation, this chapter demonstrates how intelligent systems can support sustainable agricultural practices and strengthen plant health surveillance.

Chapter 14: Antimicrobial Resistance Surveillance Using Artificial Intelligence and Big Data

This chapter examines the transformative potential of artificial intelligence and big data analytics in advancing antimicrobial resistance surveillance through a comprehensive One Health framework. It explores how AI can integrate microbiological, genomic, clinical, environmental, agricultural, and public health data streams to create more predictive and responsive surveillance systems. The chapter emphasizes that AI functions as an analytical enhancement rather than a replacement for microbiological expertise, enabling faster interpretation, broader data integration, and improved decision support. It also addresses key implementation challenges, including data bias, interoperability limitations, algorithmic transparency, privacy concerns, and inequities in surveillance capacity, highlighting the importance of responsible AI governance in combating global antimicrobial resistance.

CONCLUSION

I extend my sincere gratitude to all contributing authors, reviewers, and the editorial and production teams at IGI Global Scientific Publishing for their dedication and expertise. Their collective efforts have produced a timely and comprehensive resource that reflects the multidisciplinary nature of modern biosurveillance. It is my hope that this volume will serve not only as a reference for current scholarship but also as a catalyst for future innovation, interdisciplinary collaboration, and evidence-based policymaking in AI-enabled biosurveillance.

Matthew Chidozie Ogwu
Appalachian State University, USA

Chapter 1
Foundations of Biosurveillance in the Age of Artificial Intelligence

Matthew Chidozie Ogwu
http://orcid.org/0000-0001-6054-1667
Appalachian State University, USA

ABSTRACT

Biosurveillance is becoming a core infrastructure for biological risk governance in an era shaped by accelerating disease emergence, antimicrobial resistance, climate change, ecological disruption, food-system complexity, and expanding human-animal-environment interfaces. This chapter establishes the conceptual and operational foundations of biosurveillance in the age of artificial intelligence by tracing the transition from traditional public health reporting to integrated, anticipatory, AI-enabled intelligence systems. It examines the historical evolution of surveillance, clarifies the distinctions among monitoring, detection, prediction, intelligence, and response, and describes the major domains of modern biosurveillance, including human health, animal health, plant health, food safety, environmental monitoring, wastewater surveillance, genomic surveillance, and biosecurity.

1.0 INTRODUCTION

Biosurveillance has become a foundational requirement for biological risk governance in a world shaped by accelerating disease emergence, ecological disruption, climate instability, global mobility, food-system complexity, antimicrobial resistance, and expanding human-animal-environment interfaces. Traditionally, surveillance systems were designed to collect, verify, and report disease events after they became

DOI: 10.4018/979-8-3373-9033-8.ch001

clinically visible. In the contemporary risk landscape, however, delayed detection is no longer sufficient. Biological threats can move rapidly across borders, sectors, and ecosystems, and by the time formal reporting systems confirm an event, the window for low-cost containment may already have narrowed. The central task of modern biosurveillance is therefore not merely to document biological events, but to detect weak signals early, interpret them in context, forecast plausible trajectories, and support timely decisions.

Artificial intelligence is reshaping this field by expanding the analytical capacity of biosurveillance systems. AI can process large, heterogeneous, and rapidly changing data streams, including clinical reports, laboratory data, genomic sequences, wastewater signals, mobility patterns, social media, news feeds, environmental sensors, remote sensing outputs, and veterinary or agricultural records. Reviews of epidemic early warning systems emphasize that AI can help detect outbreak signals from high-volume open-source and formal data streams, especially where conventional health-system reporting is slow, fragmented, or under-resourced (MacIntyre et al., 2023).

Yet AI does not eliminate the fundamental challenge of biosurveillance: transforming uncertain observations into trusted action. Biological signals are noisy. A spike in search queries may reflect anxiety rather than disease incidence. A wastewater signal may indicate community infection trends but not individual risk. Genomic detection may identify a pathogen but not necessarily its transmission dynamics. Remote sensing may identify environmental suitability but not pathogen presence. A strong biosurveillance system must therefore combine AI-enabled pattern recognition with epidemiological reasoning, ecological knowledge, laboratory validation, institutional coordination, and accountable decision-making.

This chapter aims to provide a foundation for understanding biosurveillance in the age of artificial intelligence. It traces the historical evolution of surveillance, defines core concepts, describes major biosurveillance domains and data streams, explains the biosurveillance intelligence cycle, and evaluates AI as an enabling technology for early warning, One Health integration, data interoperability, analytical modelling, ethical governance, and operational response. The chapter's central argument is that AI-enhanced biosurveillance should be understood as a socio-technical system rather than a purely computational system. Its effectiveness depends not only on algorithms, but also on data quality, institutional trust, workforce capacity, governance arrangements, equity, communication, and the ability to convert alerts into timely interventions.

2.0. HISTORICAL EVOLUTION OF BIOSURVEILLANCE

The origins of biosurveillance lie in public health surveillance: the systematic collection, analysis, interpretation, and dissemination of health-related data for action. Early systems depended heavily on clinical observation, laboratory confirmation, mandatory reporting, and institutional communication between local, national, and international health authorities. These systems were essential for identifying outbreaks of notifiable diseases, monitoring endemic infections, tracking mortality, and supporting vaccination or containment programs. However, they were often reactive and slow, especially where laboratory capacity, reporting infrastructure, or political transparency were limited. The late twentieth and early twenty-first centuries brought a series of biological events that exposed the need for more agile systems: HIV/AIDS, SARS, pandemic influenza, Ebola, Zika, COVID-19, recurring foodborne outbreaks, antimicrobial resistance, zoonotic spillovers, and biosecurity incidents. Morse (2012) emphasized that emerging infectious diseases and deliberate biological threats revealed persistent vulnerabilities in global surveillance systems and the need for strengthened detection capacity.

Syndromic surveillance expanded biosurveillance beyond laboratory-confirmed diagnoses by tracking symptom patterns, emergency visits, pharmacy sales, absenteeism, and other early indicators that may precede formal case confirmation. This approach showed that surveillance could operate upstream of diagnosis, creating opportunities for earlier detection and response. However, it also introduced persistent challenges, including high false-positive rates, weak specificity, difficulty interpreting non-diagnostic signals, and uncertainty about when alerts should trigger action. Digital disease detection and internet-based biosurveillance further widened the surveillance landscape. Systems such as ProMED-mail, HealthMap, GPHIN, BioCaster, MedISys, and related platforms demonstrated that news reports, informal communications, and digital sources can reveal outbreaks before formal notification. Systematic reviews of internet-based biosurveillance emphasize the value of combining formal and informal sources for earlier detection and situational awareness, while also noting challenges related to data reliability, duplication, language bias, verification, and interpretation (Barboza et al., 2013; Velasco et al., 2014).

One Health and planetary health surveillance then pushed biosurveillance beyond human clinical systems. Biological risks do not respect administrative or disciplinary boundaries. Human disease emergence is shaped by wildlife reservoirs, livestock systems, land-use change, climate variability, food supply chains, antimicrobial use, environmental contamination, and ecological disturbance. Surveillance therefore increasingly includes integrated monitoring of animals, plants, food systems, water, soil, air, vectors, microbial communities, and ecosystems. This transition marks the shift from surveillance as a health-sector activity to biosurveillance as a cross-sectoral intelligence function.

Predictive and AI-enhanced biosurveillance now defines the current frontier. Unlike earlier systems that primarily detected events after they occurred, AI-enabled systems aim to integrate multiple data streams, identify anomalies, forecast risk, detect patterns at scale, and support anticipatory action. The strongest lesson from the history of biosurveillance is that each stage has expanded the signal space: from clinical cases to syndromes, from syndromes to digital events, from digital events to One Health data streams, and now from fragmented signals to AI-assisted biological intelligence. Figure 1 summarizes the historical transition from traditional disease reporting to AI-enabled biosurveillance. It shows how surveillance expanded from confirmed clinical cases to syndromic signals, digital disease detection, One Health data streams, and finally integrated AI-assisted biological intelligence.

Figure 1. Evolution of biosurveillance toward AI-enabled biological intelligence

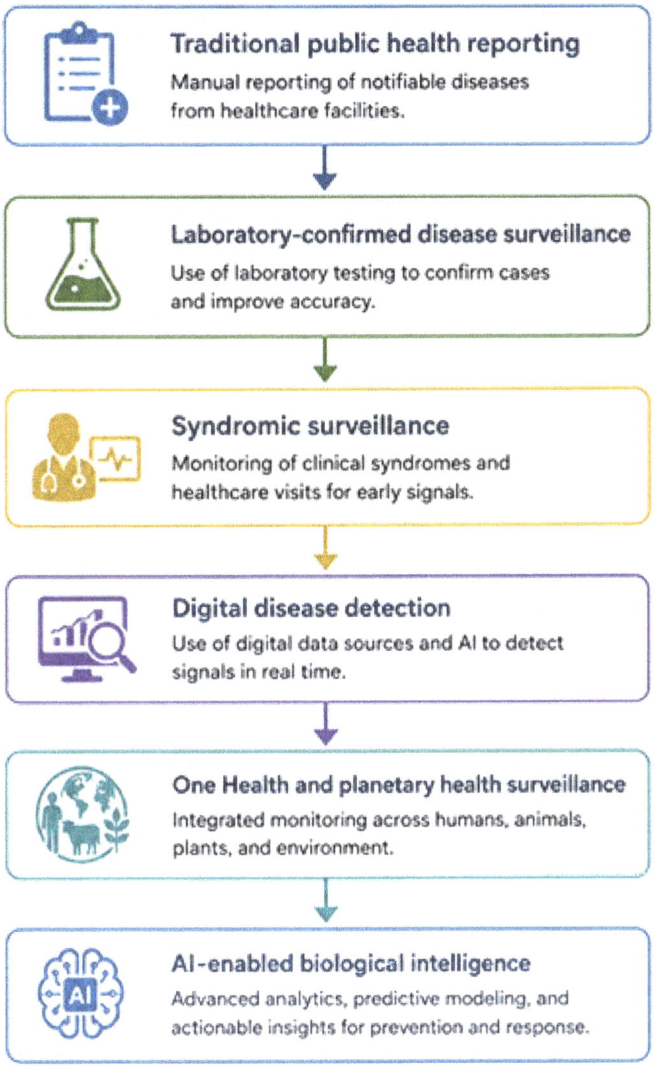

3.0. CONCEPTUAL FOUNDATIONS OF BIOSURVEILLANCE

Biosurveillance is often used broadly, but conceptual precision matters. Surveillance refers to structured observation for public action. Monitoring refers to repeated measurement of indicators over time. Detection refers to identifying signals that may indicate a biological event. Intelligence refers to the interpretation of signals in relation to risk, uncertainty, context, and decision needs. Response refers to the actions taken to prevent, contain, mitigate, or recover from a biological threat. Biosurveillance integrates these elements into an actionable cycle. The distinction between detection, prediction, and prevention is especially important. Detection identifies that something may be happening. Prediction estimates what may happen next. Prevention acts before harm escalates. AI is often strongest at detection and prediction, but prevention depends on institutions, resources, legitimacy, communication, and political will. A model may detect an anomaly, but it cannot by itself mobilize testing, isolate cases, regulate food systems, vaccinate livestock, restrict contaminated products, repair wastewater infrastructure, or build community trust.

Biosurveillance systems commonly combine event-based and indicator-based approaches. Event-based surveillance uses unstructured or semi-structured information, such as media reports, rumors, online signals, and community alerts, to detect unusual health events. Indicator-based surveillance uses predefined variables, such as laboratory results, hospital admissions, confirmed cases, mortality counts, antimicrobial resistance profiles, or vector density. The most robust systems combine both. Event-based approaches may detect unusual signals early, while indicator-based systems provide greater specificity and continuity. Sentinel surveillance, syndromic surveillance, genomic surveillance, environmental surveillance, and digital surveillance each contribute differently. Sentinel systems monitor selected sites or populations in depth. Syndromic systems prioritize early signal detection. Genomic systems reveal pathogen evolution, variant emergence, transmission pathways, and antimicrobial resistance markers. Environmental systems, including wastewater and air monitoring, provide population-level signals. Digital systems expand situational awareness through open-source and informal data.

A foundational concept in biosurveillance is the signal-to-noise problem. Biological systems are variable, data streams are incomplete, and social behavior influences reporting. False positives can create alarm fatigue, waste resources, and erode credibility. False negatives can delay response and allow escalation. Effective biosurveillance requires calibrated thresholds, uncertainty communication, validation pathways, and feedback loops. Reviews of early warning systems show that these systems are valuable but vary in effectiveness depending on design, data quality, timeliness, evaluation criteria, and operational integration (Meckawy et al., 2022).

4.0. SCOPE AND DOMAINS OF BIOSURVEILLANCE

Modern biosurveillance is inherently multidomain. Human health biosurveillance remains central, tracking infectious diseases, syndromes, hospital burden, mortality, vaccination gaps, antimicrobial resistance, and emerging health threats. However, human health signals are often late-stage indicators of broader biological processes. Zoonotic risk may emerge first in wildlife, livestock, vector populations, land-use change, or environmental conditions before appearing as human illness. Animal health surveillance includes livestock diseases, wildlife mortality, zoonotic reservoirs, companion animal signals, veterinary diagnostics, and production system indicators. It is essential for detecting spillover risk, protecting food security, and preventing economic losses. Plant health and agricultural biosurveillance track crop pathogens, invasive pests, plant stress, soil microbial changes, and climate-sensitive disease pressures. These systems are vital because plant disease can destabilize food systems, livelihoods, trade, and ecological resilience.

Food safety biosurveillance monitors contamination, adulteration, foodborne pathogens, supply-chain disruptions, antimicrobial residues, toxins, and production risks. Environmental biosurveillance tracks biological and chemical hazards in wastewater, surface water, drinking water, soil, air, sediments, and built environments. Wastewater surveillance, for example, provides community-level detection of pathogens and can reveal trends that precede clinical reporting. Genomic biosurveillance cuts across all domains. Sequencing can identify pathogens, track variants, distinguish outbreak clusters, monitor resistance genes, detect novel organisms, and connect human, animal, food, and environmental samples. However, genomic surveillance is only useful when linked to metadata, epidemiological interpretation, laboratory quality control, and decision pathways.

Biosecurity and biodefense represent a further domain, focusing on deliberate biological threats, laboratory accidents, dual use risks, and national or global security implications. This domain requires special attention to ethics and safeguards because biosurveillance can protect populations while also raising concerns about privacy, civil liberties, data misuse, and geopolitical tensions. The unifying framework is One Health. Human, animal, plant, and environmental health are interconnected through shared ecosystems, mobility networks, food systems, microbial flows, and climate processes. AI-enhanced biosurveillance is strongest when it integrates these domains rather than treating them as separate reporting silos. Food safety biosurveillance is central to biological risk governance because contamination, adulteration, antimicrobial residues, microbial hazards, and supply-chain failures can produce public-health threats that require early detection and coordinated response (Ogwu et al., 2024).

Modern biosurveillance extends far beyond traditional disease reporting to encompass interconnected human, animal, plant, environmental, and biosecurity systems. Each surveillance domain contributes complementary information that strengthens early warning, risk assessment, and coordinated response. Table 1 summarizes the principal domains of contemporary biosurveillance, their primary surveillance objectives, representative data sources, and their contributions to integrated AI-enabled One Health intelligence. Plant health biosurveillance is also essential for food-system resilience, particularly as fungal pathogens, crop disease pressure, and emerging control technologies increasingly intersect with public health, ecological sustainability, and agricultural productivity (Ogwu & Izah, 2025).

Table 1. Major domains of modern biosurveillance and their roles in AI-enabled one health intelligence

Biosurveillance Domain	Primary Focus	Representative Data Sources	Contribution to AI-Enabled Biosurveillance
Human health	Infectious diseases, syndromes, AMR	Clinical records, hospitals, laboratories	Disease detection and public health response
Animal health	Livestock, wildlife, zoonoses	Veterinary reports, animal diagnostics	Spillover detection and food security
Plant health	Crop diseases, invasive pests	Field surveillance, remote sensing	Agricultural resilience and food protection
Food safety	Foodborne hazards, contaminants	Inspection records, supply chains	Food system surveillance
Environmental surveillance	Water, wastewater, soil, air	Environmental sensors, eDNA	Population-level early warning
Genomic surveillance	Pathogen evolution and resistance	Whole-genome sequencing, metagenomics	Variant detection and transmission analysis
Biosecurity	Deliberate or accidental biological threats	Intelligence systems, laboratory networks	National and global biological security

5.0. CORE DATA STREAMS IN MODERN BIOSURVEILLANCE

Biosurveillance depends on data diversity. No single stream is sufficient. Clinical and hospital reporting data provide structured information on symptoms, diagnoses, admissions, intensive care use, mortality, and healthcare utilization. Laboratory diagnostics provide pathogen confirmation, resistance profiles, and test positivity. Genomic data provide evolutionary and transmission information. Wastewater data provide population-level signals independent of individual healthcare-seeking behavior. Veterinary, livestock, and wildlife data reveal risks at the animal-human interface. Plant disease, crop yield, pest surveillance, and agricultural production data reveal food-system and ecological vulnerabilities. Food inspection data, supply-chain records, and contamination reports identify hazards in production, distribution, and consumption systems. Climate and environmental data—temperature, rainfall, humidity, flooding, drought, vegetation, land-use change, and habitat disruption—help explain biological risk drivers. Environmental biosurveillance depends on sustained biomonitoring because pollutants, pathogens, and ecological stressors can interact across water, soil, air, sediments, organisms, and human exposure pathways (Izah et al., 2024).

Mobility, travel, trade, and demographic data are critical for modelling spread. Social media, search trends, news reports, and informal online signals can provide early clues, especially where formal reporting is delayed. Systematic reviews of social media and internet-based surveillance show that these sources can support public health monitoring, but also require careful validation because digital data are shaped by attention, media cycles, platform bias, and population access (Charles-Smith et al., 2015; Velasco et al., 2014). Remote sensing, satellite imagery, drones, and sensor networks add spatial intelligence. They can identify environmental conditions associated with vector expansion, crop disease, flooding, algal blooms, habitat disruption, urban heat, vegetation stress, and ecological change. When paired with field surveillance and AI models, these data streams allow systems to move from simple reporting to risk mapping and forecasting.

The major challenge is integration. Data streams differ in format, scale, latency, reliability, ownership, and meaning. A clinical diagnosis, a genome sequence, a satellite pixel, a wastewater concentration, and a social media post are not directly equivalent. AI can help integrate them, but only if metadata, standards, quality control, and interpretive frameworks are robust. The detection, risk analysis, and monitoring of contaminants across agro-aqua food production systems illustrate how biosurveillance must integrate laboratory evidence, environmental exposure pathways, food-chain data, and One Health interpretation (Ogwu et al., 2024).

6.0. THE BIOSURVEILLANCE INTELLIGENCE CYCLE

The biosurveillance intelligence cycle begins with problem framing. A system must first define the biological risk it is designed to detect, whether that risk involves a known pathogen, an unknown syndrome, antimicrobial resistance, food contamination, crop disease, vector expansion, wildlife mortality, or broader biological instability. Poorly framed systems collect data without clear decision relevance, whereas strong systems begin with risk questions that determine which data are needed, which signals matter, and which actions should follow.

Data acquisition follows from this problem framing. Relevant information may come from routine reporting, targeted sampling, automated data scraping, sensor capture, sequencing, remote observation, laboratory systems, or community reporting. Once collected, these data require validation and harmonization. They must be cleaned, standardized, geocoded, deduplicated, quality-checked, and linked to appropriate metadata. This work is often the least visible part of biosurveillance, but it is also one of the most important. Weak data governance can make even advanced AI models unreliable.

Signal detection then converts harmonized data into potential alerts. Statistical methods, anomaly detection, machine learning, natural language processing, and expert review can identify unusual patterns that may indicate emerging biological risk. These signals must then undergo risk assessment, where they are interpreted according to severity, plausibility, uncertainty, exposure, vulnerability, and response capacity. Forecasting and scenario generation extend this process by estimating possible spread, escalation, or impact.

Decision support translates interpreted signals into practical action. Depending on the threat, this may include further sampling, public advisories, clinical alerts, vector control, food recalls, quarantine, vaccination, movement restrictions, environmental interventions, or escalation to emergency response. Communication is equally important because biosurveillance findings must be conveyed clearly to policymakers, practitioners, affected communities, and the public.

Feedback and learning complete the cycle. Biosurveillance systems must ask whether an alert predicted a real event, whether action was timely, whether false alarms were excessive, whether vulnerable communities were protected, whether signals were missed, and whether data biases were identified. Without feedback, biosurveillance systems become static and brittle. AI models especially require continuous evaluation because biological systems, human behavior, and reporting practices change over time. Effective biosurveillance is a continuous intelligence process that transforms diverse biological observations into timely public health action. Artificial intelligence enhances each stage by improving data integration, signal detection, predictive analytics, and adaptive learning while maintaining expert oversight and institutional accountability (Table 2)

Table 2. Stages of the AI-enabled biosurveillance intelligence cycle

Intelligence Stage	Primary Activities	AI Contribution	Expected Output
Problem framing	Define surveillance objectives	Prioritize risks using historical evidence	Surveillance strategy
Data acquisition	Collect multisource data	Automated data integration	Harmonized datasets
Data processing	Cleaning, validation, harmonization	Data quality assessment	Reliable analytical inputs
Signal detection	Identify anomalies and trends	Machine learning and anomaly detection	Early warning signals
Risk assessment	Evaluate severity and uncertainty	Predictive modelling	Risk characterization
Forecasting	Project future trajectories	Scenario modelling	Short- and long-term forecasts
Decision support	Recommend interventions	Decision-support systems	Public health actions
Feedback and learning	Evaluate outcomes	Model updating and continual learning	Improved surveillance performance

7.0. ARTIFICIAL INTELLIGENCE AS A BIOSURVEILLANCE ENABLER

AI contributes to biosurveillance by extending human analytical capacity. Machine learning can detect anomalies, classify signals, forecast trends, and integrate complex datasets. Deep learning can analyze images, genomic sequences, satellite data, acoustic records, and clinical text. Natural language processing can extract disease events from news reports, clinical notes, social media, and public health documents. Computer vision can identify vectors, pests, livestock lesions, crop disease symptoms, wildlife, or environmental changes. AI is particularly useful when biosurveillance systems must process high-volume, high-velocity, high-variety data. Epidemic early warning systems increasingly depend on automated collection and analysis of open-source information, and AI can reduce the burden of manual scanning while improving the speed of signal recognition (MacIntyre et al., 2023).

However, AI should be treated as augmentation, not replacement. Expert interpretation remains essential because biological signals are context-dependent. An algorithm may detect a cluster of fever-related terms, but local experts may know whether it reflects disease, seasonal behavior, media attention, migration, conflict, or a reporting artifact. AI can prioritize signals, but humans must interpret consequences, ethics, feasibility, and response trade-offs. AI can also support genomic surveillance. Models can classify sequences, identify variants, detect resistance markers, infer phylogenetic relationships, and support rapid pathogen characterization. In environmental and agricultural biosurveillance, AI can analyze imagery for crop stress, vector habitat, pest outbreaks, or ecological disturbance. In One Health antimicrobial resistance surveillance, AI is increasingly being explored as a way to integrate human, animal, and environmental data streams, although challenges remain in data interoperability, representativeness, and implementation. AI-enhanced biosurveillance is especially relevant for tropical disease systems, where environmental variability, vector ecology, limited diagnostic access, and fragmented health infrastructure often delay detection and response (Ogwu & Izah, 2025).

The strongest AI-enabled biosurveillance systems are hybrid systems. They combine automated detection, statistical inference, domain expertise, laboratory confirmation, field intelligence, and institutional protocols. The danger is techno-solutionism: assuming that better algorithms automatically produce better preparedness. They do not. AI improves biosurveillance only when embedded in trustworthy systems capable of action.

8.0. AI-ENHANCED EARLY WARNING SYSTEMS

Early warning systems are designed to identify emerging threats before they become large-scale crises. In biosurveillance, early warning depends on detecting weak signals, distinguishing meaningful anomalies from background variation, validating signals quickly, and connecting alerts to response protocols. AI can strengthen each step by scanning large datasets, identifying unusual patterns, generating forecasts, and prioritizing signals for expert review. Epidemic early warning systems can use AI to detect unusual symptom patterns, changes in healthcare utilization, online reports, mobility-linked risk, climate-sensitive vector suitability, wastewater signals, or genomic anomalies. MacIntyre et al. (2023) argue that AI-enabled epidemic early warning has the potential to be transformative, particularly because it can analyze vast open-source data with less dependence on slow formal reporting systems.

However, early warning is not simply early detection. A warning must be meaningful to decision-makers. It must indicate what may be happening, where, with what confidence, and what should be done. The practical architecture of early warning therefore includes thresholds, triggers, escalation levels, verification protocols, and communication channels. A weak signal may trigger intensified sampling; a stronger signal may trigger laboratory confirmation; a confirmed signal may trigger public health intervention.

AI can also strengthen climate-sensitive disease forecasting. Vector-borne diseases are influenced by temperature, rainfall, humidity, vegetation, hydrology, land use, and human mobility. Machine learning can combine these drivers with surveillance data to identify areas of rising risk. In food systems, AI can detect contamination signals, supply-chain disruptions, or inspection anomalies. In crop systems, AI can identify pest and pathogen risk based on remote sensing, climate data, and field reports. The central challenge is calibration. Overly sensitive systems generate too many alerts. Overly conservative systems miss early signals. Effective early warning must balance sensitivity, specificity, timeliness, interpretability, and actionability. Systematic reviews show that early warning systems can support outbreak detection, but their effectiveness depends strongly on data source, evaluation design, alert verification, and integration with response systems (Meckawy et al., 2022).

9.0. INTEGRATION WITH ONE HEALTH AND PLANETARY HEALTH

AI-enhanced biosurveillance must be grounded in One Health because biological risk emerges across human, animal, plant, and environmental systems. Zoonotic spillover can be influenced by deforestation, wildlife trade, livestock intensification, climate change, urban expansion, agricultural practices, and human mobility. Antimicrobial resistance is shaped by antimicrobial use in humans, animals, agriculture, aquaculture, wastewater, and environmental reservoirs. Foodborne disease links production, processing, distribution, consumption, and environmental contamination. Planetary health expands this logic further by recognizing that human health depends on Earth-system stability. Climate change can expand vector ranges, alter pathogen ecology, intensify disasters, and disrupt food and water systems. Biodiversity loss can change host-pathogen dynamics. Land-use change can increase contact between humans, livestock, and wildlife. Pollution can alter microbial communities and weaken ecosystem resilience. Biosurveillance therefore must track not only disease outcomes but also ecological drivers.

One Health biosurveillance requires cross-sectoral data sharing among ministries of health, agriculture, environment, wildlife, food safety, water, trade, transportation, and security. This is difficult because institutions often operate with different mandates, data systems, legal frameworks, and cultures of evidence. AI can help integrate data, but integration is primarily an institutional problem before it is a technical one. A strong One Health biosurveillance architecture includes shared risk definitions, interoperable data standards, joint investigation teams, cross-sector dashboards, common alert thresholds, coordinated communication, and mutual accountability. Without these elements, AI may only automate existing silos. A One Health biosurveillance framework is also necessary for tracking biological and chemical risks associated with agrochemicals, because agricultural inputs can affect human health, animal systems, microbial communities, food safety, and environmental quality simultaneously (Ogwu & Izah, 2023).

Air-quality surveillance also belongs within One Health biosurveillance because airborne pollutants interact with respiratory vulnerability, microbial exposure, ecological degradation, and environmental justice concerns (Izah et al., 2024). AI applications in fisheries and aquatic systems further illustrate how biosurveillance can integrate ecological monitoring, animal health, food security, environmental change, and decision support across aquatic production systems (Izah et al., 2026).

10.0. DATA ARCHITECTURE AND INTEROPERABILITY

The most advanced biosurveillance systems are only as strong as their data architecture. Interoperability refers to the ability of different systems to exchange, interpret, and use data meaningfully. This requires standardized formats, metadata, ontologies, controlled vocabularies, identifiers, geospatial referencing, laboratory standards, and governance protocols. Biosurveillance data are difficult to integrate because they are generated at different scales. A hospital record may be individual-level and time-stamped. A wastewater sample may represent a community catchment. A genome sequence may require laboratory metadata. A satellite image may describe environmental conditions at spatial resolutions ranging from meters to kilometers. A veterinary report may be farm-level. A news item may be ambiguous in location and time. AI models need these data to be connected, but careless integration can produce false precision.

FAIR data principles—findable, accessible, interoperable, reusable—are important, but biosurveillance also requires responsible restriction. Some data are sensitive: human health records, genomic information, farm locations, wildlife poaching risk, pathogen sequences, infrastructure vulnerabilities, and community-level signals. Cybersecurity is therefore central. Biosurveillance systems must protect against breaches, misinformation, ransomware, adversarial manipulation, and misuse. Cloud computing enables scalable analysis and centralized dashboards, while edge computing allows data processing near the point of collection, useful in remote or low-bandwidth settings. Distributed architectures may be preferable when data sovereignty, privacy, or institutional constraints prevent centralized storage. Real-time dashboards are valuable, but they must communicate uncertainty, not simply produce attractive visualizations.

The unspoken weakness of many biosurveillance systems is that they are built as pilot projects rather than durable infrastructure. Sustainable architecture requires maintenance funding, trained personnel, governance agreements, evaluation protocols, and long-term institutional ownership.

11.0. ANALYTICAL METHODS AND MODELLING APPROACHES

Biosurveillance uses a spectrum of analytical methods. Descriptive analytics summarize trends, distributions, and anomalies. Statistical outbreak detection methods identify deviations from expected baselines. Spatiotemporal models map risk across geography and time. Network analysis can represent transmission pathways, mobility flows, trade routes, food supply chains, or animal movement. Genomic epidemiology links sequence variation to pathogen evolution and transmission. Ecological niche models and species distribution models can estimate environmental suitability for vectors, hosts, pathogens, or invasive pests. Agent-based models simulate individual or institutional behavior under different scenarios. Compartmental models estimate disease dynamics. Ensemble models combine multiple approaches to improve robustness. Scenario planning explores plausible futures under climate change, land-use change, or policy interventions.

AI methods increasingly sit across these approaches. Machine learning can detect nonlinear relationships and high-dimensional patterns. Deep learning can analyze images, sequences, and audio. Natural language processing can extract events from unstructured text. Reinforcement learning may support adaptive decision systems, although its use in public health must be carefully governed. Model validation is essential. A model that performs well retrospectively may fail under new conditions. Outbreaks, ecological disturbances, climate extremes, policy changes, and social behavior can shift relationships among variables. Model drift is therefore a major concern. Strong biosurveillance systems must evaluate models across space, time, populations, and data sources.

Uncertainty quantification is not optional. Decision-makers need to know whether an alert is strong, weak, speculative, or urgent. Confidence intervals, probability ranges, sensitivity analyses, scenario comparisons, and transparent assumptions improve trust. AI models that cannot explain why they generated an alert may be less useful in high-stakes biosurveillance than simpler models with clearer interpretation. Risk interpretation in biosurveillance requires robust statistical reasoning because surveillance models must distinguish meaningful signals from background noise, uncertainty, confounding, and uneven sampling intensity (Ogwu et al., 2025).

12.0. GOVERNANCE, ETHICS, AND RESPONSIBLE AI

Biosurveillance is a public good, but it is also a governance challenge. It involves observing populations, environments, animals, food systems, and digital behavior. This raises questions about privacy, consent, trust, equity, sovereignty, accountability, and dual-use risk. AI intensifies these issues because it can scale observation, automate classification, and generate decisions from data that individuals or communities may not know are being used. Privacy concerns are especially important in clinical, mobility, social media, and genomic surveillance. Wastewater surveillance may not identify individuals, but it can stigmatize communities if poorly communicated. Agricultural or wildlife surveillance can affect livelihoods, trade, and land-use decisions. Social media surveillance can blur the boundary between public communication and behavioral monitoring.

Responsible AI requires transparency, explainability, fairness, accountability, and human oversight. Algorithms should be evaluated for bias across geography, language, income, race, gender, species, ecosystem type, and data availability. Many AI systems are trained on data from well-resourced settings, which may make them less reliable in low-resource environments where surveillance needs are often greatest. Data sovereignty is also central. Indigenous communities, local communities, and nations may have rights over biological, genomic, environmental, and cultural data. Extractive data practices can reproduce colonial patterns in which samples and data are collected from vulnerable places but interpreted and monetized elsewhere. Ethical biosurveillance must include benefit-sharing, consent, governance participation, and respect for local knowledge.

Dual-use risks must be acknowledged carefully. Biosurveillance systems can protect societies from biological threats, but some data and analytical capacities could be misused. Governance frameworks must therefore balance openness with safeguards. The strongest systems are transparent enough to support trust and scientific scrutiny, but secure enough to prevent harmful misuse. Responsible AI-enabled biosurveillance requires governance frameworks that balance technological innovation with ethical, legal, scientific, and societal considerations (Table 3). These principles promote trustworthy surveillance systems capable of supporting equitable, transparent, and evidence-based decision-making while protecting individual rights and public confidence.

Table 3. Core principles for responsible AI governance in biosurveillance

Governance Principle	Purpose	Practical Implementation
Scientific validity	Ensure analytical robustness	Model validation, benchmarking, external evaluation
Transparency and explainability	Improve trust and interpretation	Explainable AI and transparent reporting
Privacy and data governance	Protect sensitive information	Secure data management and regulatory compliance
Fairness and equity	Reduce algorithmic bias	Inclusive datasets and equity assessments

continued on following page

Table 3. Continued

Governance Principle	Purpose	Practical Implementation
Interoperability	Enable cross-sector integration	Standardized data formats and FAIR principles
Human oversight	Maintain expert accountability	Human-in-the-loop decision-making
Sustainability	Support long-term operation	Workforce development and infrastructure investment
Actionability	Translate intelligence into practice	Integration with public health response systems

13.0. OPERATIONAL CHALLENGES IN AI-ENHANCED BIOSURVEILLANCE

The operational challenges of AI-enhanced biosurveillance are substantial. Institutional fragmentation is one of the largest barriers. Human health, animal health, food safety, environmental monitoring, agriculture, emergency management, and security agencies often operate separately. Their data systems may be incompatible, their incentives misaligned, and their legal authorities unclear. Data quality remains a persistent problem. Missing data, delayed reporting, uneven sampling, inconsistent definitions, laboratory variation, biased digital traces, and poor metadata can undermine model performance. AI does not solve poor data quality; it can amplify it. The apparent sophistication of AI can hide weak foundations.

Capacity gaps are equally important. Many regions lack sufficient laboratory infrastructure, sequencing capacity, trained epidemiologists, data scientists, field staff, veterinarians, environmental health specialists, and information systems. AI systems that require high computational resources and expert maintenance may fail if deployed without local capacity building.

False positives and false negatives create operational risk. Frequent false alarms can lead to decision fatigue and loss of credibility. Missed signals can lead to delayed response. Alerts must therefore be linked to graded response protocols rather than all-or-nothing decisions.

Model drift and changing baselines are major challenges. Climate change, urbanization, migration, land-use change, vaccination, pathogen evolution, and behavioral adaptation can alter patterns. Systems must be updated, tested, and recalibrated.

The implementation gap is perhaps the most important challenge. Many systems detect signals but fail to trigger action. Detection is not preparedness. Preparedness requires authority, resources, logistics, communication, public trust, and political willingness.

14.0. APPLICATIONS ACROSS KEY BIOSURVEILLANCE CONTEXTS

Pandemic preparedness is the most visible application. AI-enhanced biosurveillance can combine clinical, genomic, wastewater, mobility, digital, and environmental data to detect and track infectious disease threats. Wastewater surveillance can provide early community-level signals, especially where testing access is limited. Genomic surveillance can identify variants and transmission pathways. Antimicrobial resistance surveillance is another critical area. AMR emerges across hospitals, farms, wastewater, soils, waterways, and food systems. A One Health AI approach can integrate resistance gene data, antimicrobial use records, clinical isolates, veterinary data, and environmental sampling. The challenge is harmonizing data across sectors and interpreting resistance signals in relation to actual exposure and disease burden.

Foodborne pathogen surveillance can benefit from AI through anomaly detection in inspection records, supply-chain tracing, genomic comparison of isolates, and risk modelling. Livestock disease surveillance can integrate veterinary reports, movement records, farm biosecurity data, climate conditions, and wildlife interfaces. Crop pest and plant pathogen surveillance can combine remote sensing, field reports, diagnostic data, and climate models. Water-system biosurveillance is particularly important in settings where microbial contamination can signal public-health risk before disease burden is fully visible through clinical reporting systems (Izah & Ogwu, 2025).

Vector-borne disease surveillance can use environmental suitability modelling, mosquito surveillance, human case data, animal host data, and climate forecasts. Environmental toxicant and microbial risk surveillance can track contamination in water, soil, air, and built environments. Disaster-related biosurveillance can detect disease risks after floods, storms, displacement, infrastructure failure, and heatwaves. Low-resource and fragile settings require special attention. AI-enhanced systems should not assume high data density, stable internet, or advanced laboratory capacity. In such settings, lightweight tools, mobile reporting, sentinel networks, community surveillance, wastewater sampling, regional laboratories, and open-source platforms may be more realistic than highly centralized systems. Equity must shape design. Surveillance of antimicrobial residues in poultry and other animal-derived food products is a critical One Health function because such residues can connect livestock production, consumer exposure, food safety, and antimicrobial resistance risk (Izah et al., 2025).

15.0 TOWARD FUTURE-READY BIOSURVEILLANCE SYSTEMS

Future-ready biosurveillance systems must move from reactive reporting to anticipatory intelligence. This requires integrated data streams, adaptive models, trained workforces, cross-sector governance, ethical safeguards, and response capacity. AI can support this transition, but only if systems are designed around decisions rather than data accumulation. A future-ready system should be modular, meaning it can add or remove data streams as risks change. It should be scalable, functioning from local sentinel sites to national and global networks. It should be interoperable, allowing health, agriculture, environment, and security data to be connected. It should be adaptive, learning from feedback and recalibrating under changing conditions. It should be trusted, with transparent governance and responsible communication.

Real-time biosurveillance observatories may become increasingly important. These would integrate wastewater monitoring, genomic sequencing, environmental sensors, remote sensing, digital disease detection, veterinary and agricultural data, and climate forecasts into decision-support platforms. However, observatories must avoid becoming passive dashboards. Their purpose should be action: identifying where to sample, when to alert, what to investigate, and how to intervene. Climate-sensitive biosurveillance is increasingly important because changing temperature, precipitation, extreme events, and rural health-system vulnerabilities can reshape exposure pathways, disease risks, and preparedness needs (Sugg et al., 2023).

Community-engaged biosurveillance is also essential. Communities often detect unusual illness, animal deaths, environmental change, food contamination, or ecological disturbance before institutions do. Participatory systems can strengthen detection, legitimacy, and response. But participation must be reciprocal; communities should receive useful information, protection, and benefits—not merely provide data. The future research agenda includes explainable AI, causal inference, uncertainty-aware forecasting, multimodal data fusion, privacy-preserving analytics, low-resource biosurveillance tools, AI-assisted genomic interpretation, climate-sensitive risk models, and ethical frameworks for biological data governance.

16.0. CONCLUSION

Biosurveillance has moved beyond the narrow function of disease reporting to become a core infrastructure for biological risk governance. In an interconnected world, threats to human health, animal health, plant health, food systems, ecosystems, climate stability, and security cannot be monitored through isolated systems. Biological risks emerge across sectors, move through ecological and social pathways, and often become visible only after weak signals have accumulated across multiple data streams. The future of biosurveillance therefore depends on systems that can connect evidence across clinical, environmental, agricultural, genomic, digital, and community-based sources. Artificial intelligence can strengthen this infrastructure by accelerating detection, integrating complex data, identifying anomalies, forecasting risk, and supporting decision-making. Its value lies in expanding the speed, scale, and analytical depth of surveillance. However, AI does not remove uncertainty, institutional fragmentation, ethical responsibility, or the need for expert judgment. Algorithms can reveal patterns, but they cannot determine public priorities, build trust, interpret local realities, or guarantee that warnings lead to timely action. For this reason, AI-enhanced biosurveillance must be understood as a socio-technical system rather than a purely computational upgrade.

A strong foundation for AI-enhanced biosurveillance requires integrated One Health thinking, interoperable and validated data, responsible governance, transparent analytics, and clear links between early warning and response capacity. Public trust, equity, accountability, and communication are not secondary considerations; they are conditions for legitimacy and effectiveness. Systems that detect biological threats but fail to communicate uncertainty, protect sensitive data, include affected communities, or support coordinated intervention will remain incomplete. The most effective biosurveillance systems of the future will not simply be those with the largest datasets or the most advanced algorithms. They will be systems capable of detecting meaningful signals early, interpreting them accurately, communicating them responsibly, and mobilizing action before biological risks become crises. In this sense, artificial intelligence does not change the ultimate purpose of biosurveillance. It sharpens that purpose: to transform biological information into timely, ethical, trusted, and coordinated action that protects life

REFERENCES

Barboza, P., Vaillant, L., Le Strat, Y., Hartley, D. M., Nelson, N. P., Mawudeku, A., Madoff, L. C., Linge, J. P., Collier, N., Brownstein, J. S., Yangarber, R., Astagneau, P., & Valenciano, M. (2013). Factors influencing performance of Internet-based biosurveillance systems used in epidemic intelligence for early detection of infectious diseases outbreaks. *PLoS One*, *8*(3), e57252. DOI: 10.1371/journal.pone.0057252 PMID: 23472077

Charles-Smith, L. E., Reynolds, T. L., Cameron, M. A., Conway, M., Lau, E. H. Y., Olsen, J. M., Pavlin, J. A., Shigematsu, M., Streichert, L. C., Suda, K. J., & Corley, C. D. (2015). Using social media for actionable disease surveillance and outbreak management: A systematic literature review. *PLoS One*, *10*(10), e0139701. DOI: 10.1371/journal.pone.0139701 PMID: 26437454

Izah, S. C., Nurmahanova, A., Ogwu, M. C., Toktarbay, Z., Umirbayeva, Z., Ussen, K., Koibasova, L., Nazarbekova, S., Tynybekov, B., & Guo, Z. (2025). Public health risks associated with antibiotic residues in poultry food products. *Journal of Agriculture and Food Research*, *21*, 101815. DOI: 10.1016/j.jafr.2025.101815

Izah, S. C., & Ogwu, M. C. (2025). Modeling solutions for microbial water contamination in the Global South for public health protection. *Frontiers in Microbiology*, *16*, 1504829. DOI: 10.3389/fmicb.2025.1504829 PMID: 40241726

Izah, S. C., Ogwu, M. C., & Hamidifar, H. (Eds.). (2024). *Biomonitoring of pollutants in the Global South*. Springer Nature., DOI: 10.1007/978-981-97-1658-6

Izah, S. C., Ogwu, M. C., & Iyiola, A. O. (2026). *Artificial intelligence in fisheries: Transformative potentials and challenges* (1st ed.). CRC Press., DOI: 10.1201/9781003506287

MacIntyre, C. R., Chen, X., Kunasekaran, M., Quigley, A., Lim, S., Stone, H., Paik, H. Y., Yao, L., Heslop, D., Wei, W., & Gurdasani, D. (2023). Artificial intelligence in public health: The potential of epidemic early warning systems. *The Journal of International Medical Research*, *51*(3), 03000605231159335. Advance online publication. DOI: 10.1177/03000605231159335 PMID: 36967669

Meckawy, R., Stuckler, D., Mehta, A., Al-Ahdal, T., & Doebbeling, B. N. (2022). Effectiveness of early warning systems in the detection of infectious disease outbreaks: A systematic review. *BMC Public Health*, *22*(1), 2216. Advance online publication. DOI: 10.1186/s12889-022-14625-4 PMID: 36447171

Morse, S. S. (2012). Public health surveillance and infectious disease detection. *Biosecurity and Bioterrorism, 10*(1), 6–16. DOI: 10.1089/bsp.2011.0088 PMID: 22455675

Ogwu, M. C., & Izah, S. C. (2023). *One Health implications of agrochemicals and their sustainable alternatives* (Sustainable Development and Biodiversity, Vol. 34). Springer Nature. https://doi.org/DOI: 10.1007/978-981-99-3439-3

Ogwu, M. C., & Izah, S. C. (2025). Nanotechnology for fungal pathogen control in crops: Innovations, public health impacts, and disease prevention. *Frontiers in Fungal Biology, 6*, 1653214. DOI: 10.3389/ffunb.2025.1653214 PMID: 40959466

Ogwu, M. C., Izah, S. C., & Ntuli, N. R. (2024). *Food safety and quality in the Global South.* Springer Nature., DOI: 10.1007/978-981-97-2428-4

Ogwu, M. C., Izah, S. C., Sawyer, W. E., & Amabie, T. (2025). Environmental risk assessment of trace metal pollution: A statistical perspective. *Environmental Geochemistry and Health, 47*(4), 94. DOI: 10.1007/s10653-025-02405-z PMID: 40019615

Sugg, M. M., Ryan, S., Spurlock, T., Duncan, S., Hege, A., Ogwu, M. C., Tyson, J., Shay, E., Renwick, K., Hendren, C., Roy, M., & Runkle, J. D. (2023). Climate change and health in rural mountain environments: Summary of a workshop on knowledge gaps, barriers, and opportunities for action. *GeoJournal, 88*(5), 5397–5409. Advance online publication. DOI: 10.1007/s10708-023-10916-4 PMID: 40291811

Velasco, E., Agheneza, T., Denecke, K., Kirchner, G., & Eckmanns, T. (2014). Social media and internet-based data in global systems for public health surveillance: A systematic review. *The Milbank Quarterly, 92*(1), 7–33. DOI: 10.1111/1468-0009.12038 PMID: 24597553

Chapter 2

Community-Centered AI Biosurveillance:
Participatory Models, Trust, and Public Engagement

Fizza Waheed
http://orcid.org/0009-0009-4972-533X
University of Agriculture, Faisalabad, Pakistan

Saman Mahmood
http://orcid.org/0009-0002-2854-5945
University of Agriculture, Faisalabad, Pakistan

Nimra Aslam
http://orcid.org/0009-0009-1787-631X
University of Agriculture, Faisalabad, Pakistan

Fatima Umer Draz
http://orcid.org/0009-0000-8679-9965
University of Agriculture, Faisalabad, Pakistan

Eiman Ashiq
http://orcid.org/0009-0007-9467-4557
University of Agriculture, Faisalabad, Pakistan

Ayesha Shabbir
http://orcid.org/0009-0000-9468-499X
University of Agriculture, Faisalabad, Pakistan

Abu Sulman
http://orcid.org/0009-0008-7194-5219
University of Agriculture, Faisalabad, Pakistan

Naima Khan
Cholistan University of Veterinary and Animal Sciences, Bahawalpur, Pakistan

Adnan Afzal
http://orcid.org/0009-0001-4359-5695
Government College University, Faisalabad, Pakistan

Babar Khan
University of Swat, Pakistan

DOI: 10.4018/979-8-3373-9033-8.ch002

ABSTRACT

This chapter explores community-centered AI biosurveillance as an ethical and effective approach to modern public health surveillance. It examines how participatory models, trust-building strategies, and public engagement enhance the design, implementation, and governance of AI-driven outbreak detection systems. By integrating citizen-generated data, local knowledge, and transparent governance structures, community-centered frameworks address limitations of traditional surveillance, including bias, delayed reporting, and public mistrust. The chapter also analyzes ethical challenges related to privacy, consent, accountability, and equity, highlighting emerging technologies such as explainable AI and federated learning. Ultimately, it argues that trust and participation are foundational to sustainable, equitable, and resilient biosurveillance systems

INTRODUCTION

Due to the intensive development of artificial intelligence (AI) and data-driven technologies, there has been an essential change in the situation in the field of public health surveillance (Pillai & Kumar, 2021). The conventional bio surveillance systems, which are mostly centralized, expert-oriented, and controlled by institutions have been instrumental in the identification of outbreaks, the tracking of disease patterns, and the provision of a response to the population health (Salardi, 2025). Nonetheless, the COVID-19 crisis and frequent outbreaks of the newly emerging and re-emerging infectious diseases have revealed serious shortcomings of these traditional models. Sluggish reporting, data silos, insensitivity to context, mistrust among the general population, and disproportionate impacts have provided reasons to adopt more inclusive, responsive, and responsive solutions (Folasole, 2023). It is against this context that community-based AI bio surveillance has become a promising paradigm that reinvents data generation, data interpretation and data action: by making communities not just sources of data, but co-producers of bio surveillance ecosystems (Schulte & Bohnet-Joschko, 2022). Community-based AI bio surveillance Community-centred AI bio surveillance are systems that combine the use of sophisticated computational techniques with local knowledge and sustained participation of the population. These models do not only depend on the top-down approach to data collection via clinical or governmental sources, but they focus on the collaboration with communities when it comes to the design, implementation, and monitoring of AI-based surveillance instruments (Hapicha, 2025; Johnson, 2025). This kind of approach acknowledges that health risks are locally and unfairly experienced, and influenced by social, cultural, economic and environmental factors that are usually not visible to a purely technical system. Participatory bio surveillance aims to enhance the effectiveness of AI-enabled bio surveillance and its legitimacy by integrating the community perspective, values, and priorities in community-oriented interventions in improving the community health conditions (Frohlich & Abel, 2014).

Surge of AI in bio surveillance has provided new capabilities as never before. Machine learning algorithms can analyse large heterogeneous datasets, such as electronic health records and laboratory reports, mobility data, social media signals, wastewater data, environmental sensors, etc. in near real time. Such systems are expected to facilitate the detection of outbreaks sooner, better prediction of risks and enhancement of public health resources allocation (Cheah et al., 2025). However, there are deeper ethical, social, and political issues associated with the application of AI in this area, as well. Trust can not be a side issue, but a prerequisite to successful AI bio surveillance. Trust determines whether or not people agree to share their data, adhere to the advice of the public health authorities and accept AI-informed choices that can influence their mobility, access to employment, or services (Fontes et al., 2022). Community-based models are models that develop trust by way of being transparent, accountable, reciprocal, and making collective decisions. This includes informing of the nature of data gathered, the way AI models are used, beneficiaries of their outputs, and mitigating risks (Kangana et al., 2024). It also entails methods through which communities express their grievances, engage in dispute processes and in the governance process. Community-based bio surveillance will seek to transcend extractive forms of data practices and establish mutualistic relationships by integrating trust-building practices into the design and working of AI systems (Haider, 2009).

Notably, community-focused methods decrease the notion that technical accuracy is sufficient to ensure the success of the health of the population. Bio surveillance systems exist in the scenario of complex social systems, in which power relations and history of injustice, as well as cultural meanings, have a significant impact on the outcome (Nicholson et al., 2013; Unert et al., 2016). To illustrate, the machine learning models deployed based on skewed or that lacks completeness might underrepresent specific groups of people in a systematic manner, causing unfair risk evaluations or late responses. Participatory models can assist in bringing out and mitigating these biases by engaging communities in data verification, interpretation and evaluation of the model. By so doing they will help to create an equitable and fairer public health infrastructure (Wang et al., 2022; Lalor et al., 2024). The fusion of AI and the community also alter the concept of expertise in bio surveillance. Data scientists, epidemiologists, and public health officials can provide essential technical expertise, but the local people will also provide valuable experience of what is going on, behaviours, and weaknesses at hand (Saadatzadeh, 2024). Such complementary types of knowledge may help to see the situation better and promote a sense of joint ownership of the surveillance products. Such pluralistic concept of expertise is very significant especially in diverse societies where localized realities are usually not well understood through one-size-fits-all models (Andrejevic, 2004).

Nonetheless, community-based AI bio surveillance should not be introduced without difficulties. Effective involvement takes time, resources, and institutional dedication, which can go against the urgency and efficiency that is mostly linked to the role of responding to an outbreak (Bedford et al., 2019).

This chapter discusses community-based AI bio surveillance against the backdrop of participatory bio surveillance models, trust, and engagement with the politics. It looks at the ways in which participatory structures can be materialized in the design and governance of AI-powered surveillance systems, with reference to the examples of various settings. Trust is prefigured in the discussion as the results and as the process, which is pre-conditioned by transparency, accountability, and sensitivity to the needs of the community (Faye et al., 2025). To achieve sustainable and equitable bio surveillance, the participation of the actors, the decision-makers, and the data-driven health security beneficiaries have to be reconsidered. The community-based models provide a way to a more intelligent, but also more just, more resilient and more responsive to the real-life experiences of the populations they seek to serve (Damiani, 2019). During a time when the populace is highly suspicious of institutions and health threats are becoming more complicated, the inclusion of participation and engagement in the very core of AI bio surveillance is not just a want, but a necessity (Fontes et al., 2022; Faye et al., 2025; Mendes et al., 2025).

THE FOUNDATIONS OF AI BIOSURVEILLANCE

AI Biosurveillance is becoming a need as time is slipping through our hands but it only works effectively when combined with public trust and participation (Chen et al., 2023; Bazzano et al., 2025).

AI Biosurveillance

"AI Biosurveillance is the use of automated systems (machine learning ML, deep learning DL, federated learning FL, etc) for monitoring of health patterns in populations to detect and predict outbreaks enabling rapid public health response" (Deiner et al., 2025; Gomes de Sousa et al., 2025).

Core Components

1. Data Collection: It is the first stage of AI Biosurveillance. Data can be collected through many ways e.g. clinically through hospitals, digitally through social media or through devices and apps, etc (McMurry et al., 2024; Giri & Gupta, 2024).
2. Modeling: After collecting data, it is processed by using models (like ML or DL) in which the algorithm detects patterns and identify them as either normal or abnormal (Gomes de Sousa et al., 2025; Deiner et al., 2025).
3. Prediction: Systems then predict the probability of outbreaks. It shows if there is a high chance or low chance of outbreak and how fast the disease will spread (Chen et al., 2023; Deiner et al., 2025).
4. Reporting: The information is reported to the public health authorities and then to the public through mobile app notifications (McMurry et al., 2024; Salathé et al., 2022).

Importance in Epidemic Prediction

Modern outbreaks spread rapidly therefore traditional systems for outbreak prediction are not enough anymore. Nowadays traveling has become easy and more people travel through air and also urbanization has increased significantly this is why the diseases spread faster than ever (Chen et al., 2023; Giri & Gupta, 2024). Traditional surveillance process takes weeks by then many more become infected. So for a earlier prediction AI systems are being used. It processes real-time data and can detect patterns early (Deiner et al., 2025; McMurry et al., 2024).

Public health datas are very complex and the volume is also large. Health departments are understaffed due to which processing data this complex, variable and large becomes a challenge. Meanwhile AI can handle larger volumes of complex data in real-time (Gomes de Sousa et al., 2025; Giri & Gupta, 2024).

AI helps in detection of threats that we have not seen before. Traditional systems need to know what it is and specific diagnosis is required while AI can detect unknown patterns without knowing the actual cause. Mostly emerging disease are zoonotic and these unknown pathogens are the biggest threat (Chen et al., 2023; Deiner et al., 2025).

Table 1. Examples of AI biosurveillance

System	Year Launched	Data Sources	Geographic Scope	Key Feature
HealthMap	2006	News, social media, reports	Global	Real-time disease alerts
BlueDot	2014	News, airline data, climate	Global	COVID-19 early detection
Flu Near You	2011	Citizen symptom reports	US/Canada	Participatory Surveillance

Community-Centered Approaches

Technology without trust and participation of community is a failure no matter how sophisticated it is. AI systems depend on data quality. If there is no trust the communities will not use the services or apps so there will not be enough data resulting in poor performance of the system (Bazzano et al., 2025; Salathé et al., 2022). As most AI systems like apps and surveys require public participation in which voluntary participation rates are often low. As we know surveys and apps require the use of smart phones and internet so the groups with lack of such resources remain underrepresented. This becomes an issue of data quality (Michener et al., 2020). Community engagement addresses such blind spots by ensuring participation by the citizens.

Example

"Google Flu Trends" was launched in 2008. At first it showed promise in prediction of flu outbreaks by using search query data. But it failed in 2013 due to overestimatimg prevalence by nearly double. This happened because it depended entirely on technology without community involvement or validation. It is a classic example of how technology without community involvement becomes a failure (Chen et al., 2023).

Key Principles

- Bidirectional communication means not just collecting data but also providing information regarding results and everything back to the participants just like sending feedback to the communities (Salathé et al., 2022).
- Co-design means involving communities in making surveillance designs and respecting knowledge and priorities of locals (Bazzano et al., 2025; Michener et al., 2020).

- Transparency and building trust by telling what data is collected and how it is collected. Accountability of actions by authorities (O'Mara-Eves et al., 2025; Salathé et al., 2022).
- Integrating local knowledge to improve quality because communities can understand the aspects AI might miss (Michener et al., 2020; O'Mara-Eves et al., 2025).

Benefits of Community-Centered Approach

1. Better data quality: When people start trusting the programs or systems they become more honest, which brings better data with good results and effective use of the system (Bazzano et al., 2025; Salathé et al., 2022).
2. Earlier detection: Locals notice unusual patterns first and this helps by filling the gaps and earlier detection becomes possible (Salathé et al., 2022; O'Mara-Eves et al., 2025).
3. More effective response: Earlier detection from more participation gives time for early and better response by the authorities and the public (Michener et al., 2020; Bazzano et al., 2025).

Figure 1. Comparison between timeline and working of traditional and AI surveillance

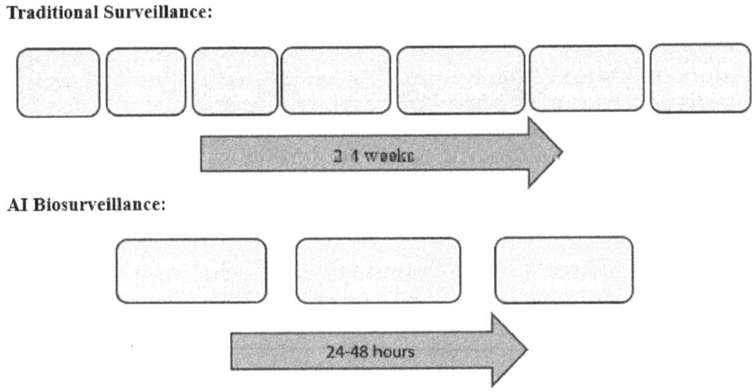

PARTICIPATORY MODELS FOR AI BIOSURVEILLANCE

Data Collection With Communities

The conventional disease surveillance system has traditionally utilized passive reporting from healthcare facilities; whereby healthcare practitioners submit reports to the health department. Although this method is useful for obtaining clinical data, it has several drawbacks, including the fact that it only identifies people who actually access healthcare services, is prone to delays from the onset of symptoms to reporting, and does not identify people with mild or asymptomatic disease at all. Participatory disease surveillance systems essentially revolutionize this process by engaging community members as data sources, thereby making citizens active participants in the surveillance process rather than mere subjects (Brownstein et al., 2009; Wójcik et al., 2014).

Citizen Science and Participatory Reporting

Citizen science models of biosurveillance utilize the strength of distributed networks of volunteers who report health data directly to the surveillance system. Participatory models of surveillance understand that members of a community are often the first to observe changes in health patterns before they are observable to the healthcare system. By creating platforms for reporting symptoms and health behaviors, participatory models of surveillance create early warning systems that work in conjunction with clinical reporting systems(Smolinski et al., 2015; Smolinski et al., 2017).

The most successful example of such an approach is Flu Near You (FNY), an online participatory surveillance system established in 2011 by the joint efforts of the American Public Health Association, HealthMap at Boston Children's Hospital, and the Skoll Global Threats Fund (Smolinski et al., 2015). FNY is an online participatory surveillance system that enables citizens in the United States and Canada to report symptoms of influenza-like illness on a weekly online survey accessible through the FNY website or mobile app. Citizens are required to register with basic demographic information and are sent weekly reminders via email to report whether they have experienced fever, cough, sore throat, or other symptoms of the flu in the past week. FNY tells even healthy people to report as it helps set a baseline for when the numbers become abnormal (Smolinski et al., 2015).

A large number of people reported through FNY and this number was close to the official numbers of CDC (centers for disease control). This showed that letting people report is as effective as the traditional system (Baltrusaitis et al., 2018).

It gives one more advantage that many people with Influenza Like Illness (ILI) never go to hospitals or clinics and such people can report at FNY. This results in a better and complete picture of the illness (Wójcik et al., 2014; Smolinski et al., 2017).

In contrast to traditional surveillance, where patients must wait to be seen by a physician, and then the physician must write a report, which is then compiled into aggregate data that takes days to weeks, FNY participants enter their symptoms days after the onset of illness, and aggregate data is available almost instantly. This is vital for early warning systems, which enable public health officials to identify outbreaks and take action before the disease becomes widespread (Brownstein et al., 2009; Santillana et al., 2015).

The success of participatory influenza surveillance has encouraged the development of similar projects for other diseases and health issues. In early 2020, when COVID-19 broke out, just like FNY they immediately developed COVID Near You. It worked similar to FNY where patients could report their test results, vaccination updates, etc. This proved that the participatory model could be changed to cope with new health challenges (Smolinski et al., 2017). Examples of such projects that have been developed worldwide include: FluTracking in Australia (launched in 2006), Influenzanet in ten European countries and many others (Wójcik et al., 2014).

Mobile Reporting Applications and Sensor Networks

In addition to web-based symptom reporting, participatory biosurveillance is increasingly incorporating mobile technology and wearable sensors that allow for more passive forms of data collection while still allowing for user control. Mobile health applications can collect a variety of data streams such as self-reported symptoms, GPS location information for tracking movement, Bluetooth proximity information for contact tracing, and smartphone/wearable sensor information for heart rate, temperature, activity, and sleep (Drew et al., 2020).

The Swiss SwissCovid app, the German Corona-Warn app, and other state-level apps in the U.S. showed the potential of decentralized architectures that support privacy-friendly systems for contact tracing. By mid-2021, Germany's Corona-Warn app had been downloaded more than 30 million times, which is roughly 36% of the population, a remarkably high level of adoption that helped control outbreaks. Yet, experience with contact tracing apps around the world has also shown that there are crucial shortcomings: many apps reached levels of adoption of less than 20%, which is ineffective since 60-75% adoption is required to have an impact at the population level (Salathé et al., 2020; Trang et al., 2020).

Smart thermometers are another example of innovative participatory biosurveillance using sensor networks. Kinsa Health's internet-connected thermometer network collects anonymized temperature data from households across the United

States. When a user takes a temperature reading, the data is automatically sent to central databases, where aggregate analysis shows fever activity at a community, state, and national level. During the 2019-2020 flu season, Kinsa's network of smart thermometers identified unusual fever activity in early March 2020, several weeks before the number of COVID-19 cases began to rise in many parts of the United States (Chamberlain et al., 2020).

ETHICAL CONSIDERATIONS: PRIVACY, CONSENT, AND DATA SECURITY

The success of participatory biosurveillance is fundamentally dependent on voluntary participation and community trust. Unlike traditional surveillance, where healthcare providers submit de-identified information to the relevant authorities without the direct consent of patients, participatory biosurveillance involves individuals voluntarily submitting their personal health information. The voluntary nature of participatory biosurveillance means that issues of privacy, consent, and security are not just legal issues but fundamental to the success of such systems. If communities do not trust these systems, they simply will not participate (Floridi et al., 2018).

Privacy Protection

Current state-of-the-art participatory biosurveillance projects use multiple levels of privacy measures (Parker et al., 2020; Vayena et al., 2018).

First, data minimization principles are used in data collection. Only information that will be used in surveillance should be collected, resisting the urge to collect as much personal information as possible "just in case" it may be useful in the future. Flu Near You, for instance, collects minimal demographic information (age group, sex, and ZIP code) that is adequate for epidemiological analysis without needing individuals' full identification. This is in contrast to COVID-19 contact tracing applications that asked for access to a vast amount of phone information, location history, and contact information, which raised privacy concerns and led to a lack of trust (Parker et al., 2020).

Then technical measures such as encryption, access controls, and secure storage of data safeguard information against unauthorized access or breaches. Data must be encrypted both in transit (between user devices and servers) and at rest (in databases). Access must be limited to authorized personnel in the public health sector with a legitimate need. Systems should be checked regularly for any problem to avoid being hacked (Vayena et al., 2018; Floridi et al., 2018).

The protection of privacy is very complicated when location tracking is involved in the systems. GPS location information discloses individuals' movements, residential addresses, work locations, and social behavior. Even when anonymized, location information can be readily re-identified by correlating it with publicly available information (Parker et al., 2020).

Informed Consent

Proper and valid consent only counts if the user truly understands what they are giving consent to. Every policy should be written in clear, simple language without using technical terms. The important information should be communicated. Visual aids such as flowcharts showing how data is transmitted through the system can be useful (Floridi et al., 2018).

Users should be able to choose what data they share and what they don't. They should also be able to remove consent any time and delete their data (Vayena et al., 2018; Parker et al., 2020).

Consent in participatory system is different from traditional system. Users can enroll in FNY and give consent for weekly symptom reporting, but their continued participation is continuous consent. If users fail to respond to weekly invitations, systems should honor this implicit withdrawal of consent rather than assuming non-response is data (falsely assuming wellness). This is a challenge in how to honor continuous consent and keep participants informed about changes in systems or data use (Smolinski et al., 2015).

Children lack the ability to give legal consent, making parental consent necessary for study participation. Yet this criterion could preclude an important subgroup of respiratory infections, which frequently spread among school-age children, from being monitored. Minors might be hesitant to submit symptoms if parental consent criteria disclose personal health information to their families. Similarly, persons who are homeless, undocumented immigrants, or come from communities that have had valid reasons to distrust the government in the past might be hesitant or unable to give consent, even if it would help them and their communities (Corburn et al., 2020).

Data Security

Despite strong privacy safeguards and informed consent, participatory biosurveillance systems are still very attractive to malicious actors. The value of healthcare data is high in black markets, where stolen healthcare data is more valuable than stolen credit card data. Attackers may attempt to break into surveillance systems to steal personal health data, find susceptible people, or use the data to deceive systems into reporting false outbreak reports (Vayena et al., 2018).

Protecting data needs more than one kind of protection. Technical controls meaning computer based protection like encryption, firewalls, and security testing. Administrative controls meaning rules and training for people who can access data (Floridi et al., 2018; Vayena et al., 2018).

Security is not only needed in central database. Mobile apps and devices can also be hacked. So every part of system needs security including apps (Trang et al., 2020).

Protection is also needed from fake or harmful submissions. It could be done by restricting how many reports can a person submit or by checking emails. But still this should not be too strict so that it will become hard for people with limited access to technology (Floridi et al., 2018).

INVOLVING COMMUNITIES IN MODEL DESIGN AND FEATURE SELECTION

There are many design choices of AI biosurveillance. They include what questions to be asked like: What symptoms should be tracked? What patterns of symptoms indicate an outbreak? How should geographic patterns be considered? What levels of risk should indicate alerts? These are not purely technical questions—they are matters of value, of what is important, what levels of risk are acceptable, and what trade-offs should be made (Floridi et al., 2018).

The traditional approach to AI development considers these design choices to be technical decisions made by developers and public health professionals. However, there are important perspectives that are missing from these technical decisions, and these are provided by the communities affected by the development of AI biosurveillance systems. They understand the local context—seasonal patterns, cultural practices, environmental conditions—that affect the spread of disease but which may not be apparent to algorithms analyzing national-level data. They understand what health concerns are most important to them (Corburn et al., 2020).

Trust as the Foundation

AI is transforming healthcare, but challenges in safety, trust, and ethics are still there. Persistent human oversight, strong regulation, and ethical governance are essential to ensure reliable and trustworthy AI systems (Astobiza et al., 2025; Mohsin Khan et al., 2025).

A hybrid approach study combined a rapid review of 32 papers with a survey of 22 stakeholders to analyze trust in medical AI. Nineteen factors were identified across human, technological, ethical/legal, and other categories, with most rated as highly relevant to trust and acceptance, while patient demographics were less significant (Shevtsova et al., 2024).

In the developing relationship between human beings and AI, trust is the only technique that governs the application and adoption of AI by clinicians. Trust is a psychological process for coping with the uncertainty that lies between the known and unknown. Various research studies have emphasized the need to improve AI-based systems and their capabilities to assist clinicians (Asan et al., 2020).

AI is revolutionizing the healthcare industry with its use in the diagnosis, development of medicines, and delivery of healthcare services, but issues related to bias, explainability, and data privacy are being raised as challenges. Even though the FDA guidelines are available to regulate the use of AI, their implementation is still limited (Mora-Cantallops et al., 2024).

The use of AI and robotics in the healthcare industry requires equal access, societal trust, and appropriate ethical regulation. The digital divide and co-regulation on a global scale are required for the appropriate use of AI and robotics in the healthcare industry (Elendu et al., 2023).

A qualitative research study was conducted in the NHS setting in the West Midlands to explore the experience of healthcare professionals utilizing AI to inform decision-making. Forty healthcare professionals, administrators, and AI developers took part in interviews and focus group discussions, which were analyzed using a thematic analysis. Although AI was perceived to enhance the speed of diagnosis and resource allocation, the group expressed concerns regarding accountability, "black box" explainability, and bias towards minority or atypical patients. The results emphasize the requirement for more stringent regulations, full data sets, and increased collaboration between healthcare professionals and developers. Future research studies should also consider the perspective of patients and provide useful ethical guidelines for the fair use of AI (Nouis et al., 2025).

Bias in AI may potentially exacerbate existing inequalities in the healthcare system. In this systematic review of 20 studies on EHR-based AI systems, six types of bias have been identified, and techniques of bias detection (fairness metrics) and bias correction (resampling, reweighting, and data transformation) have been explored. There is a need for standardized and understandable methods to guarantee ethical and fair AI use in healthcare (Chen et al., 2024).

The study discusses accountability in healthcare AI systems by pointing out the main stakeholders, patient roles, and regulatory bodies. Difficulties are overcome through explainability and a three-tier system, encouraging mutual accountability for AI-supported decisions to guarantee responsible and fair use (Bagave et al., 2025).

Healthy AI system development calls for national plans with local policies, community engagement for fairness, and international cooperation to overcome governance gaps and inequalities in resources, achieving a balance between the potential of technology and public confidence (Gao et al., 2026).

Figure 2. Impact of trust on adoption of the programs

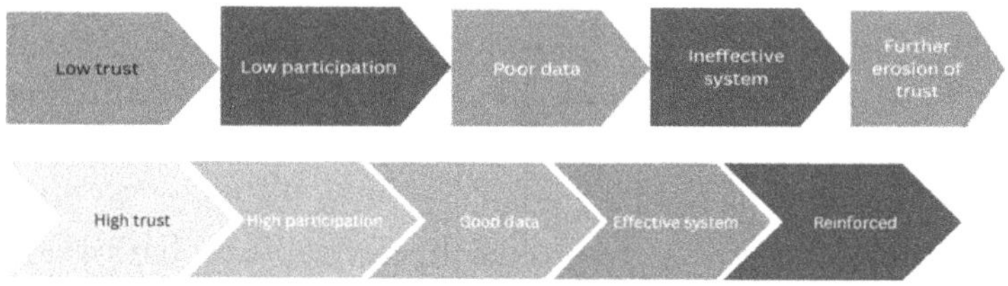

Engaging the Public Effectively

Digital community based surveillance systems now function worldwide for early detection of health incidents across human, animal, and environmental domains. A review has identified 60 operational systems that use digital technology to collect community-submitted data at a national, subnational, and multinational level. Many systems allow the data to be accessed by the user through various channels, which has led to increased public awareness and earlier detection of emerging health threats (McNeil et al., 2022).

This surveillance system improves early outbreak detection by involving local stakeholders in surveillance efforts. Outcomes from 15 countries show that success factors include engagement, trust, local ownership, communication, collaboration, skilled staff, well-defined cases, quality assurance, timely use of data, and accessible technology. Participatory and local ownership are essential for effective surveillance systems (McGowan et al., 2022).

A prospective study (2011-2021) examined factors affecting influenza vaccination. People living with children and using local transport were less likely to be vaccinated, whereas university-educated individuals and those with chronic states were more likely. Age group and regional influenza incidence also influence vaccination, and the COVID-19 pandemic was timed to coincide with the increase in vaccination (Kelley et al., 2025)

A rapid review of 22 studies revealed that web-based, visual, and mapping online platforms are associated with increased confidentiality, efficiency, and flexibility in the development of health interventions, but challenges include internet access and skills. Most of the strategies are modifications of traditional approaches, which indicate the need for standardized terminology and reporting (Doerwald et al., 2024).

Community health workers (CHWs) in 25 resource-poor countries help public health surveillance by participation, data collection, screening, and health education. Their roles enhance the control and reporting of epidemics but are affected by challenges such as inadequate training, resource constraints, and logistics. The combination of CHWs with proper training and infrastructure enhances surveillance, especially in resource-poor countries (Alhassan & Wills, 2024).

Digital surveillance systems like Influenzanet involve citizens in reporting public health information but are associated with ethical concerns. Main concerns include electronic consent, privacy protection, justice, and strengthening the ethics committee capacity. A systemic framework is needed to ensure responsible and ethical use of digital public health data. (Geneviève et al., 2019).

CHALLENGES AND CONSIDERATIONS

AI biosurveillance holds promise but faces many obstacles including:
Technical Challenges:

- Data quality issues span incomplete datasets, biased sampling and inconsistent measurement standards. So the models trained on such datasets produce unreliable results (Radanliev, 2025).
- Interoperability (communication between different systems) problems occur because of the institutes having systems different from each other resulting in delayed outbreak detection (Shenoy, Bhat, & Krishna Prakasha, 2025).
- Cybersecurity is one of the main challenges due to the sensitive data used in AI biosurveillance. So it becomes an easy target for privacy breach. It may result in employment discrimination due to the leaked data (Saheb, 2022; Shachar, Gerke & Adashi, 2020).
- Systems become unable to handle increasing data resulting in scalability limitation. Real-time processing becomes difficult at population scale (Ali, Snášel, & Platoš, 2025).

Ethical and Legal Challenges:

- Privacy concerns by the people is a valid issue. People may fear that their privacy is being invaded due to full-time surveillance like location tracking, health status etc. Laws like HIPAA (health insurance portability and accountability act) set rules and boundaries. Anonymization can also solve this problem (Saheb, 2022;Shachar, Gerke & Adashi, 2020).

- Obtaining consent becomes an issue during public health emergencies. During COVID-19, some countries made contact tracing apps mandatory while others made them voluntary but that result in less adoption by the population then those who made it mandatory ultimately resulting in less effective surveillance (Saheb, 2022; Wheatley, 2024).
- Biasness can occur with the systems trained with overrepresented groups. Algorithm may perform poorly on underrepresented groups. This amplifies existing health disparities (Radanliev, 2025; Shachar, Gerke & Adashi, 2020).

Social Challenges:

- Public mistrust due to history of unethical medical research (Tuskegee, etc.) still remains even after decades (Saheb, 2022).
- Cultural differences and literacy gaps affect understanding and adoption (Shachar, Gerke & Adashi, 2020).
- Poor understanding of AI and spread of false information (rumors) undermine acceptance. For example, during covid-19 rumors spread rapidly about contact tracing apps claiming that they contain microchips installed by the government for permanent citizens surveillance. These conspiracy theories fueled by social media resulted in less adoption of the app (Bazzano, Mantsios, Mattei, Kosorok, & Culotta, 2025).

FUTURE DIRECTIONS

Despite of all the challenges listed above, emerging technologies hold promising pathways leading to more effective, ethical and socially acceptable AI biosurveillance.

Emerging Technological Trends

- Explainable AI (XAI) surpasses traditional AI. It gives transparency and provides explanations for predictions. It also bills public trust enables auditing and identifies bias. This helps the medical professionals to explain the decisions to the general public (Rahman et al., 2022; López-Ramos et al., 2024).
- Federated learning reduces privacy risk. In this, model goes to data not data to model, meaning the data is not sent to a central service rather the device trains on its own local data and only sends the model improvements back to the central service (Rahman et al., 2022; Bharati, Hossain Mondal, Podder, & Prasath, 2022; Ali, Snášel, & Platoš, 2025).

- Blockchain is like a ledger that contains all the information about who accessed the data and when. It enhances security and trust. It also combines well with federated learning (Bhardwaj & Sumangali, 2025; Tawfik et al., 2025).

Table 2. Emerging technologies

Technology	What it does	Key advantage	Remaining challenges	Maturity level
Explainable AI (XAI)	Makes AI decisions interpretable	Builds trust through transparency	Trade-off with accuracy	Medium
Federated learning	Trains models without centralizing data	Preserves privacy	Scalability at population level	Medium-High
Blockchain	Permanent unalterable record	Accountability and consent management	Complexity, energy use	Low-Medium

Sustainable, Ethical and Participatory Programs

- Communities should be involved from the very first step and be given power in decision making and not just one time consultation. This can be done by forming community advisory boards having authority to make decisions representing the whole community and also be responsible for participatory programs (Rice et al., 2025; Abejirinde et al., 2025).
- It should be understandable for layperson and should be explained in accessible language. Transparency is the key for trust building. Data collection, it's use and every documentation should be transparent (Rice et al., 2025).
- Acknowledge historical harms and remove biasness by training systems with more data from underrepresented groups. Then test algorithms for different demographic groups (Radanliev, 2025; Shachar, Gerke, & Adashi, 2020).

Research Gaps

- Many technical gaps or challenges need further research. More attention towards security and privacy and improving real-time processing speed and many other aspects need further intense research for more accurate future use (Rahman et al., 2022; Shenoy, Bhat, & Krishna Prakasha, 2025).
- Further research is needed to understand different social aspects like which participatory model works best?, how can trust be rebuilt in communities after all the historical trauma?, what is the psychological effect of constant

health surveillance?, how to measure if a system reduces health disparities or not?, etc (Bazzano, Mantsios, Mattei, Kosorok, & Culotta, 2025; Abejirinde et al., 2025).

Evaluation of Community-Centered Programs

- Evaluation needs to be done all the time not just once. Regular monitoring before, during and after the launch of a program helps identify any problem or risk timely (Rice et al., 2025).
- Technical, ethical, social and every other aspect needs monitoring and constant evaluation. If the system does well in one aspect it doesn't mean the system is perfect (Rice et al., 2025).

CONCLUSION

This chapter has made it clear that participatory approaches, trust and meaningful public engagement are not optional components of AI biosurveillance systems but are essential requirements for success. The evidence from variety of case studies and context shows that there is a clear trend that surveillance systems that involve communities in their design, implementation and governance are more effective than systems that are purely technological.

It is clear that the four components (data collection, modelling, prediction and reporting) of AI biosurveillance are linked with each other and all benefit from community participation

The participatory approaches that have been explored throughout this chapter from citizen science initiative collaborative design of contact tracing apps show how community engagement can be facilitated in practice these the purchase show that community

Trust as the Indispensable Foundation

It is shown that trust is essential building block for all other elements of community driven biosurveillance. Without trust, community engagement will fail, the quality of data will decrease and even the most advanced AI systems will be unable to provide the benefits they promise. The Google flu trends failure is a warning example, that a system that is highly advanced from a technical point of view but lacks validation and trust from the community will fail and consequences will be dire for public health.

The three pillars of trust in AI biosurveillance systems are transparency, explainability, and accountability. But trust is not an easy or rapid process to attain. The medical traumas of the past, like the Tuskegee Syphilis study, still cause challenges, especially in the case of underrepresented groups. The modern-day issues of surveillance capitalism, data breaches, and biasness of algorithms are added to these wounds.

Navigating Complex Challenges

Technical problems such as data quality and interoperability have ethical implications because they produce biased results. Ethical problems such as privacy and consent create social distrust, which hinders technical ability. Social problems such as distrust and cultural differences require technical solutions (privacy-protecting algorithms) and social strategies (cultural adaptation of communication).

Technical, ethical, and social problems are interrelated and require integrated solutions. Federated learning solves the technical problem of data distribution and satisfies the ethical need for privacy protection. Explainable AI solves the technical problem of algorithmic opacity and meets the social need for trust. Community oversight boards meet the ethical need for governance and promote social trust.

THE PATH FORWARD: INTEGRATION AND INNOVATION

New technologies offer very promising solutions to the challenges that arise from the conflict between surveillance and privacy. Federated learning enables collaborative model building without the need to centralize privacy-sensitive data. Explainable AI enables black-box systems to be more transparent and trustworthy while maintaining efficiency. Blockchain and decentralized systems enable a secure and community-run way of monitoring access to data, thus providing a technological basis for accountability.

Constructing Resilient, Ethical, and Inclusive Systems

As we face the reality of continued pandemic threats and look to the future of infectious disease, the consequences of failure in biosurveillance have never been more dire. The next pandemic will likely come with little warning, and it must be detected and acted upon immediately.

Community-based AI biosurveillance provides a way forward to build systems that are at once more effective and more ethical. By placing communities as partners, not objects of study, these systems will provide better data, more accurate predictions, higher rates of adoption, and more equitable outcomes. By placing a priority on trust as well as effectiveness, they will build systems that are sustainable and functional not only during times of crisis but as a regular part of public health infrastructure.

A Vision for the Future

Let's think about a biosurveillance system in which the community is engaged and empowered to protect their own health. The community provides data through apps that they themselves were involved in developing, and they get feedback on disease activity in their region on a regular basis.

As we move forward in the development of the next generation of biosurveillance systems, we are at a crossroads. We can choose to follow a path in which we optimize our algorithms while viewing our communities as data sources to be exploited. This path may give us a short-term advantage but will ultimately be the cause of our failure as the lack of trust builds up and resistance grows. Or we can choose to follow a path of developing systems with our communities.

The question is not whether we will use AI in biosurveillance, because it is already clear that we will. The question is whether we will use it well, with a recognition of the limitations of technology, a commitment to the autonomy of communities, a commitment to equity, and an understanding that protecting public health is ultimately a matter of public trust. Community-centered approaches provide a roadmap for this. The future of biosurveillance (effective, ethical, and inclusive) waits for our choice to follow it

REFERENCES

Abejirinde, I.-O. O., Itanyi, I. U., Kornas, K., Zaim, R., Guha, S., Chui, V., Lipscombe, L., Rosella, L. C., & Shaw, J. (2025). Principles and practices of community engagement in AI for population health: Formative qualitative study of the AI for diabetes prediction and prevention project. *Journal of Participatory Medicine*, *17*, e69497. DOI: 10.2196/69497 PMID: 41037755

Alhassan, J. A. K., & Wills, O. (2024). Public health surveillance through community health workers: A scoping review of evidence from 25 low-income and middle-income countries. *BMJ Open*, *14*(4), e079776. DOI: 10.1136/bmjopen-2023-079776 PMID: 38582533

Ali, A., Snášel, V., & Platoš, J. (2025). Health-FedNet: A privacy-preserving federated learning framework for scalable and secure healthcare analytics. *Results in Engineering*, *27*, 106484. DOI: 10.1016/j.rineng.2025.106484

Andrejevic, M. (2004). The work of watching one another: Lateral surveillance, risk, and governance. *Surveillance & Society*, *2*(4), 479–497. DOI: 10.24908/ss.v2i4.3359

Asan, O., Bayrak, A. E., & Choudhury, A. (2020). Artificial Intelligence and Human Trust in Healthcare: Focus on Clinicians. *Journal of Medical Internet Research*, *22*(6), e15154. DOI: 10.2196/15154 PMID: 32558657

Astobiza, M. A., Alonso, M., & Ortega Lozano, R. (2025). Trust and AI in healthcare: A systematic review. *Monash Bioethics Review*. Advance online publication. DOI: 10.1007/s40592-025-00272-z

Bagave, P., Westberg, M., Janssen, M., & Ding, A. Y. (2025). *Accountability framework for healthcare AI systems: Towards joint accountability in decision making* (Preprint). arXiv. https://doi.org//arXiv.2509.03286DOI: 10.48550

Baltrusaitis, K., Brownstein, J. S., Scarpino, S. V., Bakota, E., Crawley, A. W., Conidi, G., Gunn, J. E., Gray, J., Zink, A., Santillana, M., & Smolinski, M. S. (2018). Comparison of crowd-sourced, electronic health records-based, and traditional healthcare-based influenza-tracking systems at multiple spatial resolutions in the United States. *BMC Infectious Diseases*, *18*(1), 403. DOI: 10.1186/s12879-018-3322-3 PMID: 30111305

Bazzano, A. N., Mantsios, A., Mattei, N., Kosorok, M. R., & Culotta, A. (2025). AI can be a powerful social innovation for public health if community engagement is at the core. *Journal of Medical Internet Research*, *27*(1), e68198. Advance online publication. DOI: 10.2196/68198 PMID: 39841529

Bedford, J., Farrar, J., Ihekweazu, C., Kang, G., Koopmans, M., & Nkengasong, J. (2019). A new twenty-first century science for effective epidemic response. *Nature*, *575*(7781), 130–136. DOI: 10.1038/s41586-019-1717-y PMID: 31695207

Bharati, S., Hossain Mondal, M. R., Podder, P., & Prasath, V. B. S. (2022). Federated learning: Applications, challenges and future directions (arXiv:2205.09513). arXiv. https://doi.org//arXiv.2205.09513DOI: 10.48550

Bhardwaj, T., & Sumangali, K. (2025). An explainable federated blockchain framework with privacy-preserving AI optimization for securing healthcare data. *Scientific Reports*, *15*(1), 21799. Advance online publication. DOI: 10.1038/s41598-025-04083-4 PMID: 40595873

Brownstein, J. S., Freifeld, C. C., & Madoff, L. C. (2009). Digital disease detection—Harnessing the web for public health surveillance. *The New England Journal of Medicine*, *360*(21), 2153–2157. DOI: 10.1056/NEJMp0900702 PMID: 19423867

Chamberlain, S. D., Singh, I., Ariza, C., Daitch, A., & Dalziel, B. D. (2020). Real-time detection of COVID-19 epicenters within the United States using a network of smart thermometers (preprint). medRxiv. https://doi.org/DOI: 10.1101/2020.04.06.20039909

Cheah, B. C., Vicente, C. R., & Chan, K. R. (2025). Machine learning and artificial intelligence for infectious disease surveillance, diagnosis, and prognosis. *Viruses*, *17*(7), 882. DOI: 10.3390/v17070882 PMID: 40733500

Chen, F., Wang, L., Hong, J., Jiang, J., & Zhou, L. (2024). Unmasking bias in AI: A systematic review of bias detection and mitigation strategies in electronic health record-based models. *Journal of the American Medical Informatics Association : JAMIA*, *31*(5), 1172–1183. DOI: 10.1093/jamia/ocae060 PMID: 38520723

Chen, X., Chughtai, A. A., & MacIntyre, C. R. (2023). Artificial intelligence in public health: The potential of epidemic early warning systems. *Infectious Disease Reports*, *15*(3), 222–236. DOI: 10.3390/idr15030024

Corburn, J., Vlahov, D., Mberu, B., Riley, L., Caiaffa, W. T., Rashid, S. F., Ko, A., Patel, S., Jukur, S., Martínez-Herrera, E., Jayasinghe, S., Agarwal, S., Nguendo-Yongsi, B., Weru, J., Ouma, S., Edmundo, K., Oni, T., & Ayad, H. (2020). Slum Health: Arresting COVID-19 and Improving Well-Being in Urban Informal Settlements. *Journal of Urban Health : Bulletin of the New York Academy of Medicine*, *97*(3), 348–357. DOI: 10.1007/s11524-020-00438-6 PMID: 32333243

Damiani, G. (2019). Data-driven decision making and stakeholders involvement in public health: A fascinating challenge for sustainable development. In 53 Days of Preventive Medicine – International Congress (Niš, Serbia). Public Health Institute Niš; Faculty of Medicine, University of Niš.

Deiner, M. S., Consoli, S., & Brownstein, J. S. (2025). AI-driven epidemic intelligence: The future of outbreak detection and response. *Frontiers in Artificial Intelligence*, *8*, 1645467. Advance online publication. DOI: 10.3389/frai.2025.1645467 PMID: 40810005

Doerwald, F., Stalling, I., Recke, C., Busse, H., Shrestha, R., Rach, S., & Bammann, K. (2024). A rapid review of digital approaches for the participatory development of health-related interventions. *Frontiers in Public Health*, *12*, 1461422. DOI: 10.3389/fpubh.2024.1461422 PMID: 39678234

Drew, D. A., Nguyen, L. H., Steves, C. J., Menni, C., Freydin, M., Varsavsky, T., Sudre, C. H., Cardoso, M. J., Ourselin, S., Wolf, J., Spector, T. D., Chan, A. T., Chan, A. T., Drew, D. A., Nguyen, L. H., Joshi, A. D., Guo, C.-G., Ma, W., Lo, C.-H., & Spector, T. (2020). Rapid implementation of mobile technology for real-time epidemiology of COVID-19. *Science*, *368*(6497), 1362–1367. DOI: 10.1126/science.abc0473 PMID: 32371477

Elendu, C., Amaechi, D. C., Elendu, T. C., Jingwa, K. A., Okoye, O. K., John Okah, M., Ladele, J. A., Farah, A. H., & Alimi, H. A. (2023). Ethical implications of AI and robotics in healthcare. *Revista de Medicina (São Paulo)*, *102*(50), e36671. DOI: 10.1097/MD.0000000000036671 PMID: 38115340

Faye, S. L. B., Nkweteyim, D., Sow, G. H. C., Diop, B., Diongue, F. B., Cisse, B., Dickson, N., Dia, N., Badu, K., Ayana, G., Ba, M. L., Mensah Gyening, R.-M. O., Ondua, M., Ndiaye, F., Diouf, M., King, R., Duclos, V., Musa, E., Goitom, M., & Kong, J. (2025). Reimagining artificial intelligence for zoonotic disease detection in Africa: A decolonial approach rooted in community engagement and local knowledge. *AI and Ethics*, *5*(5), 5327–5354. DOI: 10.1007/s43681-025-00779-5

Floridi, L., Cowls, J., Beltrametti, M., Chatila, R., Chazerand, P., Dignum, V., Luetge, C., Madelin, R., Pagallo, U., Rossi, F., Schafer, B., Valcke, P., & Vayena, E. (2018). AI4People—An ethical framework for a good AI society. *Minds and Machines*, *28*(4), 689–707. DOI: 10.1007/s11023-018-9482-5 PMID: 30930541

Folasole, A. (2023). Data analytics and predictive modelling approaches for identifying emerging zoonotic infectious diseases: Surveillance techniques, prediction accuracy, and public health implications. *International Journal of Engineering Technology Research & Management*, *7*(12), 292–300.

Fontes, C., Hohma, E., Corrigan, C. C., & Lütge, C. (2022). AI-powered public surveillance systems: Why we (might) need them and how we want them. *Technology in Society*, *71*, 102137. DOI: 10.1016/j.techsoc.2022.102137

Frohlich, K. L., & Abel, T. (2014). Environmental justice and health practices: Understanding how health inequities arise at the local level. In *From Health Behaviours to Health Practices* (pp. 43–56). Palgrave Macmillan. DOI: 10.1002/9781118898345.ch5

Gao, Q., Chen, L., & Huang, Z. (2026). Opportunities and challenges of artificial intelligence in public health: A systematic review on technological efficacy, ethical dilemmas, and governance pathways. *Frontiers in Public Health*, *13*, 1748797. DOI: 10.3389/fpubh.2025.1748797 PMID: 41613098

Geneviève, L. D., Martani, A., Wangmo, T., Paolotti, D., Koppeschaar, C., Kjelsø, C., Guerrisi, C., Hirsch, M., Woolley-Meza, O., Lukowicz, P., Flahault, A., & Elger, B. S. (2019). Participatory Disease Surveillance Systems: Ethical Framework. *Journal of Medical Internet Research*, *21*(5), e12273. DOI: 10.2196/12273 PMID: 31124466

Giri, P. A., & Gupta, P. (2024). Transforming disease surveillance through artificial intelligence. *Indian Journal of Community Medicine : Official Publication of Indian Association of Preventive & Social Medicine*, *49*(5), 663–664. DOI: 10.4103/ijcm.ijcm_459_24 PMID: 39421515

Gomes de Sousa, J., Marques, A. D., & Silva, N. (2025). Harnessing artificial intelligence for enhanced public health surveillance: A narrative review. *Frontiers in Public Health*, *13*, 1601151. Advance online publication. DOI: 10.3389/fpubh.2025.1601151 PMID: 40809756

Haider, H. (2009). Community-based approaches to peacebuilding in conflict-affected and fragile contexts.

Hapicha, E. M. H. (2025). Integrating predictive analytics and community engagement in Kenya: A framework for technology-driven wildlife conservation in private conservancies. National Security: A Journal of the National Defence University–Kenya, 3(1), 50–67.

Johnson, B. (2025). Artificial intelligence as an anti-corruption tool (AI-ACT): Potentials and pitfalls for top-down and bottom-up approaches.

Kangana, N., Kankanamge, N., De Silva, C., Goonetilleke, A., Mahamood, R., & Ranasinghe, D. (2024). Bridging community engagement and technological innovation for creating smart and resilient cities: A systematic literature review. *Smart Cities*, *7*(6), 3823–3852. DOI: 10.3390/smartcities7060147

Kelley, K., Gozzi, N., Mazzoli, M., & Paolotti, D. (2025). Exploring influenza vaccination determinants through digital participatory surveillance. *BMC Public Health, 25*(1), 1345. DOI: 10.1186/s12889-025-22496-8 PMID: 40211245

Lalor, J. P., Abbasi, A., Oketch, K., Yang, Y., & Forsgren, N. (2024). Should fairness be a metric or a model? A model-based framework for assessing bias in machine learning pipelines. *ACM Transactions on Information Systems, 42*(4), 99. Advance online publication. DOI: 10.1145/3641276

López-Ramos, L. M., Leiser, F., Rastogi, A., Hicks, S., Strümke, I., Madai, V. I., Budig, T., Sunyaev, A., & Hilbert, A. (2024). Interplay between federated learning and explainable artificial intelligence: A scoping review (arXiv:2411.05874v1). arXiv. https://arxiv.org/abs/2411.05874v1

McGowan, C. R., Takahashi, E., Romig, L., Bertram, K., Kadir, A., Cummings, R., & Cardinal, L. J. (2022). Community-based surveillance of infectious diseases: A systematic review of drivers of success. *BMJ Global Health, 7*(8), e009934. DOI: 10.1136/bmjgh-2022-009934 PMID: 35985697

McMurry, A. J., Zipursky, A. R., Geva, A., Olson, K. L., Jones, J. R., Ignatov, V., Miller, T. A., & Mandl, K. D. (2024). AI-enabled diagnostic prediction within electronic health records to enhance biosurveillance and early outbreak detection. *PLoS Computational Biology, 20*(6), e1012177. Advance online publication. DOI: 10.1371/journal.pcbi.1012177

McNeil, C., Verlander, S., Divi, N., & Smolinski, M. (2022). The Landscape of Participatory Surveillance Systems Across the One Health Spectrum: Systematic Review. *JMIR Public Health and Surveillance, 8*(8), e38551. DOI: 10.2196/38551 PMID: 35930345

Mendes, V. I., Mendes, B. M., Moura, R. P., Lourenço, I. M., Oliveira, M. F., Ng, K. L., & Pinto, C. S. (2025). Harnessing artificial intelligence for enhanced public health surveillance: A narrative review. *Frontiers in Public Health, 13*, 1601151. DOI: 10.3389/fpubh.2025.1601151 PMID: 40809756

Michener, L., Aguilar-Gaxiola, S., Alberti, P. M., Castaneda, M. J., Castrucci, B. C., Harrison, L. M., Hughes, L. S., Richmond, A., & Wallerstein, N. (2020). Engaging with communities — Lessons (re)learned from COVID-19. *Preventing Chronic Disease, 17*, E125. Advance online publication. DOI: 10.5888/pcd17.200250 PMID: 32678059

Mohsin Khan, M., Shah, N., Shaikh, N., Thabet, A., Alrabayah, T., & Belkhair, S. (2025). Towards secure and trusted AI in healthcare: A systematic review of emerging innovations and ethical challenges. *International Journal of Medical Informatics*, *195*, 105780. DOI: 10.1016/j.ijmedinf.2024.105780 PMID: 39753062

Mora-Cantallops, M., García-Barriocanal, E., & Sicilia, M.-Á. (2024). Trustworthy AI Guidelines in Biomedical Decision-Making Applications: A Scoping Review. *Big Data and Cognitive Computing*, *8*(7), 73. DOI: 10.3390/bdcc8070073

Nicholson, C., Jackson, C., & Marley, J. (2013). A governance model for integrated primary/secondary care for the health-reforming first world–results of a systematic review. *BMC Health Services Research*, *13*(1), 528. DOI: 10.1186/1472-6963-13-528 PMID: 24359610

Nouis, S. C., Uren, V., & Jariwala, S. (2025). Evaluating accountability, transparency, and bias in AI-assisted healthcare decision- making: A qualitative study of healthcare professionals' perspectives in the UK. *BMC Medical Ethics*, *26*(1), 89. DOI: 10.1186/s12910-025-01243-z PMID: 40629303

Parker, M., Fraser, C., Abeler-Dörner, L., & Bonsall, D. (2020). Ethics of instantaneous contact tracing using mobile phone apps in the control of the COVID-19 pandemic. *Journal of Medical Ethics*, *46*(7), 427–431. DOI: 10.1136/medethics-2020-106314 PMID: 32366705

Pillai, S. V., & Kumar, R. S. (2021). The role of data-driven artificial intelligence on COVID-19 disease management in public sphere: A review. *Decision (Washington, D.C.)*, *48*(4), 375–389. DOI: 10.1007/s40622-021-00286-8

Radanliev, P. (2025). Privacy, ethics, transparency, and accountability in AI systems for wearable devices. *Frontiers in Digital Health*, *7*, 1431246. Advance online publication. DOI: 10.3389/fdgth.2025.1431246 PMID: 40599875

Rahman, A., Hossain, M. S., Muhammad, G., Kundu, D., Debnath, T., Rahman, M., Khan, M. S. I., Tiwari, P., & Band, S. S. (2022). Federated learning-based AI approaches in smart healthcare: Concepts, taxonomies, challenges and open issues. *Cluster Computing*, •••, 1–41. Advance online publication. DOI: 10.1007/s10586-022-03658-4 PMID: 35996680

Rice, B. T., Rasmus, S., Onders, R., Thomas, T., Day, G., Wood, J., Britton, C., Hernandez-Boussard, T., & Hiratsuka, V. (2025). Community-engaged artificial intelligence: An upstream, participatory design, development, testing, validation, use and monitoring framework for artificial intelligence and machine learning models in the Alaska Tribal Health System. *Frontiers in Artificial Intelligence*, *8*, 1568886. DOI: 10.3389/frai.2025.1568886 PMID: 40260415

Saadatzadeh, R. (2024). The future of public health: Integrating artificial intelligence in disease surveillance and prevention. *Journal of Foresight and Health Governance*, *1*(2), 1–15.

Saheb, T. (2022). Ethically contentious aspects of artificial intelligence surveillance: A social science perspective. *AI and Ethics*, *3*(2), 369–379. DOI: 10.1007/s43681-022-00196-y PMID: 35874304

Salardi, S. (2025). What is the role of institutional bioethics in addressing new technological challenges? The case of the ecological and carbon footprint of the healthcare sector. In *EACME Conference 2025 Book of Abstracts*.

Salathé, M., Althaus, C. L., Anderegg, N., Antonioli, D., Ballouz, T., Bugnon, E., Čapkun, S., Jackson, D., Kim, S. I., Larus, J. R., Low, N., Lueks, W., Menges, D., Moullet, C., Payer, M., Riou, J., Stadler, T., Troncoso, C., Vayena, E., & von Wyl, V. (2022). Participatory disease surveillance systems: Engaging communities in public health monitoring. *JMIR Public Health and Surveillance*, *8*(4), e28159. Advance online publication. DOI: 10.2196/28159

Salathé, M., Althaus, C. L., Neher, R., Stringhini, S., Hodcroft, E., Fellay, J., Zwahlen, M., & Low, N. (2020). COVID-19 epidemic in Switzerland: On the importance of testing, contact tracing, and isolation. *Swiss Medical Weekly*, *150*(1112), w20225. DOI: 10.4414/smw.2020.20225 PMID: 32191813

Santillana, M., Nguyen, A. T., Dredze, M., Paul, M. J., Nsoesie, E. O., & Brownstein, J. S. (2015). Combining search, social media, and traditional data sources to improve influenza surveillance. *PLoS Computational Biology*, *11*(10), e1004513. DOI: 10.1371/journal.pcbi.1004513 PMID: 26513245

Schulte, T., & Bohnet-Joschko, S. (2022). How can big data analytics support people-centred and integrated health services: A scoping review. *International Journal of Integrated Care*, *22*(2), 23. Advance online publication. DOI: 10.5334/ijic.5543 PMID: 35756337

Shachar, C., Gerke, S., & Adashi, E. Y. (2020). AI Surveillance during Pandemics: Ethical Implementation Imperatives. *The Hastings Center Report*, *50*(3), 18–21. DOI: 10.1002/hast.1125 PMID: 32596887

Shenoy, D., Bhat, R., & Krishna Prakasha, K. (2025). Exploring privacy mechanisms and metrics in federated learning. *Artificial Intelligence Review*, *58*(8), 223. DOI: 10.1007/s10462-025-11170-5

Shevtsova, D., Ahmed, A., Boot, I. W. A., Sanges, C., Hudecek, M., Jacobs, J. J. L., Hort, S., & Vrijhoef, H. J. M. (2024). Trust in and acceptance of artificial intelligence applications in medicine: Mixed methods study. *JMIR Human Factors*, *11*, e47031. Advance online publication. DOI: 10.2196/47031 PMID: 38231544

Smolinski, M. S., Crawley, A. W., Olsen, J. M., Jayaraman, T., & Libel, M. (2017). Participatory disease surveillance: Engaging communities directly in reporting, monitoring, and responding to health threats. *JMIR Public Health and Surveillance*, *3*(4), e62. DOI: 10.2196/publichealth.7540 PMID: 29021131

Smolinski, M. S., Crawley, A. W., Olsen, J. M., Jayaraman, T., Libel, M., & Cummings, D. A. T. (2015). Flu Near You: Crowdsourced symptom reporting spanning 2 influenza seasons. *American Journal of Public Health*, *105*(10), 2124–2130. DOI: 10.2105/AJPH.2015.302696 PMID: 26270299

Tawfik, M., Abu-Ein, A. A., Noaman, H. M., Abdelhaliem, A. H., & Fathi, I. S. (2025). FedMedSecure: Federated few-shot learning with cross-attention mechanisms and explainable AI for collaborative healthcare cybersecurity. *Scientific Reports*, *15*(1), 40050. DOI: 10.1038/s41598-025-25107-z PMID: 41238815

Trang, S., Trenz, M., Weiger, W. H., Tarafdar, M., & Cheung, C. M. K. (2020). One app to trace them all? Examining app specifications for mass acceptance of contact-tracing apps. *European Journal of Information Systems*, *29*(4), 415–428. DOI: 10.1080/0960085X.2020.1784046

Unertl, K. M., Schaefbauer, C. L., Campbell, T. R., Senteio, C., Siek, K. A., Bakken, S., & Veinot, T. C. (2016). Integrating community-based participatory research and informatics approaches to improve the engagement and health of underserved populations. *Journal of the American Medical Informatics Association : JAMIA*, *23*(1), 60–73. DOI: 10.1093/jamia/ocv094 PMID: 26228766

Vayena, E., Haeusermann, T., Adjekum, A., & Blasimme, A. (2018). Digital health: Meeting the ethical and policy challenges. *Swiss Medical Weekly*, *148*(304), w14571. DOI: 10.4414/smw.2018.14571 PMID: 29376547

Wang, A., Ramaswamy, V. V., & Russakovsky, O. (2022). Towards intersectionality in machine learning: Including more identities, handling underrepresentation, and performing evaluation. In *Proceedings of the 2022 ACM Conference on Fairness, Accountability, and Transparency* (pp. 336–349). Association for Computing Machinery. https://doi.org/DOI: 10.1145/3531146.3533101

Wheatley, M. C. (2024). *Ethics of surveillance technologies: Balancing privacy and security in a digital age*. Premier Journal of Data Science., DOI: 10.70389/PJDS.100001

Wójcik, O. P., Brownstein, J. S., Chunara, R., & Johansson, M. A. (2014). Public health for the people: Participatory infectious disease surveillance in the digital age. *Emerging Themes in Epidemiology*, *11*(7), 1–7. DOI: 10.1186/1742-7622-11-7 PMID: 24991229

Chapter 3
Genomic and Metagenomic Approaches to Biosurveillance

Olugbemiga Ojo Aliu
http://orcid.org/0000-0003-4942-7454
Auchi Polytechnic, Auchi, Nigeria

ABSTRACT

Genomic and metagenomic technologies have become central to modern biosurveillance because they allow biological threats to be detected, characterized, compared, and tracked at molecular resolution. Whole-genome sequencing can reconstruct pathogen transmission, identify variants and lineages, and support outbreak attribution, while metagenomic sequencing can detect known and unexpected organisms directly from complex clinical, animal, food, wastewater, and environmental samples. In AI-enhanced biosurveillance systems, sequencing data are no longer valuable only as laboratory outputs; they become part of a larger intelligence cycle that includes sampling design, metadata capture, bioinformatics, machine learning, geospatial analysis, dashboard visualization, risk interpretation, and public health or biosecurity action. This chapter reviews the conceptual foundations, sequencing platforms, sampling strategies, laboratory workflows, bioinformatics pipelines, AI applications, One Health use cases, governance challenges, and future directions of genomic and metagenomic biosurveillance.

1.0. INTRODUCTION

Biosurveillance is the systematic collection, analysis, interpretation, and communication of information related to biological threats. Historically, many surveillance

DOI: 10.4018/979-8-3373-9033-8.ch003

systems depended on clinical reporting, culture-based laboratory confirmation, targeted diagnostic assays, syndromic signals, or ecological observations. These approaches remain important, but they are often limited by diagnostic delay, narrow target specificity, incomplete reporting, and limited ability to distinguish related strains or infer transmission pathways. Genomic and metagenomic technologies have altered this landscape by enabling surveillance systems to observe biological threats through their genetic signatures. Rather than asking only whether a known pathogen is present, sequence-informed biosurveillance can ask what lineage is circulating, how it is changing, whether it carries antimicrobial resistance or virulence determinants, where it may have originated, and how it is moving across populations and environments (Gardy & Loman, 2018; Lo & Jamrozy, 2020).

The COVID-19 pandemic demonstrated the operational value of pathogen genomics at global scale. SARS-CoV-2 sequencing enabled lineage naming, variant tracking, investigation of transmission clusters, and public communication through open analytical platforms such as Nextstrain and international data repositories such as GISAID (Hadfield et al., 2018; Rambaut et al., 2020; Shu & McCauley, 2017). However, the promise of genomic surveillance was also unevenly distributed. Countries and regions with limited sequencing infrastructure, reagent supply chains, bioinformatics capacity, and data governance systems were less able to contribute to, or benefit from, real-time genomic intelligence. This inequality remains one of the central challenges for AI-enhanced biosurveillance (World Health Organization [WHO], 2022). A One Health approach is essential for genomic and metagenomic biosurveillance because biological risks often emerge through linked human, animal, food, agricultural, and environmental pathways rather than through isolated clinical systems (Ogwu & Izah, 2023).

Genomic and metagenomic biosurveillance should be understood as complementary rather than interchangeable. Whole-genome sequencing generally focuses on recovering the genome of a specific organism and is particularly useful when a pathogen has already been isolated or detected. Metagenomics, by contrast, sequences mixed genetic material directly from complex samples and can identify microbial communities, known pathogens, unexpected organisms, antimicrobial resistance genes, and ecological disturbance signals without requiring prior culture or a single target assay (Chiu & Miller, 2019; Quince et al., 2017). This makes metagenomics especially valuable for threat-agnostic surveillance, wastewater monitoring, environmental biosurveillance, and early investigation of unexplained disease clusters. Genomic biosurveillance is particularly important for tropical disease systems, where pathogen diversity, vector ecology, environmental variability, and uneven diagnostic infrastructure create persistent challenges for early detection and response (Ogwu & Izah, 2025). The same breadth that makes metagenomics powerful also makes it difficult: host background, contamination, low-abundance organisms, database incompleteness, and ambiguous biological interpretation can all generate uncertainty.

Artificial intelligence (AI) adds another layer to this transformation. Machine learning can assist with pathogen classification, anomaly detection, antimicrobial resistance prediction, variant prioritization, literature mining, and integration of genomic, epidemiological, geospatial, and environmental data. Yet AI cannot compensate for weak sampling, poor metadata, unvalidated pipelines, or biased reference databases. A central argument of this chapter is that genomic biosurveillance is not simply a matter of sequencing more samples or applying more complex algorithms. It is a full decision-support ecosystem linking field sampling, laboratory quality, computational analysis, data governance, and responsible action.

This chapter aims to examine how genomic and metagenomic technologies are transforming biosurveillance from conventional pathogen detection into molecularly informed, AI-enhanced biological intelligence. It explains the conceptual foundations, sequencing platforms, sampling strategies, laboratory workflows, bioinformatics pipelines, machine-learning applications, One Health use cases, ethical challenges, and future directions that shape genomic and metagenomic biosurveillance across human, animal, plant, food, wastewater, and environmental systems. The chapter contributes an advanced framework for understanding how sequence data can support pathogen discovery, outbreak reconstruction, antimicrobial resistance monitoring, wastewater-based surveillance, environmental DNA analysis, food and agricultural biosecurity, and threat-agnostic early warning when combined with representative sampling, validated pipelines, interoperable metadata, responsible governance, and decision-ready interpretation.

2.0 CONCEPTUAL FOUNDATIONS OF GENOMIC AND METAGENOMIC BIOSURVEILLANCE

2.1 Genomics, Metagenomics, and Biosurveillance Intelligence

A genome is the complete genetic material of an organism or virus, while a metagenome is the collective genetic material recovered from all organisms present in a sample. In biosurveillance, genomic data are used to identify biological agents, compare related isolates, infer evolutionary relationships, detect variants, and evaluate features such as virulence or antimicrobial resistance. Metagenomic data are used to profile communities, discover organisms, detect multiple targets simultaneously, and characterize the functional potential of microbial assemblages. When linked with epidemiological and environmental metadata, these molecular data can become actionable intelligence for outbreak response, infection prevention, environmental monitoring, food safety, agricultural biosecurity, and pandemic preparedness.

The distinction between data, information, intelligence, and action is essential. Raw reads from a sequencer are data. A classified organism list or consensus genome is information. A finding that a resistant pathogen is increasing in a hospital, wastewater catchment, livestock corridor, or food supply chain becomes intelligence only when placed in context. The final step is action: issuing alerts, changing treatment guidance, intensifying sampling, recalling contaminated food, implementing infection control, or informing community risk communication. The value of genomic biosurveillance therefore depends on interpretation and governance as much as on sequencing. Figure 1 shows how genomic and metagenomic biosurveillance converts biological samples into actionable intelligence. It emphasizes that sequencing alone is not surveillance. Molecular data become useful only when they are processed, interpreted with metadata, translated into risk intelligence, and connected to public health, environmental, agricultural, or biosecurity action.

Figure 1. From biological sample to genomic biosurveillance intelligence

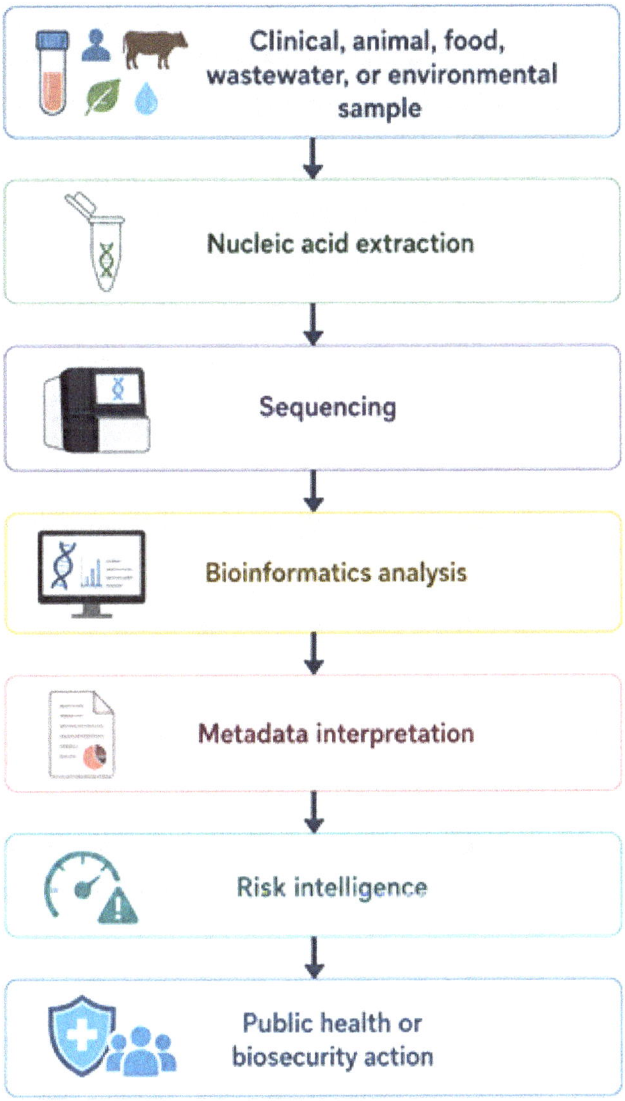

2.2 From Targeted Detection to Threat-Agnostic Surveillance

Targeted assays such as PCR, qPCR, and digital PCR remain indispensable for rapid, sensitive, and cost-effective detection of known organisms or genes. Their limitation is that they only detect what the assay is designed to detect. Whole-genome sequencing extends targeted detection by providing high-resolution characterization of a known organism. Amplicon sequencing profiles selected marker genes, such as 16S rRNA for bacteria and archaea or ITS regions for fungi. Shotgun metagenomics sequences nucleic acids more broadly, enabling organism detection, community profiling, genome reconstruction, and functional analysis. Metatranscriptomics focuses on RNA and can provide evidence of active gene expression or RNA viruses, although it is more sensitive to sample degradation and requires additional laboratory care (Chiu & Miller, 2019; Quince et al., 2017).

Threat-agnostic surveillance aims to detect biological signals without restricting analysis to one preselected pathogen. This approach is particularly important for emerging infections, novel variants, unexplained syndromes, environmental reservoirs, and complex microbial ecosystems. However, threat-agnostic does not mean assumption-free. Every workflow includes assumptions about sampling location, nucleic acid extraction, sequencing depth, host read depletion, reference databases, taxonomic thresholds, and reporting criteria. The scientific strength of metagenomic surveillance depends on making these assumptions explicit and validating them where possible.

2.3 One Health Framing

The One Health framing is central to genomic and metagenomic biosurveillance because many biological threats emerge or circulate across human, animal, plant, food, and environmental systems. Zoonotic spillover, antimicrobial resistance, foodborne outbreaks, crop disease, wastewater signals, and vector-borne pathogens cannot be understood adequately from a single-sector perspective. Genomic tools allow comparisons among isolates or genetic signals collected from humans, livestock, wildlife, vectors, foods, water, soil, and built environments. When linked with ecological and social data, they can help identify transmission interfaces, reservoirs, and intervention points.

3.0 SEQUENCING TECHNOLOGIES USED IN BIOSURVEILLANCE

3.1 Whole-Genome Sequencing

Whole-genome sequencing (WGS) is one of the most established genomic approaches for pathogen surveillance. For bacteria, WGS can identify outbreak clusters, track source attribution, detect resistance and virulence determinants, and support infection control. For viruses, WGS enables lineage assignment, variant tracking, phylogenetic analysis, and monitoring of mutations that may influence transmissibility, immune escape, diagnostics, or therapeutics. Early genomic studies of Ebola virus showed how sequencing could clarify origin, transmission, and evolution during an outbreak (Gire et al., 2014). Portable sequencing later demonstrated that genome generation could be moved closer to the field during Ebola surveillance, shortening the distance between sample collection and epidemiological interpretation (Quick et al., 2016).

3.2 Amplicon Sequencing

Amplicon sequencing targets specific genomic regions and is commonly used for microbial community profiling. In environmental and agricultural biosurveillance, 16S rRNA sequencing supports bacterial community analysis, ITS sequencing supports fungal profiling, and targeted viral or pathogen amplicons can support variant tracking. Amplicon approaches are cost-effective and scalable, but they provide limited resolution compared with shotgun sequencing and may be influenced by primer bias, amplification artifacts, and taxonomic limitations of marker genes. Error-correction tools such as DADA2 improved the resolution of amplicon sequence analysis by inferring exact sequence variants rather than relying only on broad operational taxonomic units (Callahan et al., 2016).

3.3 Shotgun Metagenomic Sequencing

Shotgun metagenomics sequences mixed DNA or RNA-derived complementary DNA from a sample. It can identify bacteria, viruses, fungi, parasites, antimicrobial resistance genes, virulence factors, plasmids, and microbial community structure in a single workflow. It can also support metagenome-assembled genomes and functional profiling. The approach is especially relevant for wastewater, environmental monitoring, food safety, and cases in which the etiological agent is unknown. Its limitations include high host background, low pathogen abundance, contamination, large computational requirements, and dependence on reference databases (Chiu & Miller, 2019; Quince et al., 2017).

3.4 Metatranscriptomics

Metatranscriptomics sequences RNA and can identify actively expressed genes, RNA viruses, and functional states of microbial communities. In biosurveillance, this can be important when the objective is to distinguish potential biological activity from residual DNA. RNA-based methods are valuable for respiratory virus surveillance, wastewater monitoring of RNA viruses, and detection of microbial stress or functional shifts. However, RNA is more labile than DNA, and field workflows must address preservation, cold-chain requirements, inhibitors, and rapid degradation.

3.5 Long-Read and Portable Sequencing

Long-read sequencing platforms, especially nanopore sequencing, have expanded the possibilities for real-time and field-deployable biosurveillance. Nanopore platforms can generate long reads, support rapid sample-to-answer workflows, and be deployed outside centralized laboratories. The MinION platform demonstrated the feasibility of portable sequencing and has been applied in outbreak settings, including Ebola and Zika virus surveillance (Jain et al., 2016; Quick et al., 2016; Quick et al., 2017). Long reads are helpful for resolving structural variation, plasmids, repetitive regions, and complete genomes. Their limitations include error profiles, sample preparation requirements, and the need for robust analysis pipelines.

3.6 Target Enrichment and Hybrid Capture

Target enrichment methods, including hybrid-capture sequencing, increase the proportion of pathogen or genomic target reads in a sample. This can be valuable when pathogen nucleic acids are overwhelmed by host or environmental background, as often occurs in clinical, vector, wildlife, or low-biomass samples. Enrichment increases sensitivity but can reduce the breadth of agnostic detection if probe sets are narrowly designed. In surveillance practice, the choice between enrichment and shotgun metagenomics depends on whether the priority is sensitivity for a defined target set or broader discovery capacity.

Table 1 compares the major sequencing approaches used in genomic and metagenomic biosurveillance, highlighting their typical applications, strengths, and operational limitations. It shows that no single sequencing method is universally sufficient: whole-genome sequencing provides high-resolution characterization of known pathogens, amplicon sequencing is scalable for targeted marker-gene profiling, shotgun metagenomics supports untargeted analysis of complex samples, and metatranscriptomics can reveal RNA viruses or active biological signals. Long-read sequencing and hybrid capture further expand surveillance capacity by improving field deployment, structural resolution, and low-abundance target recovery. However, each method also carries important constraints, including primer bias, host-background interference, contamination, database dependence, RNA instability, sequencing-error profiles, and reduced agnostic breadth in enriched workflows. The choice of method should therefore be driven by the surveillance question, sample type, required resolution, available infrastructure, and intended decision pathway.

Table 1. Major sequencing approaches used in biosurveillance

Approach	Typical use	Strengths	Limitations
Whole-genome sequencing	Characterization of known pathogens and isolates	High resolution for lineage, outbreak, and AMR analysis	Usually requires target organism enrichment, culture, or prior detection
Amplicon sequencing	Marker-gene profiling and targeted variant surveillance	Cost-effective, scalable, suitable for many samples	Primer bias, limited functional information, lower taxonomic resolution
Shotgun metagenomics	Untargeted analysis of complex samples	Detects multiple organisms and genes in one workflow	Host background, contamination, database dependence, higher cost
Metatranscriptomics	RNA viruses and active functional signals	Can identify expressed genes and active RNA signals	RNA instability and stricter preservation requirements
Long-read sequencing	Field sequencing, structural variation, complete genomes	Portable, rapid, useful for long-range context	Higher error rates than short-read sequencing in some workflows
Hybrid capture	Low-abundance target recovery	Improves sensitivity for selected pathogen groups	Less fully agnostic than untargeted metagenomics

4.0. SAMPLING STRATEGIES FOR GENOMIC AND METAGENOMIC BIOSURVEILLANCE

Sampling is often the most important and least visible determinant of genomic surveillance quality. A sophisticated model cannot correct a surveillance system that samples only convenient locations, omits high-risk populations, ignores seasonal dynamics, or lacks metadata. In genomic biosurveillance, sampling decisions determine what can be detected, what can be compared, and how confidently results can be interpreted.

4.1 Clinical and Public Health Sampling

Clinical sampling supports diagnosis, outbreak investigation, hospital infection control, variant surveillance, and antimicrobial resistance monitoring. Samples may include respiratory swabs, blood, cerebrospinal fluid, stool, wound swabs, or isolates from diagnostic laboratories. In routine surveillance, the goal may not be to sequence every positive case but to design representative or risk-based sampling that captures geography, time, disease severity, travel history, vaccine breakthrough, unusual presentations, or suspected clusters. For hospital outbreaks, genomic data can distinguish related transmission events from coincidental infections caused by similar organisms.

4.2 Animal, Wildlife, and Vector Sampling

Animal and vector sampling is essential for zoonotic and One Health surveillance. Livestock, wildlife, companion animals, mosquitoes, ticks, flies, and other vectors can serve as sentinels, reservoirs, or bridges between ecosystems and human populations. Metagenomic analysis of pooled vectors can detect known pathogens and characterize vector-associated viromes or microbiomes. However, pooled sampling complicates prevalence estimation and requires careful metadata on collection site, species identification, pool size, date, and environmental context.

4.3 Plant, Food, and Agricultural Sampling

In agricultural biosurveillance, genomic methods can detect crop pathogens, soil-borne disease agents, seed contamination, invasive pests, and microbial shifts linked to plant health. Food-system sampling can support outbreak source attribution and trace contamination across farms, processing plants, distribution networks, and retail environments. The value of sequencing in food and agricultural systems depends on chain-of-custody documentation, harmonized sampling protocols, and coordination between public health, agriculture, and regulatory agencies.

4.4 Wastewater and Environmental Sampling

Wastewater-based surveillance provides population-level information that can complement clinical surveillance, especially when many infections are asymptomatic, untested, or underreported. Wastewater sequencing can track SARS-CoV-2 variants, detect poliovirus, monitor influenza and other pathogens, and characterize antimicrobial resistance genes in community sewage (Hendriksen et al., 2019; Lo & Jamrozy, 2020). Environmental sampling can also include surface water, sediment, aerosols, soil, animal habitats, markets, and built environments. Interpretation requires caution because environmental signals may represent viable organisms, degraded nucleic acids, transient contamination, or upstream mixture from many sources. Metagenomic surveillance of water systems is valuable because microbial contamination can provide early evidence of public-health risk, especially in settings where clinical detection is delayed or sanitation infrastructure is uneven (Izah & Ogwu, 2025). Environmental biosurveillance should also include airborne exposure contexts because air pollutants, microbial aerosols, respiratory vulnerability, and ecological degradation can interact in ways that influence human and environmental health risk (Izah et al., 2024).

4.5 Sampling Design Considerations

Effective genomic biosurveillance requires explicit attention to spatial representativeness, temporal frequency, sample size, risk stratification, metadata completeness, and the action threshold that will follow detection. Sentinel sites can provide consistent trend data, while adaptive sampling can be intensified around outbreaks, ports of entry, high-risk interfaces, hospitals, farms, wastewater catchments, or ecological hotspots. Metadata should include date, location, sample type, host or matrix, collection method, preservation conditions, sequencing platform, laboratory protocol, and relevant clinical, epidemiological, ecological, or environmental variables.

5.0. LABORATORY WORKFLOWS AND QUALITY ASSURANCE

The credibility of genomic biosurveillance depends on laboratory quality. Because sequencing can detect minute quantities of nucleic acid, it can also detect contamination introduced during sampling, extraction, library preparation, sequencing, or data handling. Quality assurance should therefore be designed into the workflow rather than added after results are generated.

5.1 Sample Preservation and Nucleic Acid Extraction

DNA and RNA extraction protocols must be suitable for the sample matrix and surveillance objective. Wastewater, soil, feces, food, and plant tissues may contain inhibitors that reduce amplification or sequencing performance. Low-biomass samples are especially vulnerable to reagent contamination. RNA workflows require preservation strategies that minimize degradation. Extraction methods also influence observed community composition, making protocol consistency essential for longitudinal surveillance.

5.2 Library Preparation and Sequencing Depth

Library preparation converts extracted nucleic acids into sequencing-ready molecules. Decisions about amplicon versus shotgun libraries, RNA versus DNA sequencing, target enrichment, multiplexing, and sequencing depth shape the sensitivity and interpretability of results. Shallow sequencing may be adequate for dominant organisms or known targets but insufficient for low-abundance pathogens, variant reconstruction, or resistome analysis. Excessive sequencing depth increases cost without always producing proportional surveillance value. The correct depth depends on the question, matrix complexity, expected pathogen abundance, and required confidence.

5.3 Contamination Control

Controls are mandatory in metagenomic and low-biomass sequencing. Negative controls, extraction blanks, positive controls, mock communities, and replicate samples help identify contamination, batch effects, and pipeline errors. Index hopping and barcode misassignment can create false signals when highly abundant samples are sequenced alongside low-abundance samples. Laboratory separation of pre- and post-amplification areas, reagent tracking, and careful documentation reduce the risk of false detections.

5.4 Biosafety and Biosecurity

Sequencing laboratories may handle infectious material, unknown pathogens, or sensitive genomic data. Biosafety protocols must match sample risk, and inactivation procedures should be validated when infectious agents may be present. Biosecurity considerations include secure storage, controlled access, responsible data sharing, and awareness of dual-use risks. Pathogen sequence data can support public health, but it can also raise concerns when linked to virulence, transmissibility, or synthetic biology applications.

6.0. BIOINFORMATICS PIPELINES FOR BIOSURVEILLANCE

Bioinformatics converts raw sequence reads into interpretable surveillance outputs. A typical pipeline includes basecalling, demultiplexing, trimming, quality filtering, host read removal, taxonomic classification, assembly or mapping, variant calling, phylogenetics, resistance gene detection, and reporting. Each step can introduce bias, and each parameter choice should be documented. Figure 2 summarizes the major analytical workflow that connects laboratory sequencing to surveillance outputs. It follows the pathway from raw sequence generation through quality control, host-read removal, classification, assembly, variant calling, resistance detection, and reporting. It is useful because it simplifies a technically dense section and helps readers see where uncertainty, bias, and quality-control decisions enter the pipeline.

6.1 Raw Data Processing

Raw reads require quality assessment before interpretation. Trimming removes adapters and low-quality regions. Host read removal reduces privacy risks and improves microbial analysis, especially for clinical and animal samples. For nanopore sequencing, basecalling models and read-quality thresholds affect downstream results. In wastewater and environmental surveillance, quality filtering must balance sensitivity to low-abundance organisms against removal of spurious reads.

6.2 Mapping, Assembly, and Genome Reconstruction

Reference-based mapping aligns reads to a known genome and is efficient when the target is well characterized. De novo assembly reconstructs contigs without requiring a close reference and is useful for novel organisms or complex communities. Hybrid assembly can combine short-read accuracy with long-read continuity. Tools such as Bowtie 2, minimap2, and Prokka illustrate common components of genome mapping and annotation workflows (Langmead & Salzberg, 2012; Li, 2018; Seemann, 2014). For metagenome-assembled genomes, quality assessment tools such as CheckM can estimate completeness and contamination (Parks et al., 2015).

6.3 Taxonomic Classification and Reference Databases

Taxonomic classification assigns reads or contigs to organisms. K-mer-based tools such as Kraken 2 are widely used because they are fast and scalable for large metagenomic datasets (Wood et al., 2019). However, classification accuracy depends on database content, taxonomic curation, genome representation, and thresholds for reporting. False positives can occur from conserved sequences, contamination, database errors, or overly permissive thresholds. False negatives can occur when organisms are absent from reference databases or too divergent from known sequences.

6.4 Variant Calling, Lineage Assignment, And Phylogenetics

Variant calling identifies genomic differences such as single nucleotide variants, insertions, deletions, and structural changes. In pathogen surveillance, these differences support lineage assignment, outbreak cluster detection, evolutionary analysis, and monitoring of variants of concern. Phylogenetic analysis places sequences in evolutionary context and can help infer introductions, transmission clusters, and geographic spread. Nextstrain demonstrated the value of integrating phylogenetics with real-time visualization for pathogen evolution and public health communication (Hadfield et al., 2018; Neher & Bedford, 2018).

6.5 Antimicrobial Resistance and Virulence Detection

Resistance gene detection is central to One Health biosurveillance. Databases such as the Comprehensive Antibiotic Resistance Database support identification of known resistance determinants and interpretation of resistance mechanisms (Alcock et al., 2020; McArthur et al., 2013). Sequencing-based AMR analysis can detect resistance genes, plasmids, mobile genetic elements, and resistome patterns across clinical, agricultural, and environmental samples (Boolchandani et al., 2019). However, genotype does not always predict phenotype perfectly. Expression, gene context, copy number, mutations, host species, and clinical breakpoints all influence resistance interpretation.Although genomic AMR surveillance focuses on resistance detection and transmission intelligence, it is also linked to broader antimicrobial governance, including the evaluation of alternative antimicrobial sources and strategies that may reduce dependence on conventional antibiotics (Ogwu & Izah, 2025).

6.6 Workflow Automation and Reproducibility

Reproducible workflows are essential for surveillance systems that must generate comparable results over time and across institutions. Workflow managers such as Snakemake and Nextflow support automated, version-controlled, and portable analyses (Di Tommaso et al., 2017; Köster & Rahmann, 2012). Containerization, shared reference databases, standardized metadata, and documented parameters reduce analytic drift. Reproducibility is not only a technical issue; it is a governance requirement when decisions affect public health, agriculture, trade, or community trust.

Figure 2. Core workflow for genomic and metagenomic biosurveillance

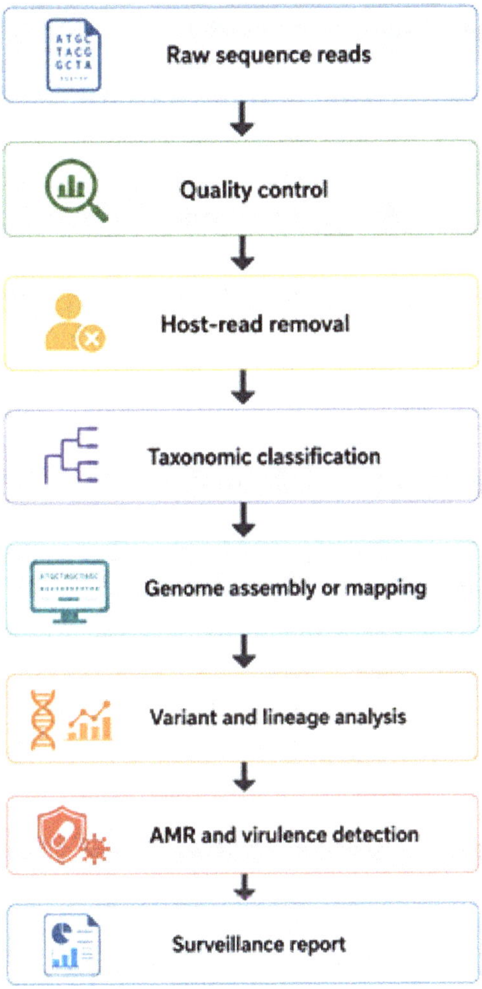

7.0. ARTIFICIAL INTELLIGENCE AND MACHINE LEARNING IN GENOMIC BIOSURVEILLANCE

AI methods can increase the speed and scale of genomic biosurveillance, but they must be deployed with careful validation. Machine learning can help classify sequences, detect anomalies, identify clusters, predict antimicrobial resistance, prioritize samples, and integrate genomic data with epidemiological and environmental covariates. Deep learning approaches have been developed to identify viral sequences from metagenomic data, including reference-free approaches designed to improve detection of divergent viruses (Ren et al., 2020).

7.1 AI-Enabled Pattern Recognition and Anomaly Detection

Anomaly detection can identify unusual genomic patterns, unexpected community shifts, emerging resistance signatures, or deviations from historical wastewater trends. These methods are useful when surveillance systems generate continuous high-volume data. However, anomaly does not equal threat. Models should distinguish technical artifacts, sampling changes, seasonal variation, and real biological shifts. Human review remains necessary, especially when action may affect communities, markets, or health systems.

7.2 Machine Learning for Classification and Prediction

Supervised learning can classify sequences, predict resistance, or estimate the probability that a sample belongs to a known outbreak cluster. Unsupervised learning can cluster samples, identify community patterns, or reveal latent structure in high-dimensional genomic data. Predictive models may integrate genomic features with mobility, climate, wastewater flow, land use, hospital admissions, veterinary reports, or trade patterns. These integrations are powerful, but they require harmonized metadata, transparent feature selection, and rigorous validation.

7.3 AMR Prediction and Risk Scoring

AMR prediction is one of the clearest applications of AI in genomic biosurveillance. Machine learning can connect genotype patterns to phenotypic resistance, identify combinations of genes and mutations, and support risk scoring across isolates or metagenomes. Nevertheless, models trained on biased clinical datasets may perform poorly in environmental, animal, or geographically underrepresented contexts. Surveillance systems should communicate confidence, uncertainty, and intended use rather than presenting predictions as definitive clinical interpretations.

7.4 Explainability and Governance

AI-enhanced genomic biosurveillance requires explainability. Decision-makers need to know why a model flagged a lineage, a wastewater catchment, a resistance pattern, or a transmission cluster. Model governance should include version control, validation datasets, bias assessment, audit trails, and procedures for human review. The most useful AI systems are not black boxes that replace expertise, but decision-support tools that help experts interpret complex signals more quickly and consistently.

8.0. APPLICATIONS ACROSS BIOSURVEILLANCE DOMAINS

8.1 Public Health and Outbreak Response

In public health, genomic surveillance supports outbreak detection, transmission reconstruction, variant monitoring, and infection prevention. During healthcare-associated outbreaks, WGS can determine whether cases share a recent common source or represent unrelated introductions. During regional epidemics, phylogenetics can identify introductions, local transmission, and spread between communities. Genomic information is strongest when combined with epidemiological context, because sequence similarity alone cannot prove direct transmission.

8.2 Wastewater-Based Genomic Surveillance

Wastewater surveillance is a major example of population-level genomic biosurveillance. It can detect pathogen signals from symptomatic and asymptomatic individuals and provide early warning when clinical testing declines. Sequencing wastewater can identify circulating variants, resistance genes, and multiple pathogens in a shared environmental matrix. However, wastewater is a mixed sample, so lineage reconstruction can be difficult when multiple variants coexist. Flow rates, rainfall, industrial inputs, sewershed structure, and sampling frequency also affect interpretation (Hendriksen et al., 2019; Lo & Jamrozy, 2020).

8.3 Antimicrobial Resistance Surveillance

AMR surveillance benefits from genomic and metagenomic approaches because resistance circulates across hospitals, communities, farms, wastewater, wildlife, and natural environments. Clinical isolate sequencing can characterize resistant pathogens and outbreaks, while metagenomics can profile resistomes in sewage, soil, livestock systems, and food environments. Global sewage metagenomics has shown that urban wastewater can provide broad information on antimicrobial resistance patterns, although interpretation must consider infrastructure, sanitation, antibiotic use, and environmental context (Hendriksen et al., 2019). Genomic and metagenomic AMR surveillance is especially important for food-animal systems because antimicrobial residues and resistant organisms can connect livestock production, consumer exposure, food safety, and broader One Health risk (Izah et al., 2025).

8.4 Zoonotic, Wildlife, and Vector-Borne Disease Surveillance

Genomic and metagenomic surveillance can detect pathogens at interfaces where humans, animals, vectors, and environments interact. Wildlife and vector metagenomics can identify known pathogens and reveal viral diversity, while livestock sequencing can track transmission and resistance in food-animal systems. These approaches are valuable for early warning but must avoid simplistic interpretations. Detection of a pathogen-like sequence in a reservoir or vector does not automatically imply imminent human disease risk. Risk assessment requires ecological, behavioral, host-range, and exposure data. The detection and monitoring of contaminants in agro-aqua food systems demonstrate why genomic and metagenomic data should be interpreted alongside environmental exposure pathways, food-chain evidence, and One Health risk assessment (Ogwu et al., 2024).

8.5 Plant Health, Food Safety, and Agricultural Biosecurity

Plant and food-system biosurveillance are increasingly important as climate change, trade, and land-use change alter pathogen movement. Sequencing can detect crop pathogens, characterize soil microbial communities, trace foodborne outbreaks, and support microbial source tracking across supply chains. In food safety, genome comparisons can connect clinical isolates with food, processing facilities, or environmental sources. In plant health, metagenomics can support detection of emerging pathogens in seeds, soils, nurseries, and traded commodities. Food safety systems increasingly require molecularly informed surveillance because microbial hazards, contaminants, adulteration, antimicrobial residues, and supply-chain failures can generate risks that move across production, processing, distribution, and consumption systems (Ogwu et al., 2024). Plant genomic biosurveillance is increasingly important for agricultural biosecurity because fungal pathogens, crop disease pressure, and emerging control technologies intersect with food security, environmental sustainability, and public-health protection (Ogwu & Izah, 2025).

8.6 Environmental and Ecosystem Biosurveillance

Environmental DNA and metagenomic approaches can monitor biodiversity, pathogens, harmful algal blooms, microbial contamination, and ecosystem disturbance. These approaches extend biosurveillance beyond human disease by recognizing that ecological disruption can precede or amplify biological risk. Environmental sequence data can reveal shifts in microbial community composition, pathogen reservoirs, and resistance genes. Yet environmental detection must be interpreted carefully because DNA can persist after organism death, be transported from distant sources, or occur at levels with unclear biological significance. Environmental metagenomics is strongest when embedded within broader biomonitoring frameworks that track pollutants, microbial indicators, ecological stressors, and exposure pathways across water, soil, air, sediments, and biological matrices (Izah et al., 2024).

9.0. DATA INTEGRATION, DASHBOARDS, AND DECISION SUPPORT

Genomic biosurveillance becomes most useful when sequence data are integrated with other information streams. These may include case counts, hospital admissions, vaccination data, antimicrobial use, travel history, animal movements, weather, hydrology, land use, trade flows, wastewater flow data, and social vulnerability indicators. Dashboards can make these signals visible to decision-makers, but dashboards should not be confused with decisions. Effective systems must define thresholds, uncertainty, user roles, and response protocols.

9.1 Integrating Genomic, Epidemiological, and Environmental Data

Integration enables richer interpretation. A variant detected in wastewater becomes more meaningful if paired with clinical testing, sewershed population, wastewater flow, and prior lineage trends. An AMR gene detected in agricultural runoff becomes more meaningful if connected to antibiotic use, livestock density, rainfall, and downstream water exposure. A wildlife pathogen signal becomes more actionable if linked to host distribution, vector abundance, land-use change, and human contact patterns.

9.2 Dashboards and Alert Systems

Dashboards should translate complex sequence data into interpretable indicators such as lineage frequency, variant growth, resistance gene abundance, detection confidence, spatial hotspots, or trend changes. Alert systems should be calibrated to avoid both alarm fatigue and delayed response. Each alert should indicate the evidence base, uncertainty, recommended follow-up, and whether the signal requires confirmatory testing. For public communication, uncertainty should be explained plainly rather than hidden.

Figure 3 illustrates how genomic and metagenomic surveillance can integrate signals across One Health domains. Human, animal, plant, food, wastewater, and environmental samples feed into a shared sequencing and interpretation platform. It reinforces the chapter's argument that genomic biosurveillance is strongest when sequence data are linked across sectors rather than interpreted in isolated laboratories or single-disease programs.

Figure 3. One health genomic surveillance network

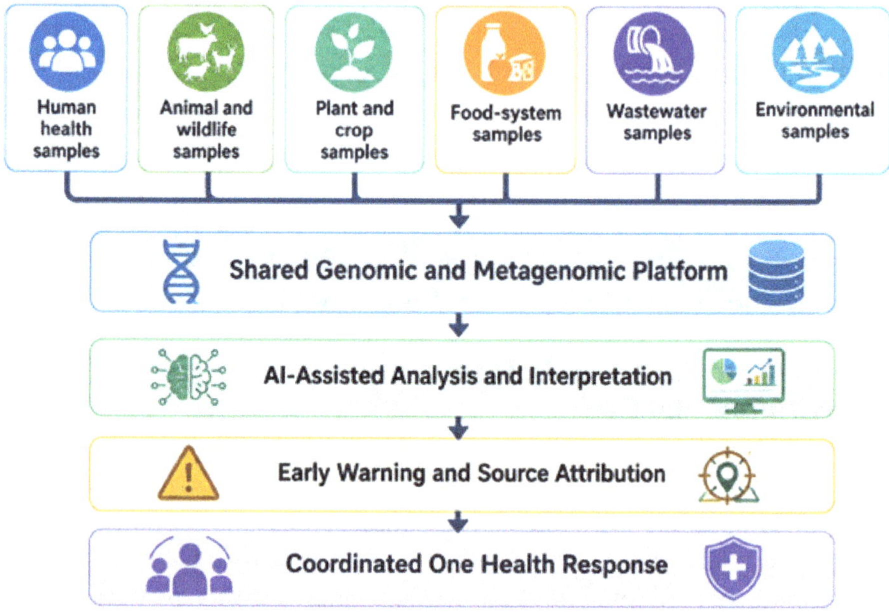

9.3 From Sequence Signals to Action

Actionable genomic intelligence may support targeted sampling, outbreak investigation, clinical guidance, public advisories, vaccination strategy, infection control, food recall, farm biosecurity, or environmental remediation. The key is to define the decision pathway in advance. If a surveillance system cannot state what action follows a positive signal, then the system risks collecting data without operational value.

Table 2 summarizes common genomic and metagenomic biosurveillance signals, their possible interpretations, and appropriate follow-up actions. It emphasizes that sequence-based detection should not be treated as automatic evidence of a confirmed outbreak or direct transmission event. A novel lineage, rising wastewater signal, high antimicrobial resistance gene abundance, foodborne genome match, wildlife-vector detection, or microbial community shift must be interpreted alongside metadata, sampling design, clinical evidence, environmental context, and confirmatory testing. This supports the chapter's central argument that genomic and metagenomic outputs become useful only when translated into risk intelligence and linked to clear decision pathways.

Table 2. Examples of genomic biosurveillance outputs and decisions

Signal	Potential interpretation	Possible follow-up action
Novel lineage detected in clinical sequences	New introduction or local evolution	Increase sequencing, review travel and cluster data, monitor growth rate
Rising pathogen RNA in wastewater	Community transmission may be increasing	Confirm with clinical data, intensify wastewater sampling, communicate trends
High AMR gene abundance in wastewater	Potential resistome burden or contamination source	Compare across sites, assess antibiotic use and sanitation context
Foodborne pathogen genome matches clinical isolates	Possible supply-chain link	Traceback investigation, targeted inspection, recall assessment
Pathogen-like sequence in wildlife vectors	Potential reservoir or vector signal	Confirm detection, identify host/vector species, assess exposure risk
Microbial community shift in soil or water	Environmental disturbance or contamination	Repeat sampling, integrate chemical and ecological indicators

10.0. ETHICAL, LEGAL, AND SOCIAL IMPLICATIONS

Genomic biosurveillance raises ethical and governance questions because molecular data can identify communities, reveal sensitive health or environmental information, influence trade, and affect public trust. Ethical design should therefore begin before sample collection.

10.1 Privacy and Community-Level Surveillance

Clinical pathogen genomics can indirectly reveal information about patients, outbreaks, travel, or social networks. Wastewater surveillance usually reports community-level signals, but small catchments can create privacy concerns if linked to schools, dormitories, prisons, workplaces, or neighborhoods. Systems should define aggregation levels, access controls, retention periods, and communication protocols that minimize stigmatization.

10.2 Data Sharing, FAIR Principles, and Pathogen Repositories

Rapid data sharing can accelerate outbreak response, variant tracking, and global risk assessment. The FAIR principles emphasize that data should be findable, accessible, interoperable, and reusable (Wilkinson et al., 2016). In pathogen genomics, repositories such as GenBank, the Sequence Read Archive, and GISAID have supported large-scale sharing and reuse of sequence data (Leinonen et al., 2011; Shu & McCauley, 2017). However, open sharing must be balanced with attribution, benefit sharing, privacy, national regulations, and security concerns.

10.3 Indigenous and Community Data Sovereignty

Environmental and pathogen sampling may occur in Indigenous territories, rural communities, farms, or ecosystems with cultural and livelihood significance. The CARE Principles for Indigenous Data Governance emphasize collective benefit, authority to control, responsibility, and ethics, complementing FAIR by centering rights and relationships rather than only data usability (Carroll et al., 2020). For biosurveillance, this means that community engagement, consent, governance, and return of results should not be treated as optional outreach. They are part of responsible surveillance design.

10.4 Equity in Sequencing Capacity

Global sequencing capacity remains uneven. The WHO genomic surveillance strategy stresses the need to strengthen laboratory networks, data systems, workforce development, and equitable access to pathogen genomic surveillance (WHO, 2022). Without sustained investment, genomic biosurveillance can reproduce global inequalities: data may be extracted from under-resourced regions while analysis, authorship, infrastructure, and benefits remain concentrated elsewhere. Capacity building should include sequencing platforms, reagents, bioinformatics training, data governance, maintenance budgets, and local decision authority.

10.5 Dual-Use and Biosecurity Concerns

Pathogen sequence data can support vaccine design, diagnostics, therapeutics, and outbreak control. It can also raise dual-use concerns if information could be misused to enhance or reconstruct biological agents. Responsible governance should include risk assessment, controlled access when appropriate, institutional review, biosafety training, and publication practices that maximize public health value while minimizing unnecessary security risk.

11.0. CHALLENGES, LIMITATIONS, AND SOURCES OF UNCERTAINTY

The major weakness in many discussions of genomic biosurveillance is technological optimism. Sequencing is powerful, but it does not automatically solve surveillance. The limitations are technical, analytical, operational, ethical, and interpretive. The interpretation of genomic and metagenomic signals requires robust statistical risk assessment because sequence detections must be distinguished from contamination, sampling bias, background environmental variation, and uncertain exposure relevance (Ogwu et al., 2025).

11.1 Technical and Analytical Limitations

Technical limitations include low pathogen abundance, host background, sample degradation, inhibitors, contamination, sequencing errors, and insufficient depth. Analytical limitations include incomplete reference databases, taxonomic misclassification, uncertain genome assembly, variant-calling errors, and poor comparability across pipelines. Metagenomic classification is especially sensitive to database quality. Unknown organisms may be missed or assigned incorrectly to the nearest known organism.

11.2 Interpretation Limits

Detection of nucleic acid does not always mean viable organism, active infection, or immediate risk. DNA can persist after organisms die, and RNA signals can degrade rapidly. Environmental signals may be transported, diluted, concentrated, or mixed across sources. Genomic similarity can suggest relatedness but does not alone prove direct transmission. AI models can identify patterns but can also overfit, amplify dataset biases, or generate confident predictions outside their valid context.

11.3 Operational Limits

Operational barriers include cost, procurement delays, workforce shortages, fragmented data systems, lack of sustained funding, and limited integration with public health decision-making. Sequencing turnaround time must match the response window. A highly accurate genomic result delivered after the decision point may have limited operational value. Surveillance systems should therefore evaluate timeliness, not only analytical accuracy.

11.4 Governance Limits

Governance challenges include unclear ownership, inconsistent sharing rules, privacy concerns, limited community participation, uneven benefit distribution, and lack of accountability when genomic signals are misinterpreted. These challenges are not peripheral; they determine whether surveillance systems are trusted, sustainable, and useful.

Table 3 summarizes key limitations that can affect genomic and metagenomic biosurveillance, the risks they create, and practical strategies for mitigation. It reinforces the chapter's central caution that sequencing and AI do not automatically produce reliable surveillance intelligence. Contamination, incomplete databases, sampling bias, weak metadata, model overfitting, privacy concerns, and unequal sequencing capacity can all distort interpretation or reduce public trust if not addressed systematically. Effective genomic biosurveillance therefore requires quality controls, representative sampling, curated databases, standardized metadata, external model validation, ethical communication, and equitable investment in local infrastructure and expertise

Table 3. Common limitations and mitigation strategies

Limitation	Risk	Mitigation strategy
Contamination	False pathogen detection	Use blanks, mock controls, replicate testing, batch tracking
Incomplete databases	Misclassification or missed organisms	Use curated databases, report confidence, update references
Sampling bias	Misleading geographic or population patterns	Use representative, sentinel, or risk-based designs
Weak metadata	Limited interpretation and integration	Require standardized metadata at collection
AI overfitting	Poor model transferability	Validate externally and monitor model drift
Privacy concerns	Community harm or stigmatization	Aggregate data, apply governance controls, communicate ethically
Unequal capacity	Data extraction without local benefit	Invest in local infrastructure, training, and authorship equity

12.0. FUTURE DIRECTIONS

12.1 Real-Time and Field-Deployable Surveillance

The future of genomic biosurveillance will involve faster, more distributed sequencing systems that bring analysis closer to the field. Portable platforms, simplified extraction protocols, cloud-enabled pipelines, and automated reporting can reduce delays. However, field deployment must not weaken quality control, biosafety, or interpretation standards.

12.2 Threat-Agnostic and Pan-Pathogen Surveillance

Future systems will increasingly move beyond one-pathogen-at-a-time surveillance. Metagenomic and metatranscriptomic methods can detect multiple organisms, characterize communities, and reveal unknown signals. The main challenge will be converting broad detection into meaningful triage. Systems will need thresholds, confidence scoring, confirmatory workflows, and decision rules that distinguish background biological diversity from actionable threat signals.

12.3 AI-Enhanced Prediction and Early Warning

AI will increasingly be used to forecast variant growth, AMR spread, spillover risk, and environmental emergence by integrating genomic data with climate, land use, mobility, wastewater, clinical, animal, and ecological data. The strongest systems will combine machine learning with domain knowledge, causal reasoning, and transparent uncertainty communication. Prediction without explanation may be insufficient for public health decision-making. Future genomic biosurveillance should account for climate-sensitive health risks because changing temperature, precipitation, extreme events, and rural health vulnerabilities can reshape pathogen exposure, environmental reservoirs, and surveillance priorities (Sugg et al., 2023).

12.4 Global Genomic Surveillance Networks

Sustainable genomic biosurveillance requires networks rather than isolated projects. Regional sequencing hubs, interoperable platforms, shared standards, local training, and stable financing are essential. The future should not be defined by emergency investment during pandemics followed by neglect during inter-pandemic periods. Genomic infrastructure must become part of routine public health, agriculture, environmental protection, and One Health systems.

12.5 Responsible Innovation

Responsible innovation means designing genomic biosurveillance systems that are scientifically valid, ethically governed, locally useful, and globally connected. The next generation of systems should embed community engagement, privacy protection, data sovereignty, FAIR and CARE-aligned governance, benefit sharing, and equitable capacity building from the beginning. Without these elements, technological sophistication may produce data abundance without trust or action.

13.0. CONCLUSION

Genomic and metagenomic approaches are reshaping biosurveillance by providing high-resolution tools for detecting, tracking, and interpreting biological threats across human, animal, plant, food, and environmental systems. Whole-genome sequencing supports lineage tracking, outbreak reconstruction, source attribution, and resistance monitoring. Metagenomics expands surveillance from targeted detection to broader, threat-agnostic analysis of complex samples. When linked with AI, geospatial analytics, epidemiology, and One Health frameworks, sequencing data can support earlier warning, more precise response, and stronger preparedness. Yet the central lesson is cautionary as well as optimistic. Sequencing is not surveillance by itself. Surveillance requires representative sampling, robust laboratory workflows, validated bioinformatics, high-quality metadata, ethical governance, equitable infrastructure, and clear pathways from data to action. AI can help detect patterns and prioritize decisions, but it cannot rescue poor data or replace expert interpretation. The future of genomic biosurveillance therefore lies not in more data alone, but in better systems: systems that are accurate, timely, inclusive, reproducible, explainable, and trusted.

REFERENCES

Alcock, B. P., Raphenya, A. R., Lau, T. T. Y., Tsang, K. K., Bouchard, M., Edalatmand, A., Huynh, W., Nguyen, A.-L. V., Cheng, A. A., Liu, S., Min, S. Y., Miroshnichenko, A., Tran, H.-K., Werfalli, R. E., Nasir, J. A., Oloni, M., Speicher, D. J., Florescu, A., Singh, B., & McArthur, A. G. (2020). CARD 2020: Antibiotic resistome surveillance with the Comprehensive Antibiotic Resistance Database. *Nucleic Acids Research*, *48*(D1), D517–D525. DOI: 10.1093/nar/gkz935 PMID: 31665441

Boolchandani, M., D'Souza, A. W., & Dantas, G. (2019). Sequencing-based methods and resources to study antimicrobial resistance. *Nature Reviews. Genetics*, *20*(6), 356–370. DOI: 10.1038/s41576-019-0108-4 PMID: 30886350

Callahan, B. J., McMurdie, P. J., Rosen, M. J., Han, A. W., Johnson, A. J. A., & Holmes, S. P. (2016). DADA2: High-resolution sample inference from Illumina amplicon data. *Nature Methods*, *13*(7), 581–583. DOI: 10.1038/nmeth.3869 PMID: 27214047

Carroll, S. R., Garba, I., Figueroa-Rodríguez, O. L., Holbrook, J., Lovett, R., Materechera, S., Parsons, M., Raseroka, K., Rodriguez-Lonebear, D., Rowe, R., Sara, R., Walker, J. D., Anderson, J., & Hudson, M. (2020). The CARE Principles for Indigenous Data Governance. *Data Science Journal*, *19*, 43. Advance online publication. DOI: 10.5334/dsj-2020-043

Chiu, C. Y., & Miller, S. A. (2019). Clinical metagenomics. *Nature Reviews. Genetics*, *20*(6), 341–355. DOI: 10.1038/s41576-019-0113-7 PMID: 30918369

Di Tommaso, P., Chatzou, M., Floden, E. W., Barja, P. P., Palumbo, E., & Notredame, C. (2017). Nextflow enables reproducible computational workflows. *Nature Biotechnology*, *35*(4), 316–319. DOI: 10.1038/nbt.3820 PMID: 28398311

Gardy, J. L., & Loman, N. J. (2018). Towards a genomics-informed, real-time, global pathogen surveillance system. *Nature Reviews. Genetics*, *19*(1), 9–20. DOI: 10.1038/nrg.2017.88 PMID: 29129921

Gire, S. K., Goba, A., Andersen, K. G., Sealfon, R. S. G., Park, D. J., Kanneh, L., Jalloh, S., Momoh, M., Fullah, M., Dudas, G., Wohl, S., Moses, L. M., Yozwiak, N. L., Winnicki, S., Matranga, C. B., Malboeuf, C. M., Qu, J., Gladden, A. D., Schaffner, S. F., & Sabeti, P. C. (2014). Genomic surveillance elucidates Ebola virus origin and transmission during the 2014 outbreak. *Science*, *345*(6202), 1369–1372. DOI: 10.1126/science.1259657 PMID: 25214632

Hadfield, J., Megill, C., Bell, S. M., Huddleston, J., Potter, B., Callender, C., Sagulenko, P., Bedford, T., & Neher, R. A. (2018). Nextstrain: Real-time tracking of pathogen evolution. *Bioinformatics (Oxford, England)*, *34*(23), 4121–4123. DOI: 10.1093/bioinformatics/bty407 PMID: 29790939

Hendriksen, R. S., Munk, P., Njage, P., van Bunnik, B., McNally, L., Lukjancenko, O., Röder, T., Nieuwenhuijse, D., Pedersen, S. K., Kjeldgaard, J., Kaas, R. S., Clausen, P. T. L. C., Vogt, J. K., Leekitcharoenphon, P., van de Schans, M. G. M., Zuidema, T., de Roda Husman, A. M., Rasmussen, S., Petersen, B., & Aarestrup, F. M. (2019). Global monitoring of antimicrobial resistance based on metagenomics analyses of urban sewage. *Nature Communications*, *10*(1), 1124. Advance online publication. DOI: 10.1038/s41467-019-08853-3 PMID: 30850636

Izah, S. C., Nurmahanova, A., Ogwu, M. C., Toktarbay, Z., Umirbayeva, Z., Ussen, K., Koibasova, L., Nazarbekova, S., Tynybekov, B., & Guo, Z. (2025). Public health risks associated with antibiotic residues in poultry food products. *Journal of Agriculture and Food Research*, *21*, 101815. DOI: 10.1016/j.jafr.2025.101815

Izah, S. C., & Ogwu, M. C. (2025). Modeling solutions for microbial water contamination in the Global South for public health protection. *Frontiers in Microbiology*, *16*, 1504829. DOI: 10.3389/fmicb.2025.1504829 PMID: 40241726

Izah, S. C., Ogwu, M. C., & Hamidifar, H. (Eds.). (2024). *Biomonitoring of pollutants in the Global South*. Springer Nature., DOI: 10.1007/978-981-97-1658-6

Jain, M., Olsen, H. E., Paten, B., & Akeson, M. (2016). The Oxford Nanopore MinION: Delivery of nanopore sequencing to the genomics community. *Genome Biology*, *17*(1), 239. Advance online publication. DOI: 10.1186/s13059-016-1103-0 PMID: 27887629

Köster, J., & Rahmann, S. (2012). Snakemake: A scalable bioinformatics workflow engine. *Bioinformatics (Oxford, England)*, *28*(19), 2520–2522. DOI: 10.1093/bioinformatics/bts480 PMID: 22908215

Langmead, B., & Salzberg, S. L. (2012). Fast gapped-read alignment with Bowtie 2. *Nature Methods*, *9*(4), 357–359. DOI: 10.1038/nmeth.1923 PMID: 22388286

Leinonen, R., Sugawara, H., & Shumway, M.International Nucleotide Sequence Database Collaboration. (2011). The Sequence Read Archive. *Nucleic Acids Research*, *39*(Database issue), D19–D21. DOI: 10.1093/nar/gkq1019 PMID: 21062823

Li, H. (2018). Minimap2: Pairwise alignment for nucleotide sequences. *Bioinformatics (Oxford, England)*, *34*(18), 3094–3100. DOI: 10.1093/bioinformatics/bty191 PMID: 29750242

Lo, S. W., & Jamrozy, D. (2020). Genomics and epidemiological surveillance. *Nature Reviews. Microbiology, 18*(9), 478. DOI: 10.1038/s41579-020-0421-0 PMID: 32690878

McArthur, A. G., Waglechner, N., Nizam, F., Yan, A., Azad, M. A., Baylay, A. J., Bhullar, K., Canova, M. J., De Pascale, G., Ejim, L., Kalan, L., King, A. M., Koteva, K., Morar, M., Mulvey, M. R., O'Brien, J. S., Pawlowski, A. C., Piddock, L. J. V., Spanogiannopoulos, P., & Wright, G. D. (2013). The Comprehensive Antibiotic Resistance Database. *Antimicrobial Agents and Chemotherapy, 57*(7), 3348–3357. DOI: 10.1128/AAC.00419-13 PMID: 23650175

Neher, R. A., & Bedford, T. (2018). Real-time analysis and visualization of pathogen sequence data. *Science, 359*(6378), 1537–1539. DOI: 10.1126/science.aaq1018 PMID: 30135232

Ogwu, M. C., & Izah, S. C. (2023). *One Health implications of agrochemicals and their sustainable alternatives* (Sustainable Development and Biodiversity, Vol. 34). Springer Nature. https://doi.org/DOI: 10.1007/978-981-99-3439-3

Ogwu, M. C., & Izah, S. C. (2025). Nanotechnology for fungal pathogen control in crops: Innovations, public health impacts, and disease prevention. *Frontiers in Fungal Biology, 6*, 1653214. DOI: 10.3389/ffunb.2025.1653214 PMID: 40959466

Ogwu, M. C., Izah, S. C., Aigberua, A. O., & Ngun, C. T. (2024). Editorial: Detection, risk analysis, and monitoring of chemical contaminants from agro-aqua food production and processing: Implications on the One Health triad. *Frontiers in Sustainable Food Systems, 8*, 1501930. DOI: 10.3389/fsufs.2024.1501930

Ogwu, M. C., Izah, S. C., Sawyer, W. E., & Amabie, T. (2025). Environmental risk assessment of trace metal pollution: A statistical perspective. *Environmental Geochemistry and Health, 47*(4), 94. DOI: 10.1007/s10653-025-02405-z PMID: 40019615

Parks, D. H., Imelfort, M., Skennerton, C. T., Hugenholtz, P., & Tyson, G. W. (2015). CheckM: Assessing the quality of microbial genomes recovered from isolates, single cells, and metagenomes. *Genome Research, 25*(7), 1043–1055. DOI: 10.1101/gr.186072.114 PMID: 25977477

Quick, J., Grubaugh, N. D., Pullan, S. T., Claro, I. M., Smith, A. D., Gangavarapu, K., Oliveira, G., Robles-Sikisaka, R., Rogers, T. F., Beutler, N. A., Burton, D. R., Lewis-Ximenez, L. L., de Jesus, J. G., Giovanetti, M., Hill, S. C., Black, A., Bedford, T., Carroll, M. W., Nunes, M., & Loman, N. J. (2017). Multiplex PCR method for MinION and Illumina sequencing of Zika and other virus genomes directly from clinical samples. *Nature Protocols*, *12*(6), 1261–1276. DOI: 10.1038/nprot.2017.066 PMID: 28538739

Quick, J., Loman, N. J., Duraffour, S., Simpson, J. T., Severi, E., Cowley, L., Bore, J. A., Koundouno, R., Dudas, G., Mikhail, A., Ouédraogo, N., Afrough, B., Bah, A., Baum, J. H. J., Becker-Ziaja, B., Boettcher, J. P., Cabeza-Cabrerizo, M., Camino-Sánchez, Á., Carter, L. L., & Carroll, M. W. (2016). Real-time, portable genome sequencing for Ebola surveillance. *Nature*, *530*(7589), 228–232. DOI: 10.1038/nature16996 PMID: 26840485

Quince, C., Walker, A. W., Simpson, J. T., Loman, N. J., & Segata, N. (2017). Shotgun metagenomics, from sampling to analysis. *Nature Biotechnology*, *35*(9), 833–844. DOI: 10.1038/nbt.3935 PMID: 28898207

Rambaut, A., Holmes, E. C., O'Toole, Á., Hill, V., McCrone, J. T., Ruis, C., du Plessis, L., & Pybus, O. G. (2020). A dynamic nomenclature proposal for SARS-CoV-2 lineages to assist genomic epidemiology. *Nature Microbiology*, *5*(11), 1403–1407. DOI: 10.1038/s41564-020-0770-5 PMID: 32669681

Ren, J., Song, K., Deng, C., Ahlgren, N. A., Fuhrman, J. A., Li, Y., Xie, X., Poplin, R., & Sun, F. (2020). Identifying viruses from metagenomic data using deep learning. *Quantitative Biology*, *8*(1), 64–77. DOI: 10.1007/s40484-019-0187-4 PMID: 34084563

Seemann, T. (2014). Prokka: Rapid prokaryotic genome annotation. *Bioinformatics (Oxford, England)*, *30*(14), 2068–2069. DOI: 10.1093/bioinformatics/btu153 PMID: 24642063

Shu, Y., & McCauley, J. (2017). GISAID: Global initiative on sharing all influenza data - from vision to reality. *Eurosurveillance*, *22*(13), 30494. DOI: 10.2807/1560-7917.ES.2017.22.13.30494 PMID: 28382917

Sugg, M. M., Ryan, S., Spurlock, T., Duncan, S., Hege, A., Ogwu, M. C., Tyson, J., Shay, E., Renwick, K., Hendren, C., Roy, M., & Runkle, J. D. (2023). Climate change and health in rural mountain environments: Summary of a workshop on knowledge gaps, barriers, and opportunities for action. *GeoJournal*, *88*(5), 5397–5409. Advance online publication. DOI: 10.1007/s10708-023-10916-4 PMID: 40291811

Wilkinson, M. D., Dumontier, M., Aalbersberg, I. J., Appleton, G., Axton, M., Baak, A., Blomberg, N., Boiten, J.-W., da Silva Santos, L. B., Bourne, P. E., Bouwman, J., Brookes, A. J., Clark, T., Crosas, M., Dillo, I., Dumon, O., Edmunds, S., Evelo, C. T., Finkers, R., & Mons, B. (2016). The FAIR Guiding Principles for scientific data management and stewardship. *Scientific Data*, *3*(1), 160018. Advance online publication. DOI: 10.1038/sdata.2016.18 PMID: 26978244

Wood, D. E., Lu, J., & Langmead, B. (2019). Improved metagenomic analysis with Kraken 2. *Genome Biology*, *20*(1), 257. Advance online publication. DOI: 10.1186/s13059-019-1891-0 PMID: 31779668

World Health Organization. (2022). Global genomic surveillance strategy for pathogens with pandemic and epidemic potential, 2022-2032. World Health Organization. https://www.who.int/publications/i/item/9789240046979

Chapter 4
Wastewater-Based Epidemiology and Environmental Genomic Monitoring

Nilotpal Das
http://orcid.org/0009-0005-0083-4093
Uttaranchal University, India

Atin Kumar
http://orcid.org/0000-0002-1653-2146
Uttaranchal University, India

Rohit Kumar
http://orcid.org/0009-0003-4713-2625
GLA University, India

Sharad Sachan
Lovely Professional University, India

ABSTRACT

Wastewater-based epidemiology (WBE) and environmental genomic surveillance are emerging population-scale tools for assessing public health and environmental risks. By analyzing wastewater and environmental samples, they enable the detection of biomarkers, pathogens, antimicrobial resistance genes, pharmaceuticals, and chemical contaminants originating from human activities. Integrated with environmental genomics approaches such as metagenomics and high-throughput sequencing, WBE provides near real-time, non-invasive insights into disease prevalence, drug use patterns, and community health trends. These methods enhance

DOI: 10.4018/979-8-3373-9033-8.ch004

early warning of infectious disease outbreaks and help assess anthropogenic impacts on wastewater systems and natural environments. However, challenges remain in data interpretation, standardization of methods, ethical concerns, and integration with conventional surveillance systems. Methodological harmonization and interdisciplinary collaboration are essential to strengthen their role in sustainable public health and environmental monitoring.

1. INTRODUCTION

Wastewater-based epidemiology (WBE) has evolved since its inception, from early uses of sewage excretion products of illicit drugs and pharmaceuticals to monitor epidemiology through human metabolism, into a more expanded population-health surveillance system that can measure chemical and biological biomarkers to estimate exposure, behavior, and prevalence of the disease on a population scale (Boogaerts et al., 2021; Boogaerts et al., 2024). In principle, WBE considers influent wastewater a pooled, composite specimen where the contribution of the whole sewered population is combined, and the concentrations of the analytes are transformed into population-adjusted mass loads in order to allow comparison over space and time (Boogaerts et al., 2024). The COVID-19 pandemic triggered a significant growth of this area: SARS-CoV-2 RNA in wastewater monitoring showed high correlations with clinical and significant epidemiological cases and early outbreak detection and transformed WBE into one of the largest environmental public health surveillance systems in the world (Parkins et al., 2023; Kumblathan et al., 2021). Simultaneously, the larger field of environmental genomics, which means mass analysis of DNA and RNA of environmental samples, including water, soil, air, and wastewater, is now becoming the focus of popular health care and environmental observation. The use of genomic techniques such as quantitative PCR, metagenomics, and next-generation sequencing (NGS) have become a standard procedure to monitor wastewater to identify viral pathogens (e.g., SARS CoV 2, poliovirus), antimicrobial resistance genes, and complex microbial communities and extend WBE beyond simple target quantification to characterize the diversity of pathogen threats, dynamics of variants, and emergencies (Malcom et al., 2025; Farkas et al., 2024). The methods based on NGS allow identifying known and novel viruses, understanding variants and recombinants, and incorporating wastewater data into One Health systems that relate human, animal, and environmental sources of infection (Malcom & Bowes, 2025; Farkas et al., 2024). This intersection forms a solid basis for a reasoned decision to make WBE and environmental genomics a formally combined conceptual tool. To begin with, genomic tools directly overcome several major weaknesses of traditional WBE, which include the inability to resolve strain-level diversity and identify novel pathogens, by offering sequence-level resolution that can be used to monitor variants, perform genome-level evolutionary analysis, and conduct genomic epidemiology (Zhu et al., 2025; Farkas et al., 2024). Second, implementing the integration of quantitative WBE data (e.g., population-normalized loads) with metagenomic or amplicon sequencing will enable the risk assessment of infectious disease and antimicrobial resistance at the community level and simultaneously determine pathogen abundance, diversity, and resistance determinants in the same samples (Malcom & Bowes, 2025; Zhu et al., 2025). Third, the WBE-genomics systems could serve as early warning systems, since wastewater sequencing has alerted to outbreaks of SARS-CoV-2 variants days to weeks before extensive clinical diagnosis, particularly in places that have limited availability of individual testing (Farkas et al., 2024; Rashid et al., 2024). Lastly, incorporating such integration into equity- and One Health-based models can facilitate more inclusive surveillance in high-, middle-, and low-income environments, even in areas with underdeveloped clinical infrastructure, and extend the benefits of the genomic public health intelligence of WBE with the benefits of non-invasiveness, cost-effectiveness, and near-real-time population coverage (Rashid et al., 2024; R. et al., 2025). Collectively, these advancements warrant a conceptual transformation of WBE as a niche environmental measurement device to a pillar of environmental genomics-enabled environmental-related public health surveillance, in which wastewater can be used as a quantitative epidemiologic matrix and an abundant source of community health genomic information.

2. PRINCIPLES OF WASTEWATER-BASED EPIDEMIOLOGY

2.1 Biomarkers in Wastewater and Their Epidemiological Relevance

Wastewater-based epidemiology (WBE) is based on biomarkers, chemical, or biological surrogates of human excretion into untreated sewage, to deduce community health, exposure, and behavior. These are exogenous products like illicit drugs, pharmacogenic drugs, and their human metabolites, and endogenous biomarkers and pathogen-specific nucleic acids (e.g., SARS-CoV-2 RNA) (Boogaerts et al., 2021; Zhang et al., 2021) (Figure 1). By taking biomarker concentrations in the influent wastewater and comparing them to accepted rates of human excretion and sewer network stability, WBE can approximate community-wide xenobiotics, infectious disease, and even chronic disease burden (Boogaerts et al., 2021; Amin et al., 2022; Zhang et al., 2021). Since sewage is a collection of urine and feces of large groups of individuals, it contains contributions of symptomatic, asymptomatic, and unreported individuals, which provides a non-invasive and anonymous, almost real-time view of population health that helps to supplement clinical and survey-based data (Boogaerts et al., 2021; Kumblathan et al., 2021; Zhang et al., 2021).

Figure 1. Biomarkers in wastewater and their epidemiological relevance

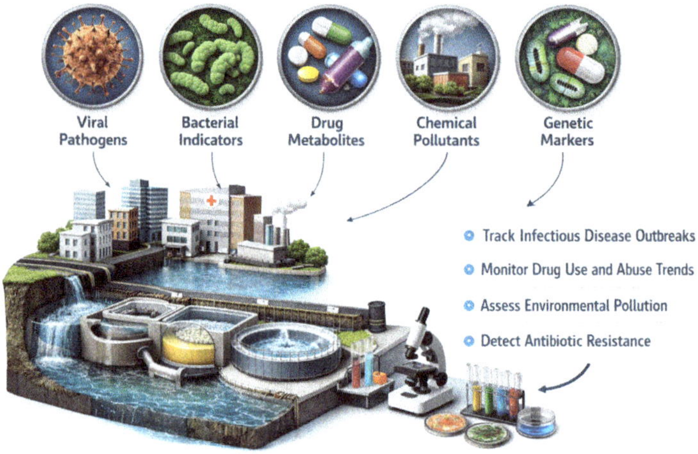

2.2 Population Normalization and Temporal Resolution

One major rule of WBE is that the mass loads of the biomarkers (e.g., mg/day/1000 inhabitants or gene copies/day/1000 inhabitants) are converted into raw biomonitoring data that is population normalized (e.g., multiplication of concentration and wastewater flow by the population size that contributes to the catchment) (Boogaerts et al., 2024; Boogaerts et al., 2021). This is necessary in order to analyze spatial and temporal patterns to compare cities or over time with changes in flow and population dynamics (i.e., commuting, tourism), which would be meaningfully analyzed (Boogaerts et al., 2024; Oloye et al., 2023). To improve on this, de facto population biomarkers (e.g., hydrochemical parameters, human-specific chemicals, or endogenous excretion products, e.g., creatinine, coprostanol, nicotine, or 5 HIAA) are being exploited more and more to estimate the actual contributing population on a particular day, but large uncertainties and a lack of an optimal external validator exist (Boogaerts et al., 2024; Shrestha et al., 2021; Oloye et al., 2023). High temporal resolution is ensured by using daily or sub-daily composite sampling or integrative methods, which allow identifying short-term changes in consumption patterns or infection dynamics and early signals of the outbreak before clinical case information (Boogaerts et al., 2021; Li et al., 2021; Shah et al., 2021).

2.3 Advantages and Limitations of WBE as a Surveillance Tool

WBE has a number of strengths as a monitoring mechanism. It is also low-cost, a single sample of composite wastewater can represent thousands to millions of individuals, and non-invasive and anonymous, which is especially useful in the low- and middle-income context (Shrestha et al., 2021; Holton et al., 2022). It offers nearly real-time and longitudinal, population-scale data, including those without symptoms and those who are not tested, allowing an early warning of the spread of infectious diseases and constant tracking of changes in the consumption of drugs, the use of antimicrobials, or biomarkers of the diseases (Boogaerts et al., 2021; Li et al., 2021; Shah et al., 2021; Holton et al., 2022). Nevertheless, there are significant constraints to accuracy and meaning. In most cases, large uncertainties arise at many steps: variable biomarker excretion rates, in-sewer degradation or dilution, sampling and storage artefacts, analytical variability, and assumption of back-calculation, and all of this can introduce relative errors of 20 per cent or more in prevalence estimates even in optimum conditions (Boogaerts et al., 2024; Li et al., 2021). Normalization of population is still a significant issue, as de jure and de facto estimates vary by up to a difference of 200 percent, making it difficult to make comparisons between studies and sites (Boogaerts et al., 2024). Other limitations are the absence of a global standardization of sampling and analytical procedures, the inability to quantitatively relate wastewater indicators to clinical outcomes at the individual scale, and ethical considerations when transitioning to small-scale or highly resolved sampling that may have stigmatization risks towards subpopulations (Boogaerts et al., 2024; Shrestha et al., 2021; Amin et al., 2022; Kumblathan et al., 2021). In turn, WBE should be viewed as a supplementary, rather than a competing, surveillance modality whose advantages are in its capacity to provide global, fast, and inclusive population surveillance, but the results of which should be put into context with clinical and other epidemiological data.

3. ENVIRONMENTAL GENOMIC MONITORING APPROACHES

3.1 Metagenomics, Metatranscriptomics, and Amplicon Sequencing

The monitoring of environmental genomics takes advantage of several high-throughput sequencing strategies that vary in the extent and extent of information they can give. Shotgun metagenomics sequences the entirety of DNA in an environmental sample, which allows culture-independent description of taxonomic composition, functional gene content, and recovery of metagenome-assembled genomes (MAGs) that disclose genomes of uncultivated or novel taxa (Rout et al., 2025; Kayani et al., 2021; Nam et al., 2023) (Table 1). Long read sequencing platforms, including nanopore and PacBio, have also enhanced continuity in assembly, complex genome organization, and detection of mobile elements and resistance genes, and can be used to perform sequencing in fields (Ciuffreda et al., 2021; Zhang et al., 2025; Kim et al., 2024). Metatranscriptomics goes further and sequences community RNA, usually mRNA, to analyse the active expression of particular genes in a specific environmental condition to correlate potential functions with real-time metabolic activity and environmental responses (Tyagi and Katara, 2024; Peng et al., 2023; Sharuddin et al., 2022). This method is effective, especially for tracing dynamic processes like nutrient cycling, degradation of pollutants, and expression of antibiotic resistance or virulence factors, but is faced with the drawbacks of RNA instability, dominance by rRNA, and difficulty in extracting it from inhibitory matrices (Peng et al., 2023; Sharuddin et al., 2022). Conversely, amplicon (metabarcoding) sequencing is based on the specificity of marker genes (ex, 16S/18S rRNA, ITS, COI, 12S) to get cost-effective, high-throughput profiles of community composition and biodiversity, a common application to eDNA-based species detection and microbiome surveys (Yoon et al., 2025; Nam et al., 2023). Although amplicon data offer limited functional resolution and may be biased based on the choice of primers and reference databases, they are still an essential tool for screening and monitoring, which can be combined with shotgun and metatranscriptomic data to balance cost, taxonomic coverage, and functional information (Yoon et al., 2025; Nam et al., 2023; Regueira-Iglesias et al., 2023).

Table 1. Comparative features of environmental genomic approaches

Approach	Main Target & Output	Typical Applications	Key Strengths / Limitations	Citations
Shotgun metagenomics	Total community DNA; taxonomic + functional profiles, MAGs	Microbial diversity, resistome, metabolic potential, pollutant degradation	High functional/ taxonomic resolution; assembly and binning demanding; large data, costlier	(Rout et al., 2025; Zhang et al., 2025; Kayani et al., 2021; Nam et al., 2023)
Metatranscriptomics	Community RNA (mRNA ± rRNA); gene expression profiles	Active functions, responses to stress, and real-time biogeochemical processes	Captures activity not just presence; RNA is unstable; complex workflows, deep sequencing	(Tyagi & Katara, 2024; Peng et al., 2023; Sharuddin et al., 2022; Aplakidou et al., 2024)
Amplicon/ metabarcoding	Marker genes (16S, ITS, COI, 12S); community composition	eDNA biodiversity, biomonitoring, basic microbiome surveys	Cost-effective, high throughput; limited functional inference; primer and database bias	(Yoon et al., 2025; Nam et al., 2023; Regueira-Iglesias et al., 2023)

3.2 Functional Gene Profiling and Microbial Community Analysis

Metagenomics has the ability to analyze communities and functional genes at the same time by annotating sequenced DNA using curated databases to determine metabolic pathways, resistance genes, as well as stress response functionalities and taxa (Rout et al., 2025; Yang et al., 2021; Nam et al., 2023). This helps in the identification of new enzymes, bioactive products, and biomarkers of contaminant degradation, climate-related reactions, or disease outbreaks in aquatic and other environments (Rout et al., 2025). Metatranscriptomics introduces the time aspect by measuring the expression of active genes, showing which pathways (e.g., nitrogen cycling, oil degradation, antibiotic resistance, rhizosphere signaling) are activated by the existing conditions and whether they change due to a change in environment or pollution (Peng et al., 2023; Sharuddin et al., 2022). Amplicon-based functional inference tools are tools that seek to infer functions of a community based on data about taxonomic markers and provide a less expensive alternative to the shotgun sequencing method, but are limited by a lack of complete reference genomes and ecosystem-specific bias (Djemiel et al., 2022). In methods, alpha/beta diversity measures, ordination, and ecological traits assignment are also a regular part of community analysis to establish a connection between microbial assemblages and gene repertoires, on the one hand, and ecological health, disturbance, or management intervention, on the other hand (Rout et al., 2025; Nam et al., 2023; Djemiel et al., 2022).

4. SAMPLING STRATEGIES AND ANALYTICAL METHODOLOGIES

4.1 Wastewater and Environmental Sample Collection Techniques

Wastewater and environmental matrix sampling should be able to include spatial and temporal variability, as well as reduce contamination and target loss. The most frequently used methods in wastewater-based surveillance are grab samples (single time point), time- or flow-weighted 24-hour composites using autosamplers, and more passive samplers which accumulate loads of analytes over days to capture more representative fluctuating signals (Kumblathan et al., 2021; Shakallis et al., 2022; Tiwari et al., 2023). In case of molecular analysis (e.g. ARGs, SARS CoV 2), the water is usually pooled in sterile glass or HDPE bottles (0.52 L), and the volume depends on the type of the matrix and the amount of solid material; large volumes are needed in low biomass effluents compared to raw influent or activated sludge (Keenum et al., 2022; Kumblathan et al., 2021). In eDNA and benthic samples, the grabs or cores are used to collect sediment, and the water is filtered either in the environment or immediately after collection, with decontamination of all tools and filters to avoid cross-contamination (Pawlowski et al., 2021; Patin and Goodwin, 2023; Lu et al., 2021). In the case of microplastics, surface waters are sampled using nets (e.g., manta trawls) and discrete samplers, whereas cores, grabs, and shovels are regularly used in sediments, and special attention should be paid to minimizing airborne and procedural contamination (Kye et al., 2023; Prata et al., 2021; Lu et al., 2021) (Figure 2).

Figure 2. Wastewater and environmental sample collection techniques

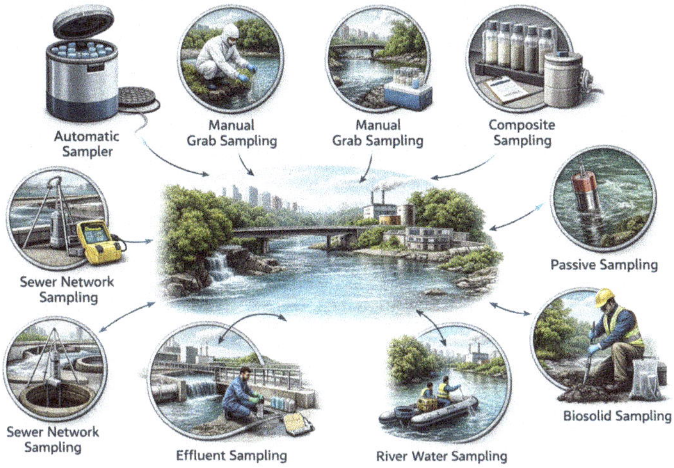

4.2 Sample Preservation, Concentration, and Extraction Methods

After collection, preservation and concentration are necessary as soon as possible to ensure the integrity of the analyte. In the case of nucleic acid targets in wastewater, it is recommended that they should be concentrated right away (typically by filtration onto 0.22 9m membranes) instead of freezing bulk water; long-term freezing before filtration will decrease DNA recovery, whereas filters can then be frozen in ethanol at -20 -80 °C until extraction (Keenum et al., 2022; Kumblathan et al., 2021). The workflows of SARS-CoV-2 WBE are usually composed of sequential stages of biosafety through pasteurization (e.g., 5660 C), solid-liquid separation, concentration of viruses, commercial RNA extraction with commercial kits, and RT qPCR (Kumblathan et al., 2021; Ciannella et al., 2023). Similar methods are also used in environmental eDNA and sediment analyses, where the sample is concentrated by filtration or centrifugation, and then DNA is extracted with different kits and protocols optimized to the inhibitor-rich matrices (e.g., humics in sediments), with different kits and protocols benchmarked on yield and inhibitor removal (Pawlowski et al., 2021; Ciannella et al., 2023). In the case of organic micropollutants and other chemical analyses, the most widespread concentration and clean up method is still solid phase extraction (SPE), whereas more recent green methods such as solid phase microextraction (SPME), fabric phase sorbent extraction, and micro QuEChERS attempt to minimize solvents usage, time, and exposure of operators and enhance selectivity (Câmara et al., 2022; Catarro et al., 2025). Use of preservation strategies such as cooling or freezing, adjusting the pH, adding some chemical preservatives, or the quick in situ extraction of the samples onto SPE cartridges followed by their storage instead of bulk water may reduce the degradation (Hernandez et al., 2023).

4.3 Quality Control and Standardization Challenges

In the fields of wastewater monitoring and environmental monitoring, quality assurance/quality control (QA/QC) and standardization of the methods are significant challenges toward the generation of similar reliable datasets. Literature reviews of AMR and SARS-CoV-2 surveillance note significant sampling (grab vs composite; sites; volumes), concentration and extraction approaches, target panels, and reporting unit differences, which inhibit cross-study analysis and meta-analysis (Liguori et al., 2022; Keenum et al., 2022; Kumblathan et al., 2021; Ciannella et al., 2023). Such aspects as field and laboratory blanks, positive controls or surrogate spikes to measure recovery, evaluation of PCR inhibition, method detection/quantification limits, and documentation of sample stability/storage conditions are critical QA/QC factors; however, most studies ignore some of these steps or present them incompletely (Liguori et al., 2022; Keenum et al., 2022; Ciannella et al., 2023; Hernandez et al., 2023; Lu et al., 2021). In the case of microplastics and eDNA, the contamination of ambient air, clothing, and plastic labware are ubiquitous, and systematic reviews reveal that the only studies that have adopted stringent control of contamination include the use of cotton clothing, filtered solutions, laminar flow working environment, covered samples, and the use of non-plastic equipment (Prata et al., 2021; Lu et al., 2021). Absence of standardized procedures in sampling, preservation, extraction, and data reporting (e.g., variation in units, incomplete metadata, etc.) also complicates interpretation and risk assessment (Liguori et al., 2022; Pawlowski et al., 2021; Ciannella et al., 2023; Lu et al., 2021). The recent consensus models thus recommend uniform target panels, uniform sampling and storage directions, reduced core quality assurance/ quality control standards, and open reporting checklists as key measures to solid and comparable wastewater and environmental monitoring.

5. DETECTION OF PATHOGENS AND INFECTIOUS DISEASES

5.1 Surveillance of Viral, Bacterial, and Parasitic Pathogens

Viral, bacterial, and parasitic pathogen surveillance in the environment and wastewater settings is becoming commonplace. Based on scoping and systematic reviews, untreated wastewater and sludge include a diverse set of human pathogens and extensive research on viruses (e.g., poliovirus, nonpolio enteroviruses, norovirus, hepatitis A, adenoviruses, and SARS-CoV-2), a lot of data on bacteria, and much less on protozoa and helminths (Corrin et al., 2024; Kilaru et al., 2021). They can be detected by molecular detection (qPCR/RT qPCR, digital PCR, sequencing), and in the case of certain enteric pathogens, they can be identified by traditional culture or infectivity assays (Ali et al., 2021; Zhang et al., 2021; Singh et al., 2024). The wastewater-based epidemiology (WBE), therefore, offers a combined cue of both symptomatic and asymptomatic shedders as an alternative to clinical surveillance and allows monitoring of various families of pathogens: viruses, bacteria, and parasites (Kilaru et al., 2021; Corrin et al., 2024; Singh et al., 2024) (Table 2).

Table 2. Examples of pathogens and surveillance outcomes

Pathogen group/ example	Typical matrix & method	Surveillance use	Citations
SARS-CoV-2	Municipal wastewater, RT-qPCR, sequencing	Track community burden, variants, and support interventions	(Shah et al., 2021; Ali et al., 2021; Tiwari et al., 2023)
Enteric viruses (polio, norovirus, HAV, adenovirus)	Sewage, RT-qPCR, genotyping	Detect circulation, preclinical outbreak signals	(Shah et al., 2021; Ali et al., 2021; Singh et al., 2024)
Bacterial pathogens (e.g., Salmonella; AMR bacteria)	Wastewater, qPCR, culture, ARG profiling	Track enteric disease and resistance trends	(Corrin et al., 2024; Zhang et al., 2021)

5.2 Early Warning Systems for Disease Outbreaks

Early warning systems (EWS) are now being fed with wastewater and other environmental data to identify outbreaks before they can explode into clinical cases. In the case of SARS-CoV-2, numerous reports revealed that a rise in viral RNA in wastewater leads to reported COVID-19 cases by several days to weeks, and a systematic review reported lead times as high as 63 days, and positivity in sewage cases preceded any clinical cases in some locations (Shah et al., 2021; Ali et al., 2021). In general, reviews of infectious disease EWS pinpoint wastewater surveillance as one of the many proactive streams of data (amongst syndromic data, social media, meteorology, and laboratory data) that are better, more timely indicators of outbreaks compared to traditional, diagnosis-based surveillance on its own (Villanueva-Miranda et al., 2025; Meckawy et al., 2022). Long-term forecasting of COVID-19 patterns in a number of sewersheds can also be enhanced by using machine learning models on longitudinal wastewater viral loads to facilitate the forecasting of risks and resource allocation (Abdeldayem et al., 2021; Villanueva-Miranda et al., 2025).

5.3 Case Studies: COVID-19 and Enteric Viruses

The most evident case study of the method of detecting pathogens and tracking outbreaks with the help of wastewater is COVID 19. Thousands of samples in dozens of countries are reported in global systematic reviews and consistently show SARS-CoV-2 RNA in wastewater tracking incidence, regularly forecasting surges, and can be used to monitor emergent variants through targeted or metagenomic sequencing (Shah et al., 2021; Ali et al., 2021; Tiwari et al., 2023). Similar evidence of enteric viruses has been found earlier than COVID 19: sustained wastewater surveillance has long been used to monitor poliovirus circulation, and has been used to warn of outbreaks of hepatitis A and norovirus earlier, as well as making warning of viral introduction to the human body months before clinical cases of poliomyelitis are observed (Shah et al., 2021; Ali et al., 2021; Singh et al., 2024). Recent narrative and scoping reviews suggest that these experiences with COVID 19 and enteric viruses show that wastewater surveillance is a valuable task that can be used as an integral part of integrated infectious disease surveillance and can provide cost effective population level intelligence of a wide range of pathogens and inform timely decision making in public health (Kilaru et al., 2021; Corrin et al., 2024; Singh et al., 2024).

6. MONITORING ANTIMICROBIAL RESISTANCE AND EMERGING THREATS

6.1 Tracking Antimicrobial Resistance Genes (ARGs)

Both human and animal populations underwent an integrated view of antimicrobial resistance circulating in wastewater and the aquatic environment due to monitoring antimicrobial resistance genes (ARGs). Quantification of dozens to hundreds of ARGs and mobile genetic elements at the same time is commonly done by high-throughput qPCR, microfluidic qPCR, and shotgun metagenomics to detect the resistome of municipal, hospital, and agricultural wastewater (Nguyen et al., 2021; Farkas et al., 2024) (Figure 3). Wastewater treatment facilities are also a significant hotspot where different microbes, antibiotics, metals, and disinfectants can coexist, which encourages the horizontal transfer of genes and the growth of antibiotic resistant bacteria (ARB); poorly treated treatment can either decrease the loads of ARGs or, in the worst case, serve as a channel of their dissemination to the receiving waters and the soils (Nguyen et al., 2021; Sassi et al., 2025; Singh et al., 2024). Monitoring panels suggested include high concentrations of both anthropogenic and clinically relevant resistance genes (e.g., sul1, tetA, intI1) and last resort resistance genes (e.g., vanA, blaCTX M) that help to track the trend and prioritize the risks (Farkas et al., 2024; Liguori et al., 2022). Nevertheless, the absence of standardized targets, reporting units, and QA/QC continues to reduce the comparison between studies and regions (Nguyen et al., 2021; Liguori et al., 2022).

Figure 3. Tracking Antimicrobial Resistance Genes (ARGs)

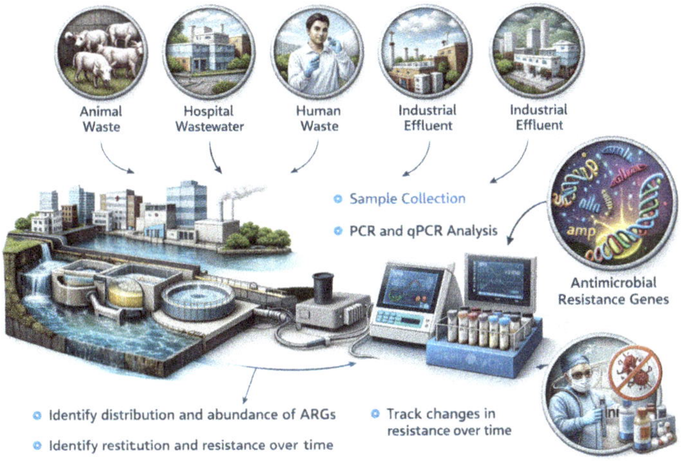

6.2 Identification of Emerging Pathogens and Variants

Detection of emerging pathogens, ARGs, and even variations is possible in wastewater with next-generation sequencing (NGS) and metagenomics, without prior selection of targets. ARG inventories through metagenomic surveying of wastewater and affected surface waters, discovery of ARG host range, and post facto screening of new resistance genes are all possible after their identification in clinics (Davis et al., 2023; Farkas et al., 2024). Initiatives like MARA and ResistoMap based on eDNA at a large scale demonstrate how years of multi-year metagenomic data over water, soil, and dust can map hundreds of ARGs, including high-profile genes like blaNDM and mcr, and be able to identify environmental hotspots that can be used to forecast imminent threats before they become clinically dominant (Sassi et al., 2025). Workflows in wastewater-based epidemiology based on similar NGS are becoming more common to monitor variants of viruses, and may be extended systematically to bacterial and fungal resistance determinants (Davis et al., 2023; Farkas et al., 2024).

7. CHEMICAL AND PHARMACEUTICAL BIOMARKERS IN WBE

7.1 Monitoring Drug Consumption, Illicit Substances, and Pharmaceuticals

Wastewater-based epidemiology (WBE) is an epidemiological methodology that relies on chemical and pharmaceutical biomarkers present in sewage to approximate community-level illicit drug and prescription drug use, as well as consumption of other xenobiotics. In the case of illicit drugs, specific human metabolites (e.g., benzoylecgonine in the case of cocaine, amphetamine-type stimulants, opioids, and cannabinoid metabolites) are determined in influent wastewater and backed up to daily per capita use, which are near-real-time and objective estimates that can be used to supplement surveys and crime statistics (Huizer et al., 2021; De Oliveira et al., 2023). The same practice has been extensively implemented on pharmaceuticals, such as antimicrobials, psychoactive, and cardiovascular drugs, by quantifying excreted parent compounds and metabolites and correcting for human excretion and sewer stability (Holton et al., 2022; Boogaerts et al., 2021; Laimou-Geraniou et al., 2022). It has been reviewed that longitudinal WBE is able to monitor spatial and temporal trends, assess the impact of policy interventions, and detect misuse or off-label use of particular medicines (Holton et al., 2022; Boogaerts et al., 2021). Emerging work can extend the use of WBE to the whole range of unintentional exposures (pesticides, PFAS, phthalates, and the composition of personal care products) by addressing the right human metabolites in wastewater (Aghaei et al., 2024; Carneiro et al., 2024).

7.2 Correlation With Community Health and Behavior

The community health status and health behavior, such as substance use patterns, medication, and lifestyle risk factors, can be related to chemical biomarker loads in wastewater. The illicit drug monitoring on a multi-year international basis has shown distinct spatio-temporal patterns, weekend/holiday peaks, and patterns associated with particular events or crises, showing that WBE is sensitive to behavioral changes (Boogaerts et al., 2024; Huizer et al., 2021; De Oliveira et al., 2023). In the case of pharmaceuticals, it has been suggested that WBE-derived consumption can be triangulated with prescription and sales data to help inform drug utilization studies and determine the effect of public health interventions or prescribing recommendations (Holton et al., 2022; Boogaerts et al., 2021). Non-communicable disease monitoring concept papers suggest that endogenous disease biomarkers (e.g., urinary proteins, cardiovascular disease, or cancer) be used with pharmaceuticals (e.g., statins, chemotherapy drugs) and lifestyle markers (alcohol, tobacco) to determine the prevalence and behavioral contributions of diseases at a population level (Amin et al., 2022).

8. APPLICATIONS IN ENVIRONMENTAL AND ECOSYSTEM HEALTH

8.1 Assessment of Wastewater Treatment Plant Performance

Wastewater treatment plants (WWTPs) play an important role in environmental and ecosystem health, as they are the key control points whereby the effectiveness of contaminant removal can be evaluated using biological, chemical, and microbiological indicators. It has been reviewed that WWTPs are large reservoirs and sources of antibiotic-resistant bacteria (ARB) and genes (ARGs); primary and secondary processes commonly reduce bulk organic material but do not eliminate ARB/ARGs, so they continue to be found in effluents and biosolids (Gao et al., 2022; Uluseker et al., 2021; Kalli et al., 2023). The tracking of ARGs, antibiotics, heavy metals, and new pollutants (e.g., microplastics, pharmaceuticals) at different levels of treatment is a sensitive indicator of plant performance and shows conditions that contribute to the proliferation of horizontal gene transfer and resistance (La Rosa et al., 2025; Gao et al., 2022; Uluseker et al., 2021) (Table 3). UV, ozone deactivation, advanced oxidation, and solar photocatalysis processes are advanced and tertiary and could significantly enhance ARB/ARGs removal, but the performance is gene-selective and depends on the composition of a matrix and operational conditions (McCorquodale-Bauer et al., 2023; Starling et al., 2021; Kalli et al., 2023).

Table 3. Examples of WWTP-related ecosystem impacts

Aspect	Key findings	Citations
ARG fate in WWTPs	Limited removal; WWTPs act as ARG hotspots and sources to receiving waters.	(La Rosa et al., 2025; Gao et al., 2022; Uluseker et al., 2021)
Advanced treatment	UV/ozone/AOPs and solar photo-Fenton enhance ARB/ARG removal versus conventional treatment.	(McCorquodale-Bauer et al., 2023; Starling et al., 2021; Kalli et al., 2023)
Environmental dissemination	Effluents and sludge spread AMR into aquatic ecosystems, posing ecological and health risks.	(La Rosa et al., 2025; Uluseker et al., 2021; Kalli et al., 2023)

8.2 One Health Perspectives Linking Humans, Animals, and the Environment

The One Health model clearly considers human, animal, and environmental health and focuses on common determinants, including pollution, antimicrobial resistance, and climate change (La Rosa et al., 2025; Ioannou et al., 2025; Banerjee and Van Der Heijden, 2022). An example of such a connection is wastewater-mediated AMR: clinical, agricultural, and community sources of antibiotics and ARGs accumulate in WWTPs, diffuse into soil and water, and eventually impact wildlife, livestock, and the human population via various routes of exposure (La Rosa et al., 2025; Uluseker et al., 2021; Werkneh and Islam, 2023). Stewbrook One Health reviews have indicated convergence in toxicological effects across species caused by contaminants of PBDEs, heavy metals, and other pollutants, and microbiomes in soil, plants, animals, and humans are interconnected networks that support ecosystem and population health (Domingo et al., 2025; Banerjee and Van Der Heijden, 2022).

9. DATA INTEGRATION, MODELING, AND RISK ASSESSMENT

9.1 Linking Genomic Data with Epidemiological Models

The integration of data in wastewater-based epidemiology, epidemiological modeling, and risk assessment is increasingly dependent on the connection between genomic (e.g., the sequence of pathogens, variants, antimicrobial resistance genes) and mechanistic and statistical epidemiological modeling. Infectious disease and COVID-19 wastewater surveillance reviews indicate that RT qPCR and sequencing results are analyzed together with the disease transmission model to estimate the prevalence of infections, reproduction numbers, and underdetection, as well as characterize the dynamics of variants (Shah et al., 2021; Ciannella et al., 2023; Kilaru et al., 2021). In general, mechanistic models integrated with artificial intelligence and machine learning can utilize genomic and other omics data to better estimate parameters, detect outbreaks, and optimize interventions to enhance the accuracy of predictions and their biological understanding (Ye et al., 2025; Singh et al., 2024).

9.2 Decision-Support Systems for Public Health Authorities

These decision-support systems are used by decision-making agencies to take up these integrated modeling and analytics pipelines. Wastewater surveillance and epidemic intelligence reviews outline the use of dashboards and GIS-based platforms to visualize predicted trends and uncertainty, as well as spatial risk, allowing them to provide early warnings, testing and vaccination priorities, and resource allocation (Cuadros et al., 2024; Ensor et al., 2024; Aryffin et al., 2025). Surveys of the wastewater surveillance literature demonstrate that, where action is explicitly designed into them, such systems have informed case finding, targeted communication, and mitigation actions, albeit that a small fraction of published literature to date incorporates explicit action pathways or decision rules (Zhu et al., 2025; Van Der Drift et al., 2025). Wastewater, clinical, mobility, and environmental data streams can be used to support scenario analysis and policy evaluation with emerging AI-enhanced epidemic intelligence tools that improve surveillance beyond providing descriptive monitoring to active and model-driven risk management (Ye et al., 2025; Aryffin et al., 2025).

10. CURRENT CHALLENGES AND KNOWLEDGE GAPS

10.1 Methodological Limitations and Uncertainty in Interpretation

The contemporary challenges and gaps in the field of wastewater-based epidemiology (WBE) are mainly caused by methodological constraints and a lack of interpretation. In the context of pathogens, variability in virus shedding, in-sewer decay, sampling design, concentration, and extraction efficiency, and RT qPCR or sequencing performance is quite high and introduces uncertainty in the estimation of prevalence and the interpretation of trends (Li et al., 2021; Kumblathan et al., 2021). There is a lack of consistency in following qPCR best-practice principles and a shortage of certified reference materials, with standard-curve parameters widely varying across studies and laboratories, making quantitative comparison between studies and across laboratories unreliable (Bivins et al., 2021; Kumblathan et al., 2021). Even with a strong detection, it is still hard to translate the wastewater concentrations into the number of cases or how much is too much to take action, as dose-response relationships, normalization of the population, and contribution of the non-human sources are not well understood (Grassly et al., 2024; Tiwari et al., 2022; Li et al., 2021). Scoping reviews thus demand more engineering characterization of sewer systems, more use of internal controls and surrogate viruses, and clear uncertainty analysis at every stage, leading to back-calculation (Li et al., 2021; Laicans et al., 2024).

10.2 Capacity Building in Low- and Middle-Income Countries

Lastly, major capacity gaps in low- and middle-income countries (LMICs) have been characterized by the risk of pandemic emergence being likely the greatest, but the weakness of wastewater and laboratory capacity (Grassly et al., 2024; Shrestha et al., 2021; R. et al., 2025). Most LMICs are characterized by the lack of centralized sewer systems, regular power, cold chains, supply of reagents, and sequencing facilities, which contribute to the long lead times, high costs, and test disruptions (Grassly et al., 2024; Shrestha et al., 2021). The COVID-19-specific reviews in LMICs remark that extensive networks of sanitation make it difficult to define catchments and population denominators, which necessitate innovative designs including subdivision into smaller sewersheds, decentralized sampling, and low-cost passive samplers (Shrestha et al., 2021; R. et al., 2025). In addition to technical problems, sustainable implementation requires program investments in the sanitation infrastructure, local laboratory training, stable supply chains, and governance systems that incorporate the outputs of WBE into the process of making decisions in the field of health, and not independent research pilots (Grassly et al., 2024; Shrestha et al., 2021; R. et al., 2025). Unless specific capacity building is made, and fair funding is received, global WBE systems run the risk of reinforcing some old inequities of surveillance rather than closing them.

11. FUTURE PERSPECTIVES AND TECHNOLOGICAL INNOVATIONS

The outlook for the future and new technologies will make wastewater-based epidemiology (WBE) and environmental genomics more sensitive, automated, and sustainable aspects of community health monitoring. Unanticipated and rapid changes in sequencing, especially high-throughput and metagenomic next-generation sequencing (NGS), can now be done at high scales, targeting untargeted detection of a broad spectrum of pathogens, variants, antimicrobial resistance genes, and other biomarkers in mixed systems such as wastewater, and well beyond single-pathogen qPCR and serves as an early warning of both known and novel threats within a One Health paradigm (Zhu et al., 2025; Farkas et al., 2024). The NGS-based environmental DNA (eDNA) methods also extend the scope of surveillance to the air and soil and aquatic ecosystems and suggest a unified ecological-population health surveillance system (Yoon et al., 2025). Simultaneously, biosensor technologies are being developed to transition out of laboratory-scale, towards portable, multiplexable and even battery-free systems, with the capability to detect contaminants and pathogens in real-time and on-site with high sensitivity and selectivity; integration with microfluidics, hybrid nanomaterials and smartphone-based display is becoming continuous or nearly real time, environmental monitoring at lower cost and power requirements in line with the Sustainable Development Goals of clean water, health and climate resilience (Huang et al., 2023; Fdez-Sanromán et al., 2025). In the case of WBE in particular, both aptamer-based biosensors and field-deployable sensor arrays are being considered as an alternative or a supplement to centralized molecular assays, where the potential to stream digital data to public health dashboards and avoid delays by transporting samples and performing batch processing could help (Zhu et al., 2025). These trends are convergent on real-time monitoring and automation. Remote or in situ sequencing platforms, autonomous samplers, and sensor networks can offer real-time wastewater and eDNA data, and machine learning and advanced bioinformatics can allow automated detection of anomalies, attribution of source, and prediction of an outbreak or pollution event (Yoon et al., 2025; Zhu et al., 2025). With a decrease in the price of sequencing and biosensing, increasingly protocols are becoming more standardized, and WBE and environmental genomics are being positioned as viable as sustainable surveillance infrastructures, providing non-invasive, population-wide coverage, the capacity to detect a variety of hazards (pathogens, resistance, chemicals) using the same samples, and capable of operating on low-power, miniaturized devices that can be deployed in a variety of environments on a long-term basis (Yoon et al., 2025; Huang et al., 2023; Farkas et al., 2024). To achieve this vision, there will be a need to implement a strong data governance policy, even-handed capacity building, and connect to traditional clinical and environmental surveillance, yet the technological trends are clearly towards automated and genomics-enabled WBE as one of the central elements of resilient and sustainable public health systems.

12. CONCLUSION

Environmental genomic monitoring of wastewater and epidemiology on wastewater are radical methods of comprehending population health dynamics and environmental integrity in a timely, non-invasive, and cost-effective approach. These tools can be used to identify infectious diseases early, detect antimicrobial resistance, and determine the level of exposure of the population to pharmaceuticals and pollutants by measuring aggregated biological and chemical signals of wastewater and environmental matrices. The combination of high-throughput sequencing, sophisticated bioinformatics, and epidemiological modeling has enhanced the resolution and predictive ability of surveillance systems to a great degree. Even though proven, there are still issues about methodological standardization, interpretation of data, ethics, and the key to their interconnection with traditional monitoring systems.

They can be overcome by standardized procedures, effective data-sharing platforms, and interdisciplinary collaboration, which will play a crucial role in maximizing their effect. An increasing global burden of emerging pathogens, urbanization, and environmental change is likely to make wastewater-based epidemiology and environmental genomic surveillance key elements of effective, sustainable public health surveillance and environmental management interventions, informative decision-making, and proactive risk reduction approaches both locally, nationally, and globally.

REFERENCES

Abdeldayem, O., Dabbish, A., Habashy, M., Mostafa, M., Elhefnawy, M., Amin, L., Al-Sakkari, E., Ragab, A., & Rene, E. (2021). Viral outbreaks detection and surveillance using wastewater-based epidemiology, viral air sampling, and machine learning techniques: A comprehensive review and outlook. *The Science of the Total Environment*, *803*, 149834–149834. DOI: 10.1016/j.scitotenv.2021.149834 PMID: 34525746

Aghaei, M., Khoshnamvand, N., Janjani, H., Dehghani, M., & Karri, R. (2024). Exposure to environmental pollutants: A mini-review on the application of the wastewater-based epidemiology approach. *Journal of Environmental Health Science & Engineering*, *22*(1), 65–74. DOI: 10.1007/s40201-024-00895-0 PMID: 38887772

Ali, W., Zhang, H., Wang, Z., Chang, C., Javed, A., Ali, K., Du, W., Niazi, N., Mao, K., & Yang, Z. (2021). Occurrence of various viruses and recent evidence of SARS-CoV-2 in wastewater systems. *Journal of Hazardous Materials*, *414*, 125439–125439. DOI: 10.1016/j.jhazmat.2021.125439 PMID: 33684818

Amin, V., Bowes, D., & Halden, R. (2022). Systematic scoping review evaluating the potential of wastewater-based epidemiology for monitoring cardiovascular disease and cancer. *The Science of the Total Environment*, *858*, 160103–160103. DOI: 10.1016/j.scitotenv.2022.160103 PMID: 36370774

Aplakidou, E., Vergoulidis, N., Chasapi, M., Venetsianou, N., Kokoli, M., Panagiotopoulou, E., Iliopoulos, I., Karatzas, E., Pafilis, E., Georgakopoulos-Soares, I., Kyrpides, N., Pavlopoulos, G., & Baltoumas, F. (2024). Visualizing metagenomic and metatranscriptomic data: A comprehensive review. *Computational and Structural Biotechnology Journal*, *23*, 2011–2033. DOI: 10.1016/j.csbj.2024.04.060 PMID: 38765606

Aryffin, H., Sahbudin, M., Pitchay, S., Halim, A., & Sahbudin, I. (2025). Technological trends in epidemic intelligence for infectious disease surveillance: A systematic literature review. *PeerJ. Computer Science*, *11*, e2874. Advance online publication. DOI: 10.7717/peerj-cs.2874 PMID: 40567698

Banerjee, S., & Van Der Heijden, M. (2022). Soil microbiomes and one health. *Nature Reviews. Microbiology*, *21*(1), 6–20. DOI: 10.1038/s41579-022-00779-w PMID: 35999468

Bivins, A., Kaya, D., Bibby, K., Simpson, S., Bustin, S., Shanks, O., & Ahmed, W. (2021). Variability in RT-qPCR assay parameters indicates unreliable SARS-CoV-2 RNA quantification for wastewater surveillance. *Water Research, 203,* 117516–117516. DOI: 10.1016/j.watres.2021.117516 PMID: 34412018

Boogaerts, T., Ahmed, F., Choi, P., Tscharke, B., O'Brien, J., De Loof, H., Gao, J., Thai, P., Thomas, K., Mueller, J., Hall, W., Covaci, A., & Van Nuijs, A. (2021). Current and future perspectives for wastewater-based epidemiology as a monitoring tool for pharmaceutical use. *The Science of the Total Environment, 789,* 148047. DOI: 10.1016/j.scitotenv.2021.148047 PMID: 34323839

Câmara, J., Perestrelo, R., Berenguer, C., Andrade, C., Gomes, T., Olayanju, B., Kabir, A., Rocha, C., Teixeira, J., & Pereira, J. (2022). Green Extraction Techniques as Advanced Sample Preparation Approaches in Biological, Food, and Environmental Matrices: A Review. *Molecules (Basel, Switzerland), 27.* Advance online publication. DOI: 10.3390/molecules27092953 PMID: 35566315

Carneiro, R., Nika, M., Gil-Solsona, R., Diamanti, K., Thomaidis, N., Corominas, L., & Gago-Ferrero, P. (2024). A critical review of wastewater-based epidemiology as a tool to evaluate the unintentional human exposure to potentially harmful chemicals. *Analytical and Bioanalytical Chemistry, 417*(3), 495–511. DOI: 10.1007/s00216-024-05596-z PMID: 39422714

Catarro, G., Pelixo, R., Feijó, M., Rosado, T., Socorro, S., Araujo, A., & Gallardo, E. (2025). Analytical Approaches Using GC-MS for the Detection of Pollutants in Wastewater Towards Environmental and Human Health Benefits: A Comprehensive Review. *Chemosensors (Basel, Switzerland), 13*(7), 253. Advance online publication. DOI: 10.3390/chemosensors13070253

Ciannella, S., González-Fernández, C., & Gómez-Pastora, J. (2023). Recent progress on wastewater-based epidemiology for COVID-19 surveillance: A systematic review of analytical procedures and epidemiological modeling. *The Science of the Total Environment, 878,* 162953–162953. DOI: 10.1016/j.scitotenv.2023.162953 PMID: 36948304

Ciuffreda, L., Rodríguez-Pérez, H., & Flores, C. (2021). Nanopore sequencing and its application to the study of microbial communities. *Computational and Structural Biotechnology Journal, 19,* 1497–1511. DOI: 10.1016/j.csbj.2021.02.020 PMID: 33815688

Corrin, T., Rabeenthira, P., Young, K., Mathiyalagan, G., Baumeister, A., Pussegoda, K., & Waddell, L. (2024). A scoping review of human pathogens detected in untreated human wastewater and sludge. *Journal of Water and Health*, *22*(2), 436–449. DOI: 10.2166/wh.2024.326 PMID: 38421635

Cuadros, D., Chen, X., Li, J., Omori, R., & Musuka, G. (2024). Advancing Public Health Surveillance: Integrating Modeling and GIS in the Wastewater-Based Epidemiology of Viruses, a Narrative Review. *Pathogens (Basel, Switzerland)*, *13*(8), 685. Advance online publication. DOI: 10.3390/pathogens13080685 PMID: 39204285

Davis, B., Brown, C., Gupta, S., Calarco, J., Liguori, K., Milligan, E., Harwood, V., Pruden, A., & Keenum, I. (2023). Recommendations for the use of metagenomics for routine monitoring of antibiotic resistance in wastewater and impacted aquatic environments. *Critical Reviews in Environmental Science and Technology*, *53*(19), 1731–1756. DOI: 10.1080/10643389.2023.2181620

De Oliveira, A., De Melo Vieira, A., & Santos, J. (2023). Trends and challenges in analytical chemistry for multi-analysis of illicit drugs employing wastewater-based epidemiology. *Analytical and Bioanalytical Chemistry*, 1 - 10. https://doi.org/DOI: 10.1007/s00216-023-04644-4

Djemiel, C., Maron, P., Terrat, S., Dequiedt, S., Cottin, A., & Ranjard, L. (2022). Inferring microbiota functions from taxonomic genes: A review. *GigaScience*, *11*, giab090. Advance online publication. DOI: 10.1093/gigascience/giab090 PMID: 35022702

Domingo, J., Souza, M., Nadal, M., & Barbosa, F. (2025). A One Health approach to polybrominated diphenyl ethers (PBDEs): Integrating human, animal, and environmental health perspectives. *Chemosphere*, *386*, 144644. DOI: 10.1016/j.chemosphere.2025.144644 PMID: 40829294

Ensor, K., Schedler, J., Sun, T., Schneider, R., Mulenga, A., Wu, J., Stadler, L., & Hopkins, L. (2024). Online trend estimation and detection of trend deviations in sub-sewershed time series of SARS-CoV-2 RNA measured in wastewater. *Scientific Reports*, *14*(1), 5575. Advance online publication. DOI: 10.1038/s41598-024-56175-2 PMID: 38448481

Farkas, K., Williams, R., Hillary, L., Garcia-Delgado, A., Jameson, E., Kevill, J., Wade, M., Grimsley, J., & Jones, D. (2024). Harnessing the Power of Next-Generation Sequencing in Wastewater-Based Epidemiology and Global Disease Surveillance. *Food and Environmental Virology*, *17*(1), 5. Advance online publication. DOI: 10.1007/s12560-024-09616-0 PMID: 39614945

Fdez-Sanromán, A., Bernárdez-Rodas, N., Rosales, E., Pazos, M., González-Romero, E., & Sanromán, M. (2025). Biosensor Technologies for Water Quality: Detection of Emerging Contaminants and Pathogens. *Biosensors (Basel), 15*(3), 189. Advance online publication. DOI: 10.3390/bios15030189 PMID: 40136986

Gao, Y., Li, X., Fan, X., Zhao, J., & Zhang, Z. (2022). Wastewater treatment plants as reservoirs and sources for antibiotic resistance genes: A review on occurrence, transmission, and removal. *Journal of Water Process Engineering, 46*, 102539. Advance online publication. DOI: 10.1016/j.jwpe.2021.102539

Grassly, N., Shaw, A., & Owusu, M. (2024). Global wastewater surveillance for pathogens with pandemic potential: Opportunities and challenges. *The Lancet Microbe, 100939*. Advance online publication. DOI: 10.1016/j.lanmic.2024.07.002 PMID: 39222653

Hernández, F., Fabregat-Safont, D., Campos-Mañas, M., & Quintana, J. (2023). Efficient Validation Strategies in Environmental Analytical Chemistry: A Focus on Organic Micropollutants in Water Samples. *Annual Review of Analytical Chemistry, 16*(1), 401–428. Advance online publication. DOI: 10.1146/annurev-anchem-091222-112115 PMID: 37068748

Holton, E., Sims, N., Jagadeesan, K., Standerwick, R., & Kasprzyk-Hordern, B. (2022). Quantifying community-wide antimicrobials usage via wastewater-based epidemiology. *Journal of Hazardous Materials, 436*, 129001. DOI: 10.1016/j.jhazmat.2022.129001 PMID: 35594673

Huang, C., Lin, C., Nguyen, M., Hussain, A., Bui, X., & Ngo, H. (2023). A review of biosensor for environmental monitoring: Principle, application, and corresponding achievement of sustainable development goals. *Bioengineered, 14*(1), 58–80. DOI: 10.1080/21655979.2022.2095089 PMID: 37377408

Huizer, M., Ter Laak, T., De Voogt, P., & Van Wezel, A. (2021). Wastewater-based epidemiology for illicit drugs: A critical review on global data. *Water Research, 207*, 117789. DOI: 10.1016/j.watres.2021.117789 PMID: 34731667

Ioannou, A., Bataka, E., Kokosis, N., Billinis, C., & Laspidou, C. (2025). One Health in Coastal and Marine Contexts: A Critical Bibliometric Analysis Across Environmental, Animal, and Human Health Dimensions. *International Journal of Environmental Research and Public Health, 22*(10), 1523. Advance online publication. DOI: 10.3390/ijerph22101523 PMID: 41154927

Kalli, M., Noutsopoulos, C., & Mamais, D. (2023). The Fate and Occurrence of Antibiotic-Resistant Bacteria and Antibiotic Resistance Genes during Advanced Wastewater Treatment and Disinfection: A Review. *Water (Basel)*, *15*(11), 2084. Advance online publication. DOI: 10.3390/w15112084

Kayani, M., Huang, W., Feng, R., & Chen, L. (2021). Genome-resolved metagenomics using environmental and clinical samples. *Briefings in Bioinformatics*, *22*(5), bbab030. Advance online publication. DOI: 10.1093/bib/bbab030 PMID: 33758906

Keenum, I., Liguori, K., Calarco, J., Davis, B., Milligan, E., Harwood, V., & Pruden, A. (2022). A framework for standardized qPCR-targets and protocols for quantifying antibiotic resistance in surface water, recycled water, and wastewater. *Critical Reviews in Environmental Science and Technology*, *52*(24), 4395–4419. DOI: 10.1080/10643389.2021.2024739

Kilaru, P., Hill, D., Anderson, K., Collins, M., Green, H., Kmush, B., & Larsen, D. (2021). Wastewater Surveillance for Infectious Disease: A Systematic Review. *American Journal of Epidemiology*. Advance online publication. DOI: 10.1101/2021.07.26.21261155 PMID: 36227259

Kim, C., Pongpanich, M., & Porntaveetus, T. (2024). Unraveling metagenomics through long-read sequencing: A comprehensive review. *Journal of Translational Medicine*, *22*(1), 111. Advance online publication. DOI: 10.1186/s12967-024-04917-1 PMID: 38282030

Kumblathan, T., Liu, Y., Uppal, G., Hrudey, S., & Li, X. (2021). Wastewater-Based Epidemiology for Community Monitoring of SARS-CoV-2: Progress and Challenges. *ACS Environmental Au*, *1*(1), 18–31. DOI: 10.1021/acsenvironau.1c00015 PMID: 37579255

Kye, H., Kim, J., Ju, S., Lee, J., Lim, C., & Yoon, Y. (2023). Microplastics in water systems: A review of their impacts on the environment and their potential hazards. *Heliyon*, *9*(3), e14359. Advance online publication. DOI: 10.1016/j.heliyon.2023.e14359 PMID: 36950574

La Rosa, M., Maugeri, A., Favara, G., La Mastra, C., Lio, M., Barchitta, M., & Agodi, A. (2025). The Impact of Wastewater on Antimicrobial Resistance: A Scoping Review of Transmission Pathways and Contributing Factors. *Antibiotics (Basel, Switzerland)*, *14*(2), 131. Advance online publication. DOI: 10.3390/antibiotics14020131 PMID: 40001375

Laicans, J., Dejus, B., Dejus, S., & Juhna, T. (2024). Precision and Accuracy Limits of Wastewater-Based Epidemiology—Lessons Learned from SARS-CoV-2: A Scoping Review. *Water (Basel), 16*(9), 1220. Advance online publication. DOI: 10.3390/w16091220

Laimou-Geraniou, M., Heath, D., & Heath, E. (2022). Analytical methods for the determination of antidepressants, antipsychotics, benzodiazepines and their metabolites through wastewater-based epidemiology. *Trends in Environmental Analytical Chemistry*. Advance online publication. DOI: 10.1016/j.teac.2022.e00192

Li, X., Zhang, S., Shi, J., Luby, S., & Jiang, G. (2021). Uncertainties in estimating SARS-CoV-2 prevalence by wastewater-based epidemiology. *Chemical Engineering Journal, 415*, 129039–129039. DOI: 10.1016/j.cej.2021.129039 PMID: 33642938

Liguori, K., Keenum, I., Davis, B., Calarco, J., Milligan, E., Harwood, V., & Pruden, A. (2022). Antimicrobial Resistance Monitoring of Water Environments: A Framework for Standardized Methods and Quality Control. *Environmental Science & Technology, 56*(13), 9149–9160. DOI: 10.1021/acs.est.1c08918 PMID: 35732277

Lu, H., Ziajahromi, S., Neale, P., & Leusch, F. (2021). A systematic review of freshwater microplastics in water and sediments: Recommendations for harmonisation to enhance future study comparisons. *The Science of the Total Environment, 781*, 146693. DOI: 10.1016/j.scitotenv.2021.146693

Malcom, H., & Bowes, D. (2025). Use of Wastewater to Monitor Antimicrobial Resistance Trends in Communities and Implications for Wastewater-Based Epidemiology: A Review of the Recent Literature. *Microorganisms, 13*(9), 2073. Advance online publication. DOI: 10.3390/microorganisms13092073 PMID: 41011405

McCorquodale-Bauer, K., Grosshans, R., Zvomuya, F., & Cicek, N. (2023). Critical review of phytoremediation for the removal of antibiotics and antibiotic resistance genes in wastewater. *The Science of the Total Environment, 161876*, 161876. Advance online publication. DOI: 10.1016/j.scitotenv.2023.161876 PMID: 36716878

Meckawy, R., Stuckler, D., Mehta, A., Al-Ahdal, T., & Doebbeling, B. (2022). Effectiveness of early warning systems in the detection of infectious disease outbreaks: A systematic review. *BMC Public Health, 22*(1), 2216. Advance online publication. DOI: 10.1186/s12889-022-14625-4 PMID: 36447171

Nam, N., Khoa, H., Trinh, K., & Lee, N. (2023). Metagenomics: An Effective Approach for Exploring Microbial Diversity and Functions. *Foods, 12*(11), 2140. Advance online publication. DOI: 10.3390/foods12112140 PMID: 37297385

Nguyen, A., Vu, H., Nguyen, L., Wang, Q., Djordjevic, S., Donner, E., Yin, H., & Nghiem, L. (2021). Monitoring antibiotic resistance genes in wastewater treatment: Current strategies and future challenges. *The Science of the Total Environment*, *783*, 146964. DOI: 10.1016/j.scitotenv.2021.146964 PMID: 33866168

Oloye, F., Xie, Y., Challis, J., Femi-Oloye, O., Brinkmann, M., McPhedran, K., Jones, P., Servos, M., & Giesy, J. (2023). Understanding common population markers for SARS-CoV-2 RNA normalization in wastewater – A review. *Chemosphere*, *333*, 138682–138682. DOI: 10.1016/j.chemosphere.2023.138682 PMID: 37201600

Parkins, M., Lee, B., Acosta, N., Bautista, M., Hubert, C., Hrudey, S., Frankowski, K., & Pang, X. (2023). Wastewater-based surveillance as a tool for public health action: SARS-CoV-2 and beyond. *Clinical Microbiology Reviews*, *37*(1), e00103-22. Advance online publication. DOI: 10.1128/cmr.00103-22 PMID: 38095438

Patin, N., & Goodwin, K. (2023). Capturing marine microbiomes and environmental DNA: A field sampling guide. *Frontiers in Microbiology*, *13*, 1026596. Advance online publication. DOI: 10.3389/fmicb.2022.1026596 PMID: 36713215

Pawlowski, J., Bruce, K., Panksep, K., Aguirre, F., Amafitano, S., Apothéloz-Perret-Gentil, L., Baussant, T., Bouchez, A., Carugati, L., Cermakova, K., Cordier, T., Corinaldesi, C., Costa, F., Danovaro, R., Dell'Anno, A., Duarte, S., Eisendle, U., Ferrari, B., Frontalini, F., & Fazi, S. (2021). Environmental DNA metabarcoding for benthic monitoring: A review of sediment sampling and DNA extraction methods. *The Science of the Total Environment*, *151783*. Advance online publication. DOI: 10.1016/j.scitotenv.2021.151783 PMID: 34801504

Peng, J., Zhou, X., Rensing, C., Liesack, W., & Zhu, Y. (2023). Soil microbial ecology through the lens of metatranscriptomics. *Soil Ecology Letters*, *6*(3), 230217. Advance online publication. DOI: 10.1007/s42832-023-0217-z

Prata, J., Reis, V., Da Costa, J., Mouneyrac, C., Duarte, A., & Rocha-Santos, T. (2021). Contamination issues as a challenge in quality control and quality assurance in microplastics analytics. *Journal of Hazardous Materials*, *403*, 123660. DOI: 10.1016/j.jhazmat.2020.123660 PMID: 33264868

R., B., N, K., Lp, T., Jt, C., Wp, H., Mt, B., & Tl, C. (2025). Advancing health equity in wastewater-based epidemiology: A global critical review and conceptual framework. *SSM - Population Health*, 30. https://doi.org/DOI: 10.1016/j.ssmph.2025.101786

Rashid, S., Rajendiran, S., Nazakat, R., Sham, N., Hasni, N., Anasir, M., Kamel, K., & Robat, R. (2024). A scoping review of global SARS-CoV-2 wastewater-based epidemiology in light of COVID-19 pandemic. *Heliyon*, *10*(9), e30600. Advance online publication. DOI: 10.1016/j.heliyon.2024.e30600 PMID: 38765075

Regueira-Iglesias, A., Balsa-Castro, C., Blanco-Pintos, T., & Tomás, I. (2023). Critical review of 16S rRNA gene sequencing workflow in microbiome studies: From primer selection to advanced data analysis. *Molecular Oral Microbiology*, *38*(5), 347–399. Advance online publication. DOI: 10.1111/omi.12434 PMID: 37804481

Rout, A., Rout, S., Panda, A., Tripathy, P., Kumar, N., Parida, S., Dey, S., Dash, S., Behera, B., & Pandey, P. (2025). Potential applications and future prospects of metagenomics in aquatic ecosystems. *Gene*, *149720*, 149720. Advance online publication. DOI: 10.1016/j.gene.2025.149720 PMID: 40789383

Sassi, A., Basher, N., Kirat, H., Meradji, S., Ibrahim, N., Idres, T., & Touati, A. (2025). The Role of the Environment (Water, Air, Soil) in the Emergence and Dissemination of Antimicrobial Resistance: A One Health Perspective. *Antibiotics (Basel, Switzerland)*, *14*(8), 764. Advance online publication. DOI: 10.3390/antibiotics14080764 PMID: 40867959

Shah, S., Gwee, S., Ng, J., Lau, N., Koh, J., & Pang, J. (2021). Wastewater surveillance to infer COVID-19 transmission: A systematic review. *The Science of the Total Environment*, *804*, 150060–150060. DOI: 10.1016/j.scitotenv.2021.150060 PMID: 34798721

Shakallis, A., Fallowfield, H., Ross, K., & Whiley, H. (2022). The Application of Passive Sampling Devices in Wastewater Surveillance. *Water (Basel)*, *14*(21), 3478. Advance online publication. DOI: 10.3390/w14213478

Sharuddin, S., Ramli, N., Yusoff, M., Muhammad, N., Ho, L., & Maeda, T. (2022). Advancement of Metatranscriptomics towards Productive Agriculture and Sustainable Environment: A Review. *International Journal of Molecular Sciences*, *23*(7), 3737. Advance online publication. DOI: 10.3390/ijms23073737 PMID: 35409097

Shrestha, S., Yoshinaga, E., Chapagain, S., Mohan, G., Gasparatos, A., & Fukushi, K. (2021). Wastewater-Based Epidemiology for Cost-Effective Mass Surveillance of COVID-19 in Low- and Middle-Income Countries: Challenges and Opportunities. *Water (Basel)*, *13*(20), 2897. Advance online publication. DOI: 10.3390/w13202897

Singh, A., Bhardwaj, P., Laux, P., Pradeep, P., Busse, M., Luch, A., Hirose, A., Osgood, C., & Stacey, M. (2024). AI and ML-based risk assessment of chemicals: Predicting carcinogenic risk from chemical-induced genomic instability. *Frontiers in Toxicology*, *6*, 1461587. Advance online publication. DOI: 10.3389/ftox.2024.1461587 PMID: 39659701

Singh, A., Pratap, S., & Raj, A. (2024). Occurrence and dissemination of antibiotics and antibiotic resistance in aquatic environment and its ecological implications: A review. *Environmental Science and Pollution Research International, 31*(35), 47505–47529. DOI: 10.1007/s11356-024-34355-x PMID: 39028459

Singh, S., Ahmed, A., Almansoori, S., Alameri, S., Adlan, A., Odivilas, G., Chattaway, M., Salem, S., Brudecki, G., & Elamin, W. (2024). A narrative review of wastewater surveillance: Pathogens of concern, applications, detection methods, and challenges. *Frontiers in Public Health, 12*, 1445961. Advance online publication. DOI: 10.3389/fpubh.2024.1445961 PMID: 39139672

Starling, M., Neto, R., Pires, G., Vilela, P., & Amorim, C. (2021). Combat of antimicrobial resistance in municipal wastewater treatment plant effluent via solar advanced oxidation processes: Achievements and perspectives. *The Science of the Total Environment, 786*, 147448. DOI: 10.1016/j.scitotenv.2021.147448 PMID: 33965817

Tiwari, A., Adhikari, S., Zhang, S., Solomon, T., Lipponen, A., Islam, M., Thakali, O., Sangkham, S., Shaheen, M., Jiang, G., Haramoto, E., Mazumder, P., Malla, B., Kumar, M., Pitkänen, T., & Sherchan, S. (2023). Tracing COVID-19 Trails in Wastewater: A Systematic Review of SARS-CoV-2 Surveillance with Viral Variants. *Water (Basel), 15*(6), 1018. Advance online publication. DOI: 10.3390/w15061018

Tiwari, A., Kurittu, P., Al-Mustapha, A., Heljanko, V., Johansson, V., Thakali, O., Mishra, S., Lehto, K., Lipponen, A., Oikarinen, S., Pitkänen, T., & Heikinheimo, A. (2022). Wastewater surveillance of antibiotic-resistant bacterial pathogens: A systematic review. *Frontiers in Microbiology, 13*, 977106. Advance online publication. DOI: 10.3389/fmicb.2022.977106 PMID: 36590429

Tyagi, S., & Katara, P. (2024). Metatranscriptomics: A Tool for Clinical Metagenomics. *Omics : A Journal of Integrative Biology, 28*(8), 394–407. DOI: 10.1089/omi.2024.0130 PMID: 39029911

Uluseker, C., Kaster, K., Thorsen, K., Basiry, D., Shobana, S., Jain, M., Kumar, G., Kommedal, R., & Pala-Ozkok, I. (2021). A Review on Occurrence and Spread of Antibiotic Resistance in Wastewaters and in Wastewater Treatment Plants: Mechanisms and Perspectives. *Frontiers in Microbiology, 12*, 717809. Advance online publication. DOI: 10.3389/fmicb.2021.717809 PMID: 34707579

Van Der Drift, A., Welling, A., Arntzen, V., Nagelkerke, E., Van Der Beek, R., & De Roda Husman, A. (2025). Wastewater surveillance studies on pathogens and their use in public health decision-making: A scoping review. *The Science of the Total Environment, 993*, 179982. DOI: 10.1016/j.scitotenv.2025.179982 PMID: 40592216

Villanueva-Miranda, I., Xiao, G., & Xie, Y. (2025). Artificial intelligence in early warning systems for infectious disease surveillance: A systematic review. *Frontiers in Public Health, 13*, 1609615. Advance online publication. DOI: 10.3389/fpubh.2025.1609615 PMID: 40626156

Werkneh, A., & Islam, M. (2023). Post-treatment disinfection technologies for sustainable removal of antibiotic residues and antimicrobial resistance bacteria from hospital wastewater. *Heliyon, 9*(4), e15360. Advance online publication. DOI: 10.1016/j.heliyon.2023.e15360 PMID: 37123966

Yang, C., Chowdhury, D., Zhang, Z., Cheung, W., Lu, A., Bian, Z., & Zhang, L. (2021). A review of computational tools for generating metagenome-assembled genomes from metagenomic sequencing data. *Computational and Structural Biotechnology Journal, 19*, 6301–6314. DOI: 10.1016/j.csbj.2021.11.028 PMID: 34900140

Ye, Y., Pandey, A., Bawden, C., Sumsuzzman, D., Rajput, R., Shoukat, A., Singer, B., Moghadas, S., & Galvani, A. (2025). Integrating artificial intelligence with mechanistic epidemiological modeling: A scoping review of opportunities and challenges. *Nature Communications, 16*(1), 581. Advance online publication. DOI: 10.1038/s41467-024-55461-x PMID: 39794317

Yoon, H., Seo, J., Shin, S., Abdelhamid, M., & Pack, S. (2025). Bioinformation and Monitoring Technology for Environmental DNA Analysis: A Review. *Biosensors (Basel), 15*(8), 494. Advance online publication. DOI: 10.3390/bios15080494 PMID: 40862956

Zhang, S., Li, X., Wu, J., Coin, L., O'Brien, J., Hai, F., & Jiang, G. (2021). Molecular Methods for Pathogenic Bacteria Detection and Recent Advances in Wastewater Analysis. *Water (Basel), 13*(24), 3551. Advance online publication. DOI: 10.3390/w13243551

Zhang, T., Jiang, M., Li, H., Gao, Y., Yousuf, S., Yu, K., Yi, X., Wang, J., Yang, L., & Liu, Y. (2025). Computational Tools and Resources for Long-read Metagenomic Sequencing Using Nanopore and PacBio. *Genomics, Proteomics & Bioinformatics, 23*(4), qzaf075. Advance online publication. DOI: 10.1093/gpbjnl/qzaf075 PMID: 40845163

Zhu, W., Wang, D., Li, P., Deng, H., & Deng, Z. (2025). Advances in Wastewater-Based Epidemiology for Pandemic Surveillance: Methodological Frameworks and Future Perspectives. *Microorganisms, 13*(5), 1169. Advance online publication. DOI: 10.3390/microorganisms13051169 PMID: 40431340

Chapter 5
Machine Learning Approaches for Outbreak Detection and Forecasting

Olugbemiga Ojo Aliu
http://orcid.org/0000-0003-4942-7454
Auchi Polytechnic, Auchi, Nigeria

Sylvester Chibueze Izah
http://orcid.org/0000-0001-5526-006X
Bayelsa Medical University, Nigeria

ABSTRACT

Outbreak detection and forecasting are central functions of biosurveillance because public-health response depends on recognizing abnormal disease activity before transmission becomes widespread and estimating how an outbreak may evolve under uncertainty. Traditional surveillance systems remain indispensable, but they are often constrained by reporting delays, incomplete case ascertainment, fragmented data streams, limited spatial resolution, and the difficulty of integrating clinical, laboratory, environmental, behavioral, and digital information in near real time. This chapter examines how machine learning can strengthen outbreak detection and forecasting by transforming heterogeneous data into actionable early-warning intelligence. It reviews core data streams, including notifiable disease reports, syndromic surveillance, electronic health records, laboratory positivity, wastewater signals, mobility data, climate and environmental predictors, genomic information, veterinary and wildlife data.

DOI: 10.4018/979-8-3373-9033-8.ch005

1.0 INTRODUCTION

Outbreaks are not merely biomedical events; they are dynamic socio-ecological processes shaped by pathogen biology, host susceptibility, mobility, health-system capacity, environmental conditions, communication networks, and policy response. Biosurveillance must therefore detect weak signals in noisy data, separate true emergence from background variation, and translate uncertainty into decisions before complete evidence is available. Machine learning has become increasingly important in this space because modern surveillance systems generate data at scales and speeds that exceed the capacity of manual analysis. The central promise of machine learning is not that it can replace epidemiology, but that it can help organize complex signals into earlier, more interpretable, and more actionable public-health intelligence.

Traditional outbreak surveillance relies on case reporting, laboratory confirmation, sentinel surveillance, syndromic monitoring, and field investigation. These systems remain the foundation of public-health response, but they are frequently delayed by care-seeking behavior, diagnostic access, reporting lags, under-ascertainment, jurisdictional fragmentation, and uneven data quality. Buckeridge (2007) noted that automated surveillance systems are evaluated not only by whether they detect outbreaks, but by determinants such as data timeliness, outbreak size, signal strength, background noise, and alert thresholds. These determinants remain central even when modern machine learning is introduced. Outbreak detection and forecasting increasingly require a One Health perspective because disease emergence is shaped by interactions among humans, animals, environmental conditions, food systems, and health infrastructure, especially in tropical and resource-limited settings where surveillance gaps can delay response (Ogwu & Izah, 2025).

The rise of digital epidemiology expanded the informational landscape. Search queries, social media, news reports, participatory surveillance, mobile-phone mobility, electronic health records, and environmental monitoring now supplement formal surveillance (Salathé et al., 2012; Althouse et al., 2015). HealthMap demonstrated that automated classification and visualization of online infectious disease reports could support global situational awareness (Freifeld et al., 2008; Brownstein et al., 2008). At the same time, Google Flu Trends showed both the appeal and danger of large digital traces: search data appeared useful for influenza detection, but model drift, algorithmic change, and overfitting exposed the limits of uncritical big-data surveillance (Ginsberg et al., 2009; Lazer et al., 2014; Yang et al., 2015).

Outbreak forecasting further extends biosurveillance from describing current signals to estimating future trajectories. Forecasts can support hospital preparedness, vaccine and therapeutic allocation, risk communication, laboratory prioritization, and decisions about public-health interventions. Yet infectious disease forecasting is difficult because surveillance data are delayed, biased, and revised; transmission processes are nonlinear; human behavior changes in response to risk; and interventions alter the future being forecast (Lauer et al., 2020). Machine learning is valuable because it can exploit high-dimensional patterns across data streams, but the same flexibility can produce models that are poorly calibrated, biologically implausible, or inequitable if not governed carefully.

This chapter treats machine learning-based outbreak detection and forecasting as an operational biosurveillance problem. The guiding question is not simply which algorithm performs best in a benchmark, but how models can support earlier detection, more reliable forecasts, transparent uncertainty, and accountable public-health action. The chapter therefore combines technical explanation with practical constraints: data quality, validation, interpretability, governance, and integration into response workflows.

2.0. CONCEPTUAL FOUNDATIONS: DETECTION, NOWCASTING, FORECASTING, AND EARLY WARNING

Outbreak detection refers to the identification of abnormal disease activity relative to an expected baseline. The baseline may represent historical seasonal patterns, endemic incidence, expected emergency department visits, laboratory positivity, or environmental pathogen concentrations. A detection system becomes useful when it distinguishes true signals from routine variation quickly enough to trigger verification and response. Statistical outbreak detection has long used threshold rules, control charts, scan statistics, cumulative sum methods, and regression-based aberration detection. Machine learning extends these approaches by allowing nonlinear relationships, complex covariates, high-dimensional features, and multimodal inputs

Nowcasting estimates current disease burden before complete reporting is available. It is especially important when surveillance reports arrive with delays or undergo revision. For example, cases occurring today may be reported days or weeks later, making the most recent data artificially low. Nowcasting models estimate the present by adjusting for reporting delays, historical revision patterns, testing patterns, and auxiliary indicators. McGough et al. (2020) demonstrated the value of Bayesian smoothing for real-time epidemic tracking, while machine learning approaches can incorporate additional predictors such as mobility, search trends, emergency visits, and test positivity. Environmental surveillance is particularly important for waterborne and sanitation-linked outbreaks because microbial contamination in water systems can act as an early indicator of public-health risk before clinical cases are fully detected (Izah & Ogwu, 2025).

Forecasting estimates future disease activity. Forecast targets may include incident cases, hospitalizations, deaths, test positivity, emergency visits, outbreak probability, spatial spread, peak timing, or reproductive number. Short-term forecasts are often most operationally useful because they align with staffing, procurement, communication, and intervention decisions. Long-term projections are more scenario-dependent and should be communicated as conditional estimates rather than predictions. Desai et al. (2019) emphasized that real-time epidemic forecasting offers public-health value only when models are evaluated, communicated, and linked to decision contexts.

Early warning combines detection, nowcasting, and forecasting. It does not require certainty; it requires timely risk intelligence sufficient to justify verification or preparedness. In outbreak work, the cost of missed detection may be severe, but excessive false alarms can erode trust and create alert fatigue. Machine learning systems must therefore be designed around thresholds, escalation pathways, human review, and explicit tolerance for false positives and false negatives. This distinguishes biosurveillance models from ordinary predictive analytics. The performance of a model is inseparable from the decision system in which it operates. Table 1 clarifies the main operational concepts used in machine learning-enabled outbreak surveillance. Although terms such as detection, nowcasting, forecasting, early warning, and situational awareness are often used interchangeably, they represent different analytical tasks with different data needs, modelling approaches, and public-health risks. Distinguishing these concepts is important because a model designed to detect abnormal disease activity may not be suitable for estimating current burden or forecasting future incidence. Similarly, an early warning system must balance timeliness with verification, while situational awareness tools must translate complex data streams into interpretable intelligence for decision-makers. This distinction helps ensure that machine learning outputs are evaluated according to their intended operational purpose rather than treated as generic prediction tools.

Table 1. Core concepts in machine learning-enabled outbreak intelligence

Concept	Meaning	Typical machine learning role	Operational risk
Detection	Identifying abnormal disease activity relative to a baseline	Anomaly detection, classification, scan statistics, clustering	False alarms or missed early signals
Nowcasting	Estimating current burden despite reporting delays	Delay adjustment, Bayesian smoothing, regression, ensemble nowcasts	Misleading real-time estimates if reporting behavior changes
Forecasting	Estimating future incidence, hospitalization, mortality, or outbreak probability	Time-series models, ensembles, deep learning, hybrid models	Overconfidence, poor calibration, long-horizon instability

continued on following page

Table 1. Continued

Concept	Meaning	Typical machine learning role	Operational risk
Early warning	Producing actionable alerts before complete evidence exists	Signal fusion, risk scoring, alert thresholds	Alert fatigue, inequitable detection, weak verification
Situational awareness	Maintaining a current picture of outbreak status and risk	Dashboards, geospatial models, NLP, multimodal integration	Data overload, poor interpretability, fragmented response

3.0. DATA STREAMS FOR MACHINE LEARNING-BASED OUTBREAK BIOSURVEILLANCE

Machine learning models are only as useful as the surveillance data they receive. Formal case notification data provide confirmed epidemiological information, but they often arrive late and depend on care-seeking, testing access, and reporting practices. Syndromic surveillance uses pre-diagnostic signals such as fever visits, respiratory symptoms, gastrointestinal complaints, absenteeism, or over-the-counter medication purchases. Such data can improve timeliness but are less specific. Emergency department-based syndromic surveillance has been used for respiratory outbreak forecasting because it can capture changes in health-care demand before laboratory confirmation is complete. Food-chain surveillance also strengthens outbreak intelligence because chemical, microbial, and processing-related contaminants can move across the food–environment–human health interface, requiring integrated monitoring within a One Health framework (Ogwu et al., 2024).

Laboratory data improve specificity through test results, diagnostic positivity, pathogen typing, and antimicrobial susceptibility. However, laboratory signals are biased by testing policies, specimen collection, diagnostic availability, and the populations who access care. Genomic surveillance adds information about pathogen lineages, introductions, clusters, and evolution. During rapidly evolving outbreaks, genomic data can help distinguish local transmission from repeated importation, identify variants, and support phylogeographic inference. Yet sequencing availability is uneven, and models trained in well-sampled regions may not generalize to settings with limited genomic infrastructure.

Digital and behavioral data offer earlier but noisier signals. Search queries and social media may capture public concern or symptoms before clinical reporting, but they also reflect media coverage, panic, misinformation, and platform changes. Lazer et al. (2014) showed that big-data surveillance can fail when models are overfit to historical correlations and not recalibrated. ARGO improved influenza estimation by combining Google searches with autoregressive information and regularized modeling, illustrating that digital traces are most useful when integrated with epidemiological baselines rather than used alone (Yang et al., 2015).

Environmental and One Health data are increasingly important. Wastewater surveillance can detect pathogen biomarkers at a population level and may provide early warning when clinical testing is limited. Climate, land cover, hydrology, vector abundance, animal movement, livestock production, wildlife mortality, and food-chain data are essential for zoonotic, vector-borne, waterborne, and agricultural outbreaks. Machine learning allows these heterogeneous data streams to be combined, but integration requires careful attention to sampling frequency, spatial alignment, missingness, and causal interpretation. Figure 1 illustrates a conceptual architecture for an AI-enabled multimodal biosurveillance system that integrates heterogeneous data streams to generate actionable outbreak intelligence within a One Health framework. The platform combines information from clinical and electronic health records, laboratory diagnostics and genomic sequencing, digital event surveillance, human mobility and behavioral data, environmental monitoring, and animal and food systems to create a unified analytical ecosystem. Through machine learning, data fusion, and predictive analytics, these diverse data sources are transformed into timely surveillance outputs, including outbreak alerts, hotspot identification, risk forecasts, uncertainty estimation, and operational recommendations for public health authorities. By integrating complementary evidence across biological, environmental, and societal domains, the framework supports earlier detection of emerging threats, improved situational awareness, and evidence-based decision-making, thereby strengthening preparedness, response, and resilience against infectious disease outbreaks.

Figure 1. Multimodal biosurveillance architecture for outbreak detection and forecasting

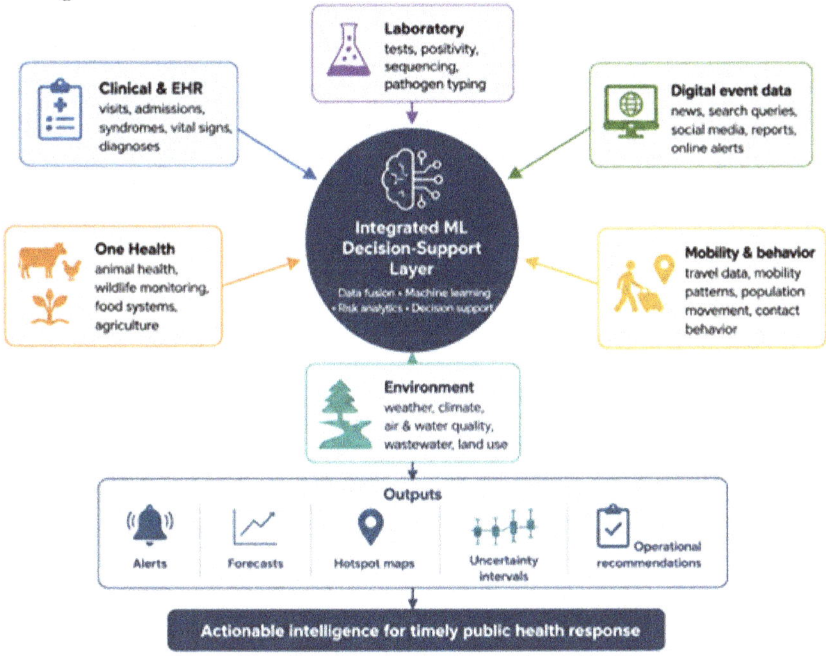

The value of machine learning lies in integrating diverse clinical, laboratory, environmental, digital, mobility, genomic, and One Health signals into a decision-support layer rather than treating any single data stream as sufficient

Table 2 summarizes the major data streams that support machine learning-based outbreak detection and forecasting. Because outbreaks rarely appear first in a single data source, effective biosurveillance depends on integrating clinical, syndromic, laboratory, digital, behavioral, environmental, and One Health signals. Each stream contributes a different form of intelligence: clinical and laboratory data provide specificity, syndromic and digital data improve timeliness, mobility and environmental data clarify transmission risk, and One Health data extend surveillance across animal, food, wildlife, and ecological interfaces. However, each source also carries limitations, including reporting delays, testing bias, misinformation, privacy concerns, sampling challenges, and institutional fragmentation. Recognizing these strengths and weaknesses is essential for designing machine learning systems that are timely, interpretable, and operationally reliable rather than simply data-rich.

Table 2. Data streams for outbreak detection and forecasting

Data stream	Examples	Strengths	Limitations
Clinical and case data	Notifiable disease reports, admissions, confirmed cases	Specific and policy-relevant	Delayed, under-ascertained, affected by testing access
Syndromic data	Emergency visits, fever, respiratory symptoms, absenteeism	Timely and sensitive	Low specificity and high background noise
Laboratory data	Test positivity, pathogen typing, sequencing	High diagnostic value	Testing-policy bias and uneven availability
Digital event data	News, search, social media, online reports	Early and broad coverage	Media effects, misinformation, duplication, platform drift
Mobility and behavior	Travel, commuting, contact patterns, population movement	Captures transmission opportunity	Privacy concerns, representativeness, proprietary access
Environmental data	Wastewater, climate, vectors, land use, hydrology	Useful for early warning and ecological risk	Sampling design and normalization challenges
One Health data	Animal, wildlife, livestock, food-chain surveillance	Critical for zoonotic and foodborne threats	Institutional fragmentation and data incompatibility

4.0. MACHINE LEARNING APPROACHES FOR OUTBREAK SIGNAL DETECTION

Outbreak detection models attempt to identify abnormal patterns before they are obvious through routine reporting. Supervised classification models can be trained to distinguish outbreak-related observations from non-outbreak observations when labelled historical data exist. Algorithms such as logistic regression, random forests, gradient boosting, support vector machines, and neural networks can incorporate temporal, spatial, clinical, environmental, and digital features. Their limitation is that labelled outbreak data are often scarce, heterogeneous, and biased toward events that were already detected.

Unsupervised and semi-supervised methods are useful when labelled outbreaks are limited. Anomaly detection identifies observations that deviate from expected patterns, while clustering can reveal unusual spatial or temporal aggregation. Change-point methods identify shifts in trends, and scan statistics detect spatial clusters. Unkel et al. (2012) reviewed statistical methods for prospective outbreak detection, many of which remain conceptually important even when embedded in modern machine learning pipelines. The key problem is not only detecting unusual patterns, but determining whether unusual patterns are epidemiologically meaningful.

Machine learning improves outbreak detection when it learns complex baselines. For seasonal diseases, expected activity depends on time of year, school calendars, climate, immunity, reporting behavior, and prior incidence. For vector-borne diseases, expected activity may depend on rainfall, temperature, land use, vector habitat, and host behavior. For hospital outbreaks, expected activity may depend on admissions, bed occupancy, antimicrobial use, device utilization, and infection-control practices. Flexible models can learn these relationships, but flexibility can also fit noise. Transparent validation and human review are therefore essential. Climate-sensitive forecasting is especially important in rural and mountain environments, where changing temperature, precipitation, land use, and health-system access can alter exposure pathways and intensify health vulnerabilities (Sugg et al., 2023).

Detection systems should be evaluated by timeliness, sensitivity, specificity, positive predictive value, false alert rate, and lead-time gain. A model that detects outbreaks early but produces constant false alarms may not be operationally useful. A model that is highly specific but detects only large outbreaks after they are obvious adds little value. Shmueli and Burkom (2010) emphasized that early detection in biosurveillance faces statistical challenges because signals are weak, data are noisy, and public-health decisions are made under uncertainty. Machine learning does not remove these challenges; it changes how they are managed.

5.0. NATURAL LANGUAGE PROCESSING AND EVENT-BASED SURVEILLANCE

Event-based surveillance uses unstructured reports from news, official bulletins, online media, and public communications to identify signals of infectious disease events. Natural language processing supports this task by classifying documents, extracting disease names, identifying locations, recognizing hosts and affected populations, detecting dates, and removing duplicate reports. Keller et al. (2009) showed that unstructured event-based reports can contribute to global infectious disease surveillance, while HealthMap demonstrated automated classification and visualization of internet media reports for global monitoring (Freifeld et al., 2008; Brownstein et al., 2008).

Modern NLP systems can use supervised classifiers, named entity recognition, topic modeling, transformer language models, and multilingual information extraction. These tools are particularly useful for early outbreak signals in places where formal reporting is delayed or limited. However, event-based surveillance is vulnerable to rumor, duplicated reporting, changes in media attention, language bias, censorship, misinformation, and uneven internet access. A surge in online reports may reflect increased concern rather than increased disease burden. Conversely, outbreaks may remain invisible if local media systems are weak or if reports are not available in languages captured by the model. ntegrated aquatic ecosystem monitoring is relevant to outbreak forecasting because environmental degradation, water quality change, and biological contamination can interact with human exposure pathways and public-health risk (Izah et al., 2023).

Human-in-the-loop review is therefore necessary. Automated NLP can triage large volumes of reports, but epidemiologists must assess credibility, context, and response relevance. A useful NLP system should preserve source traceability, confidence scores, geolocation uncertainty, and document provenance. It should also avoid reducing outbreak intelligence to keyword detection. The goal is structured situational awareness: what disease is reported, where, when, in whom, with what severity, from which source, and with what level of verification. The translation of surveillance signals into public-health action requires robust statistical interpretation, because risk models must distinguish meaningful environmental or epidemiological patterns from noise, uncertainty, and confounding (Ogwu et al., 2025).

6.0. TIME-SERIES FORECASTING AND ENSEMBLE FORECASTS

Time-series forecasting estimates future disease activity from historical observations and covariates. Classical methods include autoregressive models, exponential smoothing, state-space models, and generalized linear models. Machine learning methods include random forests, gradient boosting, support vector regression, recurrent neural networks, long short-term memory networks, temporal convolutional networks, and transformers. Deep learning may capture nonlinear temporal dependencies, but it usually requires large, stable datasets and careful regularization. Emerging outbreaks often have limited historical data, making simpler models or hybrid approaches more reliable.

Influenza forecasting has been a major testbed for operational infectious disease prediction. Multi-model and ensemble approaches have often performed better than individual models because they combine diverse assumptions and reduce instability. Reich et al. (2019) reported a collaborative, multiyear, multimodel assessment of seasonal influenza forecasting in the United States, demonstrating the value of standardized targets, prospective evaluation, and ensemble comparison. Johansson et al. (2019) similarly used an open challenge format to advance probabilistic forecasting for dengue epidemics. These efforts show that forecasting capacity improves when models are compared transparently against shared targets.

COVID-19 intensified the use of real-time forecasting. The COVID-19 Forecast Hub demonstrated the value of aggregating forecasts from multiple teams to support public-health decision-making, while Cramer et al. (2022) evaluated individual and ensemble probabilistic forecasts of COVID-19 mortality in the United States. These experiences reinforced several lessons: forecasts should be probabilistic, targets should be clearly defined, evaluation should be prospective, and ensemble approaches often provide stability when individual models fail. They also showed that forecasts degrade when testing policies change, variants emerge, reporting systems shift, or behavior changes in response to interventions.

Forecasting systems must communicate uncertainty. Point forecasts may be attractive to decision-makers, but they are often misleading. Probabilistic forecasts express a range of plausible outcomes and allow decision-makers to consider risk tolerance. Calibration matters: a 90% prediction interval should contain the observed outcome about 90% of the time over repeated forecasts. Sharpness also matters: intervals should be as narrow as possible while remaining calibrated. A forecast that is precise but wrong can be more harmful than a forecast that is uncertain but honest.

7.0. HYBRID EPIDEMIOLOGICAL-MACHINE LEARNING MODELS

Purely statistical or machine learning models can forecast well when data patterns are stable, but outbreaks are often shaped by mechanisms that change quickly. Mechanistic epidemiological models, including SIR and SEIR frameworks, represent transmission processes explicitly. Agent-based and network models can represent heterogeneity in contacts, mobility, and interventions. Hybrid epidemiological-machine learning models combine mechanistic structure with data-driven flexibility. This approach can improve predictive performance while reducing the risk of biologically implausible forecasts.

Machine learning can support mechanistic models by estimating parameters, learning residual patterns, detecting regime shifts, calibrating contact rates, and incorporating external predictors such as mobility, weather, wastewater, and search behavior. Conversely, mechanistic models can constrain machine learning by embedding epidemiological structure. For example, a model may use machine learning to estimate time-varying transmission while preserving constraints on susceptible, exposed, infectious, and recovered compartments. This is especially useful when forecasts must remain interpretable to epidemiologists and policymakers.

Hybrid methods are also useful during data-sparse emergence. Liu et al. (2020) proposed a machine learning methodology for real-time COVID-19 forecasting that combined mechanistic model estimates with internet searches and news alerts. Although early-pandemic models were often unstable, this type of architecture illustrates the broader principle: digital traces and mechanistic estimates can complement one another when carefully validated. Figure 2 illustrates the analytical pipeline underpinning AI-enabled outbreak detection and forecasting within a modern biosurveillance system. The workflow begins with the acquisition of diverse epidemiological, clinical, laboratory, environmental, genomic, and digital data streams, followed by data engineering processes that ensure quality, standardization, and analytical readiness. Machine learning and predictive models are then applied to detect anomalous patterns, identify emerging threats, and generate risk intelligence, including outbreak forecasts, hotspot identification, and uncertainty estimates. These analytical outputs inform timely public health actions such as laboratory confirmation, field investigation, surveillance prioritization, and response coordination. A continuous feedback loop incorporating model performance, outbreak investigations, laboratory confirmation, and intervention outcomes enables ongoing recalibration, adaptive learning, and performance improvement, ensuring that the surveillance system remains accurate, responsive, and capable of supporting evidence-based public health decision-making under evolving epidemiological conditions.

Figure 2. AI-enabled outbreak detection and forecasting pipeline

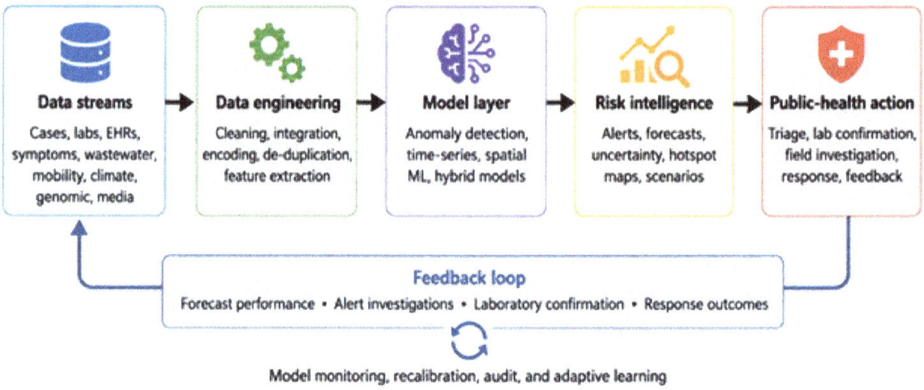

Reliable biosurveillance requires a complete workflow from data integration and model development to alert generation, epidemiological review, public-health action, and feedback-based recalibration

8.0. GEOSPATIAL MACHINE LEARNING AND OUTBREAK RISK MAPPING

Outbreaks unfold in space as well as time. Geospatial machine learning identifies where outbreaks are emerging, where risk is increasing, and where interventions may be needed. Models may use population density, mobility, land cover, climate, health-care access, vector habitat, sanitation, livestock density, travel networks, and historical incidence. Spatial clustering and hotspot detection can support rapid investigation, while predictive risk maps can guide prevention before cases appear.

Vector-borne and zoonotic diseases are especially suited to geospatial models because transmission depends on environmental suitability, reservoirs, vectors, and human exposure. Temperature, rainfall, humidity, vegetation, land-use change, and hydrology can influence vector abundance and pathogen development. Remote sensing provides covariates that can be updated at broad scales, making it useful for risk mapping. However, ecological correlation is not causation. Spatial models can overstate confidence when covariates are proxies, when reporting varies geographically, or when high-risk areas are undersampled.

Spatial validation is essential. Random cross-validation can inflate performance when nearby observations are similar. Models should be tested across time, geography, and populations to determine whether they generalize. Spatial uncertainty should be communicated visually and statistically. A hotspot map without uncertainty may encourage misplaced confidence and inequitable resource allocation. Geospatial machine learning is most useful when combined with local knowledge, field verification, and public-health capacity.

9.0. WASTEWATER AND ENVIRONMENTAL EARLY WARNING

Wastewater-based epidemiology has become an important complement to clinical surveillance because it can capture population-level pathogen signals independent of individual testing behavior. Wastewater monitoring can provide early warning for respiratory viruses, enteric pathogens, antimicrobial resistance markers, and other biomarkers. It is particularly valuable when clinical testing declines or when asymptomatic and mild infections are underreported. The same logic applies to other environmental systems, including air sampling, surface water, animal waste, and vector surveillance.

Machine learning can improve environmental surveillance by detecting anomalous trends, integrating wastewater signals with clinical outcomes, adjusting for flow and population variation, and forecasting downstream health-care demand. However, wastewater data are not straightforward. Signals depend on sampling location, catchment population, wastewater flow, decay rates, laboratory methods, normalization targets, industrial inputs, rainfall, and sewer infrastructure. A wastewater signal may precede clinical data, but interpretation requires environmental and epidemiological expertise.

The broader lesson is that environmental data expand outbreak detection beyond health-care encounters. This is important for equity because clinical surveillance often misses populations with poor health-care access. Yet environmental surveillance can also reproduce inequities if sampling infrastructure is concentrated in wealthy cities or institutions. Machine learning systems should therefore treat wastewater and environmental data as complementary evidence streams, not independent truth.

10.0 VALIDATION, BENCHMARKING, AND PERFORMANCE EVALUATION

Model validation is the dividing line between promising analytics and reliable biosurveillance. Outbreak detection models should be evaluated by sensitivity, specificity, precision, recall, false alarm rate, detection delay, and lead-time gain. Forecasting models should be evaluated by proper scoring rules, calibration, sharpness, mean absolute error, root mean squared error, weighted interval score, and continuous ranked probability score. Operational evaluation should also ask whether model outputs changed decisions, improved preparedness, or reduced harm.

Prospective validation is especially important. Retrospective performance can be inflated by data leakage, hindsight bias, changing definitions, and overfitting. Forecasting challenges in influenza, dengue, Ebola, and COVID-19 demonstrated the importance of standardized targets, archived forecasts, and comparison against baselines (Viboud et al., 2018; Johansson et al., 2019; Reich et al., 2019; Cramer et al., 2022). A model should be compared not only against other machine learning models but also against simple baselines such as persistence, seasonal averages, and autoregressive models. Complex models that fail to outperform simple baselines should not be operationalized.

Generalizability should be tested across locations, seasons, pathogen types, and surveillance regimes. A model trained in a high-income urban setting may not perform in a rural or low-resource setting. A model trained under one testing policy may fail after testing criteria change. A model trained before a variant, vaccine, or intervention may fail afterward. Evaluation should therefore include temporal validation, spatial validation, subgroup performance, fairness assessment, and model drift monitoring.

Interpretability is part of validation. Public-health users need to know why an alert was generated, which data streams contributed, how uncertain the forecast is, and what action is recommended. Black-box models may be acceptable for low-stakes triage if they are accurate and monitored, but high-stakes outbreak decisions require explainability, audit trails, and human accountability. Table 3 presents the key evaluation domains for machine learning-based outbreak detection and forecasting systems. Model performance should not be judged only by statistical accuracy, because an outbreak model that performs well retrospectively may still fail operationally if it is late, poorly calibrated, difficult to interpret, or unusable by public-health teams. Effective evaluation must therefore combine technical metrics, timeliness measures, external validation, decision impact, and governance safeguards. This broader assessment helps determine whether a model can detect signals early, forecast risk reliably, transfer across settings, support real decisions, and avoid reinforcing surveillance inequalities.

Table 3. Evaluation criteria for outbreak detection and forecasting models

Evaluation domain	Key metrics	Why it matters	Common failure mode
Detection accuracy	Sensitivity, specificity, precision, recall, false alarm rate	Determines whether signals are found without overwhelming users	High false alarm burden or missed small outbreaks
Timeliness	Detection delay, lead-time gain, reporting lag adjustment	Early signals only matter if they arrive before action windows close	Accurate but late detection
Forecast quality	MAE, RMSE, WIS, CRPS, calibration, sharpness	Assesses probabilistic and point forecast performance	Overconfident or poorly calibrated forecasts
Generalizability	Temporal, spatial, and subgroup validation	Shows whether models transfer across settings	Good retrospective fit but poor external performance
Operational usefulness	Decision impact, actionability, interpretability, user trust	Connects model output to public-health response	Technically impressive but unused dashboard
Equity and governance	Bias audit, privacy protection, transparency, accountability	Prevents unequal detection and harmful automation	Models amplify data inequalities

11.0. OPERATIONAL INTEGRATION INTO BIOSURVEILLANCE SYSTEMS

Machine learning outputs become valuable only when they are embedded in workflows. A risk score, anomaly alert, or forecast must be linked to defined actions: review, verification, laboratory testing, field investigation, risk communication, resource allocation, or intervention. Without operational pathways, a model becomes a dashboard rather than a biosurveillance system. Effective integration requires alert thresholds, escalation rules, user-centered design, data provenance, and feedback from investigations.

Human-in-the-loop systems are essential. Epidemiologists can interpret local context, recognize data artifacts, evaluate plausibility, and decide whether an alert warrants action. Laboratory and field teams provide confirmation and ground truth. Decision-makers determine proportional responses under uncertainty. Machine learning can accelerate triage, but responsibility remains human and institutional. The most trustworthy systems make model assumptions visible and preserve the ability to challenge automated outputs.

Alert fatigue is a major operational risk. If a system produces too many alerts, users may ignore them. If thresholds are too strict, early outbreaks may be missed. Thresholds should therefore be calibrated to the disease, data source, setting, and response capacity. A hospital infection-control system may tolerate more false positives for a high-consequence pathogen than a routine respiratory dashboard. Public-health agencies should define alert thresholds in relation to actionability rather than statistical significance alone.

Figure 3. Human-in-the-loop governance framework for machine learning outbreak intelligence

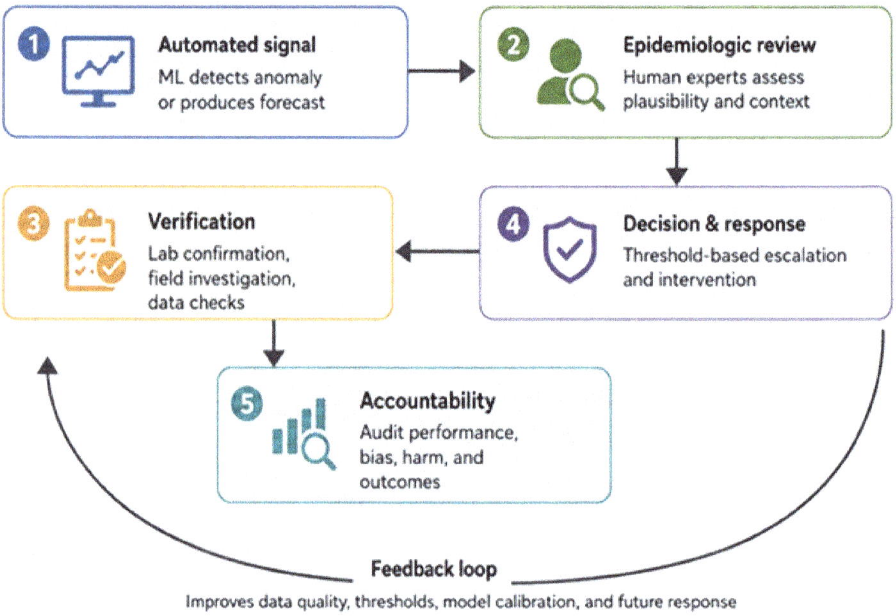

Automated alerts must be interpreted, verified, acted upon, and audited within accountable public-health systems

12.0. ETHICS, GOVERNANCE, EQUITY, AND LOW-RESOURCE SETTINGS

AI-enabled biosurveillance raises ethical and governance issues because it often uses sensitive health, mobility, behavioral, and location data. Privacy risks are particularly important when models integrate electronic health records, mobile-phone mobility, social media, or small-area geospatial data. Data minimization, aggregation, secure access controls, governance agreements, and privacy-preserving methods such as federated learning can reduce risk. However, technical privacy measures cannot replace transparent governance and public trust. Bias is a central concern. Surveillance data are not neutral reflections of disease burden; they reflect who seeks care, who is tested, who is reported, who is online, whose wastewater is sampled, and whose data systems are digitized. If machine learning models learn from biased data, they may under-detect outbreaks in marginalized populations or over-police communities with greater surveillance intensity. Fairness auditing should therefore be part of outbreak forecasting, especially when forecasts inform resource allocation, restrictions, or public messaging.

Low- and middle-income settings face distinct challenges and opportunities. Data may be sparse, delayed, paper-based, or fragmented across sectors. Laboratory and genomic capacity may be limited. Connectivity, computing infrastructure, and trained personnel may constrain deployment. At the same time, early-warning systems may be especially valuable where health-system capacity is fragile. Models imported from high-income settings should not be assumed to transfer. Locally calibrated, context-sensitive, and capacity-building approaches are necessary. One Health governance is equally important. Zoonotic and foodborne outbreaks often cross institutional boundaries among human health, animal health, agriculture, wildlife, food safety, and environmental agencies. Data sharing may be limited by jurisdiction, ownership, commercial sensitivity, or incompatible standards. Machine learning systems should therefore be designed as coordination tools as much as analytic tools. The strongest systems will integrate human, animal, and environmental intelligence while respecting institutional responsibilities and community rights. Foodborne outbreak surveillance also depends on monitoring sanitation, hygiene, and microbial contamination risks across processing environments, especially in settings where food safety infrastructure remains uneven (Ovuru et al., 2023).

13.0. CASE DOMAINS AND USE PATHWAYS

Respiratory disease forecasting remains one of the most developed applications. Influenza and COVID-19 forecasting systems illustrate the value of shared targets, real-time evaluation, probabilistic forecasts, and ensemble modeling. These systems can support hospital surge planning, vaccine campaigns, staffing, and public communication. However, respiratory models are sensitive to variant emergence, vaccination, prior immunity, testing policy, and behavior change. Vector-borne disease forecasting depends strongly on environmental and ecological covariates. Temperature, precipitation, vegetation, standing water, vector abundance, and human movement can inform risk models for dengue, malaria, West Nile virus, Rift Valley fever, and other vector-borne infections. Machine learning can integrate these predictors, but ecological plausibility and spatial validation are essential. Forecasting vector-borne disease without vector or environmental data often produces fragile predictions. Although outbreak forecasting focuses on detection and response rather than therapeutic discovery, antimicrobial biosurveillance remains connected to broader questions of pathogen control, resistance pressure, and the search for safer antimicrobial strategies (Ogwu & Izah, 2025).

Foodborne and waterborne outbreak detection benefits from anomaly detection in laboratory reports, emergency visits, pharmacy sales, consumer complaints, food-chain data, and environmental testing. Machine learning can identify clusters that cross jurisdictions or products, but outbreak confirmation still depends on epidemiological investigation, laboratory typing, and exposure assessment. Genomic surveillance increasingly strengthens the ability to link dispersed cases to common sources. Zoonotic spillover forecasting requires integrating ecological, veterinary, wildlife, land-use, climate, and human exposure data. Machine learning can identify high-risk interfaces, but the rarity of spillover events creates severe class imbalance and uncertainty. Forecasting spillover should therefore be framed as risk mapping and preparedness rather than precise prediction of when and where the next emergence will occur.

14.0. FUTURE DIRECTIONS

The next generation of outbreak detection and forecasting will be multimodal, probabilistic, explainable, privacy-preserving, and operationally embedded. Multimodal systems will integrate clinical, laboratory, genomic, environmental, mobility, and digital data streams. Probabilistic systems will communicate uncertainty rather than only point estimates. Explainable systems will show which signals drive alerts and forecasts. Privacy-preserving systems will allow learning across institutions without unnecessary exposure of sensitive data. Operational systems will connect model outputs to response actions. Foundation models and large language models may contribute to outbreak intelligence through document triage, multilingual event extraction, summarization, and evidence synthesis. However, they should be used cautiously because hallucination, source ambiguity, and weak epidemiological grounding can produce harmful misinformation. For biosurveillance, language models must preserve provenance, cite sources, and remain subordinate to verification workflows.

A major priority is building trustworthy evaluation ecosystems. Forecast hubs, open challenges, shared targets, archived predictions, and prospective benchmarking improve accountability. Future systems should evaluate not only accuracy but fairness, calibration, timeliness, interpretability, and response impact. This is how outbreak forecasting can mature from experimental modeling into reliable public-health infrastructure.

15.0. CONCLUSION

Machine learning is transforming outbreak detection and forecasting by enabling biosurveillance systems to integrate heterogeneous, high-volume, and rapidly changing data streams. Its value lies in detecting weak signals, estimating current burden despite reporting delays, forecasting near-term trajectories, mapping risk, and supporting timely public-health decisions. Yet the strongest conclusion of this chapter is that machine learning is not an autonomous replacement for epidemiology. Outbreak intelligence remains a human-centered, evidence-based, and institutionally accountable process. Models must be validated prospectively, benchmarked against simple baselines, calibrated probabilistically, interpreted transparently, and governed ethically. They must also be embedded in workflows that connect alerts to verification, investigation, communication, and response. The future of AI-enhanced biosurveillance will therefore depend less on any single algorithm than on the design of trustworthy systems. Those systems must combine epidemiological theory, machine learning, laboratory confirmation, environmental intelligence, One Health integration, and public-health judgment. When developed responsibly, machine learning can help societies detect outbreaks earlier, forecast risks more clearly, and act before small signals become large crises.

REFERENCES

Althouse, B. M., Scarpino, S. V., Meyers, L. A., Ayers, J. W., Bargsten, M., Baumbach, J., Brownstein, J. S., Castro, L., Clapham, H., Cummings, D. A. T., Del Valle, S., Eubank, S., Fairchild, G., Finelli, L., Generous, N., George, D., Harper, D. R., Hébert-Dufresne, L., Johansson, M. A., & Vespignani, A. (2015). Enhancing disease surveillance with novel data streams: Challenges and opportunities. *EPJ Data Science*, *4*(1), 17. DOI: 10.1140/epjds/s13688-015-0054-0 PMID: 27990325

Brownstein, J. S., Freifeld, C. C., Reis, B. Y., & Mandl, K. D. (2008). Surveillance sans frontières: Internet-based emerging infectious disease intelligence and the HealthMap project. *PLoS Medicine*, *5*(7), e151. DOI: 10.1371/journal.pmed.0050151 PMID: 18613747

Buckeridge, D. L. (2007). Outbreak detection through automated surveillance: A review of the determinants of detection. *Journal of Biomedical Informatics*, *40*(4), 370–379. DOI: 10.1016/j.jbi.2006.09.003 PMID: 17095301

Cramer, E. Y., Ray, E. L., Lopez, V. K., Bracher, J., Brennen, A., Rivadeneira, A. J. C., Gerding, A., Gneiting, T., House, K. H., Huang, Y., Jayawardena, D., Kanji, A. H., Khandelwal, A., Le, K., Mühlemann, A., Niemi, J., Stark, A., Shah, A., Wattanachit, N., & Reich, N. G. (2022). Evaluation of individual and ensemble probabilistic forecasts of COVID-19 mortality in the United States. *Proceedings of the National Academy of Sciences of the United States of America*, *119*(15), e2113561119. DOI: 10.1073/pnas.2113561119 PMID: 35394862

Desai, A. N., Kraemer, M. U. G., Bhatia, S., Cori, A., Nouvellet, P., Herringer, M., Cohn, E. L., Carrion, M., Brownstein, J. S., Madoff, L. C., & Lassmann, B. (2019). Real-time epidemic forecasting: Challenges and opportunities. *Health Security*, *17*(4), 268–275. DOI: 10.1089/hs.2019.0022 PMID: 31433279

Freifeld, C. C., Mandl, K. D., Reis, B. Y., & Brownstein, J. S. (2008). HealthMap: Global infectious disease monitoring through automated classification and visualization of Internet media reports. *Journal of the American Medical Informatics Association : JAMIA*, *15*(2), 150–157. DOI: 10.1197/jamia.M2544 PMID: 18096908

Ginsberg, J., Mohebbi, M. H., Patel, R. S., Brammer, L., Smolinski, M. S., & Brilliant, L. (2009). Detecting influenza epidemics using search engine query data. *Nature*, *457*(7232), 1012–1014. DOI: 10.1038/nature07634 PMID: 19020500

Izah, S. C., & Ogwu, M. C. (2025). Modeling solutions for microbial water contamination in the Global South for public health protection. *Frontiers in Microbiology*, *16*, 1504829. DOI: 10.3389/fmicb.2025.1504829 PMID: 40241726

Izah, S. C., Richard, G., Stanley, H. O., Sawyer, W. E., Ogwu, M. C., & Uwaeme, O. R. (2023). Integrating the One Health approach and statistical analysis for sustainable aquatic ecosystem management and trace contamination mitigation. *ES Food and Agroforestry, 14*, 1012. DOI: 10.30919/esfaf1012

Johansson, M. A., Apfeldorf, K. M., Dobson, S., Devita, J., Buczak, A. L., Baugher, B., Moniz, L. J., Bagley, T., Babin, S. M., Guven, E., Yamana, T. K., Shaman, J., Moschou, T., Lothian, N., Lane, A., Osborne, G., Jiang, G., Brooks, L. C., Farrow, D. C., & Biggerstaff, M. (2019). An open challenge to advance probabilistic forecasting for dengue epidemics. *Proceedings of the National Academy of Sciences of the United States of America, 116*(48), 24268–24274. DOI: 10.1073/pnas.1909865116 PMID: 31712420

Keller, M., Blench, M., Tolentino, H., Freifeld, C. C., Mandl, K. D., Mawudeku, A., Eysenbach, G., & Brownstein, J. S. (2009). Use of unstructured event-based reports for global infectious disease surveillance. *Emerging Infectious Diseases, 15*(5), 689–695. DOI: 10.3201/eid1505.081114 PMID: 19402953

Lauer, S. A., Brown, A. C., & Reich, N. G. (2020). Infectious disease forecasting for public health. arXiv preprint arXiv:2006.00073. DOI: 10.1093/oso/9780198853244.003.0004

Lazer, D., Kennedy, R., King, G., & Vespignani, A. (2014). The parable of Google Flu: Traps in big data analysis. *Science, 343*(6176), 1203–1205. DOI: 10.1126/science.1248506 PMID: 24626916

Liu, D., Clemente, L., Poirier, C., Ding, X., Chinazzi, M., Davis, J. T., Vespignani, A., & Santillana, M. (2020). A machine learning methodology for real-time forecasting of the 2019-2020 COVID-19 outbreak using Internet searches, news alerts, and estimates from mechanistic models. arXiv preprint arXiv:2004.04019.

McGough, S. F., Johansson, M. A., Lipsitch, M., & Menzies, N. A. (2020). Nowcasting by Bayesian smoothing: A flexible, generalizable model for real-time epidemic tracking. *PLoS Computational Biology, 16*(4), e1007735. DOI: 10.1371/journal.pcbi.1007735 PMID: 32251464

Ogwu, M. C., & Izah, S. C. (2025). Honey as a natural antimicrobial. *Antibiotics (Basel, Switzerland), 14*(3), 255. DOI: 10.3390/antibiotics14030255 PMID: 40149066

Ogwu, M. C., Izah, S. C., Aigberua, A. O., & Ngun, C. T. (2024). Editorial: Detection, risk analysis, and monitoring of chemical contaminants from agro-aqua food production and processing: Implications on the One Health triad. *Frontiers in Sustainable Food Systems, 8*, 1501930. DOI: 10.3389/fsufs.2024.1501930

Ogwu, M. C., Izah, S. C., Sawyer, W. E., & Amabie, T. (2025). Environmental risk assessment of trace metal pollution: A statistical perspective. *Environmental Geochemistry and Health*, *47*(4), 94. DOI: 10.1007/s10653-025-02405-z PMID: 40019615

Ovuru, K. F., Izah, S. C., Ogidi, O. I., Imarhiagbe, I., & Ogwu, M. C. (2023). Slaughterhouse facilities in developing nations: Sanitation and hygiene practices, microbial contamination, and sustainable management system. *Food Science and Biotechnology*. Advance online publication. DOI: 10.1007/s10068-023-01406-x PMID: 38274182

Reich, N. G., Brooks, L. C., Fox, S. J., Kandula, S., McGowan, C. J., Moore, E., Osthus, D., Ray, E. L., Tushar, A., Yamana, T. K., Biggerstaff, M., Johansson, M. A., Rosenfeld, R., Shaman, J., & Viboud, C. (2019). A collaborative multiyear, multimodel assessment of seasonal influenza forecasting in the United States. *Proceedings of the National Academy of Sciences of the United States of America*, *116*(8), 3146–3154. DOI: 10.1073/pnas.1812594116 PMID: 30647115

Salathé, M., Bengtsson, L., Bodnar, T. J., Brewer, D. D., Brownstein, J. S., Buckee, C., Campbell, E. M., Cattuto, C., Khandelwal, S., Mabry, P. L., & Vespignani, A. (2012). Digital epidemiology. *PLoS Computational Biology*, *8*(7), e1002616. DOI: 10.1371/journal.pcbi.1002616 PMID: 22844241

Shmueli, G., & Burkom, H. (2010). Statistical challenges facing early outbreak detection in biosurveillance. *Technometrics*, *52*(1), 39–51. DOI: 10.1198/TECH.2010.06134

Sugg, M. M., Ryan, S., Spurlock, T., Duncan, S., Hege, A., Ogwu, M. C., Tyson, J., Shay, E., Renwick, K., Hendren, C., Roy, M., & Runkle, J. D. (2023). Climate change and health in rural mountain environments: Summary of a workshop on knowledge gaps, barriers, and opportunities for action. *GeoJournal*, *88*(5), 5397–5409. Advance online publication. DOI: 10.1007/s10708-023-10916-4 PMID: 40291811

Unkel, S., Farrington, C. P., Garthwaite, P. H., Robertson, C., & Andrews, N. (2012). Statistical methods for the prospective detection of infectious disease outbreaks: A review. *Journal of the Royal Statistical Society. Series A, (Statistics in Society)*, *175*(1), 49–82. DOI: 10.1111/j.1467-985X.2011.00714.x

Viboud, C., Sun, K., Gaffey, R., Ajelli, M., Fumanelli, L., Merler, S., Zhang, Q., Chowell, G., Simonsen, L., & Vespignani, A.RAPIDD Ebola Forecasting Challenge Group. (2018). The RAPIDD Ebola forecasting challenge: Synthesis and lessons learnt. *Epidemics*, *22*, 13–21. DOI: 10.1016/j.epidem.2017.08.002 PMID: 28958414

Yang, S., Santillana, M., & Kou, S. C. (2015). Accurate estimation of influenza epidemics using Google search data via ARGO. *Proceedings of the National Academy of Sciences of the United States of America, 112*(47), 14473–14478. DOI: 10.1073/pnas.1515373112 PMID: 26553980

Chapter 6
Explainable and Multi-Source AI Frameworks for Adaptive Biosurveillance in Public Health and Environmental Systems

Aman Suresh Tharayil
http://orcid.org/0009-0000-5377-7428
Dr. MGR Educational and Research Institute, India

Anmol S. Budhewar
http://orcid.org/0009-0002-9112-107X
Sandip Institute of Technology and Research Centre, Nashik, India

Adinarayana Andy
http://orcid.org/0000-0002-5163-1094
Weatherwax Family Pharmacies Inc., USA

Gnanavel Velu
http://orcid.org/0000-0002-7081-9017
College of Pharmacy, Catholic University in Erbil, Kurdistan, Iraq

Amareswarapu V. Surendra
http://orcid.org/0000-0002-6269-1535
KL College of Pharmacy, Koneru Lakshmaiah Educational Foundation, India

Venkateswarlu Sunkari
College of Engineering and Architecture, University of Nizwa, Oman

Devi Thamizanban
http://orcid.org/0000-0002-0068-9800
Chettinad School of Pharmaceutical Sciences, Chettinad Hospital and Research Institute, Chettinad Academy of Research and Education, Kelambakkam, India

DOI: 10.4018/979-8-3373-9033-8.ch006

K. C. Gayathri
Chettinad School of Physiotherapy (CSP), Chettinad Hospital and Research Institute (CHRI), Chettinad Academy of Research and Education (CARE), Kelambakkam, India

Praveen Kumar T.
Dr. MGR Educational and Research Institute, India

Dhanalakshmi Perumal
http://orcid.org/0009-0004-2804-140X
Faculty of Pharmacy, Dr. MGR Educational and Research Institute, India

Sayali Somwanshi
http://orcid.org/0009-0000-3296-4347
Cognizant, India

Darshan Huraliganganahalli Balaramegowda
http://orcid.org/0009-0001-3001-3254
Norwich Medical School, UK

K.Blessi Priyanka
University College of Pharmaceutical Sciences, Kakatiya University, Warangal, India

Balkrishna Patil
http://orcid.org/0009-0000-5136-2157
Sandip Institute of Technology and Research Centre, Nashik, India

ABSTRACT

The growing complexity of biological threats necessitates advanced biosurveillance systems capable of three core functions. Traditional surveillance faces challenges due to fragmented data, delayed information sharing, and weak coordination between public health and environmental sectors. Advances in artificial intelligence, particularly in machine learning, deep learning, and data fusion, enable integration of diverse data sources such as epidemiological, environmental, remote sensing, genomic, and digital health data for active risk assessment. Explainable AI improves transparency and supports informed decision-making, while multi-source analytics strengthen early warning, situational awareness, and risk forecasting for emerging threats linked to zoonoses and climate change. Developing AI-driven biosurveillance systems with ethical, transparent, and resilient One Health approaches is essential to address issues of data quality, bias, privacy, and governance.

1. INTRODUCTION

Biosurveillance systems now need to create proactive systems that can adjust and use their intelligence in both public health and environmental areas, as biological threats have become more complicated and widespread than in the past. The traditional methods of surveillance system operation create obstacles which make it impossible for these systems to address the challenges presented by emerging infectious diseases and zoonotic spillovers and climate-driven ecological risks (Abba et al., 2025). Artificial intelligence technology development allows organizations to perform data-based surveillance operations because they can access more operational power than previous systems yet users face difficulty using these systems because they lack transparency and trust while users cannot understand their system functions (Abueg et al., 2021). The solution to these problems requires biosurveillance systems that use multiple data sources together with AI systems that provide transparent and responsible operational methods. The conceptual framework that supports the upcoming sections needs users to understand the core principles that govern adaptive systems that use explainable biosurveillance methods (Alsaedi et al., 2025). The One Health paradigm which connects human health with animal health and environmental health serves as the foundation for this approach which improves public health decision-making through enhanced early warning systems and data-driven decision-making.

1.1 Biosurveillance as a Socio-Technical System

Biosurveillance functions as necessary infrastructure which supports ongoing biological threat monitoring and detection and response efforts across public health and environmental protection and agricultural operations and ecological research domains (Artzrouni, 2020). The standard biosurveillance systems depend on three main sources of data which include epidemiological reporting and laboratory confirmation and retrospective analysis. The current methods of disease identification face limitations that demonstrate their ineffectiveness when dealing with new infectious diseases and zoonotic disease transmission and antimicrobial resistance and climate related ecological changes (Bernardo et al., 2013). The existing health threat surveillance systems require immediate transformation into more advanced systems which can process multiple data sources to deliver fast results during health threat developments. The implementation of One Health initiatives, which combine surveillance systems for human, animal, and environmental health, serves as an essential solution to address fragmented systems while enhancing the ability to predict and handle potential threats (Das & Rad, 2020).

Recent progress in artificial intelligence technology together with data analytics advancements has created new biosurveillance systems which detect emerging threats through adaptive prediction capabilities instead of monitoring existing threats. The systems function as sociotechnical constructs which combine data infrastructure systems with computational modeling tools together with human decision-making abilities and institutional policy frameworks and governance mechanisms. Public health surveillance encounters a major problem because AI models use "black-box" systems which create trust issues and ethical challenges and make it difficult to establish policies (Garcia-Vozmediano et al., 2025). The Explainable AI system needs to be implemented in biosurveillance systems because it will make AI prediction systems accessible through complete human understanding of their decision-making processes.

People today view Explainable Artificial Intelligence (XAI) as an essential tool which organizations must implement for their biosurveillance activities. The system needs explanation because it requires operational success to achieve its goals but its present predictive capabilities fall short of both real-world implementation needs and regulatory standards and public confidence requirements (Ghaffarian et al., 2023). The "black-box" design of most machine learning algorithms creates this requirement because their prediction mechanisms remain hidden from view which makes it difficult to use these systems for epidemiological research and their data handling capabilities.

1.2 Limitations of Conventional Surveillance Architectures

Traditional biosurveillance systems depend on multiple data sources which operate through slow reporting systems and lack the ability to share information between different fields. The research study shows that researchers can study human environmental and animal system connections only through their research of epidemiological data and environmental sensor data and genomic surveillance information and digital health data. The specialized data collection techniques which these companies use to gather information become unworkable because their operations require them to process growing data volumes and to handle rising complexity of biological research challenges. The absence of standardized data formats and interoperability standards creates additional limitations which prevent organizations from achieving effective biosurveillance through proper data integration and comprehensive operational assessment (Lazer et al., 2014).

Machine learning models have shown strong performance in outbreak forecasting and disease classification. However, many of these models function as black boxes which prevent users from understanding their internal operations and the reasons behind their output results. Public health officials and domain experts who use these outputs for resource allocation and policy decisions lose trust in the system because of its lack transparency. Medical professionals need to understand AI prediction methods because this knowledge helps them find root causes and develop proper treatment plans especially when treating patients in healthcare settings. The lack of transparent operation in these AI models creates major ethical problems together with governance issues that include algorithmic bias and fairness and accountability and data privacy. These problems with public health system adoption and implementation of the technology (Lekadir et al., 2025).

The model performance becomes worse because high-dimensional surveillance data contains duplicate features that show connections to each other. The combination of correlation-aware feature selection with post-hoc explainability techniques improves biosurveillance systems because they enhance predictive accuracy together with user understanding. The research shows that both statistical methods and explainable AI methods need to work together in biosurveillance systems.

1.3 Emergence of Multi-Source Data Ecosystems

Modern biosurveillance systems maintain their operations through data collection from multiple sources which function at different operational levels. The systems utilize multiple data categories which encompass epidemiological records and environmental monitoring data and remote sensing imagery and unmanned aerial vehicle (UAV) observations and Internet of Things (IoT) sensor networks and digital health signals from social media and electronic health records. IoT Based Patient Health Monitoring by using Biomedical Device. The platform achieves its goal through the combination of different data sources which help to identify unusual behavior patterns at an earlier stage while enabling organizations to assess potential threats. The combination of remote sensing technology with UAV surveillance systems has become vital for monitoring environmental changes and conducting ecosystem research because these systems deliver accurate measurements of space and time which enhance current public health studies. Environmental monitoring now occurs in dangerous or remote regions because the combination of UAV system advancements and edge AI technology and sensor fusion systems has made this possible (Nangunoori, 2025).

The integration of visual imagery with environmental and epidemiological datasets creates new biosurveillance challenges. The internal decision-making processes of deep learning architectures which achieve high accuracy in image recognition and land-use classification remain opaque to users. Recent investigations highlight the imperative for explainable mechanisms in multimodal fusion models, as these enable performance tracking and accountability of surveillance outputs.

1.4 Explainable AI as a Core Requirement for Biosurveillance

Explainable Artificial Intelligence mitigates the shortcomings of opaque predictive models by furnishing transparent interpretations of model behavior, feature importance, and decision rationales. In healthcare and pharmacovigilance contexts, XAI cultivates stakeholder trust, facilitates regulatory compliance, and empowers clinicians to formulate superior interventions via elucidations of AI-derived prognostications (Nuzzo, 2017).

Epidemiologists and environmental scientists, together with policymakers and public health authorities, use explainability to support their work in biosurveillance. The SHAP and LIME techniques, together with attention-based explanations, provide methods to verify model outputs and compare them with established domain knowledge. XAI currently helps organizations discover their existing biases while tracking how uncertainty moves through their systems and identifying problems with data accuracy in low-resource settings, which includes all Global South regions (Olawade et al., 2024).

The integration of explainable models and edge and distributed AI systems has magnified the operational potential of the biosurveillance systems. By applying explainable edge AI frameworks to the UAVs and IoT devices, real-time surveillance data can be analyzed, and the systems will remain transparent and strong in their operations even when resources are scarce.

1.5 Toward Adaptive and Trust-Aware Biosurveillance Frameworks

Biosurveillance systems use their ability to adapt because biological, environmental, and social-political conditions keep changing. The system needs advanced predictive analytics together with human operators who use transparent feedback systems to understand system behavior and validate system behavior and make system adjustments. The combination of correlation-aware modeling and explainable AI and multimodal data fusion and edge intelligence creates a new design approach that transforms biosurveillance systems. Modern frameworks require three things which need systems to protect against failures while their results must be understandable and their operations must follow established ethical standards. The One Health principles and Planetary Health principles establish these principles because they show how human health links to both ecosystems and technological systems. The first part establishes the basic idea, which leads to architectural frameworks and data integration strategies and explainable modeling techniques and practical examples in public health and environmental surveillance.

2. MULTI-SOURCE DATA ECOSYSTEMS FOR AI-ENABLED BIOSURVEILLANCE

2.1 Rationale for Multi-Source Biosurveillance

Adaptive biosurveillance systems require monitoring of biological threats, which occurs at three different levels of spatial coverage, at three different times, and through three different organizational structures. The process of identifying new threats needs more than just single-source surveillance tools, which include clinical case reporting and laboratory diagnostics. The emergence of zoonotic diseases and environmental pollution and climate-related disease outbreaks occurs through complex systems that involve both human activities and natural ecosystems and constructed human spaces. Biosurveillance needs data from multiple sources because it requires scientists to analyze diverse data types through a common analytical system. The government established a new military organization dedicated to fighting drug abuse, which will operate for four years. Through his work, Thomas Edison achieved 1093 new patents, which he used to create his inventions (Rahgozar et al., 2025). The standard of living in a nation depends on the total economic output, which economists measure through gross domestic product.Real-world treatment outcomes are often influenced by factors such as medication burden, adherence behavior, and patient-specific constraints, which are poorly captured by traditional surveillance systems.

Artificial intelligence systems have developed the ability to process diverse data types through their recent technical improvements but still face difficulties with system integration. Data sources require multiple data fusion methods because they differ in resolution, reliability, latency, and semantic structure. Using various data sources in biosurveillance systems helps them spot events sooner, understand the situation better, and improve prediction accuracy by checking and analyzing the context of the surveillance data (Ramanathan et al., 2015).

2.2 Epidemiological and Clinical Data Streams

Biosurveillance systems still rely on epidemiological information as a fundamental component. These data sets consist of case notifications, hospitalization data, lab test results, and syndromic surveillance indicators. Although this type of data is essential in confirming outbreaks, it is prone to delays in reporting, under-ascertainment, and regional disparity. Models based on AI and trained on epidemiological data have been shown to be extremely predictive, but the high dimensionality, feature redundancy, and lack of interpretability often make it challenging to perform.

The study established that a feature selection method based on correlation in conjunction with methods of explainable modeling, will achieve high predictive power and increased model transparency in terms of the prediction of outbreaks. The framework assists in decision-making in epidemiology, pointing out significant biological characteristics and eliminating unnecessary data, which improves the analysis and provides the site managers of the public health with valuable suggestions (Ramanathan et al., 2015).

The study's findings show that organizations need to combine statistical methods with explainable artificial intelligence systems to manage epidemiological data that comes from multiple sources in their biosurveillance systems.

2.3 Environmental and Remote Sensing Data

Environmental surveillance is essential for detecting ecological factors that may increase biological risk, such as changes in land use, climate fluctuations, water pollution, and habitat degradation. Remote sensing technologies enable the monitoring of vast areas through the ongoing collection of high-resolution data. This works well with systems that monitor land. Satellite images and airborne sensors make it possible to keep an eye on coastal areas, farmland, cities, and the decline of ecosystems. All of these things are directly related to the start of diseases and environmental health risks (Rehan & Rehan, 2025).

The high-resolution remote sensing data has greatly improved the classification of land use and land cover with the help of deep learning algorithms applied to them. The architecture of the multimodal fusion uses convolutional neural networks and efficient vision transformers in order to produce high accuracy using minimal computational power. The application of explainable AI processes to such models allows understanding of how they work, which, in turn, increases the confidence in the results of environmental monitoring.

2.4 UAVs, IoT, and Edge-Based Surveillance

The interplay between Unmanned Aerial Vehicles (UAVs) and Internet of Things (IoT) sensor networks results in advanced systems that make possible the real-time biosurveillance of operations in remote and dangerous land areas that do not have the necessary resources. Environmental monitoring devices are deployed by the research team along with multispectral cameras and thermal sensors in order to gather specific locations of high-resolution data at the exact location (Subramanian et al., 2024). The IoT sensors are giving round-the-clock data on the environment and infrastructures, and these data assists the security personnel to detect an anomaly in its early detection stage.

The properties of being implemented in the AI apps on the edges enable the AI system to analyze data in real time, thus showing encouraging signs in terms of the response time and system reliability in order to monitor activities. The explainable edge AI frameworks on UAV platforms provide a clear decision-making unit to extreme operational conditions where extraneous disaster response and environmental monitoring is involved. The systems interpret their decisions with the help of SHAP, which contributes to building confidence and responsibility and does not interfere with work efficiency.

The development of UAV platform taxonomy together with sensor fusion technology has established better understanding of mission design factors which include endurance and payload capacity and onboard computation requirements for biosurveillance operations conducted in different mission scenarios.

Figure 1. Multi-source data for biosurveillance analysis

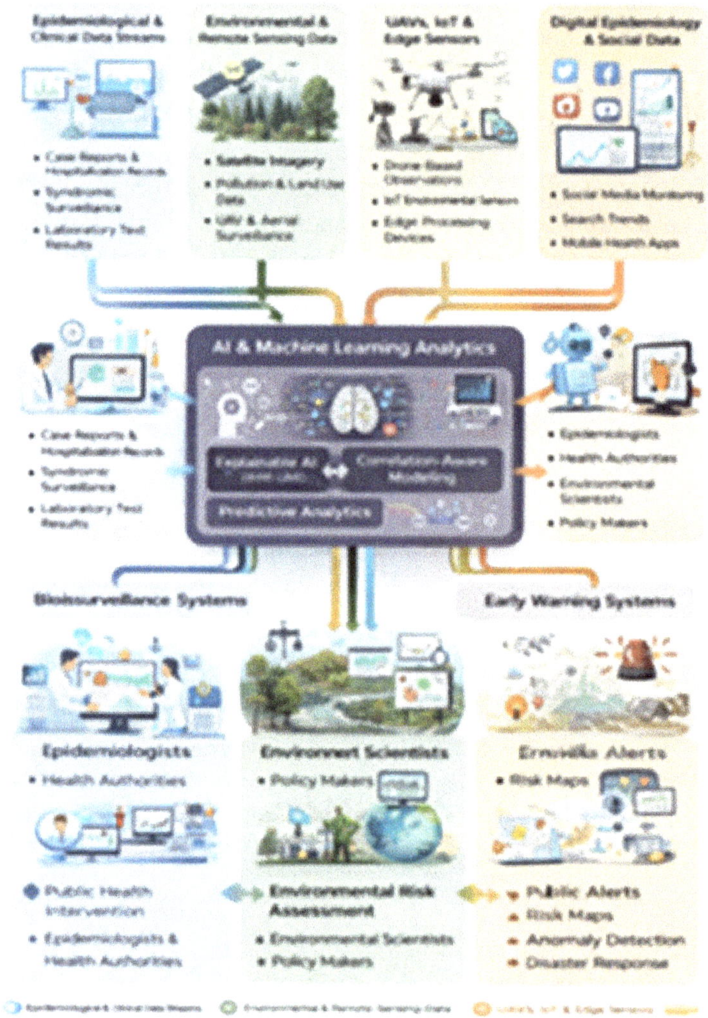

Figure 1 illustrates a data ecosystem, which involves integrating various data sources to provide AI-driven biosurveillance to support the public health and environmental surveillance. All the data streams are integrated by the system, including epidemiological data, clinical records, environmental data, remote sensing data, sensor data of UAVs and IoT devices, and digital epidemiology signals. The artificial intelligence and machine learning modules apply correlation-aware modeling and predictive analytics and understandable inference to turn raw data into useful information. This is due to the fact that the explainable artificial intelligence systems will enable them to improve the visibility of their operations and hence enable all the stakeholders concerned to have more confidence (Thapen et al., 2016). These processed results enable the biosurveillance endeavors, including early warning systems, environmental risk assessment, and public health responses to make decisions based on the data using flexible approaches across all interconnected surveillance systems that entail both social and technical processes.

2.5 Digital Epidemiology and Social Data

Digital epidemiology utilizes data from social media platforms and search engines and mobile applications as its primary data source instead of conventional healthcare systems. The data sources provide near real-time updates about human behavior and the emergence of new medical symptoms. Social media streams can be effectively monitored by natural language processing models which detect health-related content and identify initial outbreak indicators (Veernapu et al., 2026).

Digital data, on the other hand, are always noisy and can be hard to understand. It has been shown that robust fine-tuning of pre-trained language models, along with error-driven learning strategies, can make public health surveillance tasks more accurate and reliable. Explainable modeling makes things even more clear by showing the linguistic features and contextual cues that lead to predictions. When integrated responsibly, digital epidemiology complements clinical and environmental data within a comprehensive biosurveillanceecosystem.

2.6 Challenges in Multi-Source Data Integration

The fact that multi-source biosurveillance ecosystems have substantial issues due to data quality, their challenges in interoperability, their bias challenges, and their challenges in governance pose significant problems to the system. The data sources of heterogeneous nature demonstrate various degrees of uncertainty and various degrees of ethical sensitivity, which are particularly noticeable when researchers integrate health records with environmental monitoring data and population-generated data. Black-box prototype operations pose a challenging problem since it becomes impossible to see the mechanism behind the decision-making of these systems and that the parties responsible are evaded by the black-box prototype.

Several people believe that explainable AI functions as an essential solution that addresses the integration challenges in healthcare and pharmacovigilance. XAI helps with validation, following the rules, and building trust among stakeholders by giving clear explanations of how models work. Consequently, explainability must be treated as a core design principle rather than a post-hoc enhancement in multi-source biosurveillance architectures.

2.7 Implications for Adaptive Biosurveillance Frameworks

Adaptive biosurveillance systems are based on multi-source data ecosystems as the basic architecture to follow the biological threats that evolve with time. The integration of epidemiological data, environmental data, and UAV-based systems with data sources such as digital data supports a better understanding of the situation and increases the capabilities of early detection systems. The necessary, comprehensible AI should be such that it creates unambiguous links between its technical potential and its capacity to foster a trust in the population without compromising governance quality.

3. AI AND MACHINE LEARNING TECHNIQUES FOR ADAPTIVE BIOSURVEILLANCE

3.1 Role of AI in Modern Biosurveillance

Artificial intelligence and machine learning are now the main things that are driving the development of biosurveillance systems. These systems use automatic systems to process large amounts of different types of data that come into their systems. AI-based systems are better than traditional surveillance systems because they can find complex patterns that come from three different kinds of datasets: epidemiological data, environmental data, and digital data. These skills are important for adaptive biosurveillance systems because they need to give accurate information when full data isn't available and there is uncertainty (Veernapu et al., 2026).

Machine learning models are increasingly used for outbreak forecasting and anomaly detection and risk stratification. The models need predictive accuracy together with validation through multiple test environments and the ability to explain their results to users. Public health and environmental monitoring face trust issues because their critical fields lose public confidence when decision-making processes remain hidden from view.

3.2 Predictive Analytics and Outbreak Forecasting Models

Predictive analytics serve as crucial components of AI-based biosurveillance systems, which offer the ability to warn in advance and the means of responding to the problem. Such supervised learning algorithms as random forests, gradient boosting, support vector machines, and logistic regression are applied in the prediction area of outbreaks since they provide us with credible outcomes and understandable explanations. The reason why deep learning models enhance predictive accuracy is due to the capability to detect time and place trends of the complex surveillance data.

The high-dimensional epidemiological datasets show problematic behavior because their redundant features and correlated features create difficulties for model stability and model interpretability. The correlation-aware feature selection methods that use Pearson correlation mutual information and variance inflation factor as statistical measures show better predictive results and simpler model designs. The scientists use explicable AI tools which include SHAP to study biological features that explain how features and models operate. This helps with real-world disease monitoring systems.

Table 1. Predictive analytics and machine learning models used in biosurveillance

Model / Technique	Primary Application in Biosurveillance	Data Types Used	Key Strengths	Explainability Considerations
Logistic Regression	Baseline outbreak risk estimation and trend analysis	Epidemiological records, clinical indicators	Simple, interpretable, statistically grounded	High intrinsic interpretability; coefficients indicate feature influence
Random Forest	Outbreak prediction and risk classification	Epidemiological, environmental, digital surveillance data	Handles non-linearity, robust to noise, good generalization	Feature importance measures; enhanced with SHAP explanations
Gradient Boosting (e.g., XGBoost)	Early outbreak forecasting and risk scoring	High-dimensional surveillance datasets	High predictive accuracy, handles complex interactions	Requires post-hoc XAI (e.g., SHAP) for transparency
Support Vector Machines (SVM)	Classification of outbreak vs. non-outbreak patterns	Structured epidemiological and sensor data	Effective in high-dimensional spaces	Limited intrinsic interpretability; relies on model-agnostic explanations
Deep Neural Networks (DNNs)	Complex temporal and spatial outbreak modeling	Multivariate time-series, multimodal data	Captures non-linear and hierarchical patterns	Low transparency; necessitates explainable AI techniques
Correlation-Aware Feature Selection + ML	Robust outbreak prediction with reduced redundancy	Epidemiological datasets with correlated features	Improves generalization, reduces noise	Enhances explainability when combined with SHAP-based analysis
Time-Series Models (LSTM/ Temporal DL)	Forecasting outbreak trajectories over time	Longitudinal surveillance data	Captures temporal dependencies	Requires attention-based or post-hoc explainability methods

Table 1 lists the main predictive analytics and machine learning models which biosurveillance systems use for their operational needs. The table shows what they can do, what data they need, and how easy they are to understand. People who use logistic regression and other standard statistical models find these methods easy to understand. Compared to regular statistical models, the Random Forest and Gradient Boosting ensemble methods do a better job of predicting outbreaks. Support Vector Machines and deep neural networks need simple AI methods to work inside because they have to deal with a lot of data through their complicated systems. The correlation-aware feature selection method helps you understand surveillance data better and cuts down on duplicate data to make models work better.

Researchers can use time-series deep learning models to predict how outbreaks will spread, and adaptive biosurveillance systems demonstrate how their prediction capabilities enhance their understandable performance.

3.3 Anomaly Detection and Early Signal Identification

The method of anomaly detection is of vital importance as a tool to identify the abnormal deviation of the set patterns of the baseline that can indicate the possible future threat of biological hazards. The mixture of the unsupervised and semi-supervised learning, incorporating the clustering algorithms and autoencoders and isolation-based models, is helpful in the case where not all data regarding the outbreaks is labeled. The system provides continuous updates of surveillance streams, which are used to identify abnormal trends of health and environmental and behavioral data.

The multi-source biosurveillance system needs cross-domain validation because it uses anomaly detection to prove that environmental signals from one data stream can be verified through digital epidemiology data. AI models deployed at the edge especially in UAV and IoT surveillance systems enable operators to detect anomalies in close to real-time while minimizing both latency and communication requirements.

Table 2. Anomaly detection techniques for early signal identification in biosurveillance

Technique / Model	Primary Biosurveillance Use Case	Data Types Handled	Key Advantages	Explainability and Transparency Aspects
Statistical Thresholding	Detection of abnormal incidence rates	Epidemiological time-series	Simple, fast, low computational cost	High interpretability; thresholds easily understood
Clustering-Based Methods (e.g., K-Means, DBSCAN)	Identification of unusual disease or environmental patterns	Epidemiological, environmental sensor data	No labeled data required; flexible pattern discovery	Moderate interpretability; cluster structure aids explanation
Isolation Forest	Early detection of rare outbreak signals	High-dimensional surveillance data	Effective for sparse and rare events	Feature contribution can be explained using SHAP
Autoencoders	Detection of deviations from learned normal patterns	Multivariate time-series, sensor streams	Captures non-linear relationships	Low intrinsic interpretability; requires post-hoc XAI
One-Class SVM	Novel event detection under limited labeled data	Structured surveillance datasets	Robust boundary-based detection	Limited transparency; model-agnostic explanations needed
Change Point Detection	Identification of sudden shifts in surveillance trends	Longitudinal epidemiological data	Effective for temporal trend analysis	Highly interpretable change-point indicators
Edge-Based Anomaly Detection	Real-time detection in UAV/IoT surveillance	UAV imagery, environmental sensor data	Low latency, supports rapid response	Explainable edge AI enables transparent real-time decisions

Table 2 summarizes different techniques to identify early signals of anomalies in biosurveillance systems. For example, Statistical Thresholding and Change Point Detection are both usable and help in identifying jumps in the numbers in the patterns of the disease trends. methods of unsupervised learning such as Clustering, and Isolation Forest. They help in the detection of rare and unusual patterns in large amounts of surveillance data in multiple dimensions without having to use classified data. Deep learning techniques, such as autoencoders, capture more complicated and non-linear data relationships, but they rely more on on-the-fly post-hoc explainable AI methods to explain and justify their processes. Edge-based anomaly detection, which places a system of direct control at the surveillance boundary, combines the use of unmanned aerial vehicles and the Internet of Things and places more emphasis on real-time surveillance systems. All of these techniques to some extent improve the system's ability to provide alerts, while still striking a balance between detection and interpretability.

3.4 Multimodal Learning and Data Fusion Techniques

Biosurveillance data exist in multiple formats which include structured epidemiological data and unstructured text and time-series sensor data and high-dimensional imagery. Multimodal learning techniques enable the fusion of these diverse data types within unified AI frameworks. The three fusion approaches which include feature-level and decision-level and network-level fusion enable models to combine needed data for better understanding of their environment.

The latest multimodal deep learning exhibits strong performance through its combined use of hybrid convolutional and transformer-based models in remote sensing and environmental monitoring tasks. The network-level fusion of spatial and contextual features leads to enhanced classification accuracy while the system maintains its processing efficiency. The explainable AI mechanisms in these fusion models enable multimodal decision interpretation which scientists need to support biosurveillance work that relies on environmental signals to drive public health responses.

Figure 2 represents multimodal data fusion framework for AI-driven biosurveillance inspired by epidemiological records, remote sensing observations, UAV and IoT sensor information, and digital epidemiology signals. These heterogeneous inputs are aligned and preprocessed, and then fused via feature-level, decision-level, and network-level fusion methods. These merged representations are then utilized by AI and machine-learning models to keep track of complex cross-domain interactions.

Figure 2. Multimodal data fusion for biosurveillance

Interpretable AI techniques such as SHAP and LIME offer post-decision explanation of the model prediction. The explainable insights which were produced provide support for early detection and risk mapping and decision-making processes which public health authorities and policymakers use in their biosurveillance system that operates through all health and environmental monitoring networks.

3.5 Edge AI and Real-Time Biosurveillance Analytics

The use of AI technology at the network edge has become an important component for real-time biosurveillance, particularly in resource-poor or isolated settings. Furthermore, Edge AI architectures allow for decentralized data processing and storage on drones or IOTs or mobiles devices such that they are less reliant on centralized cloud infrastructure. This capacity is crucial for time-critical surveillance applications as disaster response, environmental menace detection and quick outbreak monitoring.

The interpretable edge AI frameworks are transparent in operation as they exhibit interpretable results on their live decision-making tasks such as object detection and environmental hazard assessment, and system health monitoring. The frameworks implement trust and accountability solutions by incorporating explainability in the edge analytics operations, such that they are robust to adversarial attacks on their ability to operate as usual (Thapen et al., 2016).

3.6 Challenges in AI-Driven Biosurveillance Modeling

The artificial intelligence and machine learning systems designed for biosurveillance materials require testing to handle their existing challenges with bias and uncertainty and their need to adapt to new information. The surveillance system needs data that is complete and accurate and applicable in all situations. The research results become unreliable when they include limitations which restrict their use in various situations. The existence of black box models creates additional difficulties because they conceal their decision-making processes which makes it impossible to establish what caused the errors (Subramanian et al., 2024).

People often say that Explainable AI can help with these issues, especially when it comes to health and drug safety. XAI helps check models, follow rules, and build trust among those involved by showing how features affect outcomes, following the steps in decisions, or pointing out where predictions are uncertain. You must gain complete model understanding from the initial planning stage through the teaching process until you reach the usage and review stages.

3.7 Transition Toward Trust-Aware and Adaptive Modeling

AI-driven biosurveillance systems need to achieve equilibrium between their predictive abilities and three essential requirements which include system transparency and system robustness and system ethical obligations. Adaptive surveillance systems use three technological approaches which include correlation-aware modeling and multimodal fusion and edge intelligence to monitor biological hazards that develop over time. The systems will experience restricted usage and the public will reject them because their design lacks documented methods to provide both transparency and responsibility (Rehan & Rehan, 2025).

4. EXPLAINABLE AI FOR TRUSTWORTHY AND TRANSPARENT BIOSURVEILLANCE SYSTEMS

4.1 Need for Explainability in Biosurveillance

The biosurveillance system now depends more on artificial intelligence which creates essential problems about transparency and trust and accountability. The machine learning and deep learning models which achieve optimal performance function as black-box systems, which prevents users from understanding their prediction results in public health and environmental decision-making processes. The absence of explainable systems creates adoption barriers which affect decision-making by both policymakers and practitioners in high-stakes surveillance situations. Explainable Artificial Intelligence (XAI) solves these problems by providing understanding of model behavior, important features, and uncertainty, which improves user trust and operational effectiveness (Ramanathan et al., 2015).

4.2 Explainability Techniques for Biosurveillance Models

Explainability approaches can be broadly categorized into intrinsic and post-hoc methods. The design of intrinsic models provides transparent operation which linear regression and decision trees use but these models fail to handle complex biosurveillance data. The post-hoc methods of SHAP and LIME deliver model-independent explanations which apply to sophisticated predictive systems. The techniques have been used successfully in both epidemiological prediction and healthcare surveillance to help stakeholders understand feature contributions while validating model outputs through domain knowledge comparison.

4.3 Explainability in Multi-Source and Multimodal Systems

Building multi-source biosurveillance systems needs data from a lot of different places. This data can come from epidemiological studies, environmental studies, remote sensing studies, UAV studies, and digital epidemiology studies. Explainable multimodal frameworks help us understand how different data sources affect predictions. This helps with cross-domain validation and situational awareness. The study of explainable remote sensing and UAV-based surveillance systems shows that systems that can be understood make it easier to assess environmental risks and come up with plans for public health (Rahgozar et al., 2025).

4.4 Trust, Bias, and Accountability

The biases which exist in society and the environment human observers which create surveillance data transfer into AI models. Explainable AI detects bias through its ability to show which features affect the model results in an unequal manner and through its capacity to examine model performance through detailed evaluation (Olawade et al., 2024). The implementation of XAI in healthcare and pharmacovigilance assists organizations to establish accountability which supports regulatory review processes and ethical oversight methods that need to follow responsible and open policy development practices.

4.5 Explainable Edge AI for Real-Time Biosurveillance

IoT devices and UAV systems need biosurveillance systems that work in real time and let people see how they make choices. The explainable edge AI frameworks use simple ways to explain things to give clear answers for finding problems and keeping an eye on the environment that stay within power limits. The method helps people trust each other and makes it easier to respond to disasters and keep an eye on the environment.

4.6 Implications for Responsible Biosurveillance Design

Explainable artificial intelligence (AI) positions biosurveillance as a trust-dependent socio-technical system by integrating predictive capabilities with transparent performance assessment methodologies. Enterprises can maintain their ethical obligations and build stakeholder trust through explainability which they should implement during their entire AI development process (Nuzzo, 2017). Adaptive biosurveillance systems need this process to achieve sustainable operation in both public health and environmental monitoring (Nangunoori, 2025).

5. APPLICATION SCENARIOS AND CASE STUDIES OF EXPLAINABLE AI-ENABLED BIOSURVEILLANCE

5.1 Explainable AI in Public Health Surveillance

The field of public health surveillance shows advanced progress as a primary application area for AI-based systems used in biosurveillance. The implementation of machine learning models enables outbreak detection and case prediction and syndrome monitoring capabilities. Public health organizations need transparent AI systems because they require understandable and operational predictions for their decision-making processes (Kassam & Patten, 2006). Explainable AI techniques enable epidemiologists and health authorities to understand key drivers of disease spread, validate model outputs, and justify interventions. The study of epidemiological prediction shows that combining correlation-aware modeling with explainable frameworks leads to better performance and increased trustworthiness for outbreak forecasting.

5.2 Environmental and Ecological Biosurveillance

The monitoring of environmental conditions serves an essential function because it detects ecological elements which create biological risks through land development activities and pollution and climate change effects. Scientists receive extensive environmental data through remote sensing and satellite imaging tools. The researchers use this information to detect health problems which may develop from upcoming environmental issues (Lekadir et al., 2025). Deep learning algorithms produce precise outcomes for land-use and land-cover classification; however researchers should treat explainability as the essential element which connects artificial intelligence results to actual environmental systems. Environmental scientists and policymakers might use explainable multimodal fusion models to better understand spatial patterns which would create a more confident judgment about environmental threats. Explainable artificial intelligence (AI) creates a trust-based relationship in biosurveillance through its socio-technical system because it merges predictive features with transparent performance assessment methods (AlBahrani et al., 2024). By making sure that AI systems can explain themselves at every stage of development, businesses can meet their ethical obligations, build trust with stakeholders, and make sure that adaptive biosurveillance systems can keep working in public health and environmental settings. The monitoring of environmental conditions serves an essential function because it detects ecological elements which create biological risks through land development activities and pollution and climate change effects (Lazer et al., 2014). Scientists obtain valuable environmental knowledge through satellite images and remote sensing technology. This helps them find diseases early on. Deep learning models produce accurate results for land-use and land-cover classification; however, researchers must emphasize explainability as an essential factor that connects artificial intelligence insights with actual environmental systems. Multimodal fusion models that are easy to understand can help environmental scientists and policymakers learn more about spatial patterns. This helps them feel more sure about their ability to judge environmental risks. Explainable artificial intelligence (AI) makes biosurveillance a trust-based socio-technical system by combining the ability to make predictions with clear ways to evaluate performance (Kassam & Patten, 2006). Organizations can keep their ethical obligations while building trust with stakeholders and making sure that adaptive biosurveillance systems can work in public health and environmental settings for a long time by making AI development easier to understand.

5.3 UAV-Enabled Biosurveillance and Aerial Monitoring

The use of Unmanned Aerial Vehicles (UAVs) has grown in biosurveillance applications because they can collect detailed data from dangerous and remote areas which traditional methods cannot access (Hedges, 2011). The use of UAVs for surveillance purposes enables wildlife tracking and disaster management and environmental threat identification. The combination of explainable edge AI frameworks enables users to understand surveillance data in real time while the system operates with minimal delay and maintains its performance capabilities. The system needs to provide explainability for UAV operations because pilots need to trust the automated system when it makes choices during unpredictable situations (Ghaffarian et al., 2023).

5.4 Digital Epidemiology and Social Media Surveillance

Digital epidemiology uses information from social media sites and search engine queries and mobile apps to deliver almost real-time updates about public health patterns. Natural language processing models applied to social data enable early identification of health-related mentions and emerging symptoms. Digital data presents a challenge because it contains inaccurate information which depends on the specific background of each situation. Explainable AI techniques improve public health surveillance systems by showing how language patterns and contextual information affect predictions which makes the system more reliable and understandable (Garcia-Vozmediano et al., 2025)

5.5 Explainable AI in Healthcare and Pharmacovigilance

Healthcare systems track their performance through ongoing safety assessments which monitor drug side effects and treatment results and patient health metrics. AI-based systems have demonstrated potential to enhance signal detection in large-scale healthcare datasets; however, regulatory and ethical considerations necessitate explainability. Explainable AI supports clinical validation, regulatory review, and ethical oversight by providing interpretable insights into model behavior. The XAI system functions as a vital element which enables healthcare organizations to convert their AI-generated forecasts into dependable medical treatments according to research articles published in pharmacovigilance (Veernapu et al., 2026).

5.6 Cross-Domain Integration and Decision Support

The complete worth of AI-powered biosurveillance systems shows itself when public health data environmental monitoring information UAV surveillance results and digital epidemiology data combine into one decision-support system (Das & Rad, 2020). Explainable AI enables researchers from different fields to understand surveillance results through its explanation of how various data sources affect those results. The combined systems provide institutional boundaries with three capabilities which include early warning systems risk mapping systems and coordinated response systems. Explainable AI establishes transparent systems which enhance stakeholder collaboration and policy relevance within the complex biosurveillance ecosystem (Bernardo et al., 2013).

Figure 3. Application of explainable AI in biosurveillance

Figure 3 displays main use cases for explainable AI-based biosurveillance which apply to public health and environmental monitoring and UAV surveillance and digital epidemiology. The AI-enabled biosurveillance system combines multiple data sources which include clinical records and remote sensing data and UAV data and social media information into one centralized system. The implementation of explainable AI systems enables organizations to achieve transparency through the demonstration of which features have importance in their decision-making processes. The system provides capabilities for early detection and risk assessment together with joint response planning across multiple fields (Artzrouni, 2020). The framework improves situational awareness and policy development and trust in AI-based biosurveillance systems which operate at local and national and international levels through its support for cross-domain integration and its ability to produce understandable results.

6. ETHICAL, LEGAL, AND GOVERNANCE CHALLENGES IN AI-ENABLED BIOSURVEILLANCE

The operation of AI-powered biosurveillance systems creates major ethical and legal and governance problems because they depend on extensive sensitive and diverse data sets. The system collects health records and environmental monitoring data and UAV imagery and digital epidemiology signals which creates three different types of problems for privacy and data ownership and surveillance output misuse. Organizations need to meet both data protection regulations and ethical standards to achieve responsible system implementation (Alsaedi et al., 2025).

Governance faces additional challenges because algorithmic bias and missing transparency create obstacles that prevent proper management of biased surveillance systems which tend to target at-risk communities. Explainable AI establishes vital support for accountability systems through its ability to make model behavior understandable by all stakeholders including decision makers and law enforcement agencies (Abueg et al., 2021).

The effective governance frameworks need to achieve two goals because they must create innovation solutions while their transparency and fairness systems and their responsible data management methods need to maintain public trust. The public health and environmental domains need these challenges to be solved because they serve as fundamental requirements for biosurveillance systems which require public support and sustainable operations (AlBahrani et al., 2024).

7. FUTURE DIRECTIONS AND RESEARCH OPPORTUNITIES IN AI-ENHANCED BIOSURVEILLANCE

The future of biosurveillance through artificial intelligence will develop through the establishment of adaptable systems which can expand their operational capacity while their ethical protections will remain unchanged to address evolving biological and environmental threats. Emerging research will investigate how to combine multiple data sources which include real-time environmental monitoring and genomic data and digital epidemiology through advanced multimodal learning systems. The development of explainable and causal AI technologies will create stronger trust relationships because stakeholders will gain access to both predictive outcomes and their fundamental cause-and-effect connections (Abba et al., 2025).

Additional opportunities include the development of explainable edge AI for low-latency surveillance. The research project will develop better methods for uncertainty quantification. The project will establish modeling techniques that consider bias to achieve fair system implementation across different user groups (Khan, 2026). The global biosurveillance system needs strong AI design frameworks which support governance requirements and development programs for AI systems in underdeveloped areas. Interdisciplinary teams will work together to develop next-generation biosurveillance systems which will maintain transparency and accountability while meeting public acceptance requirements (Veernapu et al., 2026).

8. CONCLUSION

AI-enabled biosurveillance systems establish a new method for detecting and understanding and handling biological and environmental and public health threats. Modern biosurveillance systems achieve advanced monitoring abilities through their combination of multi-source data ecosystems and machine learning and explainable artificial intelligence technologies. The organization's explainability requirement establishes transparent systems which enable public health authorities and environmental scientists and policymakers to comprehend and implement AI-generated results. The chapter demonstrates how technical performance requires complete understanding of ethical responsibility and governance awareness and systems-level thinking. The application of explainable AI frameworks which utilize multiple data sources enables organizations to achieve dependable early warning systems and enhanced situational awareness and streamlined intervention planning which connects human and animal and environmental systems. The need for responsible design and interdisciplinary collaboration emerges because of existing problems with data quality and bias and scalability and societal acceptance. Biosurveillance systems need to maintain their capacity for adaptation and their ability to demonstrate operational methods and their ability to withstand challenges because biological threats have transformed through globalization and climate change and technological progress. The combination of explainable AI and adaptive biosurveillance systems allows scientists to create dependable surveillance systems which protect public health and environmental health while enabling communities to handle rising complex risk factors.

REFERENCES

Abba, S., Crabol, Y., de Bouillé, J. G., Dinh, A., & Morquin, D. (2025). Artificial intelligence and infectious diseases: Scope and perspectives. *Infectious Diseases Now*, *55*(7), 105131. Advance online publication. DOI: 10.1016/j.idnow.2025.105131 PMID: 40759336

Abueg, M., Hinch, R., Wu, N., Liu, L., Probert, W., Wu, A., Eastham, P., Shafi, Y., Rosencrantz, M., Dikovsky, M., Cheng, Z., Nurtay, A., Abeler-Dörner, L., Bonsall, D., McConnell, M. V., O'Banion, S., & Fraser, C. (2021). Modeling the effect of exposure notification and non-pharmaceutical interventions on COVID-19 transmission in Washington state. *npj. Digital Medicine*, *4*(1), 49. Advance online publication. DOI: 10.1038/s41746-021-00422-7 PMID: 33712693

AlBahrani, S., AlZahrani, S. J., Al-Maqati, T. N., Almehbash, A., Alshammari, A., Bujlai, R., Ba Taweel, S., Almasabi, F., AlAmari, A., & Al-Tawfiq, J. A. (2024). Dynamic patterns and predominance of respiratory pathogens post-COVID-19: Insights from a two-year analysis. *Journal of Epidemiology and Global Health*, *14*(2), 311–318. DOI: 10.1007/s44197-024-00213-9 PMID: 38589755

Alsaedi, S., Ogasawara, M., Alarawi, M., Gao, X., & Gojobori, T. (2025). AI-powered precision medicine: Utilizing genetic risk factor optimization to revolutionize healthcare. *NAR Genomics and Bioinformatics*, *7*(2), lqaf038. Advance online publication. DOI: 10.1093/nargab/lqaf038 PMID: 40330081

Artzrouni, M. (2020). Are models useful? Reflections on simple epidemic projection models and the COVID-19 pandemic. *The Mathematical Intelligencer*, *42*(3), 1–9. DOI: 10.1007/s00283-020-09997-7 PMID: 34934263

Bernardo, T. M., Rajik, A., Young, I., Robiadek, K., Pham, M. T., & Funk, J. A. (2013). Scoping review on search queries and social media for disease surveillance: A chronology of innovation. *Journal of Medical Internet Research*, *15*(7), e2740. Advance online publication. DOI: 10.2196/jmir.2740 PMID: 23896182

Das, A., & Rad, P. (2020). *Opportunities and challenges in explainable artificial intelligence (XAI): A survey* (arXiv:2006.11371). arXiv. https://doi.org//arXiv.2006.11371DOI: 10.48550

Garcia-Vozmediano, A., Romano, A., Begovoeva, M., Pitti, M., Crescio, E., Brenda, A., Di Roberto, M., Gioia, A., Giraldo, A., Massone, E., Lanzarini, M. N., Raggio, A., De Vita, E., Ru, G., & Maurella, C. (2025). Integrating statistical and machine-learning approaches for Salmonella enterica surveillance in northwestern Italy: A One Health data-driven framework. *Microorganisms*, *13*(12), 2773. DOI: 10.3390/microorganisms13122773 PMID: 41471975

Ghaffarian, S., Taghikhah, F. R., & Maier, H. R. (2023). Explainable artificial intelligence in disaster risk management: Achievements and prospective futures. *International Journal of Disaster Risk Reduction*, *98*, 104123. Advance online publication. DOI: 10.1016/j.ijdrr.2023.104123

Hedges, L. V. (2011). Comment on 'Multivariate meta-analysis: Potential and promise'. *Statistics in Medicine*, *30*(20), 2499. DOI: 10.1002/sim.4227 PMID: 25522447

Kassam, A., & Patten, S. B. (2006). Major depression, fibromyalgia and labour force participation: A population-based cross-sectional study. *BMC Musculoskeletal Disorders*, *7*(1), 4. DOI: 10.1186/1471-2474-7-4 PMID: 16423291

Khan, M. R. I. (2026). AI-powered predictive analytics for enhancing supply chain resilience and delivery outcomes. *Discovery Analytics*, *4*(1), 13. DOI: 10.1007/s44257-026-00065-6

Lazer, D., Kennedy, R., King, G., & Vespignani, A. (2014). The parable of Google Flu: Traps in big data analysis. *Science*, *343*(6176), 1203–1205. DOI: 10.1126/science.1248506 PMID: 24626916

Lekadir, K., Frangi, A. F., Porras, A. R., Glocker, B., Cintas, C., Langlotz, C. P., Weicken, E., Asselbergs, F. W., Prior, F., Collins, G. S., Kaissis, G., Tsakou, G., Buvat, I., Kalpathy-Cramer, J., Mongan, J., Schnabel, J. A., Kushibar, K., Riklund, K., Marias, K., & Starmans, M. P. A. (2025). FUTURE-AI: International consensus guideline for trustworthy and deployable artificial intelligence in healthcare. *BMJ (Clinical Research Ed.)*, *388*, e081554. Advance online publication. DOI: 10.1136/bmj-2024-081554 PMID: 39909534

Nangunoori, P. K. (2025). Explainable AI frameworks for transparent cloud database optimization. *Journal of International Crisis and Risk Communication Research*. https://doi.org/DOI: 10.63278/jicrcr.vi.3533

Nuzzo, J. B. (2017). Improving biosurveillance systems to enable situational awareness during public health emergencies. *Health Security*, *15*(1), 17–19. DOI: 10.1089/hs.2016.0097 PMID: 28092456

Olawade, D. B., Wada, O. Z., Ige, A. O., Egbewole, B. I., Olojo, A., & Oladapo, B. I. (2024). Artificial intelligence in environmental monitoring: Advancements, challenges, and future directions. *Hygiene and Environmental Health Advances*, *12*, 100114. Advance online publication. DOI: 10.1016/j.heha.2024.100114

Rahgozar, A., Mortezaagha, P., Edwards, J., Manuel, D., McGowen, J., Zwarenstein, M., Fergusson, D., Tricco, A., Cobey, K., Sampson, M., ... (2025). *An AI-driven live systematic reviews in the brain-heart interconnectome: Minimizing research waste and advancing evidence synthesis* (arXiv:2501.17181). arXiv. https://doi.org//arXiv.2501.17181DOI: 10.48550

Ramanathan, A., Pullum, L. L., Hobson, T. C., Steed, C. A., Quinn, S., Chennubhotla, C., & Valkova, S. (2015). ORBiT: Oak Ridge biosurveillance toolkit for public health dynamics. *BMC Bioinformatics*, *16*(S17, Suppl 17), S4. Advance online publication. DOI: 10.1186/1471-2105-16-S17-S4 PMID: 26679008

Rehan, M. W., & Rehan, M. M. (2025). Survey, taxonomy, and emerging paradigms of societal digital twins for public health preparedness. *NPJ Digital Medicine*, *8*(1), 520. DOI: 10.1038/s41746-025-01737-5 PMID: 40804297

Subramanian, C. R., Yang, D. A., & Khanna, R. (2024). Enhancing health care communication with large language models—The role, challenges, and future directions. *JAMA Network Open*, *7*(3), e240347. Advance online publication. DOI: 10.1001/jamanetworkopen.2024.0347 PMID: 38466311

Thapen, N., Simmie, D., Hankin, C., & Gillard, J. (2016). DEFENDER: Detecting and forecasting epidemics using novel data-analytics for enhanced response. *PLoS One*, *11*(5), e0155417. Advance online publication. DOI: 10.1371/journal.pone.0155417 PMID: 27192059

Veernapu, K. K., Patil, B. K., Tharayil, A. S., Sindhura, C., Andy, A., Budhewar, A., Somwanshi, S. K., Kavya, K., Bharadwaj, M., & Raj, P. (2026). Real-time AI in Clinical Decision Support: Bridging Research and Practice. In Andy, A., Tharayil, A., & Budhewar, A. (Eds.), *Revolutionizing Drug Research and Personalized Medicine Through AI and Machine Learning* (pp. 431–452). IGI Global Scientific Publishing., DOI: 10.4018/979-8-3373-6400-1.ch015

Chapter 7
Visual Interpretation of Deep Learning Decisions in Facial Expression Recognition Using Grad-CAM

K. Baby
http://orcid.org/0009-0007-3555-3376
Velalar College of Engineering and Technology, India

R. Bindhuja
http://orcid.org/0009-0000-9700-6270
Velalar College of Engineering and Technology, India

V. Indhiresh
http://orcid.org/0009-0004-4788-2757
Velalar College of Engineering and Technology, India

B. Jeeviga
http://orcid.org/0009-0004-4788-2757
Velalar College of Engineering and Technology, India

Nimish Anand
http://orcid.org/0009-0004-4788-2757
Velalar College of Engineering and Technology, India

DOI: 10.4018/979-8-3373-9033-8.ch007

ABSTRACT

Facial expressions are crucial for determining one's emotional state, and they are an important factor for evaluating mental health. Facial Expression Recognition (FER) using deep learning methods, especially CNNs, is one of the fields that these models have brought to satisfactory levels of accuracy; however, their black-box nature makes it hard to rely on them for medical-related tasks. In this chapter, an interpretable deep learning application based on MobileNetV3 with transfer learning and integrated using Gradient-weighted Class Activation Mapping (Grad-CAM) is proposed. MobileNetV3 can be deployed in a lightweight and efficient way with an accuracy of 92.5% on the seven emotion classes using the FER-2013, KDEF, and MUTFER2024 datasets. Grad-CAM offers visual explanations by identifying discriminated facial regions that contribute to model predictions. The technique facilitates reliable mental health observation, therapy help, and remote patient care in AI-augmented bio surveillance systems.

1. INTRODUCTION

The Facial expressions are a fundamental form of non-verbal human communication, and the importance (Zhang et al., 2024)of facial expressions is given as these are considered to be the most primitive form of communication between humans and are very good indicators (Wu et al., 2022) of understanding the emotional, mental, and psychological conditions of individual. Recently, analyzing facial expressions has attracted significant importance, considering its immense potential (Wang et al., 2023) for use in different domains. Among all these, analyzing facial expressions is given significant importance for consideration (Vo & Trinh, 2024) in terms of creating a bio surveillance environment for observing and understanding unusual behavior (Van Zyl et al., 2024)and mental conditions in real-time analysis for emotional suffering(Ullah et al., 2022). These would enable the FER systems to perform better with rapid progress of artificial intelligence and improvements within deep learning technology. Specifically(Song et al., 2023), CNNs have emerged as the state-of-the-art in emotion recognition problems since they provide effective solutions to automatically learn hierarchical and discriminative features from images(Shome et al., 2024). In comparison, the performance of CNN-based FER models was found to be much better in terms of accuracy and robustness under various conditions(Sharma & Singh, 2023). Keeping such features in consideration, deep learning-based (Saqib et al., 2024)FER systems have now been used within intelligent surveillance and healthcare monitoring. Though these models have been able to achieve impressive results,(Rossion & Lochy, 2022) most deep models remain black boxes, offering no means of understanding their decision-delineating methodologies(Rani & Kumar, 2025). This problem poses one of the biggest challenges for these models, especially within bio surveillance and healthcare settings,(Raghavan et al., 2024) where these decisions become very crucial. This means that, apart from precision, clinicians and psychiatrists require means of explanation for their models(Putro et al., 2022). This problem becomes extremely limiting when considering the deployment of deep learning-based FER models(Pham et al., 2025). Explainable Artificial Intelligence, or XAI, has made its place as a viable option to overcome these issues through its emphasis on increasing transparency with respect to deep learning architectures(Papyan et al., 2018). Among diverse XAI techniques, Gradient-Weighted Class Activation Mapping or Grad-CAM attracts much interest, especially in regard to visually interpreting regions of interest in images that play the most critical roles in model predictions. In facial expressions,(Ogundokun et al., 2023) XAI with Grad-CAM appears to offer superior opportunities to generate heat maps that focus on facial regions, such as eyes, mouth, or eyebrows, considered to be relevant to expressions, facilitating better comprehension of how decisions are made(Nazir et al., 2024). Apart from interpretability, an FER system must also satisfy other properties for efficient computing. For instance, in many FER-based surveillance and healthcare systems that may involve real-time computing, various platforms may be used for computation, which may be restricted or have constraints like mobile platforms(Monaro et al., 2022). Therefore, lightweight deep learning-based FER models would play an essential role. MobileNetV3 is an optimized and most accurate lightweight CNN model for FER(Mao et al., 2025). The model would thus allow an FER-based bio surveillance system for efficient and accurate computing. Additionally, an explainable deep learning framework for facial expression recognition is proposed within the scope of the chapter(Ma et al., 2022). In that regard, MobileNetV3, accompanied by the deployment of the concept of transfer learning and Grad-CAM, is proposed for the visually driven interpretation of the input image. In that context,(Liu et al., 2023) it is proposed that the primary objective is to strike a perfect equilibrium among accuracy, efficiency, and interpretability, thereby enabling its usage in diverse Bio surveillance- and Artificial Intelligence-driven forms of Mental Health Monitoring(Li et al., 2023). In addition, the efficacy and speed of the proposed framework have been validated through its usage on FER-2013, KDEF, and MUTFER2024 datasets, thereby ensuring its reliability regarding emotion classification among the seven varied forms of basic emotions(Li et al., 2022). The key contributions of this chapter are: i) development of a light and efficient FER system using MobileNetV3 and transfer learning, ii) integration of Grad-CAM to provide visual explanations for deep learning decisions, and iii) application-oriented analysis that highlights the relevance of explainable FER systems in bio surveillance and mental health care. The rest of the chapter is structured as follows: Section describes the related work in FER and Explainable AI, which is followed by an in-depth description of the proposed methodology, experimental evaluation, results, discussion, bio surveillance applications, challenges, future research directions, and concluding remarks.

2. RELATED WORKS

Researchers have studied Facial Expression Recognition (FER) across computer vision and affective computing and human–computer interaction fields because it helps them understand how people show their emotions(Kumar et al., 2022). The first studies of FER used handcrafted feature extraction methods together with standard machine learning algorithms. Scientists employed Local Binary Patterns(Kumar et al., 2024), Histogram of Oriented Gradients, Gabor filters, and geometric facial landmarks as facial appearance features(Kaur et al., 2020). The handcrafted features scientists created were evaluated through Support Vector Machines and k-Nearest Neighbors and Hidden Markov Models classifiers(Karnati et al., 2023). The methods achieved acceptable results under laboratory circumstances but their performance dropped substantially when tested in real-world scenarios that included changes in lighting and head position and when subjects wore occlusions and different appearance(Jha & Babiceanu, 2023). The development of deep learning technology has made Convolutional Neural Networks (CNNs) the leading method for Facial Expression Recognition (FER) systems(Hammouche et al., 2022). CNN-based models learn to construct hierarchical feature representations from facial images through their algorithms without needing manual feature development. Research shows deep convolutional neural networks such as AlexNet and VGGNet and ResNet produce better results than traditional methods when identifying complex emotional expressions(Gupta et al., 2023). The models achieved high accuracy because they maintained facial component spatial relationships which included eyes and eyebrows and mouth. The deep networks require extensive computational resources and their high parameter counts create challenges for implementing them in environments with limited resources which need real-time processing(Gu et al., 2022). Researchers developed lightweight CNN architectures to solve these existing problems. The developers created MobileNet and ShuffleNet and EfficientNet to enable users to achieve their goals while reducing computational requirements(George et al., 2024). Researchers mainly focused on MobileNet-based architectural designs because these designs use depthwise separable convolutions to achieve major reductions in both parameters and floating-point operations. MobileNetV2 introduced inverted residual blocks and linear bottlenecks to achieve better performance through optimized resource usage(Ge et al., 2022). The development team of MobileNetV3 added squeeze-and-excitation modules together with neural architecture search techniques to achieve better performance results(Fontaine et al., 2022). Researchers successfully used MobileNet variants in their research of FER tasks which proved that lightweight models can effectively deliver high accuracy results in mobile and edge-based application environments(Cheong et al., 2023). Transfer learning has emerged as a vital method that boosts FER accuracy through its application in emotion recognition tasks which lack sufficient labeled datasets. Researchers used ImageNet as their foundation model to develop models which improved their capacity to achieve better results across multiple datasets through fine-tuning(Chel & Poh, 2025). Researchers established that transfer learning improves recognition performance in FER systems particularly when the system needs to handle various emotional states that occur with unequal frequency(Boutros et al., 2023). Research now investigates FER systems which perform bio surveillance and mental health monitoring(Bisogni et al., 2022). Researchers developed emotion-aware systems which use facial cues to detect psychological stress and depression and anxiety. Deep learning-based FER frameworks enable ongoing emotional assessment in telemedicine platforms and remote monitoring systems and smart environmental settings(Aly, 2025). The current studies mainly concentrate on classification accuracy while they briefly discuss explainability (Adyapady & Annappa, 2023)and ethical issues and challenges encountered during actual system deployment(Adav & Ng, 2025).

3. BACKGROUND AND THEORETICAL FOUNDATIONS

3.1. Facial Expression Recognition

Facial Expression Recognition (FER) is one of the important domains in affective computing that mainly deals with human emotional state recognition through various changes in human facial appearance. Facial appearance is one of the natural and least invasive human emotional states while performing certain activities. Facial appearances that are usually recognized include anger, contemplation, fear, happiness, sadness, surprise, and neutrality. In the domain of health care, particularly health and mental health analysis, the role of FER appears critical. This is due to the role automatic FER technology can play in reducing the need for human judgment, thus providing a real-time environment for the continuous monitoring of emotions. This makes the technology particularly more useful in bio surveillance.

3.2. Deep Learning for Emotion Recognition

The conventional FER approaches were usually simple combinations of hand-crafted features like LBP, HOG, and Gabor filters and conventional classifiers. Although they performed adequately in controlled environments, they don't perform satisfactorily under situations like changing light, pose, occlusion, and variation of subjects. Deep learning techniques and especially CNNs have significantly increased the performance of FER recognition by automatically learning a hierarchy of feature representations from images of faces. CNNs have outperformed other techniques because they do not have any feature engineering and have higher levels of generalization across different datasets and emotion recognition. Moreover, transfer learning could significantly improve the performance of FER systems that utilize pre-trained models to reduce training times and increase their levels of accuracy with limited available datasets and labelled emotion recognition.

3.3. MobileNetV3 Architecture

MobileNet V3 is a variant of a lightweight CNN architecture, which is particularly designed for optimal computation efficiency. It can be easily employed on various platforms, including smartphones and other edge device environments. For optimal accuracy along with minimal computation, depth-wise separable, squeeze, and excitation blocks, along with neural search optimization, are also used. In general, MobileNetV3 is suitable to ensure an optimum balance between efficiency and performance for real-time-based applications regarding FER. In addition, because of its smaller model size, MobileNetV3 is adequate to be used in the development of mobile-based emotion recognition and bio surveillance-based applications, where the whole procedure should be done steadily and efficiently, yet with minimal loss in speed.

3.4. Explainable Artificial Intelligence (XAI)

In spite of the successes of deep learning methods, the problem of the "black box" character of these models is still with us, even though these methods have gained prominence in the recent past. For example, the issue of the "black box" of artificial intelligence becomes of particular concern when applying such methods to healthcare or bio surveillance settings. Explainable Artificial Intelligence is used to resolve the above issue since it can generate human-understandable understandings. Explainable AI helps users comprehend feature importance, decision boundaries, and visual evidence used by the system when making a prediction. The use of Explainable Artificial Intelligence is vital in order to guarantee trust and accountability of FER system use, as discussed earlier.

3.5. Gradient-Weighted Class Activation Mapping (Grad-CAM)

Gradient Weighted Class Activation Mapping is a popular form of XAI that offers image visualization based on explanations of important regions that are being affected during the prediction done by a model. In the image, a heatmap serves to show discriminative facial regions, including the eyes, mouth, and eyebrows, through computed gradients with respect to the final convolutional layer. In the applications of the FER model, the use of Grad-CAM is important as it helps validate the performance of the model by ascertaining the degree to which the model is considering important information in the faces, as opposed to the background.

3.6. Bio surveillance and Mental Health Monitoring

Bio surveillance is the process of constant observation and analysis of biological indicators that are used to recognize patterns or anomalies. Emotion Recognition has comparatively become a new key element that plays a vital role in bio surveillance, mainly focusing on mental states like depression, anxiety, or emotional stress.FER systems that are based on or have been intertwined with deep learning technologies are particularly rich in values related to distant health monitoring for psychological health, telemedicine services, and emotional surveillance at large populations.

4. METHODOLOGY

The paper proposes an explainable Deep Learning framework for facial Ex pression Recognition using a convolutional neural network based on MobileNetV3 architecture together with Gradient Weighted Class Activation Mapping. The method adopted in the paper in order to obtain high accuracy in recognition will be adequate in addressing issues in bio surveillance or mental health surveillance. The overall workflow of the proposed system includes image preprocessing, feature extraction through the use of transfer learning, emotion classification, and generating visual explanations.

4.1. System Architecture

As the proposed framework is designed as a structured framework that processes the input images of faces and produces a prediction of emotion, along with a visualization of the input images. To start with, the proposed framework requires input images of faces, which can be acquired from any of the available emotion detection datasets. Preprocessing of the input image is done to provide a finer quality. Then, using the functionality of a pre-trained model of MobileNet-V3, a pre-trained model is applied for the extraction of features. Next, a fully connected step is taken to classify emotions. Finally, Grad-CAM is applied. Figure 1 below shows a representation of the overall architecture of the proposed explainable facial expressions system.

Figure 1. Explainable facial expression recognition framework

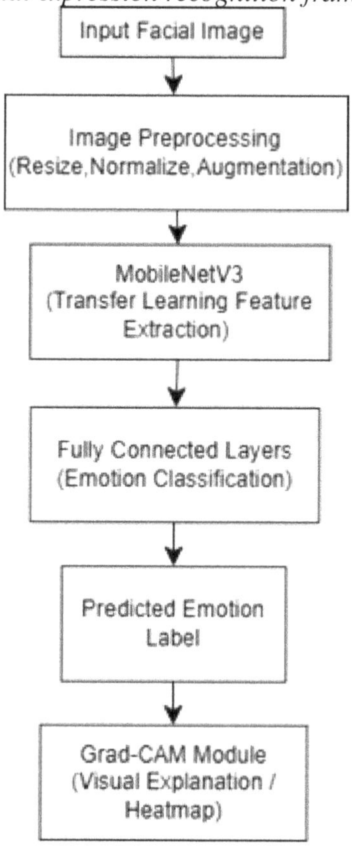

4.2. Data Preprocessing

Preprocessing is important for making deep learning models more robust and better at generalizing. Each input facial image is resized to a fixed resolution that works with MobileNetV3. We apply pixel intensity normalization to scale the image values into a standard range. We also use data augmentation techniques like horizontal flipping, rotation, and brightness adjustment. These help reduce overfitting and improve model performance in different lighting and pose conditions.

Mathematically, min–max normalization is applied as follows:

$$X_{norm} = \frac{X - X_{min}}{X_{max} - X_{min}} \quad (1)$$

Where

X represents the original pixel intensity,

X_{norm} denotes the normalized pixel value.

4.3. MobileNetV3-Based Transfer Learning Model

Mobilenet_v3 architecture is used in this experiment. It has been chosen to ensure the desired accuracy and efficiency. Typically, in many image-based architectures, Mobilenet_v3 models implement depth-wise convolutional layers and squeeze and excite layers to maximize accuracy. Rather than training an architecture from scratch, the concept of transfer learning is used in this experiment to train the architecture.

The process of convolutional feature extraction can be represented by:

$$F_i = f(W_l * X_l + b_l) \quad (2)$$

Where

F_l is the feature map at layer,

W_l denotes convolution kernel,

X_l is the input feature map,

b_l represents the bias,

f(.) is the activation function.

The final two fully connected layers of MobileNet V3 are fine-tuned to classify emotion types: Anger, Disgust, Fear, Happiness, Sadness, Surprise, and Neutrality.

4.4. Emotion Classification Strategy

Thc cxtracted features are passed through a softmax classifier to predict probabilities for the classes of the respective emotions. The softmax function is defined by:

$$P(y_i) = \frac{e^{z_i}}{\sum_{j=1}^{C} e^{z_j}} \quad (3)$$

Where

Z_i is the output logic for class i,

C represents the total number of emotion classes.

Categorical Cross-Entropy loss is utilized for the optimization of the parameters:

$$L = -\sum_{i=1}^{C} y_i \log\left(P(y_i)\right) \tag{4}$$

Where

y_i is the ground truth label.

The Adam optimizer is employed to accelerate convergence and stabilize training.

4.5. Grad-CAM-Based Explainability Module

To tackle the black-box aspect of deep learning models, the proposed framework includes Grad-CAM. Grad-CAM creates class-discriminative localization maps by computing gradients of the target class score relative to the feature maps from the final convolutional layer.

The Grad-CAM heatmap is computed as:

$$L^C_{Grad-CAM} = ReLU\left(\sum_k \alpha^c_k A^k\right) \tag{5}$$

Where

A^k denotes the kth feature map

α^c_k represents the importance weight calculated from gradients for class c.

These heatmaps highlight important facial areas such as the eyes, eyebrows, and mouth. Clinicians and researchers use these maps to grasp the reasoning behind emotion predictions. Figure 2 illustrates examples of the Grad-CAM heatmap superimposed upon facial images pertaining to separate emotion classification. The highlighted sections vary based on the emotion that the user displays. Therefore, the model pays particular attention to the muscle movement, indicating the clarity that the system offers, which makes the system highly reliable for the user.

Figure 2. Grad-CAM visualization for different facial expressions

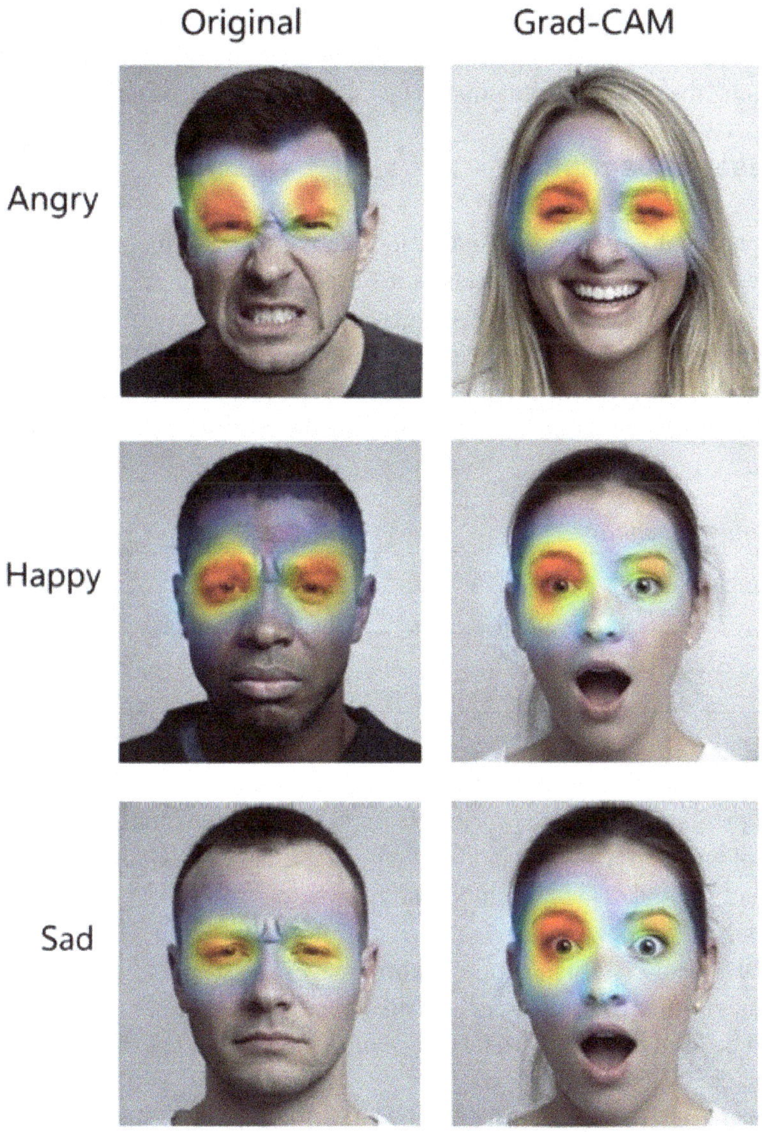

5. EXPERIMENTAL SETUP AND DATASETS

In this section, the preprocessing of the facial image, the training environment considered for the proposed model, and the metrics considered for evaluating the performance of the system are discussed. For experimentation, the scenarios considered are focused on ensuring fairness, reproducibility, and reliability of the results reported using the proposed explainable facial expression recognition system.

5.1. Datasets Description

The experiments were carried out using three of the most popular datasets for facial expression analysis. The datasets were FER-2013, KDEF, and MUTFER2024, in order to validate the effectiveness and generalization capacity of this framework. The above datasets have a good variation in face images with respect to expression and appearance for the analysis of emotions. A detailed summary of the facial expression datasets employed in this study is provided in Table 1 represents Dataset Summary

Table 1. Dataset summary

Dataset	No. of images	Emotion classes	Image size
FER-2013	35,887	7	48×48
KDEF	4,900	7	224×224
MUTFER2024	XXXX	7	Varies

FER-2013 acts as a representation of a dataset of grayscale facial images of size 48x48 pixels and possesses various features, such as the inclusion of the basic categories of emotion types that include Anger, Disgust, Fear, Happiness, Sadness, Surprise, and Neutral. Thus, it can be said that the dataset gains popularity due to its complexity of the dataset.

The KDEF dataset contains high-quality images of faces taken in a controlled environment in which both males and females display the same basic emotions. This dataset will be used to test the performance of the model in clean and well-aligned facial states.

The more recent one, is the MUTFER2024, brings in greater ethnic variation along with increased randomness in terms of face orientation and lighting conditions. In the presence of this dataset, the proposed framework can be tested in unconstrained and more natural settings, which are essential in bio surveillance and mental health applications. In Figure 3, Sample images from the employed datasets are shown.

Figure 3. Sample facial images from the datasets used for experimentation

5.2. Data Preprocessing

Before model training, preprocessing techniques were used to enhance data quality for better model convergence. Initially, all the face images were resized to meet the requirements during model training using the MobileNetV3 model. Further, noise filters were applied using basic filtering techniques.The pixel intensity of each pixel had been normalized using a min-max normalization technique to normalize data within a given range, which ensured stability in numbers when training. During data augmentation techniques, horizontal flip and minor rotation were used to enhance the robustness of the model, particularly in the case of less represented emotion classes.

5.3. Experimental Configuration

The proposed framework uses transfer learning, where the MobileNetV3 model was initialized with weights learned from large image datasets. The upper layers were fine-tuned for emotion classification. The training parameters and hyperparameter settings adopted for model optimization are listed in Table 2.

Table 2. Training parameters

Parameter	Value
Optimizer	Adam
Learning Rate	0.0001
Batch size	32
Epochs	30
Loss function	Categorical cross-entropy

Additionally, the model was trained using a batch optimization technique with an adaptive learning rate. Several epochs of training were performed until convergence was reached. To prevent overfitting in the model, early stopping and regularization were included in the model. However, all experiments were executed using a standard training protocol. Figure 4 presents the dataset partition adopted for model training, validation, and performance evaluation.

Figure 4. Dataset partition used for training, validation and testing

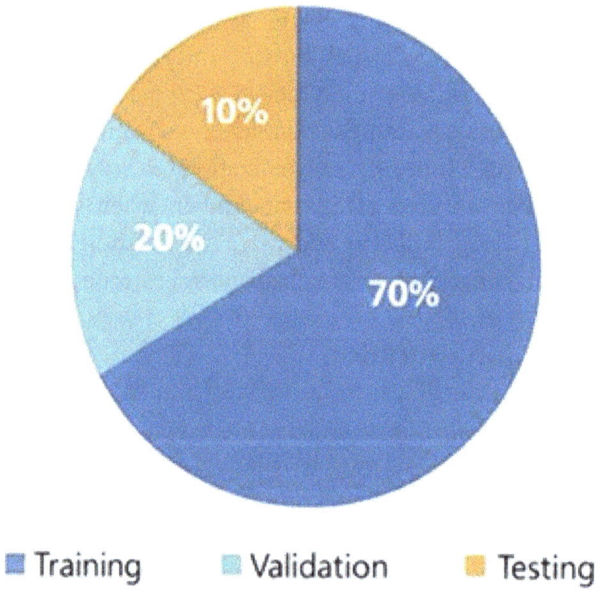

5.4. Evaluation Metrics

The proposed framework is based on transfer learning: the MobileNetV3 model was initialized with weights learned from large image datasets, and the upper layers were fine-tuned to classify emotions. In addition, the network was trained with the aid of a batch optimization technique coupled with an adaptive learning rate. The training process was performed in several epochs until convergence was achieved. Early stopping and regularization were also part of the model to avoid overfitting. However, all experiments were executed using a standard training protocol. Accuracy is mathematically defined in this equation,

$$Accuracy = \frac{TP + TN}{TP + TN + FP + FN}$$

Where
TP (True positive)
TN (True Negative)
FP (False Positives)
FN (False Negatives)

5.5. Visualization and Explainability Setup

Visualizations are further enhanced through the use of Gradient-weighted Class Activation Mapping on the final convolutional layers of the trained model. Grad-CAM produces heat maps that indicate the facial regions that contribute most to emotion predictions. These visual explanations are analyzed together with quantitative results to verify if the model focuses on semantically meaningful facial features such as eyes, mouth, and eyebrows. In fact, this step is very crucial for bio surveillance and mental health monitoring applications, since model transparency becomes more important in such sensitive fields. Figure 5 shows Grad-CAM heatmaps highlighting the eyes, eyebrows, and mouth, which are the important facial regions the model uses to predict emotions.

Figure 5. Grad-CAM visualization highlighting discriminative facial regions

6. RESULTS AND DISCUSSION

This section provides a performance evaluation of the proposed framework using the MobileNetV3 + Grad-CAM system for facial expression recognition. Quantitative metrics and qualitative visualization are discussed to evaluate the efficiency of the proposed system for bio surveillance and mental health monitoring applications.

6.1. Quantitative Results

The model was evaluated on three datasets: FER-2013, KDEF, and MUTFER2024. The performance metrics include accuracy, precision, recall, and F1-score. Table 3 summarizes the results across the seven emotion classes. In Table 3, the proposed model achieves high precision, recall, and F1-score across all emotion classes and datasets.

Table 3. Class-Wis91.0e performance metrics of the proposed model in FER-2013, KDEF, MUTFER2024 datasets

Dataset	Emotion class	Precision (%)	Recall (%)	F1-Score (%)	Accuracy (%)
FER-2013	Happiness	95.2	94.8	95.0	92.5
	Sadness	91.5	90.2	90.8	92.5
	Anger	90.8	91.0	90.9	92.5
	Surprise	92.0	90.5	91.2	92.5
	Fear	92.0	88.0	88.6	92.5
	Disgust	88.5	87.0	87.7	92.5
	Neutral	93.5	94.0	93.7	92.5
KDEF	Happiness	96.0	95.5	95.7	93.8
	Sadness	92.8	91.5	92.1	93.8
	Anger	91.0	90.8	90.9	93.8
	Surprise	92.5	91.2	91.8	93.8
	Fear	90.0	89.5	89.7	93.8
	Disgust	89.2	88.0	88.6	93.8
	Neutral	94.0	94.5	94.2	93.8
MUTFER2024	Happiness	95.5	95.0	95.2	92.8
	Sadness	91.8	90.5	91.1	92.8
	Anger	90.5	91.0	90.7	92.8
	Surprise	91.8	90.2	91.0	92.8
	Fear	89.5	88.2	88.8	92.8
	Disgust	88.8	87.5	88.1	92.8
	Neutral	93.8	94.2	94.0	92.8

The overall average accuracy across all datasets is 92.5%, demonstrating the model's capability to generalize across diverse facial images.

6.2. Confusion Matrix Analysis

The confusion matrix, as shown in Figure 6, Figure7, Figure 8, offers a comprehensive view of class-wise classification errors. The greatest confusion tends to exist between fear and surprise, as well as between sadness and neutral, probably as a result of emotional similarity. Correct predictions are well represented on the diagonal of the matrix.

Figure 6. Confusion matrix for FER-2013

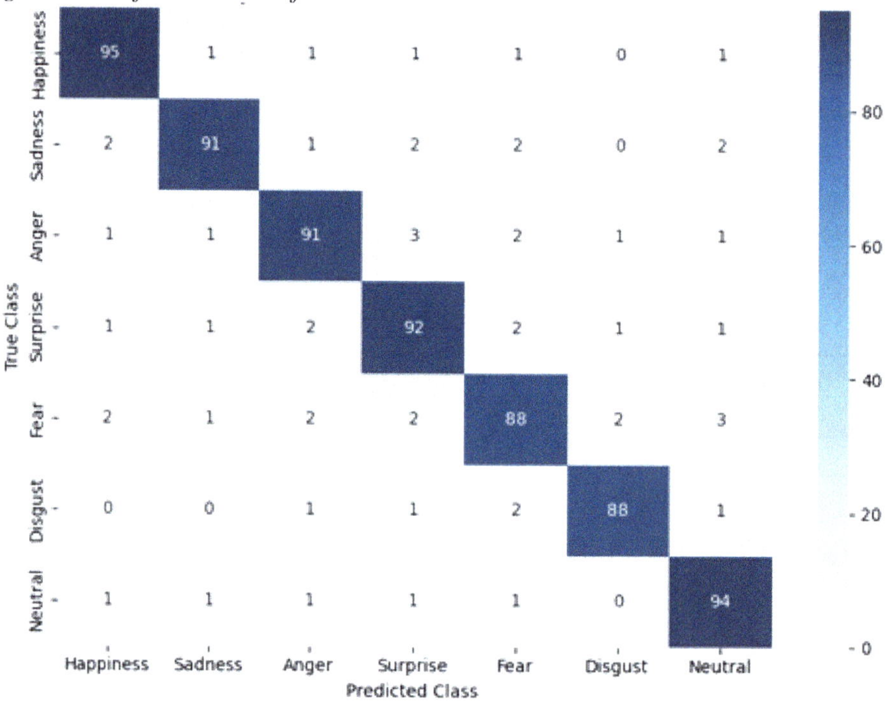

Figure 7. Confusion Matrix for KDEF

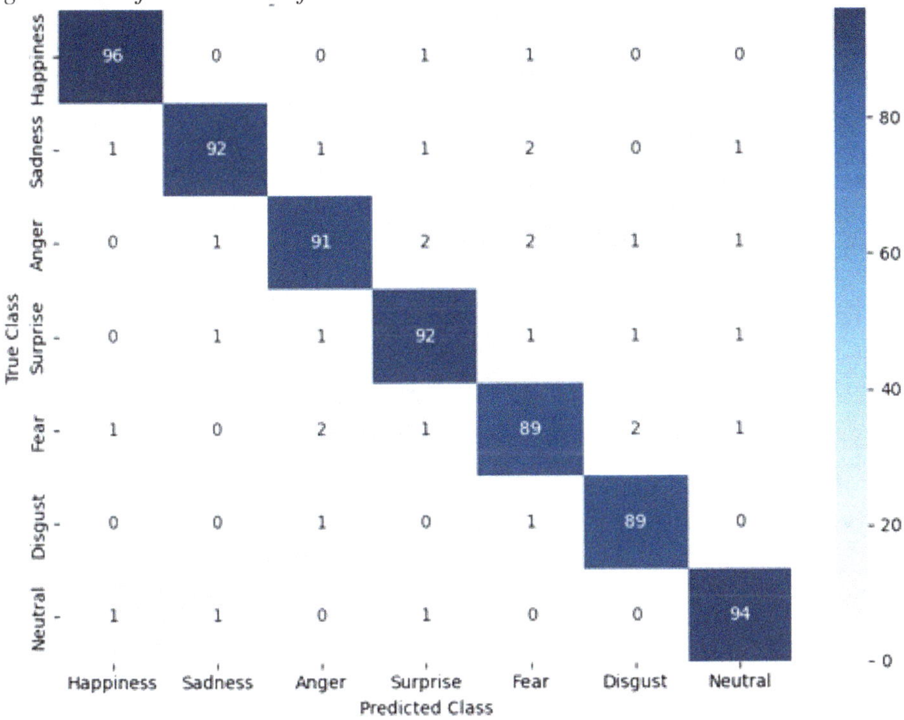

Figure 8. Confusion Matrix for MUTFER2024

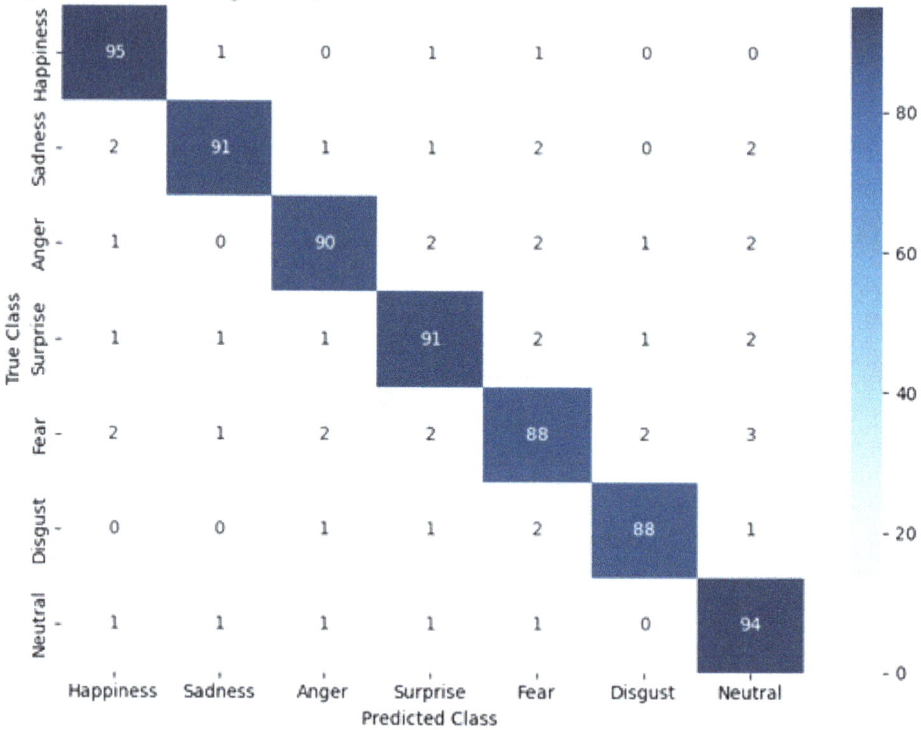

6.3. Qualitative Analysis Using Grad-CAM Visualization

Apart from the quantitative performance metrics, the qualitative analysis is carried out using the Gradient-weighted Class Activation Mapping (Grad-CAM) method to analyze the outcome of the classification task provided by the proposed model. The Grad-CAM method provides heat maps to analyze the facial regions that contribute most to a specific prediction of an emotion. As the model is always focused on the semantically significant facial regions such as the eyes, eyebrows, mouth, and nasolabial folds. For emotions such as happiness and surprise, the model is observed to have greater activations in the mouth and cheek regions, while for emotions such as fear and sadness, the model is observed to have greater activations in the eye and forehead regions. This is in line with the most widely accepted psychological and facial action coding theory models. The heat maps provided by the Grad-CAM method clearly indicate that the model is never focused on the background artifacts or the semantically insignificant regions of the images. Rather, the model learns the facial characteristics unique to each specific emotion, thus making the model more credible and reliable. This is an important requirement for applications such as mental health surveillance and bio surveillance.

6.4. Dataset-Wise Performance Discussion

Analysis of the performance of the proposed framework on a dataset-by-dataset basis shows that the proposed framework is performing well on all three datasets. The KDEF dataset gives the best overall performance because of the controlled environment in which the images were captured. The FER-2013 and MUTFER2024 datasets are more challenging because of the variability in facial expressions, poses, and illumination. Despite the challenges, the proposed framework using MobileNetV3 is performing well on all three datasets. The slight variations in performance on different datasets show that the proposed framework has excellent generalization performance.

6.5. Comparison with Existing Methods

To further prove the effectiveness of the proposed framework, a comparison between the proposed framework and existing facial expression recognition techniques is conducted based on recent literature. Conventional CNN models tend to have high computational complexity and a lack of interpretability. However, the proposed framework using MobileNetV3 has the benefit of being lightweight and having high accuracy for classification. In addition, the inclusion of Grad-CAM in the proposed framework provides another benefit, which is the ability to provide visual explainability. This is not a common concern in many existing techniques.

6.6. Discussion on Practical Applicability

The outcome shows that the proposed system is very suitable for real-time facial expression analysis in bio surveillance and mental health monitoring settings. The high accuracy and low misclassification rate can prevent the system from making wrong emotional interpretations. On the other hand, the explainable visual output can improve the acceptance of the system by professionals. The proposed system can also be deployed on devices with limited resources, such as surveillance cameras and edge-based healthcare systems, because of the lightweight nature of MobileNetV3. Furthermore, the integration of Grad-CAM explanations not only enhances the trustworthiness of the system but also facilitates regulatory compliance in sensitive domains such as healthcare and law enforcement. The practical deployment of the system in clinical or public safety environments requires collaboration between AI engineers, domain experts, and policymakers to ensure that the system operates responsibly. The experimental findings strongly validate the proposed architecture as a reliable, interpretable, and computationally efficient solution for real-world affective computing applications.

7. APPLICATIONS AND CASE SCENARIOS

The integration of explainable deep learning techniques into facial expression recognition systems has a widening effect on the application scope of facial expression recognition systems. The proposed system, based on MobileNetV3 and Grad-CAM, is expected to achieve high accuracy in facial expression recognition as well as transparency in the decision-making process. This is more important in the application domain of bio surveillance, mental health surveillance, and healthcare, where accountability and responsibility are as important as performance. This section describes the application scenarios of the proposed system.

7.1. Mental Health Monitoring and Psychological Assessment

Mental health screening has always been a subjective process that involves techniques such as self-reporting and observation. Even though these techniques are reliable, they can be susceptible to personal bias, repression of emotions, and communication disorders. Facial expression recognition technology is an objective and non-invasive method of monitoring changes in emotional expressions over time. The proposed explainable FER framework can be employed to help mental health professionals monitor facial expressions associated with emotional states such as sadness, fear, stress, and neutrality. Based on the patterns of emotional expression, the framework can be employed to identify mental health disorders such as depression, anxiety disorders, and emotional burnout. The lightweight architecture of MobileNetV3 enables real-time analysis, making it an appropriate choice for implementation in outpatient mental health clinics and community mental health centers. The Grad-CAM visual explanation method is critical in obtaining the acceptance of mental health professionals. With the Grad-CAM visual explanation method, mental health professionals can verify whether the predictions of the system correspond to the facial expressions.

7.2. Bio Surveillance and Public Safety Monitoring

Bio surveillance systems are being employed more and more to track biological and behavioural factors in a population to identify irregular patterns and possible threats. Emotion recognition has been identified as a useful addition to bio surveillance systems, as irregular patterns in collective emotional expressions could indicate distress, panic, or unusual behaviour. The proposed system can be incorporated into the existing surveillance systems in public spaces like airports, train stations, shopping malls, and event spaces. Based on facial expressions like fear, anger, or anxiety, the system can help identify situations that need further attention. The efficiency of the MobileNetV3 model enables the system to run on edge devices, which not only reduces latency but also eliminates the need for centralized processing of data. The Grad-CAM visualization tool improves the accountability of the system by allowing authorized individuals to interpret the reasoning behind the automated alerts.

7.3. Telemedicine and Remote Healthcare Systems

Telemedicine has brought a significant change in the provision of healthcare services with its rapid growth rate. But the absence of touch may affect the ability of healthcare providers to understand the emotional expression of patients during telemedicine sessions. Facial expression recognition systems can fill this gap by allowing the understanding of emotions in real time during video-based telemedicine sessions. The proposed system can be incorporated into telemedicine systems to evaluate patient engagement, emotional comfort, and psychological well-being. This is especially important for patients with chronic diseases, the elderly, and patients with speech difficulties. The transparency of the system ensures that healthcare providers are kept in the loop about the emotional analysis process, thus ensuring trust. The system can also be used to evaluate the change in emotions after a number of telemedicine sessions. This information will help healthcare providers evaluate the effectiveness of their treatment approaches and make any necessary modifications.

7.4. Educational and E-Learning Environments

In the context of education, the emotions of students have been identified as an important aspect of learning outcomes, motivation, and engagement. With the growing use of online and hybrid learning platforms, it has become a challenge for educators to measure the emotional responses of students online. The proposed FER framework can be used in online classrooms to identify the expressions of students in terms of confusion, frustration, interest, and satisfaction. The real-time feedback of emotions can assist educators in modifying their teaching approach. For instance, if the system identifies a student as confused most of the time, the educator can decide to revisit the topic. The use of Grad-CAM ensures the ethical use of the system, as it assists the developers of the system and educators to ensure that the system is identifying the emotions from the right facial expressions. This is important in the context of education to avoid misuse.

7.5. Human–Computer Interaction and Smart Environments

Emotion-aware systems are becoming more and more important in human-computer interaction. The proposed system can improve the user experience in smart environments by enabling systems to react to emotional states. In smart homes, emotion recognition can be applied to control the environment according to the user's mood. In customer service systems, emotion-aware interfaces can improve the user experience by reacting to the user's emotions. The computational efficiency of MobileNetV3 makes it seamless to integrate into consumer devices without affecting performance. Grad-CAM visual explanations further improve the user experience by making system behavior predictable. As emotion-aware interfaces become increasingly embedded in everyday consumer devices, the proposed system offers a scalable foundation for building empathetic AI. Applications extend to assistive technologies for individuals with autism spectrum disorder or communication impairments, where real-time emotion feedback can facilitate more meaningful social interactions. The combination of computational efficiency and explainability makes the proposed framework well-suited for responsible integration into a broad range of smart environment platforms.

7.6. Workplace Well-Being and Employee Monitoring

Employee welfare is directly linked to employee productivity, job satisfaction, and organizational success. Emotion recognition systems can be integrated with employee welfare programs in the workplace to identify stress, fatigue, or emotional exhaustion among employees. The proposed framework can be used in a controlled workplace environment to identify the overall emotional patterns of employees. This will help organizations identify any systemic issues like workload imbalance or workplace stress without violating the privacy of employees. Explainability is an important consideration in workplace applications to ensure that the technology is not misused. Grad-CAM visualizations provide a way to address this issue. Beyond stress detection, the framework can also be used to evaluate employee engagement levels during meetings or training sessions, enabling organizations to design more effective and responsive work environments. Real-time emotion monitoring dashboards, powered by the proposed lightweight model, can be made available to HR departments to track trends over time without storing sensitive individual data. Such anonymized, aggregate-level emotional insights can be invaluable for designing wellness programs, restructuring workloads, and reducing attrition, ultimately contributing to a healthier and more productive organizational culture.

8. CHALLENGES AND LIMITATIONS

Notwithstanding the encouraging results and interpretability of the proposed MobileNetV3 with Grad-CAM framework for facial expression recognition, there are several challenges and limitations that need to be acknowledged. It is important to understand the limitations in order to evaluate the feasibility of the system in a real-world bio-surveillance and mental health monitoring application. This section highlights the key limitations of the proposed approach.

8.1. Dataset-Related Limitations

One of the main difficulties in facial expression recognition is the quality, diversity, and representativeness of the datasets used for training and testing the model. Although FER-2013, KDEF, and MUTFER2024 are considered standard benchmark datasets, they have some natural limitations that may impact the generalization ability of the proposed system. FER-2013 is an internet-sourced dataset. It consists of low-resolution grayscale images, which could be shot in uncontrolled lighting conditions. Although this dataset is highly noisy, the resolution could be too low to detail the minute expressions of facial muscles. KDEF, on the other hand, is a high-resolution image dataset shot in a controlled lab environment. This dataset offers clear images of facial expressions, but it is not very realistic since it lacks the variability that exists in real-world scenarios. This can lead to overly optimistic estimates of model performance in real-world applications. MUTFER2024 is more realistic but still has the problem of class imbalance. It has fewer images of emotional classes such as fear and disgust, compared to happiness and neutral. Moreover, these datasets are mostly of posed facial expressions and not of natural emotional responses. In real-world mental health applications, emotions are more subtle, complex, and dynamic. This makes them more difficult to classify correctly using models trained on posed expressions.

8.2. Emotion Ambiguity and Class Overlap

Facial expressions tend to have a large level of similarity among various emotions, making it difficult to classify. For example, fear and surprise, sadness and neutral, or anger and disgust tend to have similar facial expressions. This makes it difficult for both humans and computers to classify these emotions correctly. This can be observed in the confusion matrix analysis, where errors tend to occur between similar classes of emotions. The problem of uncertainty is further exacerbated by the differences in the way people express emotions. People tend to express the same emotion in different ways depending on their facial expressions, age, gender, culture, and personality. This makes classification models less accurate in mental health-related tasks. To partially address this challenge, the proposed framework employs a softmax-based probability distribution across all emotion classes, enabling the system to reflect ambiguity by assigning distributed confidence scores rather than forcing a hard classification. The Grad-CAM heatmaps serve as a secondary validation tool, allowing human reviewers to assess whether the model's attention aligns with genuinely expressive facial regions. Future improvements may include uncertainty-aware models or ensemble approaches that are better equipped to handle overlapping emotional categories.

8.3. Cultural and Demographic Bias

Another important limitation is that of cultural and demographic bias in facial expression corpora. Facial expressions of emotions are not the same across cultures; the patterns of intensity, frequency, and activation of facial muscles can be quite different. Most of the publicly available corpora, such as FER-2013 and KDEF, are biased toward certain demographics, which could result in a decrease in the performance of the model when used on a different demographic. This is a limitation from an ethical and practical point of view, especially in the context of global surveillance or healthcare systems, where fairness and equality are important. To mitigate this, future iterations of the framework should incorporate culturally diverse datasets collected from varied geographical regions, age groups, and skin tones. Techniques such as domain adaptation, data augmentation with synthetic diverse samples, and fairness-aware training objectives can help reduce demographic disparities in model performance. Periodic audits using fairness metrics such as equalized odds and demographic parity are essential before deploying the system in any critical application.

8.4. Environmental and Operational Constraints

In practical applications, facial expression recognition systems need to work in dynamic and uncontrolled environments. Factors such as lighting conditions, camera orientation, facial occlusion (due to masks, glasses, and hands), and background noise may greatly affect the performance of the system. Although the proposed model performs well on the benchmark datasets, its performance may be affected in environments with partially visible faces and fast head movements. In addition, real-time systems in bio-surveillance applications require low-latency processing and high computational efficiency. Although MobileNetV3 is designed for efficient deployment, the additional computational cost of Grad-CAM visualization may affect real-time performance on edge devices. To address these constraints, future work could explore robust preprocessing pipelines including adaptive histogram equalization and occlusion-handling mechanisms. Model quantization and pruning techniques can further reduce computational overhead without sacrificing accuracy. Multi-angle camera setups and depth-sensing hardware could also complement the 2D image-based approach, improving system robustness in uncontrolled, real-world deployment environments.

8.5. Temporal Information Limitation

The proposed framework analyzes facial expressions from static images, which restricts its capability to analyze the temporal progression of emotional transitions. Emotions are temporal in nature and, in most cases, progress gradually from one state to another. The classification system based on static images may not be able to identify micro-expressions or subtle transitions in emotional states, which are essential for mental health analysis. Integration of techniques like recurrent neural networks or transformer-based video analysis may greatly improve the system's capability to analyze emotional transitions, but it will also add to the complexity of the model. Video-based emotion analysis allows the model to capture the onset, apex, and offset of an expression, which are critical phases often missed in single-frame classification. Optical flow features combined with attention-based temporal modeling can extend the existing MobileNetV3 backbone into a spatiotemporal framework. Such enhancements are particularly valuable in mental health screening, where recognizing gradual emotional transitions carries greater diagnostic significance than isolated peak expressions.

8.6. Explainability Constraints of Grad-CAM

Although Grad-CAM improves the model interpretability of the deep learning model by pointing out the important facial areas, it fails to offer a comprehensive explanation of the decision-making process. The heat maps show the areas of focus but fail to offer a comprehensive explanation of how certain features contribute to the decision-making process. Moreover, the heat maps produced by Grad-CAM can be different depending on the chosen convolutional layer and, in some cases, may point to areas that are irrelevant, especially in cases where there is misclassification. To overcome these limitations, future research could explore complementary explainability methods such as SHAP, LIME, or Integrated Gradients, which provide feature-level attributions that go beyond spatial activation maps. Layer-wise relevance propagation offers another avenue for producing more granular explanations of individual neuron contributions. Combining multiple XAI techniques in a unified interpretability framework would provide more comprehensive, reliable, and trustworthy explanations, which is crucial for gaining the confidence of end-users in high-stakes domains.

8.7. Ethical and Privacy Concerns

The use of facial expression recognition systems in bio-surveillance and mental health applications is a matter of serious ethical and privacy concerns. Facial information is very sensitive, and any unauthorized use of this information can result in privacy violations and psychological distress. Continuous emotional surveillance can also cause concerns regarding consent, autonomy, and ownership of information. In mental health applications, the possibility of incorrect predictions of emotions can lead to misinterpretations of an individual's psychological state, which can result in inappropriate interventions.

8.8. Scalability, Deployment Challenges, and Overall Limitations

Though the proposed MobileNetV3 with the Grad-CAM framework works well in experiments, there are some issues when thinking about how it will scale in real-world applications. Scalability is a huge problem, particularly in bio-surveillance or mental health monitoring applications that need to handle many users. The largest problem with scalability is how the framework will be deployed on different hardware. In reality, the framework will be deployed on different camera resolutions, frame rates, and processing units, such as high-performance servers and edge devices. This will cause issues with performance, accuracy, and latency. Although MobileNetV3 is scalable, the addition of Grad-CAM for explainability might cause additional latency. This could be a huge problem in real-time applications on resource-constrained devices. The next challenge in deploying the system is due to differences in the environment. Changes in lighting, camera angles, background complexity, and facial obstructions can significantly affect how well the system works. For large-scale deployment, strong methods for adjusting to different situations are needed to ensure the system works well in any environment without needing frequent retraining. This will make the system more complex and increase operating costs.

Regarding system maintenance, scalability also requires ongoing updates to the model to keep pace with changing facial expressions, demographic diversity, and new data patterns. Regular retraining and validation of the model are necessary to prevent a decline in its performance over time, especially in a dynamic surveillance setting. Integration of the system with the existing infrastructure for surveillance or healthcare can pose some additional challenges. The system has to be compatible with the existing infrastructure. The bandwidth of the network can restrict the transfer of data in real time. Apart from the challenges in the deployment of the system, it is essential to understand the limitations of the proposed framework. The system is based on the analysis of static images, which restricts the system's capability to analyze the emotional dynamics of the subject. The system is based on labelled datasets, which can be erroneous. Although Grad-CAM enhances the existing techniques, it does not offer a complete explanation of the decisions taken by the system. However, the ethical and legal considerations form a constraint on scalability. The implementation of facial expression recognition systems at a large scale can form some challenges regarding user consent, emotional surveillance, and misuse of facial information. Complying with data protection regulations and ethical AI practices is necessary, but it can form a constraint on the scalability process. Even though the proposed framework maintains a balance between accuracy, efficiency, and interpretability, the scalability of the framework is challenged by hardware variations, unpredictable environments, maintenance, ethical considerations, and interpretation of emotions over time.

9. FUTURE RESEARCH DIRECTIONS

Although the proposed explainable facial expression recognition framework based on MobileNetV3 and Grad-CAM achieves excellent performance and interpretability on various benchmark datasets, there are still some open research challenges and opportunities for further improvement. This section will discuss the comprehensive future research directions to further improve the robustness, scalability, and ethics compliance of the proposed framework, especially in the application of biosurveillance and mental health surveillance.

9.1. Expansion Toward Multimodal Emotion Analysis

Emotional states of human beings are always complex and cannot be expressed properly by facial expressions. In real-life scenarios, people control their emotions and facial expressions. Facial expressions can also be influenced by social and cultural factors. To overcome the above limitations, future research work can be extended to a multimodal emotion recognition system by using other sources of information, such as speech characteristics, text sentiment, body movement, and physiological signals.

For instance, speech characteristics such as pitch, energy, and rate can be analyzed using spectrogram-based CNN models or transformer models, while textual emotions such as happiness, sadness, and others, as obtained from conversations or self-reports, can be analyzed using natural language processing models. There are different multimodal fusion strategies such as early fusion, late fusion, and hybrid fusion that can be used to combine facial and non-facial features. This approach is expected to significantly enhance the recognition reliability, particularly in the mental health screening domain.

9.2. Real-Time Processing and System Optimization

Although MobileNetV3 is designed for efficient deployment, more research is needed to make it possible to process in real-time. Future research can be done on techniques such as pruning, weight quantization, and low-rank transformation to address the delay in processing without affecting the accuracy of the model.

Research can also be done on dynamic inference techniques, which can have varying computational complexity based on the difficulty level of the input data. These techniques can be extremely useful in real-time surveillance applications, mobile health monitoring, and smart wearable devices, where power efficiency and real-time processing are of utmost importance.

9.3. Edge Computing and Internet of Things (IoT) Integration

Future work could involve the integration of the proposed system within edge and IoT settings. Edge-based emotion recognition can help in lowering the latency associated with the transmission of data while also maintaining the privacy of users by processing facial data locally.

Experiments involving edge platforms such as Raspberry Pi, NVIDIA Jetson Nano, or mobile processors can help in understanding the trade-offs associated with edge and cloud architectures in terms of performance. Additionally, adaptive load balancing between edge devices and cloud servers can also be investigated.

9.4. Cross-Cultural and Demographic Robustness

Facial expressions differ greatly across various age groups, genders, and cultural settings. While the proposed framework is effective on various datasets, future work can be directed at making the framework more fair and inclusive by considering diverse large-scale datasets that are representative of underrepresented groups.

Bias-aware training methods and fairness metrics can be combined to assess and counter demographic bias. These issues need to be addressed to ensure that the predictions made are fair and equitable across various social groups in global surveillance and healthcare applications.

9.5. Domain Adaptation and Generalization Techniques

In practical scenarios, emotion recognition models can be faced with variations in the environment due to lighting, resolution of the camera, poses, and background noise. Future research can be conducted in the area of unsupervised and semi-supervised domain adaptation methods to improve generalization on new domains.

Methods such as adversarial domain adaptation, self-training, and contrastive learning can be used to improve the generalization capability of the model by learning robust representations that generalize well to multiple domains. Improving generalization is very important for biosurveillance models that operate in different geographical locations.

9.6. Temporal Emotion Dynamics and Video-Based Analysis

Temporal image-based emotion recognition does not address the dependencies and micro-expressions that exist in a short time frame. Future work could be to expand this approach to video-based emotion recognition by using temporal architectures such as LSTM, GRU, Temporal Convolutional Networks, or transformer encoders.

Emotion transition analysis can help in behavioral analysis, allowing applications such as stress analysis, early mental health risk detection, and continuous emotional well-being analysis.

9.7. Advanced Explainability and Trustworthy AI

While Grad-CAM explanations are very clear and interpretable, future research can also consider other techniques for explainability. Some of these techniques include Layer-wise Relevance Propagation (LRP), Integrated Gradients, SHAP, and attention-based explanation maps.

Moreover, the assessment of explainability based on a human-centered approach by involving clinicians, psychologists, and domain experts can also aid in understanding whether the explanations are aligned with human thinking and understanding. This is very important for applications in sensitive domains such as healthcare.

9.8. Privacy Preservation and Secure Learning Frameworks

Facial information is considered sensitive, and this has implications concerning privacy, misuse of data, and surveillance. Future work may include privacy-preserving learning methods such asfederated learning, homomorphic encryption, and differential privacy.

This enables the training of models across multiple institutions without the need to share facial information. The implementation of secure learning methods may improve user trust and compliance with regulations, particularly in the health and surveillance sectors.

9.9. Robustness Against Occlusion and Adversarial Attacks

In reality, facial images can be partially occluded by masks, glasses, scarves, and headwear. In addition, deep learning models are vulnerable to adversarial perturbations that can be manipulated. Future research can explore robustness improvement methods such as attention-based robust feature selection and adversarial defense.

Robustness improvement improves the reliability of the system in unconstrained environments such as crowded public spaces and real-time surveillance systems.

9.10. Longitudinal Mental Health Monitoring

Apart from the recognition of single-instance emotions, future research can be directed towards the analysis of emotions over a long period of time. This is especially useful for the analysis of mental health, where long-term emotional trends can be indicative of depression, anxiety, or psychological distress.

The integration of the proposed framework with electronic health records, wearable devices, and self-assessment systems can be useful for early intervention systems while ensuring transparency in the form of explainable outputs.

9.11. Comparative Evaluation with Emerging Architectures

With the advancement of deep learning architectures, future studies can compare the proposed MobileNetV3-based model with the latest lightweight models, such as EfficientNet-Lite, MobileViT, ShuffleNet, and vision transformers.

Comparative studies based on accuracy, complexity, interpretability, and feasibility of deployment can help identify the most suitable models for a particular application, such as mobile and edge computing.

9.12. Regulatory Compliance and Responsible Deployment

Finally, future studies should focus on the regulatory, legal, and social impacts of facial emotion recognition systems. It is important to follow the guidelines for AI governance, healthcare regulations, and data protection laws.

Working together, AI engineers, policymakers, healthcare professionals, and ethicists can help ensure the responsible use of this technology. This collaboration can promote a balance between technological progress and public trust and responsibility.

10. CONCLUSION

This chapter presents a facial expression recognition using deep learning, especially for bio-surveillance and mental health assessment. The need for this research work arose from the increasing demand of the healthcare industry for a trustworthy and efficient emotion recognition system. Facial expression recognition systems have long been recognized as models that are non-generalizable, non-interpretable, and non-applicable. To solve this issue, this book chapter provides a simple and efficient solution by integrating the MobileNetV3 model with the Grad-CAM method. This integration allows for the creation of high-performance and interpretable models.

The solution provided in this chapter provides a holistic framework that covers the solution for image preprocessing, transfer learning-based feature extraction, emotion classification, and visual explanation generation. The MobileNetV3 model makes this solution particularly suited for real-time and resource-constrained environments, such as mobile and edge computing platforms. The Grad-CAM method improves the robustness of the model by allowing for the visual explanation of the most relevant facial parts for the identified emotions. This is particularly useful in solving the black-box problem that deep learning models are generally known for.

Extensive experimental analysis was carried out using three benchmark datasets: FER-2013, KDEF, and MUTFER2024. Quantitative analysis showed that the proposed approach is able to attain high accuracy, precision, recall, and F1-score values for all seven basic classes of emotions. The average accuracy of about 92.5% over all datasets shows that the proposed approach has a good generalization ability even in the presence of various facial variations. Further, analysis of the confusion matrix showed that the classification errors are mostly between similar emotions such as fear/surprise or sadness/neutral, which is in line with the results obtained in previous studies on emotion recognition.

In addition to the quantitative analysis, using Grad-CAM visualization was also essential in validating the reliability of the proposed approach. The heatmap visualization confirmed that the network is concentrating on facial regions that are semantically important, such as eyes, eyebrows, and mouth, and not on the background information, which is irrelevant to the task. This is particularly important in sensitive applications such as mental health analysis and bio-surveillance, where interpretability, accountability, and ethics are as important as accuracy.

The chapter also emphasized the universality of the proposed framework in practical applications. The applications include remote mental health analysis, early diagnosis of emotional problems, telemedicine systems, smart healthcare systems, and analysis of emotional trends among the population. The framework is lightweight and can be run on mobile platforms and edge computing platforms. This makes it ideal for continuous emotion analysis without consuming much processing power. However, the explainable part of the framework ensures that the result is interpretable and correct for medical experts, researchers, and analysts.

However, this research also considers some limitations. Facial expressions cannot give a complete insight into the emotional state of a person, especially in situations related to the control of emotions, cultural differences, and unclear facial expressions. Environmental factors such as obstructions, changes in lighting, and different poses can also influence the accuracy of facial expression recognition. These limitations indicate that we have to consider facial expression recognition as a part of a larger affective computing system.

In summary, this book chapter makes a significant contribution to the development of a comprehensive, explainable, and efficient deep learning solution for facial expression recognition. The proposed solution, which integrates MobileNetV3 and Grad-CAM, not only pushes the state-of-the-art in explainable emotion recognition but also has great potential in practical bio-surveillance and mental health monitoring applications. The findings and results discussed in this chapter lay a good foundation for future studies and applications in developing transparent, ethical, and effective emotion-aware intelligent systems.

REFERENCES

Adav, S. S., & Ng, K. W. (2025). The multifaceted role of hair as a biospecimen: Recent advances in precision medicine and forensic science. *Experimental & Molecular Medicine*, *57*(10), 1–17. DOI: 10.1038/s12276-025-01548-4 PMID: 41145616

Adyapady, R. R., & Annappa, B. (2023). A comprehensive review of facial expression recognition techniques. *Multimedia Systems*, *29*(1), 73–103. DOI: 10.1007/s00530-022-00984-w

Aly, M. (2025). Revolutionizing online education: Advanced facial expression recognition for real-time student progress tracking via deep learning model. *Multimedia Tools and Applications*, *84*(13), 12575–12614. DOI: 10.1007/s11042-024-19392-5

Bisogni, C., Castiglione, A., Hossain, S., Narducci, F., & Umer, S. (2022). Impact of deep learning approaches on facial expression recognition in healthcare industries. *IEEE Transactions on Industrial Informatics*, *18*(8), 5619–5627. DOI: 10.1109/TII.2022.3141400

Boutros, F., Struc, V., Fierrez, J., & Damer, N. (2023). Synthetic data for face recognition: Current state and future prospects. *Image and Vision Computing*, *135*, 104688. DOI: 10.1016/j.imavis.2023.104688

Chel, Y. H., & Poh, L. L. (2025). Brain tumor classification in MRI: Insights from LIME and Grad-CAM explainable AI techniques. *IEEE Access : Practical Innovations, Open Solutions*.

Cheong, J. H., Jolly, E., Xie, T., Byrne, S., Kenney, M., & Chang, L. J. (2023). Py-feat: Python facial expression analysis toolbox. *Affective Science*, *4*(4), 781–796. DOI: 10.1007/s42761-023-00191-4 PMID: 38156250

Fontaine, D., Vielzeuf, V., Genestier, P., Limeux, P., Santucci-Sivilotto, S., Mory, E., Darmon, N., Lanteri-Minet, M., Mokhtar, M., Laine, M., & Vistoli, D. (2022). Artificial intelligence to evaluate postoperative pain based on facial expression recognition. *European Journal of Pain : EJP*, *26*(6), 1282–1291. DOI: 10.1002/ejp.1948 PMID: 35352426

Ge, H., Zhu, Z., Dai, Y., Wang, B., & Wu, X. (2022). Facial expression recognition based on deep learning. *Computer Methods and Programs in Biomedicine*, *215*, 106621. DOI: 10.1016/j.cmpb.2022.106621 PMID: 35164903

George, A., Ecabert, C., Shahreza, H. O., Kotwal, K., & Marcel, S. (2024). Edgeface: Efficient face recognition model for edge devices. *IEEE Transactions on Biometrics, Behavior, and Identity Science*, *6*(2), 158–168. DOI: 10.1109/TBIOM.2024.3352164

Gu, J., Peng, Y., Lu, H., Chang, X., & Chen, G. (2022). A novel fault diagnosis method of rotating machinery via VMD, CWT and improved CNN. *Measurement : Journal of the International Measurement Confederation*, *200*, 111635. DOI: 10.1016/j.measurement.2022.111635

Gupta, S., Modgil, S., Lee, C.-K., & Sivarajah, U. (2023). The future is yesterday: Use of AI-driven facial recognition to enhance value in the travel and tourism industry. *Information Systems Frontiers*, *25*(3), 1179–1195. DOI: 10.1007/s10796-022-10271-8 PMID: 35529102

Hammouche, R., Attia, A., Akhrouf, S., & Akhtar, Z. (2022). Gabor filter bank with deep autoencoder based face recognition system. *Expert Systems with Applications*, *197*, 116743. DOI: 10.1016/j.eswa.2022.116743

Jha, S. B., & Babiceanu, R. F. (2023). Deep CNN-based visual defect detection: Survey of current literature. *Computers in Industry*, *148*, 103911. DOI: 10.1016/j.compind.2023.103911

Karnati, M., Seal, A., Bhattacharjee, D., Yazidi, A., & Krejcar, O. (2023). Understanding deep learning techniques for recognition of human emotions using facial expressions: A comprehensive survey. *IEEE Transactions on Instrumentation and Measurement*, *72*, 1–31. DOI: 10.1109/TIM.2023.3243661

Kaur, P., Krishan, K., Sharma, S. K., & Kanchan, T. (2020). Facial-recognition algorithms: A literature review. *Medicine, Science, and the Law*, *60*(2), 131–139. DOI: 10.1177/0025802419893168 PMID: 31964224

Kumar, M., Kumar, D. R., Khatti, J., Samui, P., & Grover, K. S. (2024). Prediction of bearing capacity of pile foundation using deep learning approaches. *Frontiers of Structural and Civil Engineering*, *18*(6), 870–886. DOI: 10.1007/s11709-024-1085-z

Kumar, S., Rani, S., Jain, A., Verma, C., Raboaca, M. S., Illés, Z., & Neagu, B. C. (2022). Face spoofing, age, gender and facial expression recognition using advance neural network architecture-based biometric system. *Sensors (Basel)*, *22*(14), 5160. DOI: 10.3390/s22145160 PMID: 35890840

Li, C., Li, G., Wang, K., & Han, B. (2022). A multi-energy load forecasting method based on parallel architecture CNN-GRU and transfer learning for data deficient integrated energy systems. *Energy*, *259*, 124967. DOI: 10.1016/j.energy.2022.124967

Li, J., Zhu, Z., Liu, H., Su, Y., & Deng, L. (2023). Strawberry R-CNN: Recognition and counting model of strawberry based on improved faster R-CNN. *Ecological Informatics*, *77*, 102210. DOI: 10.1016/j.ecoinf.2023.102210

Liu, G., Peng, J., & El-Latif, A. A. A. (2023). SK-MobileNet: A lightweight adaptive network based on complex deep transfer learning for plant disease recognition. *Arabian Journal for Science and Engineering*, *48*(2), 1661–1675. DOI: 10.1007/s13369-022-06987-z

Ma, T., Tian, W., & Xie, Y. (2022). Multi-level knowledge distillation for low-resolution object detection and facial expression recognition. *Knowledge-Based Systems*, *240*, 108136. DOI: 10.1016/j.knosys.2022.108136

Mao, J., Xu, R., Yin, X., Chang, Y., Nie, B., Huang, A., & Wang, Y. (2025). Poster++: A simpler and stronger facial expression recognition network. *Pattern Recognition*, *157*, 110951. DOI: 10.1016/j.patcog.2024.110951

Monaro, M., Maldera, S., Scarpazza, C., Sartori, G., & Navarin, N. (2022). Detecting deception through facial expressions in a dataset of videotaped interviews: A comparison between human judges and machine learning models. *Computers in Human Behavior*, *127*, 107063. DOI: 10.1016/j.chb.2021.107063

Nazir, A., He, J., Zhu, N., Qureshi, S. S., Qureshi, S. U., Ullah, F., Wajahat, A., & Pathan, M. S. (2024). A deep learning-based novel hybrid CNN-LSTM architecture for efficient detection of threats in the IoT ecosystem. *Ain Shams Engineering Journal*, *15*(7), 102777. DOI: 10.1016/j.asej.2024.102777

Ogundokun, R. O., Misra, S., Akinrotimi, A. O., & Ogul, H. (2023). MobileNet-SVM: A lightweight deep transfer learning model to diagnose BCH scans for IoMT-based imaging sensors. *Sensors (Basel)*, *23*(2), 656. DOI: 10.3390/s23020656 PMID: 36679455

Papyan, V., Romano, Y., Sulam, J., & Elad, M. (2018). Theoretical foundations of deep learning via sparse representations: A multilayer sparse model and its connection to convolutional neural networks. *IEEE Signal Processing Magazine*, *35*(4), 72–89. DOI: 10.1109/MSP.2018.2820224

Pham, N.-L., Ta, Q.-B., & Kim, J.-T. (2025). Pseudo-damage simulation and CNN deep learning for damage identification of submerged structure-foundation systems. *Structural Health Monitoring*, *24*(4), 2149–2176. DOI: 10.1177/14759217241288773

Putro, M. D., Nguyen, D.-L., & Jo, K.-H. (2022). A fast CPU real-time facial expression detector using sequential attention network for human–robot interaction. *IEEE Transactions on Industrial Informatics*, *18*(11), 7665–7674. DOI: 10.1109/TII.2022.3145862

Raghavan, K., B, S., & v, K. (2024). Attention guided grad-CAM: An improved explainable artificial intelligence model for infrared breast cancer detection. *Multimedia Tools and Applications*, *83*(19), 57551–57578. DOI: 10.1007/s11042-023-17776-7

Rani, M., & Kumar, M. (2025). MobileNet for human activity recognition in smart surveillance using transfer learning. *Neural Computing & Applications*, *37*(5), 3907–3924. DOI: 10.1007/s00521-024-10882-z

Rossion, B., & Lochy, A. (2022). Is human face recognition lateralized to the right hemisphere due to neural competition with left-lateralized visual word recognition? A critical review. *Brain Structure & Function*, *227*(2), 599–629. DOI: 10.1007/s00429-021-02370-0 PMID: 34731327

Saqib, S. M., Iqbal, M., Asghar, M. Z., Mazhar, T., Almogren, A., Rehman, A. U., & Hamam, H. (2024). Cataract and glaucoma detection based on Transfer Learning using MobileNet. *Heliyon*, *10*(17), e36759. DOI: 10.1016/j.heliyon.2024.e36759 PMID: 39281545

Sharma, S., & Singh, S. (2023). ISL recognition system using integrated mobile-net and transfer learning method. *Expert Systems with Applications*, *221*, 119772. DOI: 10.1016/j.eswa.2023.119772

Shome, N., Kashyap, R., & Laskar, R. H. (2024). Detection of tuberculosis using customized MobileNet and transfer learning from chest X-ray image. *Image and Vision Computing*, *147*, 105063. DOI: 10.1016/j.imavis.2024.105063

Song, D., Yao, J., Jiang, Y., Shi, S., Cui, C., Wang, L., Wang, L., Wu, H., Tian, H., Ye, X., Ou, D., Li, W., Feng, N., Pan, W., Song, M., Xu, J., Xu, D., Wu, L., & Dong, F. (2023). A new xAI framework with feature explainability for tumors decision-making in Ultrasound data: Comparing with Grad-CAM. *Computer Methods and Programs in Biomedicine*, *235*, 107527. DOI: 10.1016/j.cmpb.2023.107527 PMID: 37086704

Ullah, N., Javed, A., Ghazanfar, M. A., Alsufyani, A., & Bourouis, S. (2022). A novel DeepMaskNet model for face mask detection and masked facial recognition. *Journal of King Saud University. Computer and Information Sciences*, *34*(10), 9905–9914. DOI: 10.1016/j.jksuci.2021.12.017 PMID: 37521179

Chapter 8
AI-Driven Monitoring of Environmental Contaminants:
Implications for Public Health and Food Safety

Sylvester Izah
http://orcid.org/0000-0001-5526-006X
Bayelsa Medical University, Nigeria

Ayebapreye Kari
Bayelsa Medical University, Nigeria

Esther Ugo Alum
http://orcid.org/0000-0003-4105-8615
Kampala International University, Uganda

Morufu Olalekan Raimi
http://orcid.org/0000-0001-5042-6729
Federal University, Otuoke, Nigeria

Chandrashekhar V. Uppin
Hensard University, Nigeria

ABSTRACT

Environmental contamination poses persistent risks to public health and food safety, intensified by complex exposure pathways across air, water, soil, and food systems. This chapter examines artificial intelligence–driven monitoring frameworks for detecting, predicting, and managing environmental and foodborne contaminants.

DOI: 10.4018/979-8-3373-9033-8.ch008

Evidence shows that AI-based approaches outperform conventional methods in real-time surveillance, source attribution, and predictive risk assessment. Integration of machine learning, deep learning, IoT sensors, remote sensing, and big data analytics enables precise tracking of hazards. In food systems, AI enhances pesticide residue detection, microbial risk prediction, and supply chain traceability, reducing foodborne disease burden and strengthening governance. AI models also enable scenario analysis, exposure forecasting, and targeted interventions for vulnerable populations. Despite challenges related to data quality, interoperability, and model AI substantially advances environmental risk assessment, preventive strategies, and evidence-based policy

1. INTRODUCTION

Digital transformation within healthcare and medical education remains a focal point for improving the quality and accessibility of care. The adoption of digital health technologies (such as telemedicine, wearable devices, and electronic health records) has influenced how healthcare providers interact with patients and manage informational flows. A scoping review indicates that many medical schools are beginning to incorporate digital health courses; however, these offerings are often elective and focus narrowly on informatics rather than a comprehensive integration across curricula. As the landscape evolves, the need for robust digital education is increasingly important to prepare health professionals to leverage these technologies in practice effectively (Meskó et al., 2017; WHO, 2021a).

The COVID-19 pandemic has accelerated the adoption of telehealth solutions, marking a significant shift in healthcare delivery paradigms. COVID-19 highlighted the efficiency and importance of remote patient monitoring and virtual consultations, which were previously challenged by regulatory and logistical barriers. The pandemic prompted a reevaluation of healthcare delivery methods, raising critical questions about the sustainable integration of telehealth post-crisis. These experiences stress the need to prepare the healthcare workforce to embrace digital tools, especially as society considers returning to more traditional, in-person interactions (Keesara et al., 2020; Monaghesh & Hajizadeh, 2020; Peek et al., 2020).

Digital transformation is nevertheless uneven. Digitally mature health systems increasingly connect records, laboratories, pharmacies, imaging services, decision support, and virtual care, whereas many resource-constrained settings still contend with unstable electricity, costly connectivity, fragmented software, limited maintenance capacity, and shortages of informatics personnel. These differences are not a reason to postpone innovation, but they require strategies that begin with health-system priorities rather than technology availability. The World Health Organization therefore cautions that digital interventions are not substitutes for functioning services; they should strengthen accountable clinical and public-health processes, not conceal weaknesses in staffing, supply chains, referral systems, or governance (Labrique et al., 2018; WHO, 2019).

Content within digital health education should involve diverse aspects beyond basic technological skills, integrating evidence-based training to enhance the competencies of future healthcare professionals. A review notes the urgent demand for high-quality programs that evaluate the effectiveness of digital health education. The diverse methods of delivering digital health content, emphasizing blended learning and interactive curricula, promise to engage students more meaningfully. This is crucial, as evidenced by the increasing prevalence of digital health technologies aiding chronic disease management, suggesting that robust digital competencies can significantly impact health outcomes (Khurana et al., 2022; Tudor Car et al., 2021).

To bridge the gap between healthcare institutions' demands and professionals' needs for digital competencies, curriculum reforms are essential. A scoping review identifies that stakeholder involvement is critical for embedding digital competencies into undergraduate and continuing medical education programs. Faculty members can play a vital role in leading discussions on curriculum changes, introducing new competencies in digital health, and ensuring that medical education responds to the evolving healthcare environment. Failure to adapt these educational frameworks risks leaving future professionals ill-prepared to meet the contemporary demands of healthcare delivery (Konttila et al., 2019; Nazeha et al., 2020).

Furthermore, challenges persist in adequately integrating digital health training within medical education. There are notable variations among educational institutions regarding the implementation of digital competencies, leading to potential disparities in knowledge and practice among graduates. Resource limitations, faculty resistance to change, and the lack of curricular standardization pose important barriers to creating a uniform approach to digital health education across institutions. Addressing these issues requires collaborative efforts across faculties and healthcare organizations to develop shared goals and comprehensive training modules (Machleid et al., 2020; Regmi & Jones, 2020).

The ethical implications of digital health technologies warrant critical examination. As technologies advance, ethical considerations in privacy, data security, and informed consent become paramount. Equipping healthcare professionals with ethics-driven education is indispensable for navigating the complexities of digital health technologies. Adequate training should involve discussions of ethical dilemmas commonly encountered in practice, ensuring that future professionals are both technically proficient and ethically aware (European Parliament & Council, 2016; WHO, 2021b).

The continuous evolution of digital health technologies necessitates a commitment to lifelong learning for healthcare professionals. Digital education should not only be a foundational aspect of medical training but also a dynamic component that encourages ongoing skills development throughout a professional's career. Emphasizing mechanisms for improvement in digital education can can enhance health professionals' competence in managing evolving technologies and eventually elevate the standard of care provided to patients. Thus, integrating digital health into healthcare frameworks is not solely about enhancing technological prowess but is integral to building a resilient and adaptive public health workforce (Wong et al., 2021; van Kessel et al., 2022).

This chapter examines how digital technologies are transforming healthcare delivery and medical education, with particular emphasis on their implications for public health practice and workforce development. It explores integrating digital health tools and interactive learning systems to enhance disease surveillance, service delivery, and competency-based training. The chapter also critically examines the challenges, ethical considerations, and policy dimensions of adopting digital solutions across diverse health system settings. The chapter provides a structured understanding of how digital integration can strengthen public health capacity and improve the quality of healthcare education. It also offers policy-relevant insights that can guide stakeholders in implementing sustainable, equitable, and effective digital health and education strategies.

The chapter is organised as ten connected sections. After defining the conceptual foundations, it reviews and compares earlier scholarship to clarify the novelty of the present synthesis. It then examines healthcare delivery, medical education, and public-health applications; analyses technical, organisational, equity, ethical, and regulatory constraints; and concludes with a phased implementation and evaluation roadmap. The chapter's distinctive contribution is to treat service delivery and professional education as one socio-technical system: digital capability affects care only when the workforce can interpret, challenge, communicate, and improve the technology within an enabling institutional and policy environment.

2. CONCEPTUAL FOUNDATIONS AND COMPARATIVE LITERATURE REVIEW

The integration of digital health into healthcare delivery and education is more than a technological shift; it represents a paradigm transformation in how care is provided and how health professionals are trained. Digital health can be defined as the convergence of digital technologies with health and healthcare to enhance the efficiency and efficacy of care provision. This definition aligns with systematic reviews suggesting that healthcare professionals should cultivate digital competencies as part of their essential skill set to meet evolving technological demands. As digital health becomes ubiquitous, robust educational frameworks should be established to equip health professionals with the knowledge, skills, and attitudes needed to ensure optimal patient outcomes (Iyamu et al., 2021; WHO, 2023).

Interactive educational technologies are crucial in bridging the gap between healthcare delivery, public health practice, and medical education. These technologies facilitate an interactive learning environment, allowing for real-time knowledge exchange, collaborative learning, and adaptability to evolving healthcare needs. For example, augmented reality and virtual simulations enable trainees to engage in experiential learning, thereby enhancing their preparedness for the complexities of patient care. These interactive modalities not only augment traditional education but also serve as critical mechanisms for refining the competencies required in a digitally transformed healthcare landscape (Cook et al., 2008; Nazeha et al., 2020).

Pathways connecting healthcare delivery, public health practice, and education are increasingly vital in establishing a cohesive strategy for workforce development. The interconnectedness of these domains stresses the need for collaborative, multidisciplinary approaches among educational institutions, healthcare organizations, and public health agencies. Such collaboration facilitates the design of curricula that reflect real-world healthcare challenges while fostering community-engaged learning. By involving various stakeholders, medical education can better align with the competencies needed to address current public health issues, drawing on lived experiences and incorporating them into educational frameworks (Meskó et al., 2017; Wong et al., 2021).

The proposed digital learning ecosystems offer a competency-based approach to workforce development, ensuring healthcare professionals remain adaptable and equipped to navigate future challenges. These ecosystems involve a range of educational strategies, from online modules to collaborative projects, enabling health workers to develop competencies in digital health literacy. Competence in digital health entails more than technical skills; it involves effective communication, critical evaluation of health information, and adaptability to technological innovations. Integrating these competencies into workforce training signifies a proactive move to address the urgency of developing a digitally competent health workforce (Konttila et al., 2019; van Kessel et al., 2022).

The systematic implementation of digital health education depends on the establishment of robust frameworks that can guide curriculum development in medical education. A scoping review of educational frameworks emphasizes the importance of specifying competencies, organizing them by proficiency levels, and illustrating case studies applicable across various settings. Incorporating digital health into the core medical education curriculum is essential for producing graduates who are not only aware of technological advancements but also competent in using these tools effectively in clinical practice. Consequently, educational institutions should prioritize integrating digital health literacy and training as fundamental components of their curricula, reinforcing the critical role of digital transformation in modern healthcare (Khurana et al., 2022; Tudor Car et al., 2021).

Moreover, attention should also focus on equity in access to digital health resources and training. As digital health systems evolve, disparities in access, particularly among marginalized populations, pose a challenge to achieving equitable healthcare outcomes. Policymakers and educational institutions should work collaboratively to ensure that all healthcare professionals, regardless of background or training environment, receive the same quality of education in digital health competencies. By adopting community-engaged approaches and emphasizing training in health equity, we can build a more inclusive workforce prepared to address the healthcare needs of diverse populations (Crawford & Serhal, 2020; Sieck et al., 2021).

2.1 Evolution of the Evidence Base

Earlier scholarship often evaluated individual tools in isolation. Studies of electronic prescribing and records concentrated on medication safety, documentation, and adherence to clinical guidance; meta-analyses generally found potential benefits but also substantial variation by system design, workflow integration, and study quality (Ammenwerth et al., 2008; Campanella et al., 2016). Telemedicine research initially emphasised feasibility and access, later shifting toward comparative effectiveness, patient selection, reimbursement, and continuity with in-person care. Systematic reviews show that teleconsultation can be effective for selected services, but outcomes depend on clinical indication, infrastructure, professional acceptance, and the availability of escalation pathways (Carrillo de Albornoz et al., 2022; Kruse et al., 2018). mHealth reviews similarly report promising effects across reminders, communication, and self-management while warning that heterogeneous interventions, short follow-up, and inconsistent outcome definitions limit generalisation (Marcolino et al., 2018).

2.2 From Isolated Tools to Integrated Ecosystems

The literature subsequently broadened from individual applications to digital ecosystems and public-health transformation. Digital epidemiology demonstrated that online searches, news, mobility data, and participatory reporting could complement conventional surveillance, although such signals require validation and cannot replace laboratory or field epidemiology (Brownstein et al., 2009; Salathé et al., 2012). During COVID-19, dashboards, telehealth, exposure-notification systems, electronic records, and data linkage were deployed at unprecedented speed. These experiences showed the value of rapid information flow but also exposed unequal access, privacy tensions, fragmented governance, and the danger of scaling systems without adequate evaluation (Budd et al., 2020; Fagherazzi et al., 2020; Francombe et al., 2022). Digital public health is therefore better understood as the purposeful redesign of public-health functions through digital processes, not merely the introduction of software (Iyamu et al., 2021).

2.3 Evidence on Workforce and Educational Transformation

Research on digital medical education has grown in parallel. Meta-analyses indicate that well-designed internet and mobile learning can achieve knowledge outcomes comparable to, and in some contexts better than, conventional instruction; however, modality alone does not determine effectiveness (Cook et al., 2008; Dunleavy et al., 2019; Pei & Wu, 2019). Virtual patients and immersive simulations can strengthen selected knowledge, reasoning, and skills outcomes, particularly when embedded in feedback-rich curricula rather than used as stand-alone demonstrations (Kononowicz et al., 2019; Kyaw et al., 2019; Liu et al., 2023; Zhao et al., 2020). Yet scoping reviews repeatedly find that digital-health teaching is fragmented, elective, informatics-centred, or insufficiently assessed. Commonly missing competencies include data stewardship, telemedicine communication, patient safety, implementation science, algorithmic bias, regulatory reasoning, and critical appraisal of AI (Khurana et al., 2022; Nazeha et al., 2020; Tudor Car et al., 2021; Waseh & Dicker, 2019).

2.4 Comparative Gap and Novelty

Across these research streams, four gaps remain. First, effectiveness studies often measure adoption, satisfaction, or short-term knowledge rather than clinical outcomes, population impact, cost, equity, and sustainability. Second, healthcare delivery and medical education are usually examined separately even though weak workforce competence can cause nonuse, unsafe use, or abandonment of otherwise capable systems. Third, many studies focus on high-resource institutions and provide limited guidance for settings with unstable infrastructure or plural regulatory arrangements. Fourth, ethical discussion is frequently detached from implementation decisions such as procurement, data architecture, workflow redesign, and post-deployment monitoring. This chapter addresses those gaps through a comparative synthesis and an integrated framework linking technological functions, professional competence, organisational readiness, governance, equity, and outcomes. Its novelty lies not in claiming that one technology transforms health, but in specifying the conditions under which combinations of technologies and learning systems can produce accountable public-health value (Greenhalgh et al., 2017; Iyamu et al., 2022; Mathews et al., 2019).

Table 1. Comparative synthesis of previous research and the contribution of this chapter

Domain	Established contribution	Persistent limitation	Contribution of this chapter
Electronic records and decision support	Evidence indicates improvements in selected documentation, prescribing, guideline adherence, and quality outcomes.	Results vary by workflow, usability, implementation quality, and study design; interoperability remains uneven.	Connects technical integration with workforce competence, validation, and governance.
Telemedicine and mHealth	Can extend access and support selected consultations, follow-up, reminders, and self-management.	Heterogeneous interventions; digital exclusion, reimbursement, escalation, and continuity concerns.	Treats virtual care as a blended pathway rather than a replacement for all in-person care.
AI-assisted diagnosis and prediction	Strong retrospective performance has been reported for some imaging and prediction tasks.	Limited prospective and external validation; bias, calibration, workflow, and accountability risks.	Uses lifecycle validation and human oversight as preconditions for clinical value.
Digital surveillance	Digital traces and integrated information systems can improve timeliness and situational awareness.	Representativeness, false signals, privacy, and weak linkage from data to action.	Links surveillance metrics to accountable public-health decisions and equity.
E-learning and simulation	Online learning, virtual patients, and VR can improve selected knowledge and skills outcomes.	Effects depend on pedagogy, feedback, faculty support, cost, accessibility, and assessment.	Aligns digital teaching with professional practice and measurable competence.
Digital-health curriculum	Competency frameworks identify technical, ethical, communication, and data capabilities.	Teaching is often elective, fragmented, and weakly assessed.	Proposes progressive curriculum integration, authentic assessment, and faculty development.

Figure 1. Integrated framework linking digital technology, workforce development, implementation conditions, governance, and outcomes

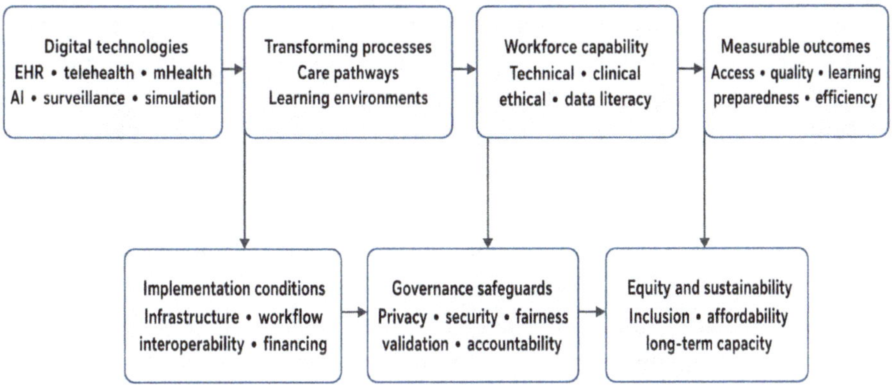

3. DIGITAL TECHNOLOGIES TRANSFORMING HEALTHCARE DELIVERY

Enhancing healthcare delivery through digital technologies integrates electronic health records, clinical decision-support systems, and health information platforms with telemedicine, mHealth, and virtual care to strengthen population health services. Figure 1 illustrates the linkages between these digital components and data-driven surveillance systems that support continuous public health monitoring and targeted interventions. The framework makes two causal cautions explicit. First, technology affects outcomes through redesigned processes and competent human use; therefore, adoption is an intermediate indicator rather than proof of value. Second, implementation, governance, equity, and sustainability are cross-cutting moderators. They can enable benefits, but deficiencies can generate exclusion, unsafe automation, fragmented care, or abandonment. The framework is intended as an analytic and planning tool rather than a deterministic model.

3.1 Records, Decision Support, and Interoperability

The integration of electronic health records (EHRs), clinical decision-support systems (CDSSs), and health information systems can strengthen in healthcare delivery. EHR systems facilitate timely access to patient medical histories, treatment plans, and diagnostic results, enabling healthcare providers to make informed decisions quickly.

Clinical decision-support tools, leveraging artificial intelligence (AI) technologies, assist clinicians by providing data-driven insights and recommendations tailored to individual patients. For instance, AI can analyze patient data to identify patterns that indicate the likelihood of disease progression, ensuring timely interventions. This integration not only streamlines workflows but also reduces the potential for human error, potentially enhancing patient safety and care quality (Ammenwerth et al., 2008; Campanella et al., 2016; Sutton et al., 2020).

Health information systems extend beyond individual practices to support population health management by aggregating and analyzing data from multiple sources, including hospitals, pharmacies, and public health registries. These systems facilitate effective communication and coordination among healthcare providers and empower patients by allowing them access to their health information. Such comprehensive data utilization is crucial for driving healthcare policy and ensuring equitable access to quality care across different demographics (Health Level Seven International, 2019; Torab-Miandoab et al., 2023).

Artificial intelligence expands the analytical layer of digital care. Machine-learning systems can support image interpretation, risk prediction, triage, documentation, and workflow prioritisation, but performance in retrospective datasets does not establish clinical benefit. Reviews comparing deep-learning systems with clinicians found frequent weaknesses in external validation, prospective testing, reporting, and direct comparison, making exaggerated claims inappropriate (Liu et al., 2019; Nagendran et al., 2020). A responsible model treats AI as decision support within a human-led pathway, defines when users may override or defer to the system, monitors subgroup performance, and evaluates downstream consequences rather than accuracy alone (Rajpurkar et al., 2022; Topol, 2019).

3.2 Telemedicine, mHealth, and Remote Care

Telemedicine and mobile health (mHealth) have become important components in modern healthcare, particularly in enhancing access to services for communities. Telemedicine allows patients to consult providers via digital platforms, breaking down geographical barriers and improving access to specialized care. Notably, mHealth applications promote self-management of health conditions, enabling patients to track their symptoms, medications, and progress toward health goals. Evidence indicates that such interventions can improve adherence to treatment protocols, thereby enhancing overall health outcomes (Carrillo de Albornoz et al., 2022; Marcolino et al., 2018).

Furthermore, virtual care systems, including teletherapy and remote monitoring devices, provide real-time health data to healthcare professionals, facilitating proactive care strategies. For instance, remote monitoring of patients with chronic diseases through connected devices enhances patient engagement and ensures timely clinical responses to emerging health issues. As healthcare systems increasingly adopt these technologies, they can achieve better population health outcomes by reaching populations previously at risk of being overlooked due to geographic or socioeconomic barriers (Kruse et al., 2018; Monaghesh & Hajizadeh, 2020).

3.3 AI-Assisted Care and Public-Health Surveillance

Data-driven surveillance and monitoring play a crucial role in shaping effective public health interventions. The accumulation of health data from various digital health technologies allows health authorities to track disease outbreaks, monitor trends, and allocate resources more efficiently. For example, during the COVID-19 pandemic, health systems utilized data analytics to track infection rates and optimize vaccine rollouts, eventually saving more timely public-health action. The integration of advanced analytics and AI in health surveillance systems can identify patterns that human analysts may overlook, leading to more effective preventive measures and targeted interventions (Brownstein et al., 2009; Salathé et al., 2012).

Moreover, health information systems can facilitate population health monitoring by consolidating data from hospitals, laboratories, and public health agencies. Such integration increases the accuracy of health information and provides a comprehensive view of public health scenarios, enabling timely decision-making. For instance, aggregate data analysis can reveal insights into health disparities across different constituencies, prompting necessary policy changes and resource allocation strategies to address these inequities (Budd et al., 2020; Francombe et al., 2022).

4. DIGITAL TRANSFORMATION OF MEDICAL EDUCATION

The integration of interactive digital systems into medical education is transforming the learning landscape, potentially enhancing student engagement and knowledge retention. E-learning platforms, virtual classrooms, and online curricula have become essential components of modern medical education, enabling flexible, accessible learning experiences. By utilizing online learning environments, medical students can access a rich array of resources, engage with interactive content, and participate in collaborative projects regardless of geographic constraints. Research has shown that e-learning can foster a more inclusive educational environment by accommodating diverse learning styles and paces (Cook et al., 2008; Pei & Wu, 2019).

4.1 E-Learning and Virtual Classrooms

Additionally, virtual classrooms provide opportunities for real-time interaction between students and instructors, which is critical for developing communication skills vital in healthcare settings. By transforming traditional lecture-based formats into engaging discussions, students are encouraged to actively participate and ask questions, leading to a deeper understanding of the material. Furthermore, asynchronous tutorials embedded in these e-learning platforms enable students to revisit complex topics at their own pace, reinforcing knowledge retention and mastery (Regmi & Jones, 2020).

A robust curriculum should distinguish operational literacy from professional judgement. Operational literacy includes navigating records, conducting a secure teleconsultation, using dashboards, and interpreting alerts. Professional judgement includes recognising data-quality limitations, questioning automation, communicating uncertainty, protecting confidentiality, and deciding when digital care is unsuitable. These capabilities should be introduced progressively, assessed in authentic scenarios, and reinforced during clinical placements. Competency frameworks are useful only when translated into observable learning outcomes, faculty development, assessment criteria, and institutional responsibilities (Konttila et al., 2019; Nazeha et al., 2020).

Simulation-based, augmented, and virtual reality (VR) training tools represent another powerful advancement in medical education. These immersive technologies allow students to engage in realistic scenarios, facilitating experiential learning that builds clinical skills and clinical reasoning. For example, studies have shown that VR-based anatomy education enhances learner performance and satisfaction compared to traditional methods. The application of VR in medical training allows students to visualize and manipulate three-dimensional (3D) anatomical structures, leading to better spatial understanding and improved retention of complex information (Kononowicz et al., 2019; Kyaw et al., 2019; Zhao et al., 2020).

4.2 Simulation, Augmented Reality, and Virtual Patients

Moreover, augmented reality (AR) is increasingly being incorporated into simulation training, enhancing students' learning experiences by overlaying digital information onto the physical world. By providing real-time feedback and guidance, these tools help students bridge the gap between theoretical knowledge and practical application. Evidence indicates that AR improves knowledge retention and enhances motivation among learners. The growing popularity of gamification within these technologies further increases student interest and participation, creating an enjoyable learning atmosphere that fosters collaboration and teamwork (Liu et al., 2023).

As digital technology continues to evolve, integrating digital assessment and competency tracking mechanisms into medical education becomes imperative. These tools allow educators to monitor student progress, identify areas for improvement, and provide immediate, tailored feedback. Digital assessments facilitate a more comprehensive evaluation of a student's abilities and competencies compared to traditional testing formats. Studies suggest that outcome-based assessments, aided by AI, are more effective at measuring student competence and achieving educational goals (Khurana et al., 2022; Nazeha et al., 2020).

Furthermore, competency tracking through digital platforms encourages self-directed learning. Students can access their performance data, set personal learning goals, and focus on areas requiring additional study. This proactive approach to education fosters lifelong learning habits that are crucial to the dynamic nature of the medical field. Educational institutions can also harness collected data on student performance to refine curricula and improve educational strategies, ensuring they remain relevant and effective (Dunleavy et al., 2019; Tudor Car et al., 2021).

4.3 Assessment, Competency Tracking, and Generative AI

Additionally, the COVID-19 pandemic has accelerated the adoption of these interactive digital systems in medical education. Institutions were prompted to quickly shift to online and distance learning, highlighting the necessity for innovative instructional approaches. This ongoing transition stresses the importance of continuous improvements in digital technologies as medical education adapts to current and future challenges. By embracing these advancements, educators can can enhance the quality of medical training and better prepare students for the complexities of patient care (Machleid et al., 2020; Wong et al., 2021).

Generative AI adds urgency to this agenda. It can support drafting, feedback, question generation, translation, and simulated dialogue, yet it may produce plausible errors, reproduce bias, expose confidential information, and obscure whether learners have achieved independent competence. Medical schools should therefore define permitted uses, require disclosure where appropriate, protect patient data, teach verification and source appraisal, and preserve assessments in which students demonstrate unaided reasoning. Human-centred educational governance is preferable to either uncritical adoption or blanket prohibition (UNESCO, 2023; WHO, 2024).

5. PUBLIC HEALTH OPPORTUNITIES ENABLED BY DIGITAL INTEGRATION

Digital integration offers transformative opportunities in public health by enhancing workforce capacity building, improving disease surveillance, and facilitating evidence-based policy-making (Table 1). The rapid adoption of digital health technologies is changing how public health professionals are trained, how data is collected and analyzed, and how health information is communicated to diverse communities (Francombe et al., 2022; Iyamu et al., 2021).

5.1 Workforce Capacity and Continuing Professional Development

Digital health technologies play a significant role in healthcare workers' continuous professional development (CPD). E-learning platforms enable targeted educational interventions, facilitating timely access to the latest knowledge and skills necessary to address evolving public health challenges. For instance, the introduction of mobile-based platforms has allowed health professionals, particularly in low- and middle-income countries (LMICs), to engage in peer-to-peer learning and mentorship from afar. This approach fosters a collaborative learning environment that enhances skill acquisition and retention while addressing workforce shortages, eventually leading to better health outcomes (Dunleavy et al., 2019; Wong et al., 2021).

Moreover, an increasing number of academic institutions and public health organizations are recognizing the importance of integrating digital competencies into training programs. Research indicates that having health workforce members proficient in digital health literacy is essential for ensuring effective service delivery and improved patient outcomes. Telemedicine training has gained prominence in medical education, enabling students to develop skills that prepare them for the realities of remote healthcare provision. Such workforce development initiatives also support an agile and responsive public health system capable of adapting to emergency scenarios, as demonstrated during the COVID-19 pandemic (Konttila et al., 2019; Waseh & Dicker, 2019).

Digital integration can enhances disease surveillance and outbreak preparedness by enabling real-time data collection and analysis. The deployment of health information systems enables public health agencies to identify emerging diseases and monitor outbreaks swiftly. The World Health Organization emphasizes that integrated digital solutions can help monitor health indicators, facilitate timely interventions, and mobilize resources to mitigate the spread of disease. For instance, during the COVID-19 pandemic, many countries leveraged digital technologies for contact tracing and effective communication of health advisories, which proved instrumental in controlling virus transmission (Brownstein et al., 2009; Budd et al., 2020).

5.2 Surveillance, Preparedness, and Community Engagement

Furthermore, engaging communities through digital platforms allows for more effective population health management. Utilizing mHealth applications can empower individuals to take charge of their health, manage chronic conditions, and report symptoms, thus creating a more participatory health ecosystem. Community engagement is enhanced as public health professionals can target messaging and interventions based on real-time data, making strategies more relevant and context-specific. Digital literacy initiatives foster greater participation from vulnerable populations, ensuring that health interventions are inclusive and equitable (Crawford & Serhal, 2020; van Kessel et al., 2022).

The integration of digital health tools into public health frameworks enables evidence-based policy-making by providing health authorities with timely and accurate data for informed decision-making. Such data facilitates the identification of health trends, resource allocation, and strategically oriented interventions aimed at improving health outcomes. The use of big data analytics and artificial intelligence further supports this process, allowing for predictive modeling that can inform health strategies before health crises escalate (Iyamu et al., 2022; WHO, 2019).

5.3 Evidence-Informed Policy and Health Communication

Moreover, effective health communication is crucial in promoting public understanding of health initiatives and policies. Digital platforms enable the swift dissemination of health information to broad audiences, leveraging social media and mobile applications to engage and educate communities on important health topics. These tools also enable feedback loops in which community members can share their concerns and experiences, fostering collaboration between public health agencies and the populations they serve (Fagherazzi et al., 2020; Francombe et al., 2022).

The central policy question is not whether more data can be produced, but whether data improve decisions without imposing disproportionate burdens or risks. Public-health dashboards should be connected to explicit action thresholds, accountable decision-makers, and feedback to reporting facilities and communities. Evaluation should examine timeliness, completeness, representativeness, false alerts, response time, resource allocation, and distributional effects. A technically impressive surveillance platform that systematically misses rural populations or produces no operational response has limited public-health value.

Table 2. Recommended indicators for evaluating integrated digital health and education

Evaluation domain	Illustrative indicators
Access and equity	Reach by geography, income, disability, language, sex/gender, age; device/connectivity barriers; completion or service-use gaps
Clinical and public-health effectiveness	Diagnostic or treatment outcomes; adherence; referral completion; timeliness; outbreak detection-to-response interval
Educational effectiveness	Knowledge, skills, clinical reasoning, communication, transfer to practice, retention, and competency progression
Safety and ethics	Adverse events, privacy incidents, cybersecurity events, inappropriate automation, subgroup error rates, complaints
Implementation	Adoption, fidelity, usability, workload, workflow disruption, abandonment, training completion, support requests
Economic and sustainability	Total lifecycle cost, cost-effectiveness, recurrent financing, maintenance capacity, vendor dependency, environmental burden

6. CHALLENGES AND BARRIERS TO DIGITAL HEALTH ADOPTION

The rapid integration of digital technologies into healthcare poses significant technical and operational challenges that need to be addressed to realize the full potential of these innovations. This section of the chapter involves various issues, including infrastructure limitations, workforce gaps, and access inequities, particularly within underserved populations (Greenhalgh et al., 2017; Iyamu et al., 2022).

6.1 Infrastructure, Interoperability, and Integration

One of the foremost challenges to the adoption of digital health technologies is inadequate infrastructure. Many healthcare facilities, especially in low-resource settings, lack the necessary technological framework to support digital initiatives. Issues such as unreliable internet connectivity, inadequate electricity supply, and outdated equipment hinder the effective use of digital health tools. Furthermore, existing healthcare information systems often suffer from interoperability issues; many systems cannot communicate effectively with one another due to heterogeneous data formats and architecture, leading to fragmented patient care. A systematic review found that fragmented health information systems pose important obstacles to the implementation of integrated care solutions, which are essential for delivering comprehensive healthcare (Labrique et al., 2018; WHO, 2019).

Additionally, the lack of standardized protocols complicates the integration of new technologies into existing practices. For example, when adopting electronic medical records (EMRs), physicians frequently encounter barriers to system interoperability across different healthcare settings, which complicate the sharing of vital patient information. This lack of interoperability not only impedes efficient workflows but also affects overall patient safety, as critical data may be lost or miscommunicated across various touchpoints in the healthcare system (Health Level Seven International, 2019; Torab-Miandoab et al., 2023).

Interoperability is not solely a technical specification. Syntactic exchange may allow systems to transmit fields, while semantic interoperability ensures that the receiving system interprets meaning consistently; organisational interoperability determines whether governance, workflow, consent, and accountability support the exchange. Standards such as FHIR provide an important foundation, but implementation profiles, terminology services, identifiers, data-quality rules, and institutional agreements are still required. Procurement that ignores these dependencies can create expensive vendor lock-in and duplicate data entry (Health Level Seven International, 2019; Torab-Miandoab et al., 2023).

6.2 Workforce Readiness and Organisational Change

Beyond technical limitations, workforce skills gaps are pronounced barriers to successful digital health adoption. Many healthcare professionals report feeling unprepared to use new technologies, and a systematic review highlights a common perception of insufficient training in digital tools. These skills gaps are exacerbated by variations in digital literacy and comfort with technology among different healthcare professionals, leading to inconsistent implementation across teams. Additionally, institutions' readiness to incorporate digital solutions into their operations is often uneven. Institutional strategies for technology adoption can significantly influence staff attitudes and acceptance levels; leadership support and investment in training are critical for fostering a culture that embraces technological change (Konttila et al., 2019; Nazeha et al., 2020).

Resistance to change is another considerable barrier, particularly among healthcare practitioners who may perceive the introduction of digital health technologies as a threat. Concerns over job security, loss of autonomy, and potential negative impacts on the patient-provider relationship contribute to skepticism surrounding digital health innovations. Physicians and healthcare workers often express apprehension about the impact of technology on their clinical practice, which, if not addressed, may result in nonadoption or abandonment of new systems. Encouragement from management, clear communication about the benefits of digital tools, and ongoing support can help alleviate these concerns and facilitate smoother transitions (Greenhalgh et al., 2017; Kruse et al., 2018).

Access inequities pose a significant challenge in the digital health landscape. The "digital divide," characterized by disparities in access to technology and the internet, disproportionately affects underserved populations, exacerbating existing health disparities. Research shows that individuals in rural or low-income urban settings often lack reliable access to necessary technologies, limiting their ability to engage with digital health solutions. This divide hinders the implementation of essential public health initiatives that effectively reach these vulnerable communities (Crawford & Serhal, 2020; Rodriguez et al., 2020; Sieck et al., 2021).

6.3 Digital Exclusion and Financial Sustainability

Furthermore, health literacy plays a crucial role in determining an individual's ability to utilize digital health resources effectively. Many individuals may lack the skills necessary to navigate digital platforms or may feel intimidated by technology, leading to underutilization of valuable health services. Public health interventions should be designed with considerations for varying levels of digital literacy, ensuring they do not inadvertently exclude or alienate those who stand to benefit the most. Targeted efforts, such as community training programs, are essential for empowering underserved populations to leverage digital health tools (van Kessel et al., 2022).

Financial sustainability must be assessed across the full lifecycle: connectivity, devices, licences, integration, cybersecurity, support, training, upgrades, evaluation, and eventual decommissioning. Pilot funding often covers development but not recurrent costs or institutional ownership. A credible business case should compare the digital intervention with realistic alternatives, identify who pays and who benefits, test affordability under scale, and avoid shifting costs to patients through data charges, devices, or travel to digitally equipped centres. These considerations are particularly important in low-resource settings, where opportunity costs can displace essential services (Labrique et al., 2018; WHO, 2019).

7. ETHICAL, LEGAL, AND GOVERNANCE CONSIDERATIONS

As digital health technologies proliferate within healthcare and education, a spectrum of ethical, legal, and governance considerations necessitates meticulous examination. Issues related to data privacy, security, informed consent, and regulatory frameworks are critical to ensuring that the benefits of digital health are realized while protecting individual rights and public trust (Vayena et al., 2018; WHO, 2021b).

7.1 Privacy, Cybersecurity, and Consent

Data privacy and security are paramount in the digital health landscape. The protection of sensitive patient information and educational data is governed by stringent legal frameworks such as the General Data Protection Regulation (GDPR) in the European Union, which mandates enhanced data privacy protections. As healthcare increasingly hinges on big data analytics and artificial intelligence (AI), there are concerns about how personal health data is utilized, stored, and shared. Failure to ensure robust data security can lead to significant breaches of confidentiality and privacy, resulting in both ethical and legal repercussions for healthcare institutions (European Parliament & Council, 2016).

Recent studies indicate a growing concern among patients regarding the use of AI and digital tools in managing sensitive health data. This anxiety highlights the need for transparent data governance practices, including explicit consent protocols for data usage. Informed consent should not only be obtained but also clearly understood by patients, particularly as technological complexity increases. Implementing strong security measures and fostering a culture of privacy protection are essential to mitigate these risks and maintain patient trust (WHO, 2021b).

Consent is necessary but rarely sufficient for complex data ecosystems. Patients and learners may have little practical ability to understand secondary uses, model training, cross-border processing, vendor access, or future linkage. Governance should therefore combine lawful processing, data minimisation, purpose limitation, access controls, audit trails, retention schedules, breach response, and independent oversight. Educational data require equal care because performance analytics and proctoring systems can affect progression, professional reputation, and psychological safety (European Parliament & Council, 2016; UNESCO, 2023).

7.2 Fairness, Transparency, and Professional Responsibility

The ethical implications of digital health interventions involve concerns, particularly related to equity and access. The deployment of digital technologies can inadvertently exacerbate health disparities if underserved populations lack access to necessary resources or digital literacy. The balance between technological advancement and ethical obligations, such as promoting equitable access and managing diverse patient populations, remains a crucial challenge (Crawford & Serhal, 2020; Rajkomar et al., 2018).

Algorithmic fairness requires empirical testing rather than assurances of neutrality. Bias can enter through unrepresentative data, historical inequities, label choice, missingness, measurement error, deployment context, and differential access to follow-up services. The discovery of racial bias in a widely used population-health algorithm illustrates how an apparently objective cost proxy can encode unequal access to care (Obermeyer et al., 2019). Developers and health systems should define relevant subgroups, report performance and error trade-offs, investigate causes of disparity, and involve affected communities in decisions about acceptable risk (Rajkomar et al., 2018; Wiens et al., 2019).

Legal considerations also come into play, particularly concerning the responsibilities of healthcare providers when implementing AI systems. Clinicians often grapple with defining accountability when AI influences clinical decision-making. Understanding the distinction between ethical and legal obligations in algorithmic contexts is critical for clinicians, particularly when navigating issues related to malpractice liability and informed consent. As AI applications continue to proliferate, regulatory frameworks should evolve to provide clear guidance on these issues, ensuring that both patients' rights and practitioners' responsibilities are clearly delineated (European Parliament & Council, 2024; WHO, 2021b).

7.3 Regulation, Validation, and Accountability

Establishing effective regulatory frameworks is crucial for the ethical governance of digital health interventions. Current regulations often lag behind technological advancements, creating gaps in oversight and accountability that could undermine the effectiveness of digital health solutions. There is an urgent need for harmonized policies that address the ethical use of AI in healthcare, ensuring that all stakeholders (patients, providers, and researchers) are protected (NIST, 2023; WHO, 2021b).

Standardization of practices within digital health is also a pressing concern. Disparate approaches to data handling and ethical considerations can lead to inconsistencies that undermine public health efforts. Developing standardized guidelines for the collection, storage, and sharing of health data can foster a more secure digital healthcare environment and facilitate compliance with existing and future regulations. Furthermore, these regulations should not only focus on compliance but also emphasize transparency and stakeholder engagement, empowering patients to make informed decisions regarding their digital health information (FDA et al., 2021; Mathews et al., 2019).

Explainability should not be treated as a universal cure for unsafe AI. Post-hoc explanations can be unstable, incomplete, or persuasive without being faithful to the model's reasoning. Depending on the clinical risk, stronger safeguards may include simpler interpretable models, prospective validation, calibrated uncertainty, human review, fail-safe escalation, restricted indications, and continuous outcome monitoring. Governance frameworks from WHO, NIST, and regulators converge on lifecycle accountability: responsibilities should be assigned before procurement and continue through deployment, modification, incident response, and retirement (Ghassemi et al., 2021; NIST, 2023; WHO, 2024).

8. POLICY IMPLICATIONS FOR PUBLIC HEALTH AND MEDICAL EDUCATION

As digital innovation rapidly transforms public health and medical education, several policy implications arise that demand strategic alignment with educational and public health standards. In particular, effective accreditation practices, quality assurance mechanisms, and institutional support are crucial for fostering sustainable digital health initiatives (WHO, 2021a).

8.1 Standards, Architecture, and Curriculum Alignment

Aligning digital innovation with recognized educational and public health standards is foundational to ensuring the effective integration of new technologies into existing systems. This alignment helps ensure that health professionals are trained to adapt to evolving technologies and that digital solutions sufficiently address pressing health challenges. A framework such as the European Digital Competence Framework for Educators can offer a structured approach for educators and health professionals to develop the competencies needed to navigate an increasingly digital landscape. Effective alignment also requires integrating competencies relevant to both sectors, fostering a more holistic view of how digital tools can enhance health education and service delivery (WHO, 2019; WHO, 2023).

National and institutional policies should be architecture-aware. They should specify core registries, identifiers, terminology, interoperability profiles, security baselines, data stewardship roles, and procurement requirements that reduce duplication. New platforms should demonstrate compatibility with approved standards and provide export, migration, and audit capabilities. Policies should also distinguish experimental tools from systems used for clinical or educational decisions, applying proportionate evidence and oversight requirements (Health Level Seven International, 2019; NIST, 2023; WHO, 2023).

Additionally, embracing a competency-based approach to medical education can enhance the relevance and quality of training programs. Competency frameworks can cultivate essential skills among healthcare providers, such as data literacy, analytical skills, and digital engagement, equipping them to work effectively in digital health environments. Understanding and integrating these competencies within curricula will create a workforce capable of navigating the complexities of modern healthcare demands, thereby reinforcing the value of digital health interventions (Khurana et al., 2022; Nazeha et al., 2020).

8.2 Accreditation, Quality Assurance, and Institutional Support

Effective accreditation and quality assurance processes are vital for maintaining the integrity of public health and medical education programs, particularly amidst rising enrollment rates and the expansion of institutions. Accreditation serves not only to uphold educational quality but also to ensure that all programs meet established standards that reflect both educational and clinical competencies. In Poland, for instance, the rapid increase in medical student admissions necessitated rigorous accreditation processes to maintain educational quality amid workforce shortages. If accreditation standards are not integrated with the evolving demands of digital health, educational programs risk producing graduates who may lack the requisite skills to succeed in a digitally transformed healthcare landscape (Mathews et al., 2019; Tudor Car et al., 2021).

Institutional support plays a crucial role in fostering an environment conducive to digital innovation. This support extends beyond funding; institutions should cultivate a culture that promotes adaptability, ongoing professional development, and the use of digital health technologies in training. Such institutional backing is essential to ensure that educators feel empowered to integrate digital elements into their curricula and that students are equipped with the skills necessary for effective practice. Policies emphasizing continuous improvement and quality monitoring further support sustainability in medical education (Greenhalgh et al., 2017).

8.3 Financing and Public-Private Partnerships

Sustainable funding models are paramount for the long-term viability of digital health and education initiatives. In many contexts, public funding is insufficient to meet rising demand for digital health services. This reinforces the need for innovative funding strategies that can enable continued investment in digital infrastructure and training resources. For instance, initiatives in low- and middle-income countries may benefit from a blend of public funds, private investment, and international grants to foster sustainable health innovation (Labrique et al., 2018; WHO, 2019).

Accreditation bodies can accelerate safe adoption by requiring programmes to demonstrate digital-health outcomes, faculty capability, authentic assessment, patient and learner protection, and quality improvement. Health facilities and universities should establish multidisciplinary digital governance committees with clinical, educational, informatics, legal, cybersecurity, procurement, ethics, and community representation. Such committees should not merely approve technology; they should examine need, alternatives, evidence, equity, workflow, cost, accountability, and exit strategy.

9. FUTURE DIRECTIONS AND STRATEGIC RECOMMENDATIONS

As the landscape of public health and medical education continues to evolve with digital technology, it is crucial to identify future directions that will enhance integration, effectiveness, and sustainability. This analysis highlights emerging innovations, research priorities, the need to bridge evidence gaps, and coordinated strategies to develop public health capacity and workforce competency (Rajpurkar et al., 2022; WHO, 2024).

9.1 Emerging Technologies and Research Priorities

Emerging digital innovations hold the potential to revolutionize public health and medical education. Technologies such as artificial intelligence, telemedicine, and mobile health (mHealth) applications facilitate more efficient patient care and educational delivery, thus enhancing overall health outcomes. Prioritizing research into these technologies is essential to determine their effectiveness, usability, and best practices for implementation. For example, understanding how AI can support diagnostic processes can inform clearer protocols for its adoption across healthcare settings (Topol, 2019; WHO, 2024).

Moreover, advances in health information systems are vital for data collection and analysis, which directly influence public health outcomes. Integrating robust monitoring systems powered by real-time data analytics can improve disease surveillance and health interventions. Research should focus on developing methodologies to evaluate the impact of these digital interventions on health practices and educational curricula, eventually guiding policymakers in their decision-making (Brownstein et al., 2009; Francombe et al., 2022).

To inform effective digital health interventions, evidence gaps should be addressed. Current literature indicates a diversity of barriers to the successful implementation of digital health technologies, including workforce education levels, infrastructural limitations, and resistance to change. Therefore, targeted research is required to evaluate the effectiveness of various digital health strategies in specific contexts, accounting for nuances across populations and geographical regions. For example, examining the challenges faced by health workers in Ethiopia regarding the usage of electronic health information systems highlights crucial areas for development (Iyamu et al., 2022; Mathews et al., 2019).

9.2 Evaluation and Phased Implementation

A practical roadmap should proceed through six linked phases. Phase 1 defines the health or learning problem, affected population, baseline performance, and plausible non-digital alternatives. Phase 2 co-designs the intervention with users and communities, maps workflow and data flows, and conducts privacy, safety, equity, and cybersecurity assessments. Phase 3 establishes technical, usability, educational, and clinical validity using representative data and explicit success criteria. Phase 4 implements in a bounded setting with training, support, incident reporting, and process evaluation. Phase 5 evaluates outcomes, cost, subgroup effects, unintended consequences, and sustainability. Phase 6 scales only when evidence and implementation capacity justify expansion, while preserving monitoring and the ability to modify or withdraw the system (Greenhalgh et al., 2017; Mathews et al., 2019; NIST, 2023).

Evaluating the effectiveness of existing digital education programs is equally important. Metrics for success should go beyond user satisfaction and include long-term impacts on health outcomes and professional competency. This evaluation should include feedback from both educators and health professionals to refine educational methodologies and align them with technological advancements continually. Such assessment frameworks could strengthen the integration of digital training into professional development programs, ensuring a digitally competent workforce prepared to meet modern healthcare demands (Agarwal et al., 2016; Mathews et al., 2019).

Implementing coordinated strategies is essential for enhancing public health capacity and workforce competency in the digital age. This necessitates the design of comprehensive training and development programs that integrate digital skills into existing curricula. Institutions should prioritize continuous professional development and digital literacy training to equip healthcare professionals to effectively use digital tools and innovations (Konttila et al., 2019; Wong et al., 2021).

9.3 Coordinated Capacity Building and Partnerships

Public-private partnerships can play a crucial role in funding and supporting these initiatives. Collaborations can leverage resources, expertise, and technology from both sectors to bolster digital health interventions, ensuring that they are sustainable and accessible across different communities. For instance, in teaching hospitals, partnerships with tech firms can facilitate the provision of advanced digital resources while fostering real-time feedback and user-experience-driven adaptations (Labrique et al., 2018).

Also, engaging stakeholders (including policymakers, educators, and health professionals) in the development of digital health strategies is vital to ensure that priorities align with community health needs. This collaboration will foster a sense of ownership and accountability, driving forward the implementation of practical digital health innovations that can expand public health capacity and improve care delivery (Greenhalgh et al., 2017; WHO, 2021a).

Research priorities follow from this roadmap. Studies should use comparative designs where feasible, report context and implementation mechanisms, include longer follow-up, and measure outcomes that matter to patients, learners, professionals, and health systems. AI studies require external and prospective validation; educational studies should connect learning to practice; public-health studies should assess representativeness and actionability; and all evaluations should examine cost and equity. Reporting standards such as mERA improve transparency, but funders and journals should also require protocols, limitations, negative findings, and implementation data to reduce selective claims (Agarwal et al., 2016; Liu et al., 2019; Nagendran et al., 2020).

10. CONCLUSION

Digital technologies are increasingly central to the transformation of healthcare delivery and medical education, with significant implications for public health systems and workforce development. This chapter has highlighted how integrated digital health tools and interactive educational platforms can enhance disease surveillance, service delivery, competency-based training, and evidence-informed decision-making. At the same time, it has stressed persistent challenges related to infrastructure, digital literacy, interoperability, equity, ethical governance, and regulatory alignment that may limit effective adoption if not adequately addressed. The discussion emphasizes that realizing the full public health potential of digital integration requires coordinated strategies that align technology, education, and policy objectives. Strengthening institutional capacity, investing in continuous professional development, promoting inclusive access, and fostering multi-stakeholder collaboration are critical for sustainability. Thus, strategic and context-sensitive integration of digital technologies can support resilient health systems, a digitally competent workforce, and improved population health outcomes.

The comparative evidence supports a cautious but constructive conclusion: digital tools can strengthen access, learning, surveillance, and decision-making, yet benefits are contingent rather than automatic. Technology, workforce competence, workflow, infrastructure, governance, financing, and public trust operate as an interdependent system. Weakness in one component can neutralise or reverse gains in another.

The most defensible strategy is therefore problem-led, standards-based, evidence-generating, and equity-oriented. Institutions should build digital competence as a core professional capability, validate tools across settings and population groups, protect data and autonomy, monitor real-world outcomes, and scale only with sustainable operational capacity.

REFERENCES

Agarwal, S., LeFevre, A. E., Lee, J., L'Engle, K., Mehl, G., Sinha, C., & Labrique, A. (2016). Guidelines for reporting of health interventions using mobile phones: Mobile health (mHealth) evidence reporting and assessment (mERA) checklist. *BMJ (Clinical Research Ed.)*, *352*, i1174. DOI: 10.1136/bmj.i1174 PMID: 26988021

Ammenwerth, E., Schnell-Inderst, P., Machan, C., & Siebert, U. (2008). The effect of electronic prescribing on medication errors and adverse drug events: A systematic review. *Journal of the American Medical Informatics Association : JAMIA*, *15*(5), 585–600. DOI: 10.1197/jamia.M2667 PMID: 18579832

Brownstein, J. S., Freifeld, C. C., & Madoff, L. C. (2009). Digital disease detection—Harnessing the Web for public health surveillance. *The New England Journal of Medicine*, *360*(21), 2153–2157. DOI: 10.1056/NEJMp0900702 PMID: 19423867

Budd, J., Miller, B. S., Manning, E. M., Lampos, V., Zhuang, M., Edelstein, M., Rees, G., Emery, V. C., Stevens, M. M., Keegan, N., Short, M. J., Pillay, D., Manley, E., Cox, I. J., Heymann, D., Johnson, A. M., & McKendry, R. A. (2020). Digital technologies in the public-health response to COVID-19. *Nature Medicine*, *26*(8), 1183–1192. DOI: 10.1038/s41591-020-1011-4 PMID: 32770165

Campanella, P., Lovato, E., Marone, C., Fallacara, L., Mancuso, A., Ricciardi, W., & Specchia, M. L. (2016). The impact of electronic health records on healthcare quality: A systematic review and meta-analysis. *European Journal of Public Health*, *26*(1), 60–64. DOI: 10.1093/eurpub/ckv122 PMID: 26136462

Carrillo de Albornoz, S., Sia, K.-L., & Harris, A. (2022). The effectiveness of teleconsultations in primary care: Systematic review. *Family Practice*, *39*(1), 168–182. DOI: 10.1093/fampra/cmab077 PMID: 34278421

Cook, D. A., Levinson, A. J., Garside, S., Dupras, D. M., Erwin, P. J., & Montori, V. M. (2008). Internet-based learning in the health professions: A meta-analysis. *Journal of the American Medical Association*, *300*(10), 1181–1196. DOI: 10.1001/jama.300.10.1181 PMID: 18780847

Crawford, A., & Serhal, E. (2020). Digital health equity and COVID-19: The innovation curve cannot reinforce the social gradient of health. *Journal of Medical Internet Research*, *22*(6), e19361. DOI: 10.2196/19361 PMID: 32452816

Dunleavy, G., Nikolaou, C. K., Nifakos, S., Atun, R., Law, G. C. Y., & Tudor Car, L. (2019). Mobile digital education for health professions: Systematic review and meta-analysis by the Digital Health Education Collaboration. *Journal of Medical Internet Research*, *21*(2), e12937. DOI: 10.2196/12937 PMID: 30747711

European Parliament and Council of the European Union. (2016). Regulation (EU) 2016/679 on the protection of natural persons with regard to the processing of personal data and on the free movement of such data (General Data Protection Regulation). *Official Journal of the European Union, L, 119*, 1–88.

European Parliament and Council of the European Union. (2024). Regulation (EU) 2024/1689 laying down harmonised rules on artificial intelligence (Artificial Intelligence Act). *Official Journal of the European Union, L 2024/1689.*

Fagherazzi, G., Goetzinger, C., Rashid, M. A., Aguayo, G. A., & Huiart, L. (2020). Digital health strategies to fight COVID-19 worldwide: Challenges, recommendations, and a call for papers. *Journal of Medical Internet Research, 22*(6), e19284. DOI: 10.2196/19284 PMID: 32501804

Francombe, J., Booth, A., & Buckley Woods, H. (2022). Digital innovations for public health functions: A scoping review of international evidence. *JMIR Public Health and Surveillance, 8*(7), e34605. DOI: 10.2196/34605 PMID: 35605152

Ghassemi, M., Oakden-Rayner, L., & Beam, A. L. (2021). The false hope of current approaches to explainable artificial intelligence in health care. *The Lancet. Digital Health, 3*(11), e745–e750. DOI: 10.1016/S2589-7500(21)00208-9 PMID: 34711379

Greenhalgh, T., Wherton, J., Papoutsi, C., Lynch, J., Hughes, G., Hinder, S., Fahy, N., Procter, R., & Shaw, S. (2017). Beyond adoption: A new framework for theorizing and evaluating nonadoption, abandonment, and challenges to the scale-up, spread, and sustainability of health and care technologies. *Journal of Medical Internet Research, 19*(11), e367. DOI: 10.2196/jmir.8775 PMID: 29092808

Health Level Seven International. (2019). *FHIR Release 4 (Version 4.0.1)*. https://hl7.org/fhir/R4/

Iyamu, I., Gómez-Ramírez, O., Xu, A. X., Chang, H.-J., Watt, S., McKee, G., & Gilbert, M. (2022). Challenges in the development of digital public health interventions and mapped solutions: Findings from a scoping review. *Digital Health, 8*, 20552076221102255. DOI: 10.1177/20552076221102255 PMID: 35656283

Iyamu, I., Xu, A. X., Gómez-Ramírez, O., Ablona, A., Chang, H.-J., McKee, G., & Gilbert, M. (2021). Defining digital public health and the role of digitization, digitalization, and digital transformation: Scoping review. *JMIR Public Health and Surveillance, 7*(11), e30399. DOI: 10.2196/30399 PMID: 34842555

Keesara, S., Jonas, A., & Schulman, K. (2020). Covid-19 and health care's digital revolution. *The New England Journal of Medicine, 382*(23), e82. DOI: 10.1056/NEJMp2005835 PMID: 32240581

Khurana, M. P., Raaschou-Pedersen, D. E., Kurtzhals, J., Bardram, J. E., Ostrowski, S. R., & Bundgaard, J. S. (2022). Digital health competencies in medical school education: A scoping review and Delphi method study. *BMC Medical Education*, *22*(1), 129. DOI: 10.1186/s12909-022-03163-7 PMID: 35216611

Kononowicz, A. A., Woodham, L. A., Edelbring, S., Stathakarou, N., Davies, D., Saxena, N., Tudor Car, L., Carlstedt-Duke, J., Car, J., & Zary, N. (2019). Virtual patient simulations in health professions education: Systematic review and meta-analysis by the Digital Health Education Collaboration. *Journal of Medical Internet Research*, *21*(7), e14676. DOI: 10.2196/14676 PMID: 31267981

Konttila, J., Siira, H., Kyngäs, H., Lahtinen, M., Elo, S., Kääriäinen, M., Kaakinen, P., Oikarinen, A., Yamakawa, M., Fukui, S., Utsumi, M., Higami, Y., Higuchi, A., & Mikkonen, K. (2019). Healthcare professionals' competence in digitalisation: A systematic review. *Journal of Clinical Nursing*, *28*(5–6), 745–761. DOI: 10.1111/jocn.14710 PMID: 30376199

Kruse, C. S., Karem, P., Shifflett, K., Vegi, L., Ravi, K., & Brooks, M. (2018). Evaluating barriers to adopting telemedicine worldwide: A systematic review. *Journal of Telemedicine and Telecare*, *24*(1), 4–12. DOI: 10.1177/1357633X16674087 PMID: 29320966

Kyaw, B. M., Saxena, N., Posadzki, P., Vseteckova, J., Nikolaou, C. K., George, P. P., Divakar, U., Masiello, I., Kononowicz, A. A., Zary, N., & Tudor Car, L. (2019). Virtual reality for health professions education: Systematic review and meta-analysis by the Digital Health Education Collaboration. *Journal of Medical Internet Research*, *21*(1), e12959. DOI: 10.2196/12959 PMID: 30668519

Labrique, A. B., Wadhwani, C., Williams, K. A., Lamptey, P., Hesp, C., Luk, R., & Aerts, A. (2018). Best practices in scaling digital health in low- and middle-income countries. *Globalization and Health*, *14*(1), 103. DOI: 10.1186/s12992-018-0424-z PMID: 30390686

Liu, K., Zhang, W., Li, W., Wang, T., & Zheng, Y. (2023). Effectiveness of virtual reality in nursing education: A systematic review and meta-analysis. *BMC Medical Education*, *23*(1), 710. DOI: 10.1186/s12909-023-04662-x PMID: 37770884

Liu, X., Faes, L., Kale, A. U., Wagner, S. K., Fu, D. J., Bruynseels, A., Mahendiran, T., Moraes, G., Shamdas, M., Kern, C., Ledsam, J. R., Schmid, M. K., Balaskas, K., Topol, E. J., Bachmann, L. M., Keane, P. A., & Denniston, A. K. (2019). A comparison of deep learning performance against health-care professionals in detecting diseases from medical imaging: A systematic review and meta-analysis. *The Lancet. Digital Health*, *1*(6), e271–e297. DOI: 10.1016/S2589-7500(19)30123-2 PMID: 33323251

Machleid, F., Kaczmarczyk, R., Johann, D., Balčiūnas, J., Atienza-Carbonell, B., von Maltzahn, F., & Mosch, L. (2020). Perceptions of digital health education among European medical students: Mixed methods survey. *Journal of Medical Internet Research*, *22*(8), e19827. DOI: 10.2196/19827 PMID: 32667899

Marcolino, M. S., Oliveira, J. A. Q., D'Agostino, M., Ribeiro, A. L., Alkmim, M. B. M., & Novillo-Ortiz, D. (2018). The impact of mHealth interventions: Systematic review of systematic reviews. *JMIR mHealth and uHealth*, *6*(1), e23. DOI: 10.2196/mhealth.8873 PMID: 29343463

Mathews, S. C., McShea, M. J., Hanley, C. L., Ravitz, A., Labrique, A. B., & Cohen, A. B. (2019). Digital health: A path to validation. *npj. Digital Medicine*, *2*(1), 38. DOI: 10.1038/s41746-019-0111-3 PMID: 31304384

Meskó, B., Drobni, Z., Bényei, É., Gergely, B., & Győrffy, Z. (2017). Digital health is a cultural transformation of traditional healthcare. *mHealth*, *3*, 38. DOI: 10.21037/mhealth.2017.08.07 PMID: 29184890

Monaghesh, E., & Hajizadeh, A. (2020). The role of telehealth during COVID-19 outbreak: A systematic review based on current evidence. *BMC Public Health*, *20*(1), 1193. DOI: 10.1186/s12889-020-09301-4 PMID: 32738884

Nagendran, M., Chen, Y., Lovejoy, C. A., Gordon, A. C., Komorowski, M., Harvey, H., Topol, E. J., Ioannidis, J. P. A., Collins, G. S., & Maruthappu, M. (2020). Artificial intelligence versus clinicians: Systematic review of design, reporting standards, and claims of deep learning studies. *BMJ (Clinical Research Ed.)*, *368*, m689. DOI: 10.1136/bmj.m689 PMID: 32213531

National Institute of Standards and Technology. (2023). *Artificial Intelligence Risk Management Framework (AI RMF 1.0)* (NIST AI 100-1). DOI: 10.6028/NIST.AI.100-1

Nazeha, N., Pavagadhi, D., Kyaw, B. M., Car, J., Jimenez, G., & Tudor Car, L. (2020). A digitally competent health workforce: Scoping review of educational frameworks. *Journal of Medical Internet Research*, *22*(11), e22706. DOI: 10.2196/22706 PMID: 33151152

Obermeyer, Z., Powers, B., Vogeli, C., & Mullainathan, S. (2019). Dissecting racial bias in an algorithm used to manage the health of populations. *Science*, *366*(6464), 447–453. DOI: 10.1126/science.aax2342 PMID: 31649194

Peek, N., Sujan, M., & Scott, P. (2020). Digital health and care in pandemic times: Impact of COVID-19. *BMJ Health & Care Informatics*, *27*(1), e100166. DOI: 10.1136/bmjhci-2020-100166 PMID: 32565418

Pei, L., & Wu, H. (2019). Does online learning work better than offline learning in undergraduate medical education? A systematic review and meta-analysis. *Medical Education Online*, *24*(1), 1666538. DOI: 10.1080/10872981.2019.1666538 PMID: 31526248

Rajkomar, A., Hardt, M., Howell, M. D., Corrado, G., & Chin, M. H. (2018). Ensuring fairness in machine learning to advance health equity. *Annals of Internal Medicine*, *169*(12), 866–872. DOI: 10.7326/M18-1990 PMID: 30508424

Rajpurkar, P., Chen, E., Banerjee, O., & Topol, E. J. (2022). AI in health and medicine. *Nature Medicine*, *28*(1), 31–38. DOI: 10.1038/s41591-021-01614-0 PMID: 35058619

Regmi, K., & Jones, L. (2020). A systematic review of the factors—enablers and barriers—affecting e-learning in health sciences education. *BMC Medical Education*, *20*(1), 91. DOI: 10.1186/s12909-020-02007-6 PMID: 32228560

Rodriguez, J. A., Clark, C. R., & Bates, D. W. (2020). Digital health equity as a necessity in the 21st Century Cures Act era. *Journal of the American Medical Association*, *323*(23), 2381–2382. DOI: 10.1001/jama.2020.7858 PMID: 32463421

Salathé, M., Bengtsson, L., Bodnar, T. J., Brewer, D. D., Brownstein, J. S., Buckee, C., Campbell, E. M., Cattuto, C., Khandelwal, S., Mabry, P. L., & Vespignani, A. (2012). Digital epidemiology. *PLoS Computational Biology*, *8*(7), e1002616. DOI: 10.1371/journal.pcbi.1002616 PMID: 22844241

Sieck, C. J., Sheon, A., Ancker, J. S., Castek, J., Callahan, B., & Siefer, A. (2021). Digital inclusion as a social determinant of health. *npj. Digital Medicine*, *4*(1), 52. DOI: 10.1038/s41746-021-00413-8 PMID: 33731887

Sutton, R. T., Pincock, D., Baumgart, D. C., Sadowski, D. C., Fedorak, R. N., & Kroeker, K. I. (2020). An overview of clinical decision support systems: Benefits, risks, and strategies for success. *npj. Digital Medicine*, *3*(1), 17. DOI: 10.1038/s41746-020-0221-y PMID: 32047862

Topol, E. J. (2019). High-performance medicine: The convergence of human and artificial intelligence. *Nature Medicine*, *25*(1), 44–56. DOI: 10.1038/s41591-018-0300-7 PMID: 30617339

Torab-Miandoab, A., Samad-Soltani, T., Jodati, A., & Rezaei-Hachesu, P. (2023). Interoperability of heterogeneous health information systems: A systematic literature review. *BMC Medical Informatics and Decision Making*, *23*(1), 18. DOI: 10.1186/s12911-023-02115-5 PMID: 36694161

Tudor Car, L., Kyaw, B. M., Nannan Panday, R. S., van der Kleij, R., Chavannes, N., Majeed, A., & Car, J. (2021). Digital health training programs for medical students: Scoping review. *JMIR Medical Education*, *7*(3), e28275. DOI: 10.2196/28275 PMID: 34287206

United Nations Educational, Scientific and Cultural Organization. (2023). *Guidance for generative AI in education and research.* UNESCO.

U.S. Food and Drug Administration, Health Canada, & Medicines and Healthcare products Regulatory Agency. (2021). *Good machine learning practice for medical device development: Guiding principles.*

van Kessel, R., Wong, B. L. H., Clemens, T., & Brand, H. (2022). Digital health literacy as a super determinant of health: More than simply the sum of its parts. *Internet Interventions : The Application of Information Technology in Mental and Behavioural Health*, *27*, 100500. DOI: 10.1016/j.invent.2022.100500 PMID: 35242586

Vayena, E., Blasimme, A., & Cohen, I. G. (2018). Machine learning in medicine: Addressing ethical challenges. *PLoS Medicine*, *15*(11), e1002689. DOI: 10.1371/journal.pmed.1002689 PMID: 30399149

Waseh, S., & Dicker, A. P. (2019). Telemedicine training in undergraduate medical education: Mixed-methods review. *JMIR Medical Education*, *5*(1), e12515. DOI: 10.2196/12515 PMID: 30958269

Wiens, J., Saria, S., Sendak, M., Ghassemi, M., Liu, V. X., Doshi-Velez, F., Jung, K., Heller, K., Kale, D., Saeed, M., Ossorio, P. N., Thadaney-Israni, S., & Goldenberg, A. (2019). Do no harm: A roadmap for responsible machine learning for health care. *Nature Medicine*, *25*(9), 1337–1340. DOI: 10.1038/s41591-019-0548-6 PMID: 31427808

Wong, B. L. H., Khurana, M. P., Smith, R. D., El-Omrani, O., Pold, A., Lotfi, A., O'Leary, C. A., & Saminarsih, D. S. (2021). Harnessing the digital potential of the next generation of health professionals. *Human Resources for Health*, *19*(1), 50. DOI: 10.1186/s12960-021-00591-2 PMID: 33853625

World Health Organization. (2019). *WHO guideline: Recommendations on digital interventions for health system strengthening*. World Health Organization.

World Health Organization. (2021a). *Global strategy on digital health 2020–2025*. World Health Organization.

World Health Organization. (2021b). *Ethics and governance of artificial intelligence for health.* World Health Organization.

World Health Organization. (2023). *Classification of digital interventions, services and applications in health: A shared language to describe the uses of digital technology for health* (2nd ed.). World Health Organization.

World Health Organization. (2024). *Ethics and governance of artificial intelligence for health: Guidance on large multi-modal models.* World Health Organization.

Zhao, J., Xu, X., Jiang, H., & Ding, Y. (2020). The effectiveness of virtual reality-based technology on anatomy teaching: A meta-analysis of randomized controlled studies. *BMC Medical Education, 20*(1), 127. DOI: 10.1186/s12909-020-1994-z PMID: 32334594

Chapter 9
Environmental Microbiomics for Public Health Safety:
Applications in Food Adulteration and Contamination Detection

Baljeet Yadav
http://orcid.org/0000-0003-0624-6784
KR Mangalam University, Gurugram, India

Anuj Kumar Bharti
http://orcid.org/0009-0009-3054-896X
KR Mangalam University, Gurugram, India

Kazi Mizan
http://orcid.org/0009-0003-8387-0756
KR Mangalam University, Gurugram, India

Garima Awasthi
KR Mangalam University, Gurugram, India

Yash Bhardwaj
http://orcid.org/0009-0009-0631-2199
KR Mangalam University, Gurugram, India

Shivani Sehgal
KR Mangalam University, Gurugram, India

ABSTRACT

Food traceability is an essential element of food safety, food quality and consumer confidence among global food supply chains. Food fraud, also known as intentional adulteration, substitution, mislabeling, or counterfeiting foodstuffs for commercial gain, is a tremendous risk to health and brand image. This review examines the latest digital technologies such as Things (IoT) sensor networks, blockchain, cloud-based data-as-a-service platforms and artificial intelligence (AI) to scale up the integration of collected data, and to detect anomalies and prevent a significant amount of

DOI: 10.4018/979-8-3373-9033-8.ch009

fraud. An emerging field, environmental microbiomics, is the ability of microbial signatures to be associated with geographic origin and events of contamination. Irrespective of these developments, there are challenges related to standardization of data, interoperability, scalability, or data privacy, as well as adoption among smallholders. Spatial-temporal analytics, ethical data governance, and inclusive implementation frameworks must be of future concern to research.

1. INTRODUCTION

Food traceability comprises tracking and documenting the production, processing, and distribution of food items to ensure safety and quality from farm to fork (Opara, 2003; Regattieri et al., 2007; Bosona & Gebresenbet, 2013; Aung & Chang, 2014; Urbano et al., 2020). The Food Safety Modernization Act of 2011 mandates traceability for high-risk foods in the United States, while the European Commission aims to extend its tracing network across the agro-food supply chain (Golan et al., 2004; Smith et al., 2005; Charlebois et al., 2014; European Commission, 2022). For food fraud prevention, the Global Food Safety Initiative encourages the implementation of a defence strategy determining the vulnerability of food products to fraud and the control measures appropriate to mitigate it (Marvin et al., 2016; Manning & Soon, 2016; Mei Soon et al., 2019). Best practices and guidance suggest aggregation of provenance data, spatial data, and time data within one format, which can be easily accessed. (Karlsen et al., 2013; Olsen & Borit, 2013; Storøy et al., 2013; Jiménez-Carvelo et al., 2023).

The work on mitigating food fraud is difficult without traceability and transparency. which is a widespread international problem that compromises confidence in the food supply chain by consumers and corporations (Verbeke & Ward, 2006; van Rijswijk & Frewer, 2008; Manning & Soon, 2016; Urbano et al., 2020). Practices such as the introduction of fair and affordable food safety, although intended to be beneficial, increase vulnerability to fraudulent activities (Rouvière & Caswell, 2012; Lindley, 2022). The dynamics of food fraud are complex, making it harder to predict and control (Spink & Moyer, 2011; Marvin et al., 2016). Such empirical data as the progressive variation in sediment levels of bottled water with storage or the presence of genetically modified organisms in the ingredient consignments reveal the pervasive nature of the fraudulent activity even in such a well-studied commodity (Fanelli et al., 2021; Guruswamy et al., 2022).

Regular checking of parameters periodically examined within food supply chains, such as temperature and humidity, chemical residues, and physicochemical properties, provides clear data supporting detection and prevention of food fraud (Wang et al., 2010; Costa et al., 2013).

The analyzed article is a broad perspective of emerging digital technologies to be deployed in food traceability and fraud prevention, which comprise blockchain, the Internet of Things, cloud-based solutions and data mining. It also examines practicalities, policy and regulation, significant concerns and ethical considerations. such technologies get them going. Lastly, the paper identifies potential research that can be done in the future to enhance transparency and integrity in the food supply chain.

2. BACKGROUND AND DEFINITIONS

Traceability is the procedure that involves tracking and tracing of a product in various different stages (Olsen & Borit, 2013; Dabbene et al., 2014; European Commission, 2022). The structure of this farm-to-fork traceability system is illustrated in Figure 1.

Figure 1. Farm-to-fork food supply chain and critical tracking events

Tracking dictates the downstream flow of a product to its destination, and tracing dictates the upstream flow of the product to its destination (Regattieri et al., 2007; Bosona & Gebresenbet, 2013; Aung & Chang, 2014; Urbano et al., 2020). Traceability information might be required by different types of stakeholders, such as suppliers, manufacturers, distributors, retailers, and consumers (Kher et al., 2010; Karlsen et al., 2013; Guruswamy et al., 2022). Data that is required by law or the company itself has to be sent at a specified point and combined at multiple levels (Jansen-Vullers et al., 2003; Dabbene et al., 2014). Traceability also closely relates to the time and temperature monitoring, hence supporting the real-time data acquisition of product life span, safety, quality, and shelf-life information (Shanahan et al., 2009; Wang et al., 2010; Costa et al., 2013). It is essential for economic sustainability tools and the development of global agri-food value chains. Food traceability, in the definition of the European Commission, is the ability to track and trace food, feed, and ingredients through the entire processes of production, processing, and distribution (European Commission, 2022). The United States Food and Drug Administration is adopting a risk-based preventive controls framework for food animals, where technologies are required to track, trace, and communicate animal food safety issues throughout the entire supply chain (Golan et al., 2004; Pearson et al., 2019). Elements such as collaboration, compliance, and transparency have become fundamental elements (Khan et al., 2018; Pearson et al., 2019).

Food fraud encompasses a range of actions like intentional adulteration, substitution, mislabeling, counterfeiting, theft, and diversion, which are done to generate a financial benefit and can pose a risk to human health, safety, and food integrity (Spink & Moyer, 2011; Mei Soon et al., 2019). The food supply chains are often complex and fragmented, and there is need to have transparency and coordination between all the participants of the food supply chain to realise end-to-end visibility of their processes and products (Zhao et al., 2019; M. Jiménez-Carvelo et al., 2023). The Global Food Safety Initiative recommends organisations to undertake a food fraud vulnerability assessment, use industry best practices and guidelines, and enforce appropriate restricting techniques, including analytical tests, trackability, and track-and-trace chain of custody, and therefore bringing their supply chain to transparency when acquiring ingredients that pose a probable fraud risk (Marvin et al., 2016; Mei Soon et al., 2019; Yadav et al., 2026). The emergent spatial-temporal data that are related to food fraud can raise an alarm about illegal activities involving fraud through event detection systems (Sharma et al., 2020; M. Jiménez-Carvelo et al., 2023).The systems supporting these practices are automated and standardised, which allows such data to be exchanged and analysed promptly using divergent and complementary data streams (Khan et al., 2018; Sharma et al., 2020; Guruswamy et al., 2022).

3. DIGITAL TRACEABILITY PLATFORMS

The digital solutions that can be used to enhance food traceability and prevent food fraud are also becoming feasible at a more rapid pace, and the opportunities available to enhance food security, protect consumer confidence and promote economic growth are numerous (Zhao et al., 2019; Guruswamy et al., 2022; Hassoun et al., 2022; Ellahi et al., 2023). But the modern applications, as well as case studies reported, are not only manifestations of those opportunities, but also raise significant points that should be considered to implement them efficiently and on a large scale.

Traceability is a method of tracking the source of an item of food up the supply chain and down the supply chain (Opara, 2003; Aung and Chang et al., 2014). The system allows quick access to the information as every step is documented (Shanahan et al., 2009; Lei et al., 2022a). The safe sourcing of food throughout the world, recall management, and tracking conditions all the way along the supply chain, regaining trust after an event, and provenance verification are essential aspects of traceability (Pearson et al., 2019; Guruswamy et al., 2022). The emergence of affordable sensors and data-collecting machines has led to the rapid digitalization of food systems (Tsang et al., 2019; Hassoun et al., 2022). Although marketing and track-and-trace functions have been provided for years, capabilities for real-time monitoring and management of conditions and incidents, combined with provenance information, remain limited. The different types of food fraud and their supply chain implications are illustrated in Figure 2.

Figure 2. Types of food fraud and supply chain vulnerabilities

4. GLOBAL TRADE

4.1. Global Trade Data Standards

Digital Global Trade Data Standards, a category under the Global Data Standards established by GS1, are increasingly accepted and deployed globally to facilitate the integrated exchange of trade data among key stakeholders of the economy (Zhang & Bhatt, 2014; Urbano et al., 2020; Guruswamy et al., 2022). Current empirical evidence demonstrates that the absence of digital Global Trade Data Standards constitutes the main hindrance to the free flow of trade data among parties and jurisdictions (M. Jiménez-Carvelo et al., 2023). Lack of universally accepted data classification and format protocols for trade data inhibits accurate data transmission between systems and devices, hence jeopardizing fast processing of such data. This condition raises significant integration costs when setting up direct data interchange with a multitude of non-standardized external parties (Khan et al., 2018; Guruswamy et al., 2022).

Digital Global Trade Data Standards consist of a data schema encompassing a collection of digital Global Trade Data Standards detailing standardized, machine-readable, and interoperable trade data used and exchanged amongst national trade repositories, such as the ASEAN Free Trade Area Trade Repository or the EU Trade Help desk (Charlebois et al., 2014; Urbano et al., 2020). They enable systems interoperability for importing, exporting, and transit trade data exchanged between public and private platforms worldwide (M. Jiménez-Carvelo et al., 2023). Such interoperability avoids disparate point-to-point integrations that do not allow country-agnostic trade data transmission. The absence of a suitable digital Global Trade Data Standards data schema leads to countries developing ad hoc data protocols that are ultimately incompatible among themselves (Kher et al., 2010). Digital Global Trade Data Standards and the associated data schema comprise GS1's recommendations to countries contemplating a data protocol to govern the cross-jurisdictional exchange of trade data (Zhang & Bhatt, 2014; M. Jiménez-Carvelo et al., 2023).

Digital traceability platforms are investable technologies enabling data-related value creation to trace food products and shipments across the supply chain. Platforms concentrate on food-related transactions, processed goods, life sciences, and logistics, and support a variety of products, including canned fish, plants, flowers, tea, and grains (Lei et al., 2022a). Even when the details of track-and-trace processing vary, the best practices, key processes, data requirements, and related implementation guidance from GS1 generate significant value for every situation (Guruswamy et al., 2022).

4.2. Blockchain and Distributed Ledger Technologies in Food Supply Chains

Implementation of a supply chain traceability solution based on blockchain is set to enhance the tracking of products and consumer safety, reduce the cost of verification, and enable compliance; nevertheless, the existing understanding of its potential benefits, drawbacks, and operational mechanisms is still insufficient (Pearson et al., 2019; Kamble et al., 2020; Ellahi et al., 2023).

Blockchain is a term used to refer to a set of technologies that can be used to establish distributed ledger systems (Pearson et al., 2019; Casino et al., 2021). The main feature of these platforms is their user network, which, in turn, logs the transactions occurring in more than one place without a central intermediary (Tian et al., 2016). Stateful permanent history created through such ledgers is extremely difficult to edit once being created (Creydt and Fischer, 2019; Pearson et al., 2019). The architectures may be very diverse as blockchains can be based on low-trust rules and basic data formats, or they can be highly complex using smart contracts and high-trust settings (Kim and Laskowski et al., 2018). The universal denominator is the facilitation of transactions without relying on a trusted third party for verification (Galvez et al., 2018; Pearson et al., 2019). Producers, intermediaries, processors, retailers, and consumers are just some examples of stakeholders that can contribute material to the ledger or retrieve information out of it (Tsang et al., 2019; Guruswamy et al., 2022). The change of the ownership can also be traced throughout the supply chain (Wang et al., 2021).

Walnut Blockchain uses the blockchain to identify auditing processes and actions within the food.supply chains (Pearson et al., 2019). This service enables scrutiny of specific products without requiring direct access to confidential commercial information (Wang et al., 2021). The food industry reports shortages of specialised blockchain developers, complicating governance arrangements and increasing consultancy costs across the supply chain (Lemeilleur et al., 2019). Many public blockchains are designed for potentially competitive enactments and prohibit private transactions, whereas a consortium governance model that does not erect membership barriers provides better insulation from the risks of opportunism and transaction costs (Pearson et al., 2019; Feng et al., 2020). All firms can track product ownership and associated documentation, allowing an audit-ready position for regulatory checks (Xu et al., 2018; Casino et al., 2021; Guruswamy et al., 2022).

Recent events such as the bankruptcy of FTX have raised concerns about the credibility of blockchain companies owing to links with cryptocurrency, even when services have no potential exposure to digital currencies or security tokens (Pearson et al., 2019; Salah et al., 2019). Communications framing services in the context of food traceability can help clarify their core mandates (M. Jiménez-Carvelo et al., 2023).

4.3. Internet of Things and Sensor Networks

The Internet of Things and sensor networks enable the collection, transmission, and processing of real-time data throughout food supply chains, complementing rather than necessarily fully integrating with traceability platforms (Tsang et al., 2019; Urbano et al., 2020; Hassoun et al., 2022). Many IoT devices may only capture non-food-safety-related data, and sophisticated data monitoring and analysis systems are required for processing large information flows (Sharma et al., 2020). Moreover, the effect of such activities on food safety is highly dependent on the extent to which the data are subsequently made visible and accessible to stakeholders. The collection of large volumes of data makes it difficult to ensure that inputs are reliable and sufficiently accurate to be useful for preparing and managing the traceability system (Ringsberg et al., 2014; Urbano et al., 2020), and because of the time necessary to collect, transmit, and prepare the data for use, the elapsed time after the incidents and conditions being reported may be already too long for detecting safety risk (Wang et al., 2010).

Three main types of sensors are commonly used in food supply chains: physical property sensors, biosensors, and micro-electromechanical systems sensors. Physical property sensors measure parameters such as moisture, temperature, light, and pH (Shanahan et al., 2009; Urbano et al., 2020); biosensors monitor processes relating to quality and contamination (Dias et al 2021) and MEMS sensors detect movement or positioning and are used for monitoring food tampering, detecting shocks and vibrations, ensuring that products remain upright, and identifying tampering at the transportation stage (Costa et al., 2013; Guruswamy et al., 2022).

Internet of Things (IoT) system consist of microcontrollers and embedded with low cost chips and have wireless communication capabilities (Tsang et al., 2019; Urbano et al., 2020). Their data can be incorporated into the food traceability systems to help increase the surveillance of both physical and event with uncertainties (Tian, 2017). Wireless sensor network are used to monitor and locate sensors in detecting pests and diseases, and transmit data in various environmental conditions as required, hence assisting in curbing food loss and waste (Hobbs, 2020; Guruswamy et al., 2022; Hassoun et al., 2022). The effectiveness of proper implementation of the IoT in agriculture greatly depends on the effectiveness of data transmission technologies, and the choice of the communication methods become dependent on the needs raised by the operational conditions.

Figure 3. Blockchain-driven IOT architecture for a food digital twin

Analytical Method	Principle / What It Detects	High-Risk Food Applications
DNA Barcoding	Verifies species origin using genetic markers (e.g, COI gene)	Fish/Seafood (species substitution), Beef/Poultry (origin verification), Plant-based ingredients
Chemical Fingerprinting	Compares molecular composition to a reference profile (e.g, spectroscopy-based profiles)	Honey (sugar syrups, dilution with cheaper oils) Spices (adulterants/dyes)
Isotopic Analysis	Measures ratios of stable isotopes to determine geographical origin or processing	Wine/Juice (geographical origin, watering – feed or origin related), Organic food products

4.4. Cloud-Based Traceability Solutions

Cloud computing offers a lot of beneficial attributes for traceability solutions, such as scalable computing resources and accessibility from multiple devices (Zhao et al., 2019; Urbano et al., 2020).Cloud based traceability solutions eliminate the necessity of storing the data locally and provide the software with the most recent updates that are distributed across users instantly. The software-as-a-service model, hosted in the cloud, has most often been used for traceability (Guruswamy et al., 2022; Hassoun et al., 2022). As with any solution, a trade-off between security and ease of use is involved (Feng et al., 2020; Urbano et al., 2020). For SaaS models, the software vendor is responsible for software maintenance and upgrades, but user data is stored off-premises on vendor servers.

Cloud computing supplies the computing capability, storage, and service required for traceability (Urbano et al., 2020); the security risk perception, the degree of security assurance, and the actual security of cloud-based traceability solutions are significant, especially for food traceability (Fenge et al., 2020).

5. EVIDENCE-BASED APPLICATIONS IN TRACEABILITY

Traceability systems that verify provenance, monitor and alert during transport, or authenticate against a standard are available (Urbano et al., 2020; Fanelli et al., 2021). Provenance systems increase awareness of events and allow sellers to demonstrate credibility (Van Rijswijk & Frewer, 2008; Guruswamy et al., 2022); studies report up to 22% sales increases for documented authenticity (Verbeke &Ward, 2006; Urbano et al., 2020). Monitoring systems automatically detect states such as temperature, send notifications, and guide responses (M. Jiménez-Carvelo et al., 2023); easy adjustments to indicator thresholds mitigate risk. Authentication solutions include tamper-proof seals, unique identifiers, and verification processes (Y. Shah et al., 2010; Mei Soon et al., 2019). The success of an application depends on the specifics of the food commodity involved (Smith et al., 2005; Guruswamy et al., 2022).

Fake products are likely to contain harmful components, and this is often a greater threat than naturally occurring components (Spink & Moyer, 2011; Marvin et al., 2016). Tamper detection services help in reducing these threats (Y. Shah et al., 2010; Mei Soon et al., 2019); retailers can restrict access to multi component or new components. Standards address counterfeit mitigation, throughout a variety of supply chains, which targets cosmetic fraud, but requires more attention to functional alternatives (Charllebois et al., 2014; Mei Soon et al., 2019).

5.1 Case Studies of Provenance Verification

The concept of food authenticity is relevant, owing to the rise in fake and counterfeit food products (Fanelli et al., 2021; Siddiqui et al., 2023). The European Union has compiled a list of the food fraud cases detected between 2006 and 2022 (European Commission, 2022; M. Jiménez-Carvelo et al., 2023). The incidents of mislabeling, substitution, undeclared allergenic ingredients, and counterfeiting of high-value products are the most common ones (Marvin et al., 2016; Mei Soon et al., 2019). Early detection can prevent crises and reduce food fraud risk.

Traceability has emerged as a solution for monitoring the food supply chain (Guruswamy et al., 2022; Ellahi et al., 2023). Two projects have demonstrated the impacts and benefits of effective traceability on food safety, food fraud, and justice (Tian, 2017; Pearson et al., 2019). Traceability systems of beef and fish based on blockchain technology have reduced fraud cases and inefficient processes and achieved financial gains in Brazil (Casino et al., 2019). The European project traceability along the milk supply chain based on low-cost IoT sensors facilitated quick access to information and track 55 traceability events to prevent dairy fraud (Tang et al., 2019; Urbano et al., 2020).

5.2 Real-Time Monitoring and Alerting Mechanisms

Besides event detection, an alerting mechanism can also facilitate rapid response towards fraud cases, by proposing alternative workflows (M. Jiménez-Carvelo et al.,

2023). The sensitivity of the alerts is an important aspect in relation to workflow proposed by a system (Guruswamy et al., 2022). An alert mechanism that detects minor deviation or abnormalities would be quickly ignored, given the rapid pace of the food supply chain. On the other hand, when the standards established in an organisation are not tight enough and warning decisions are not made, the risks of fraud during unexpected events remains unaddressed (M. Jiménez-Carvelo et al., 2023). Organisations adopt alternative methods, namely event detection on a purely statistical basis or on methods comprising of an unobserved component model (Sharma et al., 2020; Guruswamy et al., 2022). One more element to be taken into consideration is the score obtained by the event detection, in case it becomes a critical one, a response to the alerting mechanism should be triggered, even proposing an alternative workflow. Organisations operating in the industry may differ in their respective scores (Urbano et al., 2020).

5.3 Authentication and Anti-Counterfeiting Measures

The authentication and anti-counterfeiting may include the employment of tamper-proof seals or a unique identifier that can be employed to secure shipments and activate authentication procedures (Y. Shah et al., 2010; M. Jiménez-Carvelo et al., 2023). The anti-counterfeiting strategies differ based on the needs, product, and value (Mei Soon et al., 2019); fraud detection puts the responsibility on the user and since the product is going through various supply chains and in many cases not transparent schemes with standard identifiers provide a practical overall solution (Zhang and Bhatt, 2014; Pearson et al., 2019).

6 FRAUD PREVENTION MECHANISMS

Fraud prevention mechanisms strengthen supply chain integrity (Mei Soon et al., 2019). Tamper-evident packaging, applied tactile sensations, and serialization deter product adulteration (Y. Shah et al., 2010; Mei Soon et al., 2019). Blockchain maintains detailed records of events, demonstrating evidence of compliance and readiness for auditing (Pearson et al., 2019; Ellahi et al., 2023; M. Jiménez-Carvelo et al., 2023). Anomaly detection assures components remain within parameters defined by historical data (M. Jiménez-Carvelo et al., 2023). DNA verification and bio-sensing determine product composition and origin, enabling detection of intentional modifications (Fanelli et al., 2021; Liu et al., 2022).

Figure 4. Analytical detection methods for food fraud and high-risk categories traceability technologies

Feature	Benefit	Barrier/Challenge
Blockchain	Immutability and transparency	High costs and lack of global protocols
Smart Tags	Real-time monitoring of temperature/humidity	High financial burden for small firms
SaaS (Software)	Easy updates and scalability	Security risks and ethical concerns

Regulatory & Industry Best Practices

- Mandatory Frameworks: European Commission Regulations, FSMA
- FSMA Modernization Act (2011 [U.S.])
- Global Standards: GS1 Global Traceability Standards (Barcodes, RFID, DataExange)

6.1 Tamper-Evident Packaging and Serialization

Tamper-evident features, such as destructible film wrappers, breakable caps, and labeled shrink bands, provide external evidence of tampering, help deter product reuse after tampering, and support tracing of tampered products back to the affected batches (Y. Shah et al., 2010; Mei Soon et al., 2019). Serialization consists of assigning a unique and visible serial number to each product item (Y. Shah et al., 2010; Zhang& Bhatt, 2014). Anti-counterfeiting or product authentication systems, based on tamper-evident seals, labelled caps, small ambient sensors, or serialised indicators, have been installed in food, beverage, and pharmaceutical supply chains for many years (Y. Shah et al., 2010; Mei Soon et al., 2019). Some of these systems are supported by international standards that specify human and machine-readable formats for expressing serial numbers and a two-level hierarchy for the associated data structure (Y. Shah et al., 2010); the first level is a globally unique code for the dispenser, and the second level is a code specific to the item (Y. Shah et al., 2010).

6.2 Blockchain-Driven Audit Trails

Aiming to enhance data integrity and prevent unauthorized alterations of tracked events, blockchain technology enables the establishment of transparent and immutable audit trails (Creydt & Fischer, 2019; Pearson et al., 2019; Casino et al., 2021). When implemented, these systems help to simplify compliance with food safety regulations, such as hazard analysis and critical control points, and to provide the documentation needed for audits, thereby preventing penalties and ensuring uninterrupted access to strategic markets (Trienekens & Zuurbier, 2008; Pearson et al., 2019).

Integrating designated link keys into a centralised distributed ledger connects multiple ledgers distributed across different locations and organizations (Xu et al., 2018). For audit-readiness purposes, two public ledger solutions are being investigated to record selected events with broad stakeholder access and no transaction costs (Lei et al., 2022a). From the perspective of the audit trail owner, event records, types of access, and pertinent information can be categorised as public, semi-public, or private based on user rights and intended disclosure (Pearson et al., 2019; Feng et al., 2020).

6.3 AI and Machine Learning for Anomaly Detection

Spatial-temporal analyses of data may also help in combating food fraud including counterfeiting and theft because suspicious behaviour in product scanning can be identified (M. Jiménez-Carvelo et al.,2023). Abnormal spatial or temporal parameters surrounding the events permit the detection of fraud (Sharma et al., 2020; M. Jiménez-Carvelo et al., 2023). As an example, counterfeited products are usually scanned in locations or at times that deviate from expected patterns(M. Jiménez-Carvelo et al., 2023). In the same way, theft or diversion generates scans at areas which are not within the network of legitimate retailers (Guruswamy et al., 2022). The situation is complicated by the unpredictability of the consumer scanning behaviour (M. Jiménez-Carvelo et al., 2023). Few knowledge-sharing platforms spread the potential anomaly patterns (Sharma et al., 2020; Guruswamy et al., 2022). The reluctance to use spatial-temporal information to detect crime is associated with the fact that food companies lack awareness of the benefits of this approach, there are no systems that can capture and process complex data in real-time, and uncertainties about returns on investment (Sharma et al., 2020; Hassoun et al., 2022; M. Jiménez-Carvelo et al., 2023).

6.4 DNA-Based Verification and Bio-Sensing Approaches

Food fraud is a major and often under appreciated issue in many sectors (Marvin et al.,2016; Fanelli et al., 2021). High-profile incidents in recent years have included stories of honey being diluted with sugar syrup and presented to unsuspecting consumers as pure honey, or water being added to olive oil and sold to the consumer as virgin olive oil (Spink & Moyer, 2011; Marvin et al.,2016; Liu et al., 2022). The concept of tracking and monitoring food products throughout the logistical process and their implementation by applying bar codes, Cellular IoT, IoT embedded devices, Blockchain Technology, and Artificial Intelligence is gaining widespread use

(Tsang et al., 2019; Guruswamy et al., 2022). However, all the above-mentioned approaches have their drawbacks (Fanelli et al., 2021).

DNA-based verification is the identification of species-specific DNA fragments in food products and their tracing throughout the entire transparent supply chain (Fanelli et al., 2021; Liu et al., 2022). The technology in question is also applicable to the agri-food industry and authenticity of cosmetic items, which verifies the presence of the mentioned active components (Liu et al., 2022; Siddiqui et al., 2023). In addition to detecting and verifying the authenticity of food, it has a positive effect on enhancing food supply chain transparency (Fanelli et al., 2021). The DNA technique is capable of tracking the food products even in a counterfeiting scenario where the original food product has been altered, thus indicating its broad applicability (Liu et al., 2022). Highly processed products like instant noodles, soy sauce, and bread, however, have not yet been extensively studied, and they would thus require further experimentation (Fanelli et al., 2021)..

Bio-sensing approaches involve detection and identification of pathogenic organisms such as carcinogenic compounds, Salmonella, and Listeria for food safety assurance (Dias, 2021). Embedded within the intention of enhancing food safety assurance, bio-sensing microphones measure the surrounding concentrations of pathogenic organisms, and thus could also be used as a proliferation detector, but remain an indirect tracking method (Costa et al., 2013; Guruswamy et al., 2022).

Figure 5. Analytical detection methods for food fraud and high-risk categories

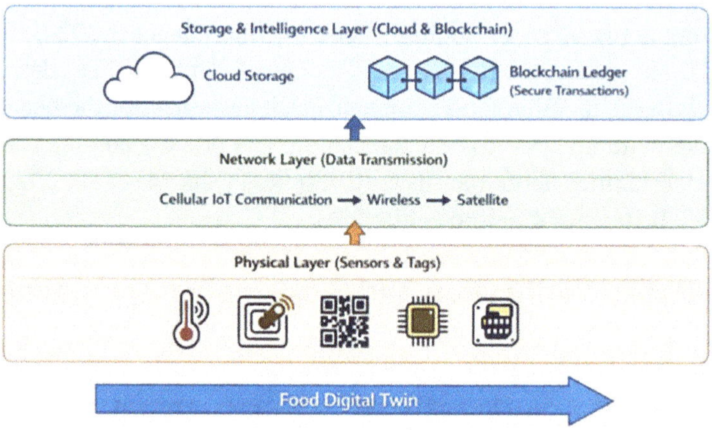

7 Food Safety and Authenticity in Environmental Microbiomics

Environmental microbiomics is an effective method for maintaining food safety by monitoring issues at an early stage (Ercolini, 2013; Bokulich et al., 2016). It includes the study of all microorganisms that are found at the places where food is grown or made (Ercolini, 2013; Quince et al., 2017). This technique involves spotting all kinds of germs at the same time, which could further help in detecting any small changes in food, which means the food is unsafe, before people get sick (Bokulich et al., 2016; Jagadeesan et al., 2019). This technique helps food companies in finding the spots where dangerous bacteria like Listeria and Salmonella hide and in cleaning them before they cause an outbreak (Franz et al., 2014).

This technique also helps in preventing the production of adulterated food. A group of specific kinds of germs is present at each place. This helps in checking whether the food is genuine or adulterated. For example, if a food label says it came from a place, but the germs present in it don't match that place, it means that the food is adulterated (Ellis et al., 2022). When this information is used along with AI, it helps track the food from the farm to the consumer's plate (Jagadeesan et al., 2019).

Food Safety Surveillance via Environmental Microbiomics

A powerful technology, environmental microbiomics, has been developed to improve food safety surveillance and to allow early detection of microbial contamination along the entire food production continuum (Ercolini, 2013; Bokulich et al., 2016). It involves extensive analysis of microbial communities in agricultural soils, irrigation water, food processing plants, storage environments, packaging materials and food products using culture independent, high-throughput sequencing technologies (Quince et al., 2017). Environmental microbiomics allows the simultaneous characterisation of whole microbial communities and therefore the detection of subtle ecological changes that often occur before the onset of food spoilage or contamination events (Bokulich et al., 2016; Jagadeesan et al., 2019). This contrasts with conventional microbiological approaches that only analyse a restricted set of culturable microorganisms.

Recent advances in next-generation sequencing (NGS), shotgun metagenomics, metatranscriptomics and long-read sequencing technologies have revolutionised the capacity to detect microbial taxa, functional genes and virulence determinants from ambient samples without the need for prior cultivation. These methods enable the thorough description of bacterial, fungal, viral and archaeal populations, giving essential information on microbial succession, disease persistence and ecosystem dynamics from farm to fork. In addition, environmental DNA (eDNA) analysis has been recognised as a fast surveillance tool for the detection of microbial fingerprints from soil, water, air and food-contact surfaces and will improve environmental monitoring and identification of sources of contamination.

Environmental microbiomics assists in identifying persistent contamination niches in food processing systems. Foodborne pathogens such as Listeria monocytogenes, Salmonella enterica, pathogenic Escherichia coli and Cronobacter sakazakii can survive in biofilms on processing equipment, drains, conveyor belts and other food-contact surfaces. This microbiome profile enables organisations to uncover these hidden reservoirs, assess sanitation effectiveness and deploy targeted hygiene interventions before foodborne disease outbreaks occur due to contamination. Leong et al. (2014) reported that *Listeria monocytogenes* can persist in food-processing environments, emphasizing the importance of continuous environmental monitoring. Hence, microbiome-based environmental monitoring is increasingly being integrated into preventive food safety management systems.

A further important application of environmental microbiomics is food authentication and adulteration detection. Each agricultural region has a characteristic microbial fingerprint, a product of its climate, soil composition, farming practices and surroundings. These microbial fingerprints are distinctive and can be utilised as biological indicators of geographical origin and production techniques. The presence of a microbial community that is not in line with the expected environmental profile of a food product can be a sign of adulteration, substitution, mislabeling or unauthorised processing (Ellis et al., 2022). Microbiome-based authentications are becoming increasingly important for high-value commodities such as dairy products, spices, fruits, wine, honey and seafood.

A recent study found that ambient microbiomics combined with artificial intelligence (AI) significantly enhances food safety surveillance. Machine learning algorithms may be trained with large metagenomic datasets to identify patterns of contamination, predict risks of spoilage, classify microbial communities and identify aberrant microbial signatures with high accuracy. AI can be used to differentiate between contamination from soil, irrigation water, processing equipment or human handling in microbial source tracing and hence, facilitate rapid trace-back investigations during food-borne disease outbreaks (Jagadeesan et al., 2019). Predictive microbiome models can enable regulatory authorities and the food sector to establish risk-based surveillance systems that can anticipate contamination hotspots before an outbreak occurs.

Environmental microbiomics is becoming more and more relevant in the framework of the One Health approach, which considers the interconnected health of humans, animals, plants and the environment. The surveillance of environmental microbial populations is of importance for understanding the dissemination of antimicrobial resistance (AMR) genes, the genesis of zoonotic illnesses and environmental reservoirs of clinically relevant bacteria. Monitoring of wastewater, agricultural runoff, animal surroundings and irrigation systems has become a prerequisite to track resistant bacteria and prevent their dissemination to the food chain. New research suggests that metagenomic sequencing combined with artificial intelligence-powered analytics can enable real-time surveillance of antibiotic resistance determinants and prompt public health decision making.

Environmental microbiomics is a paradigm shift strategy for modern food safety surveillance by the merging of high-throughput sequencing, environmental DNA analysis, advanced bioinformatics and artificial intelligence. These integrated systems provide early warning of contamination events, improve pathogen source attribution, enable verification of food authenticity, facilitate antimicrobial resistance surveillance and allow data-driven decisions across the whole farm-to-fork chain. As sequencing technologies become cheaper and computational methods more advanced, environmental microbiomics is anticipated to become a component of next-generation food safety systems and sustainable public health protection.

8 REGULATORY AND GOVERNANCE CONSIDERATIONS

A line of inquiry has emerged aimed at addressing regulatory, institutional, and technological challenges in food fraud verification (Lindley, 2022). Inspections, sampling, and laboratory analyses of food products are important regulatory control measures (Rouviere & Caswell, 2012; Lindley, 2022). However, they are resource-intensive, meaning that for many food products in Australia, there is limited oversight and confidence in the safety and authenticity of food products (Lindley, 2022). Under current conditions with many food fraud cases remaining undetected, the need for better and complementary verification methodologies has become more pressing (M. Jiménez-Carvelo et al., 2023). Emerging detection technologies have intersected regulatory frameworks worldwide, including food supply chains (Lindley, 2022). Science-based food fraud control technologies can support the prevention of food fraud through cultivar and chemometrics fingerprinting of food products (Fanelli et al., 2021). The regulatory framework for event-driven and information-based food fraud prevention covers digital soil, carbon farming, blockchain technology, and artificial intelligence (Guruswamy et al., 2022; Lindley, 2022).

Food is one among the many products affected by fraud, and sometimes illegal or unregulated substances are added to the food (Lei et al., 2022b). The attention of regulatory agencies will be on tracing back the ingredients used to manufacture the food and whether the ingredient is natural or synthetic (Golan et al., 2004; Lindley, 2022).

8.1 International Standards and Compliance

Food safety legislation mandates tracing from the end product backward to the initial supplier, and full traceability from the raw material source to the final consumer ensures compliance with regulations (Charlebois et al., 2014; Guruswamy et al., 2022). Certain technologies permit cross-validating traceability data with other available information through either an authorized channel or a comprehensive public ledger (Casino et al., 2021; Lei et al., 2022b).

To promote enhanced traceability, the Global Health Organization aids in establishing mobile and web-based communication solutions (Guruswamy et al., 2022). The global framework (ISO 22005) suggested by the International Organization for Standardization provides the key components of information exchange to trace food and feed products (Guruswamy et al., 2022). International data sharing standards applicable to matters pertaining to trade are essential with the increasing number of nations and organisations involved in cross-border business enterprises (Kher et al., 2010; Guruswamy et al., 2022).

8.2 Data Privacy and Ownership

To create resilient and trusted food systems, it is essential to share information among a large group of stakeholders (Manning and Soon, 2016). Nevertheless, food traceability-related transparency can be at the expense of data privacy and ownership (Feng et al.,2020; M. Jiménez-Carvelo et al., 2023). Data privacy is the right to manage third party access to personal information, to the purposes of its use, to the permissibility of its sharing, and to its time limitations. Individual privacy protection consists of restricting and regulating the disclosure of personal information (Feng et al., 2020; M. Jiménez-Carvelo et al., 2023). Within the context of food traceability, an organization gathers many types of data of which some can be considered personal regarding its suppliers and customers (Lei et al., 2022b). This can be related to identity, demographics, buying behaviors and the information related to loyalty programs among others. Customers usually anticipate the privacy of such data, depending on the laws of the country they live in (M. Jiménez-Carvelo et al., 2023).

Alongside the issue of privacy, there might be a requirement to consider the ownership of data, particularly, when an organization delivers data to a third party or a broader audience (Feng et al., 2020). Organizations generally desire to own data that they gather, utilize or generate (Pearson et al., 2019). The technology that facilitates traceability may need solutions to safeguard the data privacy and ownership rights while retaining the advantages of enhanced traceability (Feng et al., 2020; M. Jiménez-Carvelo et al., 2023).

8.3 Traceability in High-Risk Sectors

The sectors that involve dangerous food items should have precise control mechanisms to prevent the occurrence of disastrous crises and losses to those involved (Manning and Soon, 2016; Lei et al., 2022b). As an example, in the United States, nitrofurazone is banned in food animals; nevertheless, aquaculture products imported from other countries without these prohibitions still find their way into the country (Smith et al., 2005; M. Jiménez-Carvelo et al., 2023). Special traceability facilities have also been incorporated to identify the origin of nitrofurazone contamination at the importation stage thereby facilitating necessary measures to prevent issues (Shanahan et al., 2009). There was also the incident when thousands of vials of the SARS-CoV-2 virus were accidentally dispensed and made accessible to the population (M. Jiménez-Carvelo et al., 2023). Viral strains were supposed to be successfully tracked down to and contained within a limited number of laboratories (Guruswamy et al., 2022).

Food fraud has also been common across a wide range of commodities and geographical contexts (Marvin et al., 2016; Mei Soon et al., 2019). In 2016, Bahrain found a high incidence of prohibited food substances and contaminants in imported fish, whereas no such cases were detected with domestic fish (M. Jiménez-Carvelo et al., 2023). Nevertheless, the findings of laboratory tests confirmed the existence of such substances in domestic fish (Lei et al., 2022b). Early detection of first suppliers and following transactions, along with the corresponding countermeasures, can greatly mitigate the economic losses and protect the health of people (M. Jiménez-Carvelo et al., 2023).

9 CHALLENGES AND LIMITATIONS

Traceability technologies collect data across agriculture, processing and sales activities (Aung and Chang, 2014; Debbene et al., 2014; Urbano et al., 2020). Credible entries help prevent fraud. Data exchange is established through Global Data Standards (Guruswamy et al., 2022). Blockchains have a cryptographic pedigree of events (Creydt & Fischer, 2019; Pearson et al., 2019). Audit-readiness makes the verification easier. Tamper-proof seals are difficult or impossible to forge, and serialised codes are used instead of bulk identifiers (Y. Shah et al., 2010). Through transformation, DNA tags remain (Fanelli et al., 2021; Liu et al., 2022). Anomaly detection detects behaviour that does not conform to the norm such as when there is an input that does not correspond to transacted products (M. Jiménez-Carvelo et al., 2023). The response bias is vital: absence of response implies a lack of engagement or capacity to respond; while suboptimal corrective actions imply awareness of the issue (Mei Soon et al, 2019; Guruswamy et al., 2022).

Data accuracy, trust, and currency matter (Karlsen et al., 2013; Urbano et al., 2020). Systems demand technical understanding and infrastructure (Pearson et al., 2019; Kamble et al., 2020). Platforms, algorithms, and data need corroboration.

Trust models may increase complexity (Pearson et al., 2019). Different foodstuffs, processes, regulations, audit expectations, and jurisdictions challenge data interchange (Charlebois et al., 2014; Guruswamy et al., 2022). Without data generally accepted as trustworthy, credential verification before attachment becomes critical (Karlsen et al., 2013; M. Jiménez-Carvelo et al., 2023). Scale, speed, and energy consumption affect mathematical models. Algorithms need tuning and feedback for improvement (Sharma et al., 2020; M. Jiménez-Carvelo et al., 2023). Only some offer widely shared datasets (Guruswamy et al., 2022). Security and ownership remain problematic.

Smart sensors track moisture, temperature, humidity, pH, shock, and gases (Costa et al., 2013; Urbano et al., 2020). Artificial intelligence and data science simplify the task but also necessitate comprehensive datasets (Salah et al., 2019; M. Jiménez-Carvelo et al., 2023). Models require assessment to avoid misleading operators about data stability or technological advantage (Guruswamy et al., 2022; Hassoun et al., 2022). Global standards assist in exchange but may lack the granularity required for transference (Urbano et al., 2020). Guidance on the level of transformation compatible with provenance retention and on financing models for maintaining records is scarce (Lemeilleur et al., 2019). Those who inspect for commitment to authenticity still look for verification responses (Rouière & Caswell, 2012; Lindley, 2022). Technical capability, cost, training, and oversight further condition uptake (Lemeilleur et al., 2019; Pearson et al., 2019). The additional burden, particularly among resource-constrained producers, deters participation (Lemeilleur et al., 2019; Hobbs, 2020). All options become valuable in the case of collective adoption (Khan et al., 2018; Guruswamy et al., 2022).

Demand responds to changing commercial and procurement practices (Mei Soon et al., 2019). The authorities want to have governance and legitimacy, and societal pressures demand responsibility and safety (Trienekens & Zuurbier, 2008; Lindley, 2022). Customers monitor responses after news reports (M. Jiménez-Carvelo et al., 2023). People who want to evade or mislead become proactive, shifting from opportunistic to professional behaviour (Mei Soon et al., 2019).

9.1 Data Quality and Interoperability

The concept of traceability, which implies tracking a product within a food supply chain, gains traction in the changing, globalised and diversified food systems

(Bosona & Gebresenbet, 2013; Urbano et al., 2020). Producers, distributors and consumers of food products have to rely on reliable, trustworthy, and explanatory information on food products to ensure food safety, combat fraud and reduce wastage (Guruswamy et al., 2022). Such dependence on data is unsafe, though, when the information is of low quality (Urbano et al., 2020). Data quality is associated with the level of data excellence that is largely judged based on the notions of accuracy, reliability, credibility, and relevance (Karlsen et al., 2013). The low quality of data can cause misinformation, misguided actions and false conclusions (M. Jimenez-Carvelo et al., 2023).

Interoperability of many traceability platforms is not realized, which causes a separation of supply chain actors. On the same platform, suppliers do not have access to data from suppliers on different platforms (Urbano et al., 2020; Ellahi et al., 2023). Closed systems do not offer the comprehensive records of the past operations of a food supply chain or sufficient information about products of other supply chains (Guruswamy et al., 2022). Systems such as food traceability also depend on the availability of key, standardised data schemas and common structures for data exchange (Urbano et al., 2020). Lack of popular data standards may be an obstacle to interoperability (Lei et al., 2022b). A low level of data standardisation is also a barrier to access because it leads to the activation of undesired logics that result in suboptimal solutions (Jansen-Vullers et al., 2003; Sharma et al., 2020; Guruswamy et al., 2022).

9.2 Adoption Barriers for Smallholders

The adoption of digital traceability technologies such as blockchain remains limited. Smallholders in developing and emerging economies face barriers in access to equipment and training, collaborative incentives are insufficient, and costs of installing hardware or buying packages or software are prohibitive (Lemeilleur et al., 2019; Pearson et al., 2019). The situation is compounded at the global level by a lack of resources for keeping data records (Pearson et al., 2019), prior digitalisation investments, and skills for advanced internet use (Lemeilleur et al., 2019).

9.3 Scalability and Energy Use

The growth of the food supply chain leads to the increase of the number of transactions, participants, and data, which further makes scaling of traditional systems harder. (Zhao et al., 2019; Hassoun et al., 2019; Urbano et al., 2020). Therefore, distributed architectures such as regional cloud that allow for horizontal scaling or hybrid architectures such as edge computing combined with regional cloud that allow for both vertical and horizontal scaling are being considered to meet such a need (Hassoun et al., 2019; Urbano et al., 2020). The rapid development of information and communication technology has enabled a variety of processes to take place simultaneously, leading to an increase in energy consumption (Tsang et al., 2019; Guruswamy et al., 2022). In particular, the emergence of technologies capable of processing large amounts of data in a short time necessitates adopting energy consumption reduction techniques (Sharma et al., 2020; Guruswamy et al., 2022). Green supply-chain environment construction and reduction of processing loads are among the techniques under consideration. In addition to improving energy efficiency, traceability is expected to achieve sustainable development by decreasing waste and the number of recalls (Guruswamy et al., 2022).

9.4 Ethical and Social Implications

Traceability is critical to ensuring genuine food safety and maintaining consumer trust in food products (Urbano et al., 2020). Society is put at higher risk not only due to a lack of availability of key information but also due to the occurrence of frauds that leads to major contaminants in food (Spink & Moyer, 2011; M. Jiménez-Carvelo et al., 2023); particularly the presence of hazardous and banned materials is a major concern (Marvin et al 2016).

Detection of when and where the fraud takes place in the supply chain of the food is extremely important in ensuring safety (Mei Soon et al., 2019; M. Jiménez-Carvelo et al., 2023). Food Fraud mainly comprises lowering of food quality either through the substitution of key raw materials, the introduction of banned hazardous materials or even additional fraud which is related to similar misbehaviour in services such as quantity fraud (Marvin et al., 2016; Mei Soon et al., 2019). There is increased interest in innovative methods that exploit food fraud detection to ensure public health safety, income loss for the consumer, degradation of trust, and financial loss of the supplier (Guruswamy et al., 2022). Traceability information represents an important and valuable data source to help in the detection of food fraud (Fanelli et al., 2021; M. Jiménez-Carvelo et al., 2023). Various platforms are being deployed to manage the food fraud traceability information; in particular, the cloud-based platforms dominate the solutions in this perspective (Zhao et al., 2019; Guruswamy et al., 2022).

10 FUTURE DIRECTIONS AND RESEARCH AGENDA

Traceability and fraud prevention continue to be the forefront issues for the agri-food sector as online shopping and globalisation make food products even more complex to track (Urbano et al., 2020; Guruswamy et al., 2022). Numerous systems have contemplated improving food traceability and, with that, substantiating food provenance, keeping track of conditional aspects like temperature and humidity throughout the food supply chain, and preventing food fraud (Lei et al., 2022b). Agri-food safety has a great economic importance, and hence traceability and fraud prevention have scaled up enforcement within regulation and legislation (Lindley, 2022). The Public Health Agency of Canada, the Canadian Food Inspection Agency, and Health Canada have identified traceability as one of the priorities for the agri-food sector in the Vision and Action Plan for the Canadian Agri-Food Sector (Lindley, 2022). Research projects deliver other directions (Guruswamy et al., 2022). Little systematic analysis, however, exists on the integration of emerging monitoring technologies in tracking the food chain (M. Jiménez-Carvelo et al., 2023). The provision of a comprehensive overview and a clearer picture of contamination prevention is needed (Fanelli et al., 2021). The analysis indicates first-order and second-order fraud prevention by relying either on procedural enforcement or actors' reputation (Pearson et al., 2019). Nothing either on the tracking side exists on survey data on the situation regarding traceability and food fraud in the agri-food sector on these topics (M. Jiménez-Carvelo et al., 2023). Some important initiatives remain isolated as to their content and their relevance, or are otherwise undocumented (Lemeilleur et al., 2019). Still in the agri-food sector, Spatial-Temporal Event Analysis enables early signalling of food fraud-related anomalies. These diagrams supplement traditional graphs strictly based on time (M. Jiménez-Carvelo et al., 2023).

Much gain in cooperation and information exchange across the supply chain is possible, and big data from credible sources allows firms to better evaluate the vulnerability of their products to fraudulent actions (Salah et al., 2016; Mei Soon et al., 2019; Sharma et al., 2020). Such an evaluation, which constitutes an environmental scan of opportunities for managerial action, constitutes first-order fraud prevention grounded in risk analysis (Marvin et al., 2016; Mei Soon et al., 2019). A large number of countries and organisations have pursued the aim of easier international trade or food safety, and a plethora of standards and good practices documents have emerged, such as ISO 22005:2007 or the CODEX Alimentarius Hygiene Principles and Practices (Charlebois et al., 2014; Guruswamy et al., 2022). A systematic study of the key aspects of representing food traceability or fraud in a manner that enables their assessment against standards or good practice documents helps capture an overall perspective without anything such as a comprehensive repository or knowledge base existing on that topic (Karlsen et al., 2013; Lindley, 2022). This also allows the degrees of freedom of alternative monitoring technologies to be approximated from the corresponding specifications (Urbano et al., 2020).

Traceability and fraud prevention continued to appear as forefront issues for the agri-food sector as online shopping or globalisation made many more and much more complex food products (Hobbs, 2020; Guruswamy et al., 2022). Agri-food safety has economic importance, yet traceability and fraud prevention have scaled up in enforcement within regulation (Trienekens & Zuurbier, 2008; Lindley, 2022). Knowledge-packing of procedures before first or second orders of alternative monitoring technologies remains scarce (Lemeilleur et al., 2019). Models capturing either traceability or fraud exist, but not both together (Ringsberg, 2014; M. Jiménez-Carvelo et al., 2023). Little understanding of co-occurring regulations or their commonalities prevails (Lindley, 2022).

11 CONCLUSION

Food traceability systems significantly contribute to food safety and quality improvements by monitoring and controlling food supply chains through widely studied digital technologies and tools, including the Internet of Things, big data, Artificial Intelligence, and Blockchain. However, food fraud and the inadvertent dissemination of dangerous food-related elements still pose serious threats, resulting in severe economic losses, tarnished brand images, and health crises. Food fraud prevention and mitigation attract increasing attention and require intensive research efforts. Within food fraud prevention, spatial-temporal event analysis provides an innovative approach to signal emerging food fraud-related anomalies in supply chains and can enhance the capability to flag suspicious food-supply-chain activities promptly. Potential sources of data include general statistics, tracking records by the Global Positioning Systems, and internet postings; however, the response remains to be the ability to capture the pertinent data. The predetermined origin, transit, and destination means the possibility of fraudulent behaviours such as mislabelling and illegal diversion.

REFERENCES

Aung, M. M., & Chang, Y. S. (2014). Traceability in a food supply chain: Safety and quality perspectives. *Food Control*, *39*, 172–184. DOI: 10.1016/j.foodcont.2013.11.007

Bokulich, N. A., Lewis, Z. T., Boundy-Mills, K., & Mills, D. A. (2016). A new perspective on microbial landscapes within food production. *Current Opinion in Biotechnology*, *37*, 182–189. DOI: 10.1016/j.copbio.2015.12.008 PMID: 26773388

Bosona, T., & Gebresenbet, G. (2013). Food traceability as an integral part of logistics management in food and agricultural supply chain. *Food Control*, *33*(1), 32–48. DOI: 10.1016/j.foodcont.2013.02.004

Casino, F., Kanakaris, V., Dasaklis, T. K., Moschuris, S., & Rachaniotis, N. P. (2021). Modeling food supply chain traceability based on blockchain technology. *IFAC-PapersOnLine*, *54*(1), 115–121. DOI: 10.1016/j.ifacol.2019.11.620

Charlebois, S., Sterling, B., Haratifar, S., & Naing, S. K. (2014). Comparison of global food traceability regulations and requirements. *Comprehensive Reviews in Food Science and Food Safety*, *13*(5), 1104–1123. DOI: 10.1111/1541-4337.12101

Costa, C., Antonucci, F., Pallottino, F., Aguzzi, J., Sarriá, D., & Menesatti, P. (2013). A review on agri-food supply chain traceability by means of RFID technology. *Food and Bioprocess Technology*, *6*(2), 353–366. DOI: 10.1007/s11947-012-0958-7

Creydt, M., & Fischer, M. (2019). Blockchain and more - Algorithm driven food traceability. *Food Control*, *105*, 45–51. DOI: 10.1016/j.foodcont.2019.05.019

Dabbene, F., Gay, P., & Tortia, C. (2014). Traceability issues in food supply chain management: A review. *Biosystems Engineering*, *120*, 65–80. DOI: 10.1016/j.biosystemseng.2013.09.006

Dias, J. S. (2021). Nutritional quality and health benefits of vegetables: A review. *Food and Nutrition Sciences*, *3*(10), 1354–1374. DOI: 10.4236/fns.2012.310179

Ellahi, R. M., Wood, L. C., & Bekhit, A. E.-D. A. (2023). Blockchain-Based Frameworks for Food Traceability: A Systematic Review. *Foods*, *12*(16), 3026. DOI: 10.3390/foods12163026 PMID: 37628025

Ellis, D. I., Muhamadali, H., & Allen, D. P.. (2022). Fingerprinting food: Current technologies for the detection of food fraud and adulteration. *Trends in Food Science & Technology*, *119*, 122–136. DOI: 10.1016/j.tifs.2021.11.016

Ercolini, D. (2013). High-throughput sequencing and metagenomics: Moving forward in the culture-independent analysis of food microbial ecology. *Applied and Environmental Microbiology*, *79*(10), 3148–3155. DOI: 10.1128/AEM.00256-13 PMID: 23475615

European Commission. (2022). Food safety: From the farm to the fork. Publications Office of the European Union. https://food.ec.europa.eu/horizontal-topics/farm-fork-strategy_en

Fanelli, V., Mascio, I., Miazzi, M. M., Savoia, M. A., De Giovanni, C., & Montemurro, C. (2021). Molecular Approaches to Agri-Food Traceability and Authentication: An Updated Review. *Foods*, *10*(7), 1644. DOI: 10.3390/foods10071644 PMID: 34359514

Feng, H., Wang, X., Duan, Y., Zhang, J., & Zhang, X. (2020). Applying blockchain technology to improve agri-food traceability: A review of development methods, benefits and challenges. *Journal of Cleaner Production*, *260*, 121031. DOI: 10.1016/j.jclepro.2020.121031

Franz, E., Gras, L. M., & Dallman, T. (2014). Significance of whole genome sequencing for food safety and public health. *Food Microbiology*, *40*, 102–108. DOI: 10.1016/j.fm.2013.12.003

Galvez, J. F., Mejuto, J. C., & Simal-Gandara, J. (2018). Future challenges on the use of blockchain for food traceability analysis. *Trends in Analytical Chemistry*, *107*, 222–232. DOI: 10.1016/j.trac.2018.08.011

Golan, E., Krissoff, B., Kuchler, F., Calvin, L., Nelson, K., & Price, G. (2004). Traceability in the U.S. food supply: Economic theory and industry studies. United States Department of Agriculture Economic Research Service. Agricultural Economic Report No. 830. DOI: 10.22004/AG.ECON.33939

Guruswamy, S., Pojić, M., Subramanian, J., Mastilović, J., Sarang, S., Subbanagounder, A., Stojanović, G., & Jeoti, V. (2022). Toward better food security using concepts from Industry 5.0. *Sensors (Basel)*, *22*(21), 8377. DOI: 10.3390/s22218377 PMID: 36366073

Hassoun, A., Aït-Kaddour, A., Abu-Mahfouz, A. M., Rathod, N. B., Bader, F., Barba, F. J., Biancolillo, A., Cropotova, J., Galanakis, C. M., Jambrak, A. R., Lorenzo, J. M., Luz, C., Mañes, J., Moltó, J. C., Nuñez, J. M. M., Ozogul, F., & Regenstein, J. M. (2022). The fourth industrial revolution in the food industry—Part I: Industry 4.0 technologies. *Critical Reviews in Food Science and Nutrition*, ●●●, 1–17. DOI: 10.1080/10408398.2022.2034735 PMID: 35114860

Hobbs, J. E. (2020). Food supply chains during the COVID-19 pandemic. *Canadian Journal of Agricultural Economics, 68*(2), 171–176. DOI: 10.1111/cjag.12237

Jagadeesan, B., Gerner-Smidt, P., Allard, M. W., Leuillet, S., Winkler, A., Xiao, Y., Chaffron, S., Van Der Vossen, J., Tang, S., Katase, M., McClure, P., Kimura, B., Ching Chai, L., Chapman, J., & Grant, K. (2019). The use of next generation sequencing for improving food safety: Translation into practice. *Food Microbiology, 79*, 96–115. DOI: 10.1016/j.fm.2018.11.005 PMID: 30621881

Jansen-Vullers, M. H., van Dorp, C. A., & Beulens, A. J. (2003). Managing traceability information in manufacture. *International Journal of Information Management, 23*(5), 395–413. DOI: 10.1016/S0268-4012(03)00066-5

Jiménez-Carvelo, A. M., Li, P., Erasmus, S. W., Wang, H., & van Ruth, S. M. (2023). Spatial-Temporal Event Analysis as a Prospective Approach for Signalling Emerging Food Fraud-Related Anomalies in Supply Chains. *Foods, 12*(1), 61. DOI: 10.3390/foods12010061 PMID: 36613277

Kamble, S. S., Gunasekaran, A., & Sharma, R. (2020). Modeling the blockchain enabled traceability in agriculture supply chain. *International Journal of Information Management, 52*, 101967. DOI: 10.1016/j.ijinfomgt.2019.05.023

Karlsen, K. M., Dreyer, B., Olsen, P., & Elvevoll, E. O. (2013). Literature review: Does a common theoretical framework to implement food traceability exist? *Food Control, 32*(2), 409–417. DOI: 10.1016/j.foodcont.2012.12.011

Khan, S., Haleem, A., Khan, M. I., Abidi, M. H., & Al-Ahmari, A. (2018). Implementing traceability systems in specific supply chain management through critical success factors. *Sustainability (Basel), 5*(1), 1515822. DOI: 10.3390/su10010204

Kher, S. V., Frewer, L. J., Jonge, J. D., Wentholt, M. T., Davies, O. H., Luijckx, N. B. L., & Cnossen, H. J. (2010). Experts' perspectives on the implementation of traceability in Europe. *British Food Journal, 112*(3), 261–274. DOI: 10.1108/00070701011029138

Kim, H. M., & Laskowski, M. (2018). Toward an ontology-driven blockchain design for supply-chain provenance. *Intelligent Systems in Accounting, Finance & Management, 25*(1), 18–27. DOI: 10.1002/isaf.1424

Lei, M., Liu, S., Luo, N., Yang, X., & Sun, C. (2022a). Trusted-auditing chain: A security blockchain prototype used in agriculture traceability. *Heliyon, 8*(11), e11477. DOI: 10.1016/j.heliyon.2022.e11477 PMID: 36406715

Lei, M., Xu, L., Liu, T., Liu, S., & Sun, C. (2022b). Integration of privacy protection and blockchain-based food safety traceability: Potential and challenges. *Foods, 11*(19), 2932. DOI: 10.3390/foods11152262 PMID: 35954029

Lemeilleur, S., Maître-d'Hôtel, É., Lepiller, O., & Hobeika, A. (2019). Uncertain impacts of new technologies: The case of digital agriculture and blockchains. FOOD SYSTEMS AT RISK, 91.

Leong, D., Alvarez-Ordóñez, A., & Jordan, K. (2014). Monitoring occurrence and persistence of Listeria monocytogenes in foods and food processing environments in the Republic of Ireland. *Frontiers in Microbiology*, *5*, 436. DOI: 10.3389/fmicb.2014.00436 PMID: 25191314

Lindley, J. (2022). Food regulation and policing: Innovative technology to close the regulatory gap in Australia. *Frontiers in Sustainable Food Systems*, *6*(2), 836264. DOI: 10.1007/s00003-022-01372-2 PMID: 35282596

Liu, K., Xing, R., Sun, R., Ge, Y., & Chen, Y. (2022). An accurate and rapid way for identifying food geographical origin and authenticity: Editable DNA-traceable barcode. *Foods*, *12*(1), 82. DOI: 10.3390/foods12010017 PMID: 36613233

Manning, L., & Soon, J. M. (2016). Building strategic resilience in the food supply chain. *British Food Journal*, *118*(6), 1477–1493. DOI: 10.1108/BFJ-10-2015-0350

Marvin, H. J. P., Bouzembrak, Y., Janssen, E. M., van der Fels-Klerx, H. J., van Asselt, E. D., & Kleter, G. A. (2016). A holistic approach to food safety risks: Food fraud as an example. *Food Research International*, *89*(1), 463–470. DOI: 10.1016/j.foodres.2016.08.028 PMID: 28460939

Mei Soon, J., Krzyzaniak, S. A., Shuttlewood, Z., Smith, M., & Jack, L. (2019). Food fraud vulnerability assessment tools used in food industry. *Food Control*, *101*, 225–232. DOI: 10.1016/j.foodcont.2019.03.002

Olsen, P., & Borit, M. (2013). How to define traceability. *Trends in Food Science & Technology*, *29*(2), 142–150. DOI: 10.1016/j.tifs.2012.10.003

Opara, L. U. (2003). Traceability in agriculture and food supply chain: A review of basic concepts, technological implications, and future prospects. *Journal of Food Agriculture and Environment*, *1*(1), 101–106.

Pearson, S., May, D., Leontidis, G., Swainson, M., Brewer, S., Bidaut, L., Freyd, J. G., Parr, G., Maull, R., & Zisman, A. (2019). Are distributed ledger technologies the panacea for food traceability? *Global Food Security*, *20*, 145–149. DOI: 10.1016/j.gfs.2019.02.002

Quince, C., Walker, A. W., Simpson, J. T., Loman, N. J., & Segata, N. (2017). Shotgun metagenomics, from sampling to analysis. *Nature Biotechnology*, *35*(9), 833–844. DOI: 10.1038/nbt.3935 PMID: 28898207

Regattieri, A., Gamberi, M., & Manzini, R. (2007). Traceability of food products: General framework and experimental evidence. *Journal of Food Engineering*, *81*(2), 347–356. DOI: 10.1016/j.jfoodeng.2006.10.032

Ringsberg, H. (2014). Perspectives on food traceability: A systematic literature review. *Supply Chain Management*, *19*(5/6), 558–576. DOI: 10.1108/SCM-01-2014-0026

Rouvière, E., & Caswell, J. A. (2012). From punishment to prevention: A French case study of the introduction of co-regulation in enforcing food safety. *Food Policy*, *37*(3), 246–254. DOI: 10.1016/j.foodpol.2012.02.009

Salah, K., Rehman, M. H. U., Nizamuddin, N., & Al-Fuqaha, A. (2019). Blockchain for AI: Review and open research challenges. *IEEE Access : Practical Innovations, Open Solutions*, *7*, 10127–10149. DOI: 10.1109/ACCESS.2018.2890507

Shah, R. Y., Prajapati, P. N., & Agrawal, Y. K. (2010). Anticounterfeit packaging technologies. *Journal of Advanced Pharmaceutical Technology & Research*, *1*(4), 368–373. DOI: 10.4103/0110-5558.76434 PMID: 22247875

Shanahan, C., Kernan, B., Ayalew, G., McDonnell, K., Butler, F., & Ward, S. (2009). A framework for beef traceability from farm to slaughter using global standards: An Irish perspective. *Computers and Electronics in Agriculture*, *66*(1), 62–69. DOI: 10.1016/j.compag.2008.12.002

Sharma, R., Kamble, S. S., Gunasekaran, A., Kumar, V., & Kumar, A. (2020). A systematic literature review on machine learning applications for sustainable agriculture supply chain performance. *Computers & Operations Research*, *119*, 104926. DOI: 10.1016/j.cor.2020.104926

Siddiqui, S. A., Bahmid, N. A., Taha, A., Khalifa, I., Khan, S., Rostamabadi, H., & Jafari, S. M. (2023). Recent advances in food applications of phenolic-loaded micro/nanodelivery systems. *Critical Reviews in Food Science and Nutrition*, *63*(27), 8939–8959. DOI: 10.1080/10408398.2022.2056870 PMID: 35426751

Smith, G. C., Tatum, J. D., Belk, K. E., Scanga, J. A., Grandin, T., & Sofos, J. N. (2005). Traceability from a US perspective. *Meat Science*, *71*(1), 174–193. DOI: 10.1016/j.meatsci.2005.04.002 PMID: 22064063

Spink, J., & Moyer, D. C. (2011). Defining the public health threat of food fraud. *Journal of Food Science*, *76*(9), R157–R163. DOI: 10.1111/j.1750-3841.2011.02417.x PMID: 22416717

Storøy, J., Thakur, M., & Olsen, P. (2013). The TraceFood framework - Principles and guidelines for implementing traceability in food value chains. *Journal of Food Engineering*, *115*(1), 41–48. DOI: 10.1016/j.jfoodeng.2012.09.018

Tian, F. (2016). An agri-food supply chain traceability system for China based on RFID & blockchain technology. In 2016 13th International Conference on Service Systems and Service Management (pp. 1-6). IEEE. DOI: 10.1109/ICSSSM.2016.7538424

Tian, F. (2017). A supply chain traceability system for food safety based on HACCP, blockchain & Internet of things. In *2017 International Conference on Service Systems and Service Management* (pp. 1-6). IEEE. DOI: 10.1109/ICSSSM.2017.7996119

Trienekens, J., & Zuurbier, P. (2008). Quality and safety standards in the food industry, developments, and challenges. *International Journal of Production Economics*, *113*(1), 107–122. DOI: 10.1016/j.ijpe.2007.02.050

Tsang, Y. P., Choy, K. L., Wu, C. H., Ho, G. T. S., & Lam, H. Y. (2019). Blockchain-driven IoT for food traceability with an integrated consensus mechanism. *IEEE Access : Practical Innovations, Open Solutions*, *7*, 129000–129017. DOI: 10.1109/ACCESS.2019.2940227

Urbano, O., Perles, A., Pedraza, C., Rubio-Arraez, S., Castelló, M. L., Ortolá, M. D., & Mercado, R. (2020). Cost-effective implementation of a temperature traceability system based on smart RFID tags and IoT services. *Sensors (Basel)*, *20*(4), 1163. DOI: 10.3390/s20041163 PMID: 32093218

van Rijswijk, W., & Frewer, L. J. (2008). Consumer perceptions of food quality and safety and their relation to traceability. *British Food Journal*, *110*(10), 1034–1046. DOI: 10.1108/00070700810906642

Verbeke, W., & Ward, R. W. (2006). Consumer interest in information cues denoting quality, traceability and origin: An application of ordered probit models to beef labels. *Food Quality and Preference*, *17*(6), 453–467. DOI: 10.1016/j.foodqual.2005.05.010

Wang, L., Kwok, S. K., & Ip, W. H. (2010). A radio frequency identification and sensor-based system for the transportation of food. *Journal of Food Engineering*, *101*(1), 120–129. DOI: 10.1016/j.jfoodeng.2010.06.020

Wang, Y., Chen, C. H., & Zghari-Sales, A. (2021). Designing a blockchain enabled supply chain. *International Journal of Production Research*, *54*(1), 6–14. DOI: 10.1080/00207543.2020.1824086

Xu, L., Chen, L., Gao, Z., Chang, Y., Iakovou, E., & Shi, W. (2018). Binding the physical and cyber worlds: A blockchain approach for cargo supply chain security enhancement. In *2018 IEEE International Symposium on Technologies for Homeland Security* (pp. 1-5). IEEE. DOI: 10.1109/THS.2018.8574184

Yadav, B., Sharma, G., Sankhla, M. S., Sehgal, S., Tomar, J. S., & Dubey, A. (2026). Integrative Forensic Methods in Food Contaminant Detection and Identification. In Ogwu, M., & Izah, S. (Eds.), *Detection, Identification, and Monitoring of Food Contaminants* (pp. 59–82). IGI Global Scientific Publishing., DOI: 10.4018/979-8-3373-3982-5.ch003

Zhang, J., & Bhatt, T. (2014). A Guidance Document on the Best Practices in Food Traceability. *Comprehensive Reviews in Food Science and Food Safety*, *13*(5), 1074–1103. DOI: 10.1111/1541-4337.12103

Zhao, G., Liu, S., Lopez, C., Lu, H., Elgueta, S., Chen, H., & Boshkoska, B. M. (2019). Blockchain technology in agri-food value chain management: A synthesis of applications, challenges and future research directions. *Computers in Industry*, *109*, 83–99. DOI: 10.1016/j.compind.2019.04.002

Chapter 10
AI-Enabled Early Warning Systems for Climate-Linked Biological Risks

Rachan Karmakar
http://orcid.org/0000-0002-3220-6287
Department of Environmental Science, Graphic Era (Deemed to be) University, Dehradun, India

Debasis Mitra
http://orcid.org/0000-0003-0525-8812
Department of Microbiology, Graphic Era (Deemed to be) University, Dehradun, India

Rajesh Kumar
Department of Chemistry, Government Degree College Chaubattakhal, Sri Dev Suman Uttarakhand University, Pauri Garhwal, India

Pratibha Naithani
Department of Environmental Science, Graphic Era (Deemed to be) University, Dehradun, India

Rahul Kumar
http://orcid.org/0009-0000-1323-2220
Department of Microbiology, Graphic Era (Deemed to be) University, Dehradun, India

Andrés Moreno
Faculty of Chemical Sciences and Technologies, University of Castilla-La Mancha, Ciudad Real, Spain

Roman Tandilch
Faculty of Pharmacy, Rhodes University, Grahamstown, South Africa

Bipin Sati
Department Science and Technology, Uttarakhand State Council for Science and Technology, Dehradun, India

Nikhil Ranjan Behera
Department of Environmental Science, Graphic Era (Deemed to be) University, Dehradun, India

Nurudeen Olatunbosun Adeyemi
http://orcid.org/0000-0001-6341-775X
Federal University of Agriculture, Abeokuta, Nigeria

Bekri Xhemali
http://orcid.org/0000-0002-3700-2259
Laboratory of Plant Protection, Kosovo Institute of Agriculture, Pejë, Kosovo

DOI: 10.4018/979-8-3373-9033-8.ch010

ABSTRACT

Climate change's accelerated biological risk to human health, agriculture, wildlife, and ecosystems has altered pathogen ecology, vector dynamics, and species interactions. This chapter examined AI-enabled predictive analytics systems for early warning of climate-related biological risk, using evidence from climate science, epidemiology, ecology, and computational models. The findings show that Machine Learning and Deep Learning frameworks with Earth Observing Technology and real-time sensing have improved lead time, spatial resolution, and risk characterization for non-linear climate conditions. The platforms improved predictions of vector-borne disease risk, zoonotic spillover risks, agricultural pest management, and ecosystem hazards. However, challenges remain in data availability, model transferability, uncertainty communication, and system integration. This chapter discusses gaps and directions for developing scalable, explainable, and policy-relevant AI-driven predictive analytics systems amid rapid climate changes.

1. INTRODUCTION

Before it was recognized as a health and ecological security issue, climate change had an impact on the structure and function of biological systems. Changes in climate with gradual warming trends, more frequent extreme weather events (i.e., droughts, floods, and storms), and changes in hydrological regimes have been shown to be associated with alterations in pathogen survival, vector competence, host susceptibility to infection, and ecosystem resilience (Sutherst, 2004). The responses to these trends are not linear and usually demonstrate either a delayed or threshold response to the observed changes; therefore, conventional surveillance systems would have difficulty detecting the impact of these changes early on (Southall et al., 2021). Consequently, the impact of climate-induced biological risks represents a multi-scalar and dynamic phenomenon that affects human health and agriculture as well as wildlife and ecosystem integrity. Additionally, it became evident that much of the existing knowledge on biological risks associated with climate forcing was based on historical trends and did not take changes that occurred into consideration for future predictions. For example, the increasing availability of high-resolution climate data combined with increased ecological observations and health record data provides an environment in which it is possible to implement AI-based systems to predict climate-related biological risks prior to their occurrence (Inam, 2025). This chapter provides a structured literature review of articles (peer-reviewed) published between 2015 and 2026 on Artificial Intelligence (AI)-powered early warning systems for biological risk due to climate change. Articles were identified through searches of

Scopus, Web of Science, PubMed, and IEEE Xplore with appropriate keywords. Studies with some relevance to this topic were included while excluding studies that were in another language, duplicates, and studies that were not relevant to the topic, leaving an exact synthesis for us to examine.

1.1. Climate Change as a Driver of Emerging and Re-Emerging Biological Risks

The biological risk associated with the development of climate-related hazards (e.g., climate change, climate-induced disasters, etc.) is shaped by the impacts of climate-related stressors (i.e., temperature and rainfall) on basic ecological properties (i.e., vector populations) that drive transmission or host susceptibility (Bacillus thuringiensis (Bt) vector populations, etc.). Temperature is the primary climatic variable that influences the development of new biological risks, and studies have consistently found that temperature drives changes in the rate of pathogen replication, extrinsic incubation time (EIT) of a pathogen, and lifetime of a vector. In a study by Mordecai et al. (2020), malaria transmission was shown to have an optimal temperature of approximately 25 °C, and both lower and higher temperatures reduced the potential for vector transmission (i.e., vectorial capacity). Mordecai et al. (2017) reported similar temperature-dependent patterns for dengue virus transmission; dengue transmission rates increased sharply as the temperature increased from 18 °C to 32 °C. Because precipitation also plays a role in shaping biological risk, it can also create new reproductive habitats for vectors through new breeding ground opportunities. As with temperature, increased rainfall creates new breeding ground opportunities, whereas prolonged droughts promote the practice of storing water, thereby promoting the proliferation of mosquitoes. A multi-country study conducted across Southeast Asia provided evidence of an 8%–12% increase in the incidence of dengue for each 10 mm increase in mean monthly rainfall experienced, if the temperature during this period was conducive to dengue transmission (Huang et al. 2021). The availability of humidity also affects the longevity of the vector and the viability of the pathogen, particularly in tropical and subtropical regions.

In addition to the existing literature on vector-borne diseases, there is an increased risk of zoonotic spillover associated with the impact of climate change on wildlife habitat destruction and wildlife movement disruptions. For example, Carlson et al. (2022) predicted that due to climate-driven range shifts, by 2070 over 15,000 new viral sharing events will take place among mammal species, greatly increasing the potential to transmit viruses to humans. The greatest risk of zoonotic disease occurs in areas that are undergoing very rapid land-use changes, wherein climate change stress is compounded by human-induced land-use change. Similar to ecosystems, agricultural and ecosystem-related biological risks are sensitive to changes in effective corridors for dispersal caused by changes in weather patterns. The rise in temperature and atmospheric carbon dioxide are driving the development of insects and increasing the distributional ranges of many invasive species (Subedi et al., 2023). A global meta-analysis estimated that insect-induced crop losses will increase by between 10% and 25% for each degree Celsius increase in temperature, especially in temperate climates. A warming climate and nutrient enrichment (nutrients through exports from agriculture) both cause an increase in harmful algal blooms (HABs), which increase the amount of toxins produced and disrupt ecosystem functions in aquatic ecosystems (Paerl, 2017). The occurrence of record high cyanobacteria bloom frequency in lakes where surface water temperatures are greater than 1 – 2 °C above the long-term average temperature is considered a direct result of a warming planet (Paerl, 2017). Together, these studies confirm that biological risks from climate change arise from a complex web of physical, ecological, and biological processes, and the nonlinear interactions between these processes limit the ability to use simple threshold-based warning systems to predict biological risks from climate change. These studies highlighted the importance of developing robust predictive and data-driven strategies for evaluating complex heterogeneous datasets.

1.2. Limitations of Conventional Bio-Monitoring Systems and The Need for Artificial Intelligence-Driven Early Warning Systems

Conventional bio-surveillance systems have traditionally been designed to support retrospective analytical monitoring rather than proactive anticipation of potential risks. Most epidemiological/ecological surveillance models are based on case count reports and periodically collected field surveys and laboratory confirmations, each of which creates significant delays in receiving and analyzing the data. For infectious diseases, the reporting lag time can be several weeks to months, particularly in areas with limited resources. Liu et al. (2025) conducted a global study and found that more than 60% of the reports of infectious disease outbreaks reached national authorities after local transmission peaked. Furthermore, the limited geographic reach of these systems limits their effectiveness. In most instances, bio-surveillance networks reportedly been established at denser levels in urbanised/economically developed areas and at lower rates in rural/forested/transboundary areas where there are frequently climate-related biological risks (Sheffield et al., 2018). Wildlife/ecosystem health bio-surveillance is even more limited in terms of creating effective monitoring systems because of the continued difficulty associated with expanding the geographic area for collecting and analyzing data associated with wildlife species population levels additionally, the difficulty associated with the continuous monitoring of pathogen prevalence and environmental conditions in most areas of the globe is logistically daunting and expensive to conduct.

The awareness of advanced analytical methods has increased in recent years owing to the limitations associated with traditional methods used for developing early warning systems, including regression and rule-based methods, which are not capable of handling complex non-linear relationships or interactions among multiple variables (Ali et al., 2024). In addition, many of the models used by epidemiologists assume that the relationship between climate and disease remains linear, which is not the case as our climate continues to change at an accelerating pace. Linear models developed to predict malaria based on climate have been shown to lose accuracy when extended beyond their original study periods, and exhibit as much as a 30% increase in prediction errors during years of unusual climatic conditions (Nissan et al., 2021). As climate data streams, remote sensing products, and digital health records have expanded, the volume and complexity of these data have exceeded the capacity for manual human analysis. High-resolution (daily to sub-daily) satellite-derived land surface temperature, vegetation indices, soil moisture, precipitation, and genomic data generated through genomic sequencing and rapid digital reporting of infectious diseases have provided many opportunities for creating new methods of extracting early warning signals from these diverse data sources using automated feature extraction and automated pattern recognition techniques.

The limitations of traditional analytical approaches have been increasingly addressed with the implementation of the developed AI methods. The capacity of machine and deep learning to deal with complex (non-linear) relationships and moderate-type integration of multi-dimensional data and dynamically respond to environmental variations (retraining) will enhance their use in this process (Rane et al., 2024). Evidence from early adopters indicates improved lead times and accuracy in predictive models. For example, the use of machine learning methods for an early warning system in Brazil for dengue fever resulted in 4 to 6 weeks lead time predictive capability and an area under the curve of >0.85, while traditional statistical methods achieved area under the curve of 0.78 or less (Costa et al., 2026). The transformation to AI-enhancing alert systems will therefore be viewed not only as a technological trend (development), but also as a functional response to the growing complexity, speed, and scale associated with climate-induced biological hazards. Thus, the evolution of the systematic integration of climate science, ecological science, and computer intelligence into unified, multidimensional, and predictive models is illustrated in later sections.

2. CLIMATE–BIOLOGY INTERACTION PATHWAYS AND RISK DYNAMICS

According to Pfenning-Butterworth et al. (2024), climate-linked biological threats arise from many interactions between atmospheric processes, ecological systems, and biological responses over a wide range of spatial and temporal scales. These influences do not act independently of each other but rather interact through climate drivers, ecological and socio-environmental contexts that affect species traits, ecosystem structure, and human activities on the planet (Figure 1). Climatic forcing has been shown by an increasing number of examples to cause biological responses that have a nonlinear, threshold-dependent, and delayed pattern, which makes it difficult for detection efforts to identify potential threats while they are emerging. The presence of spatial variability also adds complexity to the dynamic nature of climate-linked biological risks, whereby identical climatic stimuli can lead to inconsistent results in terms of the impacts produced within local ecological and socio-environmental settings. Because of the many pathways along which ecological systems, biological beings, and the climate are interconnected, it is critical that any design process for building/selling artificial intelligence (AI)-enabled systems to support early warning efforts include adequate time to understand how these pathways exist.

Figure 1. Climate-linked biological risk emergence and AI-enabled early warning framework

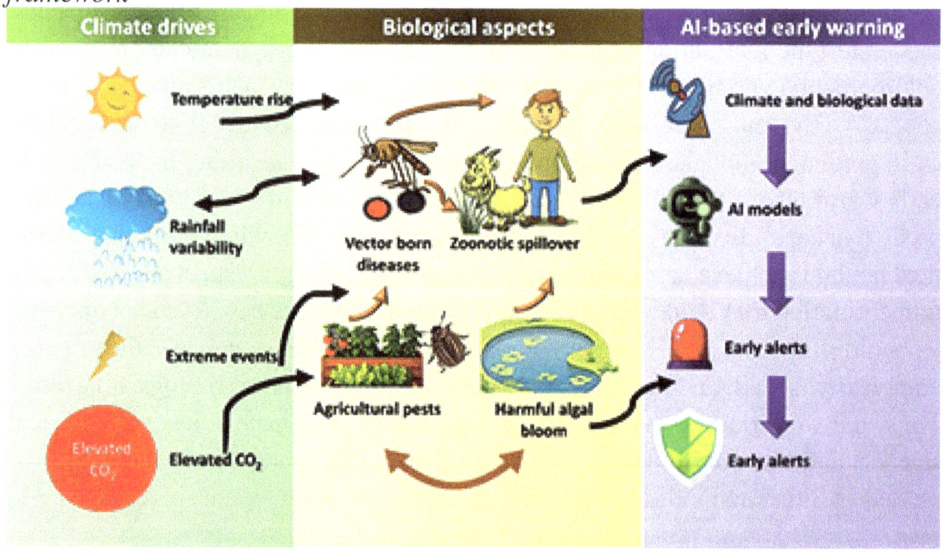

2.1. Climate Variables are Affecting Biotic Hazards

Temperature remains the most important climate variable affecting the dynamics of biological hazards (Yu, 2025). Temperature has been shown to directly affect metabolic rates, replication rates of pathogens, and competence of vectors through laboratory and field research. For vector-borne diseases, temperature-related development rates determine the length of the extrinsic incubation period of the pathogens in the vector. The synthesis by Tesla et al. (2018) indicated that Dengue Fever transmission efficiencies increased exponentially from 20 °C to 30 °C. Further increases in temperature above 30 °C led to a dramatic decrease in transmission efficiency owing to vector mortality. Thus, the temperature window for optimal transmission was not linear. The quantitative relationships between important climatic variables and their biological responses are presented in Table 1. Precipitation can have both direct and indirect effects on biotic hazards by increasing water sources for mosquitoes and altering river flow to transmit waterborne diseases. However, flooding from heavy precipitation can also eliminate breeding grounds and decrease the number of vectors present (Benedum et al., 2018). Longitudinal data throughout East Africa demonstrated that moderate rainfall anomalies (as defined by previous year averages) increased malaria by 9-14%, while very heavy rainfall events decreased malaria incidence by disrupting habitat for larvae. The discovery of the aforementioned non-linear relations to assess risk deterministically constituted a significant obstacle for predicting risk (risk assessment).

Table 1. Impacts of climate drivers on biological risk parameters

Climate variable	Biological system	Quantified response (as reported)	Reference
Temperature (+1 °C)	Insect crop pests	Global crop yield losses to insect pests projected to increase by **10%–25% per °C warming** for major staples (wheat, rice, maize)	Deutsch et al. (2018)
Rainfall (+10 mm/month)	Dengue incidence	Reported studies found **positive association between rainfall and dengue cases,** with increased rainfall correlated with higher dengue incidence (e.g., highest relative risk observed at ~60 mm rainfall; reported significant positive correlations between rainfall and dengue risk)	Kakarla et al. (2019)
Relative humidity (>60%)	Aedes mosquito survival and incidence	Studies show increased **Aedes mosquito abundance with higher relative humidity,** correlated with increased dengue incidence; humidity effects on habitat suitability and transmission risk were statistically significant (e.g., mean relative humidity correlated positively with dengue transmission)	Islam et al. (2021)

continued on following page

Table 1. Continued

Climate variable	Biological system	Quantified response (as reported)	Reference
Flood events	Leptospirosis incidence	Flooding events were associated with a **significant increase in leptospirosis incidence and mortality** (e.g., cases rose markedly following flood events in Rio Grande do Sul)	Michelon et al. (2025)
Surface water warming (+1–2 °C)	Cyanobacterial blooms	Cyanobacterial bloom occurrences were reported to **increase in frequency and intensity under higher water temperatures,** exceeding a **>20% increase** across warming scenarios	Goble et al. (2017)

Climate factors, such as relative humidity, temperature, and precipitation, work together to impact vector survival and persistence of pathogens. Relative humidity levels exceeding 60% improve the survival and lifespan of adult mosquitoes, thereby prolonging the time they could infect other animals with diseases. In laboratory studies, *Aedes albopictus* adults lived nearly 20% longer in humid environments than in dry environments with similar temperatures. Climate stress (such as rising CO_2 levels) also affects the interaction of plants with pathogens and insect pests. Increased CO_2 levels have an effect on the physiology of many plants by decreasing the nitrogen content of their tissues and changing the chemistry of their defensive compounds. These physiological changes result in changes in the feeding habits of herbivores and the growth rates of pest populations. Zvereva and Kozlov stated that insect herbivory increased by an average of 17% in response to an increase in atmospheric CO_2 across several crop species (Kalasariya et al., 2024). The rise in insect herbivory due to elevated levels of CO_2 indirectly increases biological risk to agricultural systems.

Extreme climate events, such as heatwaves, floods, and droughts, created an environment of stress that affected biological systems in an acute way.Heatwaves caused mass deaths in wildlife populations, while flooding produced an increase in the amount of pathogens that were transmitted via density-independent mechanisms. Flooding has been shown throughout history to be associated with outbreaks of leptospirosis and cholera due to contamination of water supplies and the displacement of human populations (Naing et al., 2019). A global meta-analysis demonstrated that for every major flood event, there was a 2-4 fold increase in the incidence of leptospirosis, suggesting that climate drivers impact biological hazards through multiple pathways that interact with each other. The extent of these effects and the direction of change were different between different biological systems, emphasizing the need for models that can account for non-linear, multivariate relationships and not rely solely on single variable thresholds (Waqas & Humphries, 2024).

2.2 Categories of Climate Related Biological Risks

Climate-related biological risks were identified across different areas, with each category being characterized by a separate means of transmission and a different sensitivity to changes in climate. One of the best studied areas of climate-related biological risks are vector-borne diseases because of their close connection with environmental first and foremost. These diseases include: Malaria, Dengue, Chikungunya, Zika, and West Nile Virus and all of these demonstrate that they respond to changes in climate through changes in the distribution, abundance and competency of their vectors. There have been many examples of the global expansion of the latitudinal and altitudinal ranges of vectors. Recent decades have also seen gradual increases in altitudinal range for malaria transmission zones within the East African Highlands; on the order of a shift of 150-300m, which has resulted in exposing populations that were previously non-endemic to malaria.

The second category of climate-related biological risks were Zoonotic (animal) diseases. Climate-induced habitat changes have also played a role in changing patterns of animal movement and interaction and thereby have resulted in greater potential for a pathogen/spillover event. Recent literature regarding animal populations as reservoirs for filoviruses demonstrated that climate anomalies had an influence on the timing and type of roosting and migration behaviour exhibited by bats; with these changes subsequently influencing how viruses are expressed in their hosts. Eby et al., 2022 have shown that climatic stress combined with forage scarcity result in increased spillover risk from bats to livestock and humans. The increase in land use change and agricultural expansion has heightened the above-mentioned risk factors. Agricultural biological risks have not only been influenced by agricultural land-use change but are also significantly influenced by temperature and rainfall variation or climate variability. Crop pathogens, insect pests, and invasive organisms respond rapidly to warming trends with increasing geographical spread (Ma et al., 2025). For example, following warming and changes in rainfall patterns, wheat rust pathogens worldwide are transported into many new geographic areas. A global assessment showed that climate-sensitive crop diseases have contributed to approximate 5-15% loss in yield over the last several decades in many major cereal-producing areas of the world. With warmer winters improving overwinter survival, the fall armyworm has expanded geographically. Livestock diseases are affected by climate change. Changes in temperature and rainfall have altered the way livestock are grazed and have affected the availability of water for livestock, as well as the vector populations that impact livestock, such as cattle, sheep, and goats. Rift Valley fever (RVF) outbreaks are known to be closely associated with the occurrence of anomalously high rainfall linked to El Niño (Kassem et al., 2025). Spatial and temporal analyses of RVF outbreaks in Eastern Africa indicate that more than 70% of the significant

RVF outbreaks in this region occurred following periods of above-average rainfall and/or vegetation greening detected using satellite-based indexes (Linthicum et al., 1999). Invasive species spread, outbreaks of plant pathogens, and algal toxins are examples of biological risks assessed at the ecosystem level. As temperature rises, the range of some invasive insects moves northward, such as the pine processionary moth, which is known to have moved up to 5 km a year in Europe (de Boer & Harvey, 2020). Increasing temperatures and nutrient enrichment in lakes have led to increased cyanobacterial growth and toxin production, resulting in degradation of aquatic systems. Cyanobacterial bloom events have been documented to be over 20% more frequent in temperate lake regions because of long-term temperature increases (Wang et al. 2025). The different categories reflect different patterns of temporal dynamics, spatial scales, and data requirements for assessment, indicating the need for flexible early warning systems that can respond to a wide variety of biological risks leading to climate change.

2.3. Spatio-Temporal Complexity and Uncertainty in Biological Risk Prediction

The effects of the environment on biological risk dynamics are complicated and unpredictable because of the latencies between environmental factors and the biological response to environmental factors, particularly the effects of the environment under climate change. Some examples of environmental factors that affect biological risks are climate, ecosystems, and human activities, which can cause different types of diseases (Chen et al., 2023). Another factor affecting biological risk dynamics is the spatiotemporal heterogeneity. In the case of vector-borne diseases, the lag between rainfall anomalies and incidence peaked between two weeks and several months and varied depending on the region and species, making it difficult to define standard early warning thresholds. The factors that because spatiotemporal heterogeneity are diverse and consist of ecological, social, and socio-economic components, which determine how quickly the biological hazards increase, how long the disease remains in an area, and the economic cost of controlling the disease. An analysis of dengue fever transmission between cities in Latin American countries showed that an area with a similar climate could transmit three times as much dengue fever as a city with a similar climate, due to differences in urban physical structure and the way in which a city manages water (Lowe et al., 2021).

Uncertainty about climate means that it is extremely difficult to accurately predict climate patterns. Over the years, the association between climate and biological factors has not always been constant; rather, as species evolve and adapt to the ecosystem's natural changes, or in some cases as humans change the way they interact —through agricultural management practices or the use of chemicals or pesticides — the relationships can change. The reliability of models developed using historical climate data often decreases when they are used to predict future climate patterns. For example, during a study to validate climate-based malaria models, the accuracy of predictions dropped by 25% to 40% when transferring models across multiple decades without recalibration. Additionally, uncertainty is present in every step of the risk prediction process because of various factors, including uncertainty arising from the use of climate model projections, measurement errors, and incomplete biological data. For example, climate change projections resulted in a greater than 50% difference in the predicted future dengue risk based on the chosen climate-model-emission scenario combination. This level of uncertainty complicates the ability to make effective decisions, particularly when developing early warning systems to initiate costly intervention strategies.

The identification of the threshold effects made the prediction even more difficult. Once temperature or climate conditions reach certain thresholds, unique responses may be exhibited, including sudden pest emergence (Skendžić et al., 2021) as a result of a short winter, sudden disease outbreaks (insects) once temperatures increase, and sudden emergence of other living organisms once their typical temperature limits have been exceeded. The inability to identify tipping points via conventional modelling methods, combined with the difficulty of detecting early signs of these events within "noise" (i.e., spatio-temporal complexity), clearly demonstrates the limitations of conventional linear (predictive) models. Accordingly, they offer a clear rationale for the use of AI models that are adaptable, learn from the data, and adapt to changing environmental conditions. These AI models can learn and produce adaptive patterns, integrate multi-scale data from multiple sources, and provide new predictions as new data become available in real time. Below, we discuss AI models in detail. (Zhong, 2025).

3. AI Architectures and Computational Frameworks for Early Warning

Over the last several years, the use of AI-assisted early warning systems for biological risks associated with climate has increased because of advances in computational architecture that allow for the analysis of large heterogeneous time-structured datasets (Ifeanyi et al. 2025). The interaction of climate with biological systems results in the creation of data streams of extremely high dimensionality, including re-analysis data produced by climatologists, satellite images taken by remote sensors, observations of ecological systems, genomic sequences obtained through biological testing and health records, and epidemiological studies. Classical statistical methods are not sufficient to perform analyses of this type of complexity, and therefore require a nonlinear/nonstationary approach for accurate assessment. As a result, AI systems have been adopted more widely for extracting patterns that are not readily observable, modelling time-dependent relationships, and generating probabilistic predictions. Early AI systems were developed from traditional machine learning to deep learning and hybrid approaches that combine machine learning with uncertainty quantification and explainability to inform decision-making in real-world applications.

3.1. Machine Learning Models for Biological Risk Prediction Related to Climate Change

Historically, AI in biological early warning systems has primarily utilized traditional machine learning techniques (owing to the inherent interpretability and moderate data needs of these algorithms) to generate predictive elements. Random Forests, Gradient Boosting Machines, and Support Vector Machines have all been used extensively to combine climate variables with biological endpoints (see Villanueva-Miranda et al., 2025). The characteristic nonlinear interaction of variables represented in the predictive models resulted in excellent performance and predictive quality when predicting above and beyond what traditional approaches have been able to generate (Table 2). Random Forests have emerged as one of the primary methodologies applied in predicting vector-borne diseases because of the robustness of the algorithms in preventing overfitting as well as their ability to accommodate highly correlated predictor variables. A current example includes a multidisciplinary, provincial-based study undertaken within China that utilized Random Forest Models to predict dengue fever during outbreaks based on environmental conditions (e.g., temperature, precipitation, humidity, and vegetation indices). The average accuracy exceeded 85 percent, while the study's predictive output was approximately one month prior to the observed outbreak occurrence. The analysis conducted to determine variable importance indicated that temperature and nighttime humidity were the most important predictors, demonstrating the ability of Random Forest models to identify primary biological drivers.

Table 2. Reported predictive performance of AI models in climate-linked biological risk studies

Biological risk	AI model	Prediction lead time	Performance metric	Study region	Reference
Dengue fever	LSTM	4–8 weeks	$R^2 = 0.80$	Brazil	Chen et al., (2025)
Dengue fever	Random Forest	1 month	AUC = 0.92	China	Ren & Xu (2024)
Malaria	XGBoost	1–2 months	RMSE reduced by ~30% vs GLM	Sub-Saharan Africa	Kamana et al., (2022)
Rift Valley fever	Random Forest	4–6 weeks	Accuracy > 90%	East Africa	Mulwa et al. (2024)
Harmful algal blooms	CNN–LSTM	2–4 weeks	Accuracy > 90%	Lake Erie	Schaeffer et al. (2024)

Various Gradient Boosting methods (Extreme Gradient Boosting (XGBoost)) have improved the prediction accuracy in complex climate environments (Ghafarian, et al., 2022). The XGBoost models performed better than generalized linear models in a multi-country African malaria forecasting study by reducing the monthly incidence root mean square error by approximately 30%. Many of the benefits arise from the capability of the algorithm to account for interaction effects and nonlinear thresholds in the climate-disease interaction.

Support Vector Machines have been applied to limited-training data scenarios, such as crop disease and pest risk predictions. For example, a wheat rust early warning system combining Support Vector Machines and integrating temperature anomalies and leaf wetness duration achieved classification accuracies greater than 80% for different climate conditions (Ruan, et al., 2021). However, the ability of these models to scale across regions is restricted owing to their vulnerability to kernel selection and parameter tuning. Classical machine learning models provide insights into dynamic climate-biology interactions, although their reliance on manually created features and limited temporal learning capacity prevents them from fully capturing these interactions. Therefore, there has been a shift to deep learning architectures.

Machine learning models give relatively accurate forecasts, use moderate amounts of computational resources, and can be interpreted comparatively easily. However, because they depend on hand-crafted features and do not have strong capabilities for recognizing long-term temporal dependencies, machine learning models have limited capacity to scale and transfer across rapidly changing climate conditions.

3.2. Temporal/Seasonal Models and Deep Learning

The increasing use of deep learning has provided new opportunities for developing models that consider the spatiotemporal complexity inherent in the many biological risks associated with Climate Change (Zeghina et al., 2024). Convolutional neural networks (CNNs) are being increasingly used as a method of analyzing spatial patterns (i.e., distance and location) in remote sensing products, such as land surface temperature, vegetation indices, and precipitation fields, using the same CNNs to build models for malaria risk mapping by learning spatial hierarchy from gridded (e.g., MODIS) data, which eliminates the need for manual feature engineering of the same for malaria risk mapping. For instance, a recent study combining land surface temperature and NDVI obtained from MODIS reported that CNN-based models performed approximately 20% better than a Random Forest model in terms of spatial prediction accuracy (Zhou et al., 2025). These CNN-based models recognize both small-scale and large-scale environmental heterogeneity, which is important when considering habitat suitability for malaria vectors.

Researchers have used advanced architectures such as Recurrent Neural Networks (specifically, Long Short-Term Memory networks) to model time-dependent relationships between climate influences and ecological processes in the prediction of dengue fever cases. The LSTM-type models offered up to eight weeks of lead time between the prediction and actual occurrence of cases, while achieving a significantly higher level of accuracy than traditional autoregressive techniques (Chen & Moraga 2026). A study in Brazil identified a significant improvement in the coefficient of determination (R^2) of the LSTM types of models, from 0.62 to 0.81, over conventional autoregressive methods (Chen & Moraga 2026).

The use of hybrid architectures, which combined convolutional neural networks and LSTMs, further improved the integration of spatial and temporal learning by providing an end-to-end modelling capability for climate-induced risk. Researchers have applied these models to forecast harmful algal blooms using both satellite imagery and historical recorded data for bloom events. They determined that a hybrid or deep learning model achieved greater than 90% accuracy in predicting algal bloom events for Lake Erie, and significantly outperformed all previous models that relied on empirical thresholds to predict bloom occurrence.

Graph neural networks (GNNs) represent the most recent advances in modelling for connected or network systems (i.e., spreads of disease). By incorporating spatial adjacency, human mobility and connectivity through ecosystems, GNN-based approaches have demonstrated improved generalizability between regions when applied specifically to COVID-19 through climate factors. There is great potential to use GNN-based approaches for modelling zoonotic infections or vector-borne diseases. GNN-based approaches and deep learning techniques also experience challenges regarding interpretability, computational cost, and data requirements, restricting direct usage in the field.

Deep learning models accurately recognize complex spatial and temporal relations, resulting in improved predictive accuracy for a wide variety of heterogeneous climate conditions than traditional statistical methods. However, they require large volumes of data, much more processing power, and are usually not interpretable by end-users, making them unsuitable for use in resource-limited settings.

3.3. Data Fusion and Integrated Computational Pipelines

AI-enabled systems for detecting early warning signs of biological events caused by extreme weather conditions are based heavily on Data Fusion Frameworks capable of bringing together disparate heterogeneous data streams located at various levels (Albahri et al 2024). An example of using multiple input types to generate reliable predictions about the likelihood of biological risk due to climate-related factors includes meteorological observations, earth observing satellite products (e.g., MODIS, Landsat, Sentinel 1,2,3, etc.), biological monitoring records, and socio-environmental data. As a method employed to generate the above-mentioned system outputs, Data Fusion was implemented at various input levels, including the Feature, Model and Decision levels. The integration of climate variables such as Temperature, Precipitation, and Ecological Indicators (e.g., NDVI, Land Cover/Vegetation Map) represents one form of feature-level fusion. Various Earth observation data products generated from a range of platforms (MODIS, Sentinel-2, ERA5 reanalysis data) were incorporated into a Data Fusion framework to generate improved performance in Early Warning and Risk Assessment capabilities associated with Rift Valley Fever outbreaks, with > 25% increase in outbreak detection sensitivity when compared to a singular source model (Van der Weyde et al 2021).

Model-level fusion employs ensemble learning strategies, in which outputs from multiple AI models are combined to reduce uncertainty and improve robustness. Ensemble approaches are particularly effective under conditions of high climatic variability. In a European West Nile virus forecasting system, an ensemble of Random Forest, LSTM, and Bayesian models reduced false-alarm rates by approximately 40% relative to individual models (Angelou et al., 2025). Decision-level fusion translates probabilistic model outputs into actionable risk categories. Bayesian decision frameworks are used to incorporate uncertainty and stakeholder-defined thresholds. These frameworks allow early warning systems to balance sensitivity and specificity, depending on public health or ecological management priorities.

Operational pipelines increasingly rely on cloud-based architectures to enable real-time data ingestion and model updates. Platforms, such as Google Earth Engine and Amazon Web Services, were used to process large-scale climate and satellite datasets. Automated retraining pipelines allow models to adapt to new data, addressing non-stationarity in climate–biology relationships. Fusing multiple types of biological and environmental data sources into one prediction increases the robustness and adaptability of predictions to different types of climate change risks. However, problems with interoperability, lack of consistency in data quality, and increased computational complexity may limit the ability of these data fusion systems to scale and to be used operationally in data-limited locations.

3.4. Explainability, Uncertainty Quantification, and Model Robustness

To support their use in high-stakes decision making, early warning systems based on artificial intelligence should be both understandable and trustworthy in data and model outputs (Visave 2025). To achieve this goal, Artificial Intelligence (XAI) was employed to make transparent how an AI model makes predictions, allowing users to understand what factors contribute to a prediction's risks. SHapley Additive exPlanations (SHAP), for example, provide a quantitative measurement of the contribution of individual variables or features to complex AI models. In a study predicting dengue fever, the SHAP approach determined that the model's variance of approximately 60% could be attributed to anomalies in the minimum temperature and the amount of rainfall that occurred prior to the outbreak, which corresponds to well-understood biological processes (Sophia et al. 2025). Local interpretable model-agnostic explanation (LIME) techniques were used to interpret single predictions of outbreak alert systems. Ultimately, the use of these techniques increases stakeholders' trust in models and improves communication between model developers and decision-makers.

Uncertainty was quantified using Bayesian neural networks, Monte Carlo Dropout and Ensemble Variance Estimation. These uncertainty quantification methods produce probabilistic, rather than deterministic, outputs, enabling early warning systems to communicate their level of confidence. The use of a Bayesian deep learning framework for the prediction of malaria reduced overconfidence in model predictions and improved the calibration of predictions made during extreme climate events (Njotto et al., 2025). A transferability evaluation was performed on the robustness of the models when applied to other geographical areas and time frames. Most studies have reported a decreased model performance when applying models outside their training domains. Therefore, domain adaptation and transfer learning were employed to enhance the generalizability of the results. The implementation of transfer learning for the prediction of dengue fever across Southeast Asian Nations demonstrated 80% accuracy when transferring the models across the regions S.B.K.U et al (2025). Overall, this development shows that AI architectures alone will not produce the required predictions without an appropriate framework for explainability, uncertainty management, and adaptive learning. This framework provides the computational support necessary to enhance early warning systems for climate-related biological risks using AI. Notwithstanding great technical developments relating to AI architectural systems, these systems are limited in their effectiveness due to a number of practical limitations. The lack of sufficient data from low-income and climate vulnerable areas continues to limit the development of models because long-term datasets from an epidemiological, ecological, and climate perspective are often absent or incomplete. In addition, real-time data capture is severely Limited by the absence of adequate sensor networks, inadequate numbers of meteorological stations, and unreliable web access. Furthermore, deep learning methods depend upon considerable computing infrastructure which requires considerable computing resources (e.g. high-performance computers and cloud-based solutions), which is not available to many in resource-constrained environments. Another significant concern is that model accuracy can deteriorate over time as the patterns associated with climate change, ecologic interactions, and disease will continue to change and develop. Therefore, continued model retraining, development of adaptive learning strategies, and substantial uncertainty quantification practices must be put in place to ensure reliable early warning capabilities are maintained while the climate is developing or changing.

Table 3. Comparative assessment of artificial intelligence methods for climate linked biological risk prediction

AI Method	Advantages	Limitations	Data Requirements	Explainability	Computational Complexity	Best Applications
Random Forest	Robust against overfitting; handles nonlinear relationships; works well with mixed variables	Limited temporal learning; requires manual feature engineering	Moderate	High	Low–Moderate	Vector-borne disease prediction, ecological risk mapping, agricultural pest forecasting
XGBoost	High predictive accuracy; captures complex interactions; efficient with structured datasets	Requires hyperparameter tuning; less interpretable than Random Forest	Moderate	Moderate	Moderate	Malaria forecasting, outbreak prediction, climate–health modelling
CNN	Excellent for extracting spatial features from satellite and remote sensing imagery	Requires large training datasets; weak temporal modelling capability	High	Low	High	Remote sensing analysis, habitat suitability mapping, spatial disease prediction
LSTM	Captures temporal dependencies and delayed climate effects; highly effective for time-series forecasting	Computationally intensive; requires long historical datasets	High	Low	High	Dengue prediction, malaria forecasting, seasonal outbreak prediction
Graph Neural Networks (GNN)	Models spatial connectivity and transmission networks; strong generalization potential	High computational cost; limited operational implementation	High	Low	Very High	Zoonotic spillover prediction, disease transmission modelling, mobility networks
Hybrid Models (CNN–LSTM/ Ensemble)	Integrates spatial and temporal learning; highest predictive accuracy; robust across multiple data sources	Complex architecture; difficult interpretation and deployment	Very High	Low–Moderate	Very High	Harmful algal bloom prediction, integrated climate-risk forecasting, multi-source early warning systems

When it comes to predicting biological risks related to climate change, no single AI approach can be considered universally best. Random Forest and XGBoost (traditional machine learning techniques) are great examples of predictive analytic approaches that provide a balance among precision, interpretability, and computation speed. They are effective solutions when data are scarce. However, while CNNs, LSTMs, and hybrid models have the potential to outperform in their ability to represent the complex spatial (geographic) and temporal (time-based) dynamics of predictions, they require substantial datasets for model building and an abundance of computing power during evaluation (Table 3). GNNs are believed to be a good choice for capturing similar links between various transmission pathways, but their actual operational implementation has yet to be fully exploited due to their computational complexity and lack of explainability.

4. AI-ENABLED EARLY WARNING & OPERATIONAL SYSTEMS

To translate AI frameworks into early warning systems, integrated platforms that can accept and process data quickly, keep their AI models updated, and communicate risk information in an actionable format are required (Tiggeloven et al., 2025). Unlike simulations produced for scientific studies, these systems were designed to function in instances where there may be delays in reporting data and uncertainty surrounding the data received from external sources as well as established limitations (constraints) of an organization. The types of biological risks associated with climate change created the need for a platform that allowed the user to analyze simultaneous sources of risk and then generate risk alerts quickly enough for operational decision-makers to act on the alerts. Various global/regional/national biosurveillance and climate–health predictions have been launched over the past decade (Zeng et al., 2021). Although the systems were different in terms of capacity and complexity, the design principles related to AI-based early warning systems remain the same; they are largely based on principles of automation, scalability, and the ability to operate as part of an existing policy and response structure. Evaluating how these system functions provides insight into the application of AI technologies in real-world environments.

4.1. AI-Based Global and Regional Biosurveillance Platforms

Numerous global-based AI-enabled biosurveillance systems have been developed that utilize climate and Biological Risk Surveillance (BRMS) (Bergquist, et al, 2024). One of the oldest and best-documented biosurveillance systems is called BlueDot, an AI-enabled Canadian System that integrates NLP, machine learning, airline ticketing data, and Climate Data to identify infectious disease threats (Table 4). BlueDot detected a potential Pneumonia Outbreak linked to Wuhan 1-3 days prior to any official notification, thereby demonstrating the power of AI-driven Anomaly Detection for this purpose. While it initially only monitored Infectious Diseases, Future versions of BlueDot began Adding Climate Sensitive Variables to optimize the ability to assess break contexts.

Table 4. AI based early warning platforms for climate linked biological risks

Platform	Risk focus	AI techniques used	Outcome	Operational scale
BlueDot	Infectious diseases	NLP, ML classifiers	Detected outbreak signals days before official alerts	Global
HealthMap	Vector-borne diseases	ML, text mining	Detected outbreaks 6–10 days earlier than traditional reporting	Global
ECDC WNV system	West Nile virus	ML risk mapping	High-risk regions identified weeks in advance (2018 heatwave)	Europe
EPIDEMIA	Malaria, RVF	ML + climate forecasting	Prediction lead time up to 2 months	Africa
Lake Erie HAB system	Harmful algal blooms	CNN–LSTM	Bloom prediction accuracy >90%	Regional

The Health Information System at Boston Children's Hospital, called HealthMap, is a comprehensive automated system for collecting, processing, and displaying health information. HealthMap uses a variety of techniques, including data mining through websites, social media, and government sources, machine learning to classify diseases by climate factors, and geospatial production of outbreak monitoring reports in near real-time (Bergquist et al., 2024). Studies assessing the performance of HealthMap showed that for several diseases affected by climate (such as dengue and chikungunya), HealthMap could identify and locate outbreaks 6–10 days before traditional health reporting methods did (Lim et al., 2025). In addition to developing and implementing the use of multi-year averages for display of seasonal and other non-seasonal patterns, HealthMap regularly integrated climate data layers into its system to better understand what affects the seasonal occurrence and/or anomalies of vector-borne diseases, for example, the use of machine learning classifiers to produce risk maps for the European Center for Disease Prevention and Control (ECDC) regarding the potential risk of outbreaks of West Nile virus (Farooq et al., 2022). The ECDC produced risk maps for West Nile virus, including satellite temperature anomalies, precipitation, and avian migration, using machine-learning classifiers to create these maps. The ECDC's risk maps successfully identified and mapped high-risk areas in Europe weeks before they peaked during the summer of 2018 owing to the European heatwave, enabling targeted intervention efforts (Boudou et al., 2025).

The EPIDEMIA project (Mulwa et al., 2025) includes early warning systems based on machine learning for malaria and Rift Valley fever, which utilize climate forecasts, vegetation indices, and entomological data. AI-based forecasts produced by a multi-country evaluation resulted in a maximum lead time of 2 months for malaria outbreaks with more than 80% predictive accuracy in many endemic countries (Kumar et al., 2025). These platforms show that AI-enabled early warnings can provide effective services in data-poor environments by employing robust data fusion approaches. The Global Animal Health Surveillance systems benefitted from AI-enabled platforms, such as FAO's EMPRES-i+, which used machine learning algorithms to monitor transboundary animal diseases. They used climate indicators to determine the suitability of areas for outbreak-related risks of climate-sensitive livestock diseases. Operating reports have demonstrated how AI enhances the ability to detect potential risk zones of the Rift Valley fever during El Niño-related rainfall anomalies. Ultimately, these examples illustrate how AI-enabled biosurveillance has progressed from an experimental concept to a highly effective component of regional and global early warning systems.

4.2. Integrating Earth Observation and Real-Time Sensing Technology

Earth observation technology serves as the core framework of many artificial intelligence (AI)-assisted early warning systems, providing up-to-date spatially dispersed environmental information. Numerous satellite platforms and climate reanalysis products, including MODIS, Sentinel-1, Sentinel-2, Landsat, and ERA5, provide the high-resolution inputs required to assess climate-related biological risks (Kotchi et al., 2019). AI algorithms have been developed to extract useful features from environmental data to identify anomalies or deviations from what would normally be expected and to relate environmental signals to biological effects. Most established applications use satellite-derived vegetation indices and rainfall estimates to forecast the likelihood of vector-borne diseases. For example, in the case of the Rift Valley fever, MODIS imagery was used to detect anomalies in the normalized difference vegetation index (NDVI), which corresponded with early evidence of increased mosquitoes in a given area. Through the use of machine learning models, the accuracy of NDVI as a predictor of future Rift Valley fever outbreaks was found to be > 85% within a 4-to 6-week lead time (Anyamba et al.,

2009). These applications have been operationalized in Eastern Africa and are used to guide vaccination and livestock movement control.

Thermal satellite data are regularly used to map the potential for the survival of vectors and the development of pathogens using temperature-based maps. In particular, Land Surface Temperature Products were able to generate precise mapping of risks in areas that did not have high-density meteorological networks (Kalluri et al., 2007). A number of articles have been published using Sentinel-3 thermal satellite images combined with artificial intelligence-based risk mapping models, leading to the generation of higher-resolution mapping for dengue fever risk assessment, from 5 km to 1 km grids, thus allowing for more effective planning for interventions at the urban level (Shen et al., 2025). The application of real-time sensors to augment our knowledge of what occurs in real life complements satellite-based approaches and provides even more accurate mapping data. Weather stations based on the Internet of Things (IoT), automated traps for capturing mosquitoes, and measuring water quality have created extensive high-frequency local datasets. AI-based models have been developed to integrate IoT-based data collected from the ground with satellite products, thus enhancing the spatial accuracy and temporal responsiveness of the mapping data. For example, in Singapore, an AI-based dengue early warning system was created to integrate temperature and humidity data from IoT-based sensors and LSTM models, providing approximately 30% less delay in detecting dengue outbreaks than the traditional surveillance methods (Shi et al., 2016).

In addition to their use in precision agriculture and ecosystem risk management, UAVs have been integrated into operational models (Xing, et al., 2026). Data from UAV-generated images were processed using Convolutional Neural Networks (CNN) and used to identify early signals of crop diseases and pest infestations. AI UAV monitoring systems in precision agriculture determine the presence of Fungal Disease Epidemics from fields as much as two weeks prior to the time of routine manual inspections, allowing time for intervention. Geospatial Remote Sensing resources hosted on cloud settlements have enabled all these data sources to be invoked functionally (Shahi et al., 2023). The Google Earth Engine concept has been applied as a tool for creating vertical stacks of large, regional (i.e., continental) datasets of open-source satellite data to feed artificial intelligence models with comprehensive and timely data streams for virtually any desired application. Automated data processing systems enable continuous updating of risk map information by capturing new data in an incremental fashion as it becomes available, providing a real-time mechanism to manage biological risk based on climatological events. As a result of the development of Automatic Terrorism Detection Systems, the combination of Earth Observation and Real-time Sensing with Artificial Intelligence supports Adaptive Capacity and provides a continuous feedback loop for monitoring from one central location to multiple satellite locations, creating a Dynamic Guiding System of data that is available for timely action (Tiggeloven, et al., 2025).

4.3. Decision-Support Systems and Policy Interfacing

The effectiveness of AI-abled early warning systems relies on their ability to provide timely and appropriate information for use by decision-makers as part of an effective response. To enable public health authorities and Agricultural and Environmental Agencies to make well-timed, effective responses to alerts generated from AI Early Warning Systems, "Decision Support Systems" were established. The emphasis was placed on developing decision support systems that met three primary criteria: usability, transparency, and clear alignment with established institutional response protocols. Comprehensive "Risk Visualization Tools" (RVT) were also established as part of the decision support framework. The RVT provided users with an interactive dashboard to generate or view probabilistic risk maps, Risk Trend Analysis maps, and Confidence Interval maps. In the Malaria Mile early warning system in southern Africa, AI-generated probabilistic risk maps are integrated into the National Health Information Systems (NHS) to enhance the targeting of Indoor Residual Spraying (IRS) and distribution of Bed Nets. Early evaluations indicated that targeted IRS and bed net strategies significantly reduced malaria incidence by 20% compared to regions using blanket IRS and distribution interventions.

Alert thresholds were developed using careful consideration to optimize the relationship between sensitivity and false alarms, taking into consideration the Bayesian framework to provide an awareness estimate of future performance and implement a cost–benefit analysis (Gogoi et al., 2025). Probabilistic alerts used to identify the potential risk of West Nile virus required sufficient confidence in the potential risk before initiating vector control measures, thereby limiting the number of unnecessary treatments and preventing the spread of an outbreak (Nasci & Mutebi, 2019). The establishment of policies involving model development and stakeholders involves close collaboration between model developers, stakeholders, and policymakers. Many platforms also utilize a co-design method whereby potential users define the indicators, formats of the alerts, and how to respond to the alerts. Using this method to develop platform designs has led to increased trust in the platform's use and uptake of AI-based systems. A study comparing climate–health Early Warning Systems found that platforms co-designed with users were twice as likely to be sustained in operational form after the pilot phase of implementation.

As of late 2022, a growing number of organizations at the international level have begun developing AI systems linked to existing frameworks, including the International Health Regulations and the UN's Sendai Framework for Disaster Risk Reduction (DRR). Several of these systems, particularly the Climate–Biological

Risk Alert System (CBRASS), have combined climate-related alerts with those of other hazards to promote the integration of climate and health information across disciplines. WHO reported that integrated climate–health early warning systems allowed several countries to better prepare for heat and vector-borne health risks. Ethical and governance considerations have been important for building decision-support systems; hence, emphasis was placed on transparency in how models were developed, where the data used in those models came from, and how uncertainty in the predictions of the models should be communicated to avoid the potential misuse of AI-based outputs (Laine et al., 2024; Pfenning-Butterworth et al., 2024). The privacy of an individual's health and movement must also be considered when using an AI system. As such, protecting user privacy while collecting and retaining data is also an important principle of AI decision-support systems. AI-enabled decision support systems provide a vital link between the application of artificial intelligence and the implementation of actions to mitigate risk. The manner in which AI-enabled decision support systems are designed and implemented will determine whether early warning information will be used to reduce risk in the context of climate-related biological threats.

AI-enabled early warning systems must be appropriately governed to ensure they operate properly. Most often, successful use of such systems depends just as heavily (if not more so) on ensuring that governance (and operational) challenges are addressed in addition to the ability to perform technically.

As a result, interoperability problems exist among climate information systems, health information systems, agricultural information systems, and environmental information systems due to different types (or genres) of data formats, incompatibilities between the platform that each operates under, and the lack of common standards for all to utilize. Furthermore, ethical issues associated with these systems, such as algorithm transparency, algorithm fairness, algorithm accountability, and protection of sensitive health or geospatial data, require careful consideration in order to support public confidence (i.e., trust) in the system. To do this will require implementing privacy-preserving technologies and use of secure data-sharing mechanisms.

In addition, the wide adoption of AI-enabled early warning systems (e.g., into public health) is usually due to a lack of institutional capacity, limited technical expertise in implementing and maintaining the software/hardware, and the long-term costs resulting from maintaining like maintaining compared to upgrading the system and recalibrating the models used. Finally, the absence of common protocols for validating and benchmarking the performance of these systems makes it harder to compare findings across studies and/or operational applications, indicating that the development of internationally accepted standards will improve reliability, reproducibility, and confidence in stakeholders.

5. EMPIRICAL CASE STUDIES OF AI-BASED CLIMATE–BIOLOGICAL RISK PREDICTION

Empirical case studies conducted on the application of AI to predict the risks associated with climate and biological factors have provided evidence as to how these AI-enabled early warning systems are performed under actual climatic conditions and real-world biological circumstances. In comparison to simulation-based studies, these case studies demonstrated proof of both operational viability and how limitations associated with data and policy application exist in multiple locations/regions as well as through a variety of biological contexts. AI models must be uniquely developed to predict the manifestation of biological climate risks that differ across human and animal health, agricultural systems, and natural ecosystems, as opposed to all-purpose models. Also noted in these case studies were how model performance characteristics were affected by data quality, spatial resolution/scale, and institutional collaboration and support regarding AI technology development for use in predicting the potential impact of climate-related biological threats (Sun et al., 2025). A review of the literature documenting AI applications that have been implemented within the context of a variety of vector-borne disease systems, zoonotic risk scenarios, agricultural production systems, and hazard assessments at the ecosystem level demonstrated both the advantages and shortcomings of AI technology when utilized to address the threat of biological damage or threats resulting from a significant climate event (Mushtaq et al., 2025).

5.1. Vector-Borne Disease Early Warning Under Climate Stress

Vector-borne disease early warning systems are an exemplary case study of how artificial intelligence (AI) can be leveraged as part of a climate-related early warning system. For example, predicting future outbreaks of dengue fever due to changing climate conditions in Brazil is one of the most thoroughly tested examples of this type of technology (Figure 2). Utilizing a combination of long short-term memory (LSTM) networks, as well as climate information (specifically temperature, precipitation, and relative humidity), the AI-driven system has been implemented for various municipalities throughout Brazil. As a result, the system was able to provide a lead time of 4-8 weeks for predicting hospitalizations related to dengue and achieved an average prediction accuracy of over 80%, exceeding the performance of traditional seasonal autoregressive models (Huang et al., 2020). Evaluation of the models indicated that previous rainfall and minimum temperature anomalies were the primary predictors of hospitalization and were in accordance with the understanding of the mosquito's life cycle and ecology. Furthermore, in Southeast Asia, machine learning (ML)-based dengue surveillance and early warning systems have combined satellite-derived climate indicators with historical epidemiological records. A study performed on a large area of Thailand and Vietnam utilized Random Forest (RF) and Gradient Boosting (GB) algorithms and achieved an area under the curve (AUC) between 0.82–0.89, for each of the provinces that were evaluated (Nguyen et al., 2025). These ML systems demonstrated the ability to accurately predict under varying climatic conditions, including anomalies associated with the El Niño phenomenon.

Two examples provided further evidence of the application of AI in low-data environments: first, through malaria early warning systems in sub-Saharan Africa, and second, through West Nile virus forecasting in Europe using machine learning approaches. Malaria early warning systems utilize an AI model that incorporates forecasts for climate change, past instances of vegetation indices, and previous incidence data to assess the likelihood that an area will experience malarial outbreaks within two months and provide the most accurate and reliable estimates, enabling pre-emptive vector control by providing a lead time before outbreaks. In addition to a case study conducted in Kenya and Uganda (Eddamiri et al., 2025), reductions in response time to outbreaks of 20% to 30% compared to traditional surveillance methods have been reported. Second, a case study conducted for West Nile viruses in Europe (Farooq et al., 2022) used machine learning to combine temperature anomaly, precipitation, and bird migration information to identify and predict the likelihood of outbreaks occurring. In 2018, during a heat wave, risk mapping generated by AI was able to accurately identify the areas of highest incidence weeks ahead of the anticipated outbreak, allowing targeted efforts to control mosquitoes and protect blood supplies. The results of these case studies collectively show that through the application of AI, improved timeframes and spatial targeting represent an enhancement of traditional methods for controlling vector-borne diseases during the time of climate-related stressors.

Figure 2. Lead time achieved by AI-enabled early warning systems

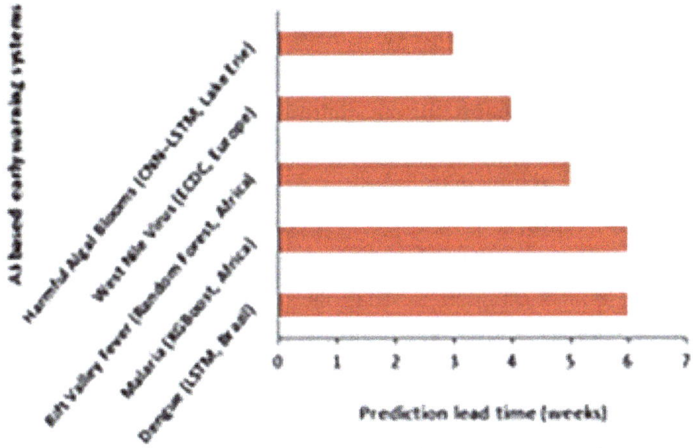

5.2. Monitoring Wildlife/Zoonotic & AI

Zoonotic diseases have emerged because of the complexity of the interactions between wild animals, the environment, and humans. To determine the best locations to prevent zoonotic diseases from spilling over into humans, AI has been used to analyze the factors of climate change (Roberts et al., 2021) and land use changes in Africa and Southeast Asia using Machine Learning (ML) models projecting how many novel viral sharing events will be shared among mammals by 2070. This global modeling study estimated over 15,000 novel viral-sharing events by 2070, with high concentrations of potential spillover across tropical Africa and South East Asia. In this modeling, one of the main contributors to increased interspecies contact was the result of climate-induced range shifts in various mammals. Bat-associated zoonotic diseases are excellent examples. ML models that utilized climate variables, vegetation cover, and bat populations were used to develop spillover risk predictions for Henipaviruses. The results of these models indicate that climatic stress combined with a lack of food in an area are the two main predictors of increased spillover. Therefore, the information obtained from bat models can be utilized to help direct surveillance efforts in high-risk areas. Systems developed for Wildlife Disease Monitoring have also incorporated AI to help analyze the images and acoustics of the wildlife being monitored. For example, in North America, systems have been developed to use AI-driven acoustic monitoring to detect changes in bat activity levels caused by the presence of a white-nose fungus. In addition, climate variables were integrated into this monitoring system to allow for a more comprehensive view of seasonal and interannual shifts in bat behavior. The early detection sensitivity of Acoustic Monitoring Systems has improved by approximately 25% compared to traditional manual survey methods (Haque et al., 2024).

Researchers used artificial intelligence (AI) in another study to predict the risk zones of avian influenza using climate data, migratory bird migration routes, and poultry density of poultry (Wang et al., 2025). Machine learning was utilized to map the risk of avian influenza outbreaks across all areas of East Asia, identifying that certain climatic anomalies are strong predictors of whether an outbreak would occur and allow for the implementation of targeted biosecurity measures. These examples illustrate how AI can enable proactive zoonotic risk monitoring through early warning systems, as the likelihood of spillover increases owing to shifts in ecology caused by climate change.

5.3. Predicting Biological Risk Due to Agriculture and Livestock

Agriculture is highly affected by climate-related biological risks, including crop diseases, insect pests, and livestock pathogens. Artificial intelligence (AI)-based Early Warning Systems have been increasingly adopted to assist climate-resilient food production. The prediction of crop disease is the clearest example. An artificial neural network model for predicting the incidence of wheat rust has been used successfully with a classification accuracy greater than 85% under different weather conditions over several growing seasons by combining temperature, relative humidity, and leaf wetness duration as input data (Ali et al 2023). By providing farmers with timely alerts, growers were able to effectively target fungicide applications, reducing yield losses and the amount of chemicals applied to control wheat rust. Farmers have benefited from the AI predictiveness of insect pest outbreak predictions. An example can be found in China. A machine learning Early Warning System was used to predict the risk of Fall Armyworm (FAW) infestations by correlating climate variables and remote sensing indicators. In this example, this system achieved greater than 80% accuracy and provided warning periods of 2-3 weeks of lead time so that effective pest management could be developed and implemented (Jiang et al., 2022).

To quickly identify diseases in livestock, early detection systems are an essential part of all nations with climates that could affect agriculture (like many in Africa). To predict where Rift Valley Fever might occur in East Africa, researchers have successfully combined rainfall anomaly data, vegetation index measurements, and livestock density count numbers, using machine learning algorithms to produce results with greater than 85% accuracy in predicting outbreaks over the course of the years before the outbreak. Additionally, these systems enabled governments to receive early alerts of potentially dangerous livestock (Rift Valley Fever) outbreaks, allowing them to implement vaccination strategies and restrict the movement of livestock during periods of increased risk (Himeidan, 2016). AI has also been used to build models to predict the potential for parasitic diseases in livestock linked to weather. The research that has been published so far in Europe found a 30% increase in the accuracy of determining where risks of fasciolosis exist by using temperature and moisture indices along with machine learning classifiers as opposed to rule-based predictive modelling (Caminade et al., 2015). The value of AI-based early warning systems in agriculture in limiting the economic impact and improving climate resilience in food systems was demonstrated through this and other examples.

5.4. Biological Risk Prediction and Monitoring at the Scale of the Ecosystem

Artificial intelligence (AI) has also been applied to identify and manage biological risks associated with ecosystems, such as invasive species and harmful algal blooms (HABs). An example of this is the widespread use of artificial intelligence models for monitoring invasive insects (Nasios, 2025). Artificial intelligence was used with regard to the northward movement of the pine processionary moth in Europe to develop predictive models based on temperature and tree cover data (Womersley et al., 2024). The predictive models closely followed the observed shifts in the species range of approximately 5 km annually and were able to accurately identify the projected movement of the species under climate warming scenarios (Womersley et al., 2024).

Another example of the use of artificial intelligence in ecosystem science is the prediction of harmful algal blooms (HABs). The application of deep learning techniques that take into account chlorophyll-a concentrations, water temperature, and other nutrient data obtained from satellite imagery is an example of how artificial intelligence has advanced ecosystem prediction capabilities. Hybrid models such as those of Liu et al. (2025) combining a convolutional neural network (CNN) structure with a long short-term memory (LSTM) structure have been successfully deployed in predicting HAB occurrences in both fresh and saltwater environments. The CNN-LSTM Hybrid model for Lake Erie produced over 90% correct predictions of algal bloom weeks in advance of toxin production (Schaeffer et al., 2024). Such predictions have provided advance warnings to water suppliers and health departments about the possible presence of toxins in drinking water.

Artificial intelligence has been used to monitor forest diseases through machine learning of satellite images to provide early warning signs of the presence of bark beetles caused by drought stress induced by global warming (Bozzini et al., 2025). AI-based monitoring methods in northern North America (USA and Canada) provide approximately 35% higher rates of early detection than traditional aerial surveys (Delplanque et al., 2024). Monitoring the health of aquatic ecosystems has also expanded into marine environments. AI models have been used to predict coral bleaching risks using a combination of sea temperature anomalies and thermal stress indices. By utilizing a global early warning system, coral bleaching events occurring in masses under extreme temperatures can be predicted with a high degree of reliability (Cornet et al., 2025). The applications of AI systems to these global-scale ecosystems are examples of how AI-enabled early warning systems can be used to identify climate-related biological risks beyond human health, and therefore support additional goals related to the management and conservation of the environment. The case studies reviewed in this paper show that AI has consistently improved both the projected lead times for biological-related hazard-related risk assessments and the assessment of spatial risk across all of the biological-related hazards examined. However, the performance of AI tools varies according to local data quality, climate variability, and institutional capability indicating that locally-adapted methodologies are critical to effectively implementing AI for operational use in a consistent and reliable manner.

6. CONCLUSION

Artificial intelligence greatly enhances forecasting climate-related biological hazards through longer ranges of dates; better spatial resolutions; and capability to utilize different kinds of environmental, epidemiological, and ecological datasets together. Use of machine and deep learning models in a real time with Earth observation and real-time sensor technology is consistently more accurate than using other traditional methods to measure the relation between multiple variable through complex nonlinear relationships. Further, the use of Explainable AI and uncertainty quantification make models more transparent and help decision-making become clearer and more effective. Despite these advantages, significant challenges to data access, difficulty transferring models across geographical areas, high compute requirements, and governance have to be resolved first for AI-based early warning systems to work effectively in an adaptive, trustworthy, and operationally scalable manner. Advancements in artificial intelligence to monitor the effects of drought stress on trees by monitoring forest diseases. Using machine learning applied to satellite imagery, early warning signals regarding bark beetle activity are provided because of the stress of drought caused by climate change. In northern North America (the USA and Canada), using AI for monitoring yields 35% earlier detection than traditional aerial surveys. The use of AI for monitoring aquatic ecosystems has expanded to include marine systems. AI was employed to project coral bleaching risks based on a combination of sea temperature anomalies and thermal stress indices. A global early warning system was developed to precisely predict mass coral bleaching events due to excessive heat. AI for monitoring allows for the identification of biological risks associated with climate change beyond those for human health; therefore, it provides benefits for management and conservation activities at a broader level.

Despite advances in technology and research, multiple gaps remain regarding the use of such technology to combat climate change as a cause of disease or public health threats. For example, the limitations on available data limit model performance in low-resource areas and ecologically complicated environments where there are very few long-term biological and health records available. In these types of smaller regions, the AI model outputs based on regionally developed datasets did not transfer well to other areas with different climates and socio-ecological systems. In addition, while there are mechanisms to quantify the uncertainty of a model in its prediction of a future outcome, there is still a very limited use of probabilistic forecasting mechanisms to inform operational decision making. This has led to organizations either being overly confident or underutilizing model output. Finally, the issue of explainability related to how deep learning architectures create model outputs is a major hurdle for public health officials and environmental managers making high-stakes decisions. When the reasoning behind a model's output is opaque, distrust is created among these entities. It is also important to point out that most AI models currently in existence focus on only one category of risk (i.e., Human Infection Disease) and do not take into account the combined effects or cascading biological risks from Human Health, Animal Health and Environmental Health. This has led to a fragmented approach to collecting data on climate-impacted biological risks in one place, without capturing how they affect different interconnected domains.

To establish AI-enabled early warning systems (EWSS), a larger investment in coordinated methodological, infrastructural, and governance-oriented advancement will be necessary. Technically, models that are both transferable and adaptive based on learning across regions and evolving climatic conditions will need to be prioritized. Utilizing federated learning and transfer learning strategies has the potential to mitigate issues related to data availability while ensuring the sovereignty of the data owned by individuals/organizations involved in the study. Improved integration of uncertainty-aware modeling and Explainable AI methodologies will be critical for improving the interpretability of data and supporting transparent decision making regarding the use of AI-generated alerts. Continuing investment in climate, ecological, and biological monitoring infrastructure will be essential to provide continuity and quality of data collected in these regions, and in some cases, may be a necessity due to the sensitivity associated with climate change. The improvement of early warning systems may also be made possible through the integration of participatory surveillance and citizen-generated data that have been carefully validated. Further investment in developing a stronger alignment between operational and policy EWSS with existing governance frameworks for public health, agriculture, and the environment, and the co-design of EWSS in collaboration with end-users, ethical safeguards in place, and clear systems for actioning alerts generated through the use of AI will be necessary for the long-term viability of EWSS in the future. Finally, an emphasis on a multi-hazard EWSS that considers the interaction between climate stressors, biological risks, and socio-environmental vulnerabilities will be important for future research efforts to reflect the complexity of the world and facilitate pro-active responses to reduce the risk caused by rapidly progressing climate change.

DECLARATION OF CONFLICT OF INTEREST

The authors declare that they have no affiliation with or involvement in any environment or business entity that has a financial interest in the topic of this paper or in the materials presented therein.

DATA AVAILABILITY STATEMENT

There were no data utilised in the current research or in the article.

ACKNOWLEDGMENT

We are thankful to Graphic Era (Deemed to be University), India for support. We would like to thanks Paperpal (https://paperpal.com/), Cactus Communications Services Pte Ltd., Singapore for their invaluable online assistance in refining the language and grammar of our manuscript.

REFERENCES

Albahri, A. S., Khaleel, Y. L., Habeeb, M. A., Ismael, R. D., Hameed, Q. A., Deveci, M., Homod, R. Z., Albahri, O. S., Alamoodi, A. H., & Alzubaidi, L. (2024). A systematic review of trustworthy artificial intelligence applications in natural disasters. *Computers & Electrical Engineering*, *118*, 109409. DOI: 10.1016/j.compeleceng.2024.109409

Ali, A., Jayaraman, R., Azar, E., & Maalouf, M. (2024). A comparative analysis of machine learning and statistical methods for evaluating building performance: A systematic review and future benchmarking framework. *Building and Environment*, *252*, 111268. DOI: 10.1016/j.buildenv.2024.111268

Ali, Y., Iqbal, S., Aatif, H. M., Naveed, K., Khan, A. A., Ijaz, M., Magsi, M. M., Ahmad, S., Syed, A. U. A., Magsi, M. A., Iqbal, R. K., Bukhari, N. A., Hatamleh, A. A., & Raza, A. (2023). Predicting Stripe Rust Severity in Wheat Using Meteorological Data with Environmental Response Modeling. *Journal of King Saud University. Science*, *35*(4), 102591. DOI: 10.1016/j.jksus.2023.102591

Angelou, A., Pappa, A., Markov, P. V., Gewehr, S., Stilianakis, N. I., & Kioutsioukis, I. (2025). Early warning system of the seasonal west nile virus infection risk in humans in northern greece, 2020–2024. *Scientific Reports*, *15*(1), 7129. Advance online publication. DOI: 10.1038/s41598-025-91996-9 PMID: 40021915

Anyamba, A., Chretien, J.-P., Small, J., Tucker, C. J., Formenty, P. B., Richardson, J. H., Britch, S. C., Schnabel, D. C., Erickson, R. L., & Linthicum, K. J. (2009). Prediction of a Rift Valley fever outbreak. *Proceedings of the National Academy of Sciences of the United States of America*, *106*(3), 955–959. DOI: 10.1073/pnas.0806490106 PMID: 19144928

Benedum, C. M., Seidahmed, O. M. E., Eltahir, E. A. B., & Markuzon, N. (2018). Statistical modeling of the effect of rainfall flushing on dengue transmission in Singapore. *PLoS Neglected Tropical Diseases*, *12*(12), e0006935. DOI: 10.1371/journal.pntd.0006935 PMID: 30521523

Bergquist, R., Zheng, J.-X., & Zhou, X.-N. (2024). Synergistic integration of climate change and zoonotic diseases by artificial intelligence: A holistic approach for sustainable solutions. *Science in One Health*, *3*, 100070. DOI: 10.1016/j.soh.2024.100070 PMID: 39077386

Boudou, M., Garvey, P., ÓhAiseadha, C., O'Dwyer, J., Burke, D. T., & Hynds, P. (2025). Effects of the 2018 European summer heatwave on the incidence of sporadic bacterial gastroenteritis in a temperate maritime climate region (Republic of Ireland). *One Health*, *20*, 101105. DOI: 10.1016/j.onehlt.2025.101105 PMID: 40547209

Bozzini, A., Huo, L., Brugnaro, S., Morgante, G., Persson, H. J., Finozzi, V., Battisti, A., & Faccoli, M. (2025). Multispectral drone images for the early detection of bark beetle infestations: Assessment over large forest areas in the Italian South-Eastern Alps. *Frontiers in Forests and Global Change*, *8*, 1532954. Advance online publication. DOI: 10.3389/ffgc.2025.1532954

Caminade, C., Van Dijk, J., Baylis, M., & Williams, D. (2015). Modelling recent and future climatic suitability for fasciolosis in Europe. *Geospatial Health*, *9*(2), 301. DOI: 10.4081/gh.2015.352 PMID: 25826311

Carlson, C. J., Albery, G. F., Merow, C., Trisos, C. H., Zipfel, C. M., Eskew, E. A., Olival, K. J., Ross, N., & Bansal, S. (2022). Climate change increases cross-species viral transmission risk. *Nature*, *607*(7919), 555–562. DOI: 10.1038/s41586-022-04788-w PMID: 35483403

Chen, X., & Moraga, P. (2025). Forecasting dengue across Brazil with LSTM neural networks and SHAP-driven lagged climate and spatial effects. *BMC Public Health*, *25*(1), 973. Advance online publication. DOI: 10.1186/s12889-025-22106-7 PMID: 40075398

Chen, X., & Moraga, P. (2026). Dengue forecasting and outbreak detection in Brazil using LSTM: Integrating human mobility and climate factors. *Infectious Disease Modelling*, *11*(1), 338–354. DOI: 10.1016/j.idm.2025.11.002 PMID: 41321683

Chen, X., Wang, Q., Cui, B., Chen, G., Xie, T., & Yang, W. (2023). Ecological time lags in biodiversity response to habitat changes. *Journal of Environmental Management*, *346*, 118965. DOI: 10.1016/j.jenvman.2023.118965 PMID: 37741191

Cornet, V. J., Cantin, N. E., Joyce, K. E., Leggat, W., Ainsworth, T. D., & Heron, S. F. (2025). Enhancing coral bleaching predictive tools through integrating sensitivity to heat exposure. *Biological Conservation*, *302*, 110958. DOI: 10.1016/j.biocon.2024.110958 da Costa, J. M. F., Costa, A. C., Silveira, C. da S., & Gonçalves, S. T. N. Marcos Junior, A. D., & Cavalcanti, L. P. de G. (2026). Forecasting and Early Warning Systems for Dengue Outbreaks: Updated Narrative Review. Revista Da Sociedade Brasileira de Medicina Tropical, 59. https://doi.org/DOI: 10.1590/0037-8682-0429-2025

de Boer, J., & Harvey, J. (2020). Range-Expansion in Processionary Moths and Biological Control. *Insects, 11*(5), 267. DOI: 10.3390/insects11050267 PMID: 32353938

Delplanque, A., Linchant, J., Vincke, X., Lamprey, R., Théau, J., Vermeulen, C., Foucher, S., Ouattara, A., Kouadio, R., & Lejeune, P. (2024). Will artificial intelligence revolutionize aerial surveys? A first large-scale semi-automated survey of African wildlife using oblique imagery and deep learning. *Ecological Informatics, 82*, 102679. DOI: 10.1016/j.ecoinf.2024.102679

Deutsch, C. A., Tewksbury, J. J., Tigchelaar, M., Battisti, D. S., Merrill, S. C., Huey, R. B., & Naylor, R. L. (2018). Increase in crop losses to insect pests in a warming climate. *Science, 361*(6405), 916–919. DOI: 10.1126/science.aat3466 PMID: 30166490

Eby, P., Peel, A. J., Hoegh, A., Madden, W., Giles, J. R., Hudson, P. J., & Plowright, R. K. (2022). Pathogen spillover driven by rapid changes in bat ecology. *Nature, 613*(7943), 340–344. DOI: 10.1038/s41586-022-05506-2 PMID: 36384167

Eddamiri, S., Bassine, F. Z., Hakam, O., Kambiet, P. L. K., & Chehbouni, A. (2025). Applications of artificial intelligence for forecasting and managing extremes. In *Climate Change and Rainfall Extremes in Africa* (pp. 91–113). Elsevier., DOI: 10.1016/B978-0-443-28867-8.00005-8

Farooq, Z., Rocklöv, J., Wallin, J., Abiri, N., Sewe, M. O., Sjödin, H., & Semenza, J. C. (2022). Artificial intelligence to predict West Nile virus outbreaks with eco-climatic drivers. *The Lancet Regional Health. Europe, 17*, 100370. DOI: 10.1016/j.lanepe.2022.100370 PMID: 35373173

Ghafarian, F., Wieland, R., Lüttschwager, D., & Nendel, C. (2022). Application of extreme gradient boosting and Shapley Additive explanations to predict temperature regimes inside forests from standard open-field meteorological data. Environmental Modelling &amp. *Environmental Modelling & Software, 156*, 105466. DOI: 10.1016/j.envsoft.2022.105466

Gogoi, G. J., Das, S., & Chakraborty, A. (2025). Bayesian Neural Network based analysis of supervised control and data acquisition of Horizontal Axis Wind Turbine for fault detection. *Wind Energy and Engineering Research, 4*, 100016. DOI: 10.1016/j.weer.2025.100016

Haque, S., Mengersen, K., Barr, I., Wang, L., Yang, W., Vardoulakis, S., Bambrick, H., & Hu, W. (2024). Towards development of functional climate-driven early warning systems for climate-sensitive infectious diseases: Statistical models and recommendations. *Environmental Research*, *249*, 118568. DOI: 10.1016/j.envres.2024.118568 PMID: 38417659

Himeidan, Y. (2016). Rift Valley fever: Current challenges and future prospects. *Research and Reports in Tropical Medicine*, *1*, 1. Advance online publication. DOI: 10.2147/RRTM.S63520 PMID: 30050334

Huang, C.-H., Lin, C.-Y., Yang, C.-Y., Chan, T.-C., Chiang, P.-H., & Chen, Y.-H. (2021). Relationship between the Incidence of Dengue Virus Transmission in Traditional Market and Climatic Conditions in Kaohsiung City. *The Canadian Journal of Infectious Diseases & Medical Microbiology*, *2021*, 1–10. DOI: 10.1155/2021/9916642 PMID: 34422144

Huang, Y., Xu, C., Ji, M., Xiang, W., & He, D. (2020). Medical service demand forecasting using a hybrid model based on ARIMA and self-adaptive filtering method. *BMC Medical Informatics and Decision Making*, *20*(1), 237. Advance online publication. DOI: 10.1186/s12911-020-01256-1 PMID: 32950059

Inam, S. A. (2025). A review of artificial intelligence for predicting climate driven infectious disease outbreaks to enhance global health resilience. *Discover Public Health*, *22*(1), 738. Advance online publication. DOI: 10.1186/s12982-025-01167-4

Islam, S., Haque, C. E., Hossain, S., & Hanesiak, J. (2021). Climate Variability, Dengue Vector Abundance and Dengue Fever Cases in Dhaka, Bangladesh: A Time-Series Study. *Atmosphere (Basel)*, *12*(7), 905. DOI: 10.3390/atmos12070905

Jiang, C., Zhang, X., Xie, W., Wang, R., Feng, C., Ma, L., Li, Q., Yang, Q., & Wang, H. (2022). Predicting the potential distribution of the fall armyworm Spodoptera frugiperda (J.E. Smith) under climate change in China. *Global Ecology and Conservation*, *33*, e01994. DOI: 10.1016/j.gecco.2021.e01994

Kakarla, S. G., Caminade, C., Mutheneni, S. R., Morse, A. P., Upadhyayula, S. M., Kadiri, M. R., & Kumaraswamy, S. (2019). Lag effect of climatic variables on dengue burden in India. *Epidemiology and Infection*, *147*, e170. Advance online publication. DOI: 10.1017/S0950268819000608 PMID: 31063099

Kalluri, S., Gilruth, P., Rogers, D., & Szczur, M. (2007). Surveillance of Arthropod Vector-Borne Infectious Diseases Using Remote Sensing Techniques: A Review. *PLoS Pathogens*, *3*(10), e116. DOI: 10.1371/journal.ppat.0030116 PMID: 17967056

Kamana, E., Zhao, J., & Bai, D. (2022). Predicting the impact of climate change on the re-emergence of malaria cases in China using LSTMSeq2Seq deep learning model: A modelling and prediction analysis study. *BMJ Open*, *12*(3), e053922. DOI: 10.1136/bmjopen-2021-053922 PMID: 35361642

Kassem, A., Khattab, M. S., Ismael, E., Yassin, A. M., Hamza, D., & Osman, A. H. (2025). Climate change and the emergence of Rift Valley fever virus: A pathological and environmental study of sheep jaundice in Egyptian slaughterhouses. *BMC Veterinary Research*, *21*(1), 539. Advance online publication. DOI: 10.1186/s12917-025-05022-1 PMID: 40963143

Kotchi, S., Bouchard, C., Ludwig, A., Rees, E., & Brazeau, S. (2019). Using Earth observation images to inform risk assessment and mapping of climate change-related infectious diseases. *Canada Communicable Disease Report = Relevé des Maladies Transmissibles au Canada*, *45*(5), 133–142. DOI: 10.14745/ccdr.v45i05a04 PMID: 31285704

Kumar, S. M., Nath, S. N., Vani, H. C., Shrinivasa, B. M., Ahmed, N., Tomer, V., Bankoti, K., Negi, G., Ali, S. S. B., Jaiswal, M. K., & Warkade, V. (2025). The role of artificial intelligence (AI) in the elimination of malaria: A systematic literature review. *Malaria Journal*, *24*(1), 427. Advance online publication. DOI: 10.1186/s12936-025-05661-7 PMID: 41291780

Laine, J., Minkkinen, M., & Mäntymäki, M. (2024). Ethics-based AI auditing: A systematic literature review on conceptualizations of ethical principles and knowledge contributions to stakeholders. Information &amp. *Information & Management*, *61*(5), 103969. DOI: 10.1016/j.im.2024.103969

Lim, A., Shearer, F. M., Sewalk, K., Pigott, D. M., Clarke, J., Ghouse, A., Judge, C., Kang, H., Messina, J. P., Kraemer, M. U. G., Gaythorpe, K. A. M., de Souza, W. M., Nsoesie, E. O., Celone, M., Faria, N., Ryan, S. J., Rabe, I. B., Rojas, D. P., Hay, S. I., & Brady, O. J. (2025). The overlapping global distribution of dengue, chikungunya, Zika and yellow fever. *Nature Communications*, *16*(1), 3418. Advance online publication. DOI: 10.1038/s41467-025-58609-5 PMID: 40210848

Linthicum, K. J., Anyamba, A., Tucker, C. J., Kelley, P. W., Myers, M. F., & Peters, C. J. (1999). Climate and Satellite Indicators to Forecast Rift Valley Fever Epidemics in Kenya. *Science*, *285*(5426), 397–400. DOI: 10.1126/science.285.5426.397 PMID: 10411500

Liu, Q., Liu, M., Liang, W., Li, X., Jing, W., Chen, Z., & Liu, J. (2025). Global distribution and health impact of infectious disease outbreaks, 1996–2023: A worldwide retrospective analysis of World Health Organization emergency event reports. *Journal of Global Health*, *15*, 04151. Advance online publication. DOI: 10.7189/jogh.15.04151 PMID: 40375732

Lowe, R., Lee, S. A., O'Reilly, K. M., Brady, O. J., Bastos, L., Carrasco-Escobar, G., de Castro Catão, R., Colón-González, F. J., Barcellos, C., Carvalho, M. S., Blangiardo, M., Rue, H., & Gasparrini, A. (2021). Combined effects of hydrometeorological hazards and urbanisation on dengue risk in Brazil: A spatiotemporal modelling study. *The Lancet. Planetary Health*, *5*(4), e209–e219. DOI: 10.1016/S2542-5196(20)30292-8 PMID: 33838736

Ma, C.-S., Wang, B.-X., Wang, X.-J., Lin, Q.-C., Zhang, W., Yang, X.-F., van Baaren, J., Bebber, D. P., Eigenbrode, S. D., Zalucki, M. P., Zeng, J., & Ma, G. (2025). Crop pest responses to global changes in climate and land management. Nature Reviews Earth &amp. *Nature Reviews. Earth & Environment*, *6*(4), 264–283. DOI: 10.1038/s43017-025-00652-3

Michelon, W., Santos, L. D., Santin, L. C., & Viancelli, A. (2025). Increased leptospirosis incidence following flooding in Rio Grande do Sul, Brazil. *International Journal of Environmental Health Research*, *35*(12), 3841–3846. DOI: 10.1080/09603123.2025.2498623 PMID: 40292553

Mordecai, E. A., Cohen, J. M., Evans, M. V., Gudapati, P., Johnson, L. R., Lippi, C. A., Miazgowicz, K., Murdock, C. C., Rohr, J. R., Ryan, S. J., Savage, V., Shocket, M. S., Stewart Ibarra, A., Thomas, M. B., & Weikel, D. P. (2017). Detecting the impact of temperature on transmission of Zika, dengue, and chikungunya using mechanistic models. *PLoS Neglected Tropical Diseases*, *11*(4), e0005568. DOI: 10.1371/journal.pntd.0005568 PMID: 28448507

Mordecai, E. A., Ryan, S. J., Caldwell, J. M., Shah, M. M., & LaBeaud, A. D. (2020). Climate change could shift disease burden from malaria to arboviruses in Africa. *The Lancet. Planetary Health*, *4*(9), e416–e423. DOI: 10.1016/S2542-5196(20)30178-9 PMID: 32918887

Mulwa, D., Kazuzuru, B., Misinzo, G., & Bett, B. (2024). An XGBoost Approach to Predictive Modelling of Rift Valley Fever Outbreaks in Kenya Using Climatic Factors. *Big Data and Cognitive Computing*, *8*(11), 148. DOI: 10.3390/bdcc8110148

Mulwa, D., Kazuzuru, B., Misinzo, G., & Bett, B. (2025). Machine Learning Approach to Predicting Rift Valley Fever Disease Outbreaks in Kenya. *Zoonotic Diseases*, *5*(3), 20. DOI: 10.3390/zoonoticdis5030020

Mushtaq, M. A., Ateeq, M., Ikram, M., Alam, S. M., Kaleem, M. M., Ashraf, M. A., Asim, M., Almutairi, K. F., Seleiman, M. F., & Shireen, F. (2025). Securing fruit trees future: AI-driven early warning and predictive systems for abiotic stress in changing climate. *Plant Stress*, *17*, 100953. DOI: 10.1016/j.stress.2025.100953

Naing, C., Reid, S. A., Aye, S. N., Htet, N. H., & Ambu, S. (2019). Risk factors for human leptospirosis following flooding: A meta-analysis of observational studies. *PLoS One*, *14*(5), e0217643. DOI: 10.1371/journal.pone.0217643 PMID: 31141558

Nasci, R. S., & Mutebi, J.-P. (2019). Reducing West Nile Virus Risk Through Vector Management. *Journal of Medical Entomology*, *56*(6), 1516–1521. DOI: 10.1093/jme/tjz083 PMID: 31549724

Nasios, I. (2025). AI-driven multi-source data fusion for algal bloom severity classification in small inland water bodies: Leveraging Sentinel-2, DEM, and NOAA climate data. *Remote Sensing Applications: Society and Environment*, *40*, 101800. DOI: 10.1016/j.rsase.2025.101800

Nguyen, T. G., Dang, D. K., Ngo, C. T., Nguyen, H. D., Pham, V. T., Bui, Q.-T., Can, T.-V., & Van Luu, N. (2025). Prediction of riverbank erosion susceptibility areas of the Vietnam Mekong Delta using machine learning and remote sensing in the context of geographical studies. *Physical Geography*, •••, 1–25. DOI: 10.1080/02723646.2025.2579104

Nissan, H., Ukawuba, I., & Thomson, M. (2021). Climate-proofing a malaria eradication strategy. *Malaria Journal*, *20*(1), 190. Advance online publication. DOI: 10.1186/s12936-021-03718-x PMID: 33865383

Njotto, L. L., Senyoni, W., Cronie, O., & Stensgaard, A.-S. (2025). Bayesian spatio-temporal modeling and prediction of malaria cases in Tanzania mainland (2016-2023): Unveiling associations with climate and intervention factors. *International Journal of Health Geographics*, *24*(1), 20. Advance online publication. DOI: 10.1186/s12942-025-00408-8 PMID: 40750887

Paerl, H. W. (2017). Controlling harmful cyanobacterial blooms in a climatically more extreme world: Management options and research needs. *Journal of Plankton Research*, *39*(5), 763–771. DOI: 10.1093/plankt/fbx042

Pfenning-Butterworth, A., Buckley, L. B., Drake, J. M., Farner, J. E., Farrell, M. J., Gehman, A.-L. M., Mordecai, E. A., Stephens, P. R., Gittleman, J. L., & Davies, T. J. (2024). Interconnecting global threats: Climate change, biodiversity loss, and infectious diseases. *The Lancet. Planetary Health*, *8*(4), e270–e283. DOI: 10.1016/S2542-5196(24)00021-4 PMID: 38580428

Rane, N., Paramesha, M., Choudhary, S., & Rane, J. (2024). Machine Learning and Deep Learning for Big Data Analytics: a Review of Methods and Applications. SSRN Electronic Journal. https://doi.org/DOI: 10.2139/ssrn.4835655

Ren, H., & Xu, N. (2024). Forecasting and mapping dengue fever epidemics in China: A spatiotemporal analysis. *Infectious Diseases of Poverty*, *13*(1), 50. Advance online publication. DOI: 10.1186/s40249-024-01219-y PMID: 38956632

Roberts, M., Dobson, A., Restif, O., & Wells, K. (2021). Challenges in modelling the dynamics of infectious diseases at the wildlife–human interface. *Epidemics*, *37*, 100523. DOI: 10.1016/j.epidem.2021.100523 PMID: 34856500

Ruan, C., Dong, Y., Huang, W., Huang, L., Ye, H., Ma, H., Guo, A., & Ren, Y. (2021). Prediction of Wheat Stripe Rust Occurrence with Time Series Sentinel-2 Images. *Agriculture*, *11*(11), 1079. DOI: 10.3390/agriculture11111079

Schaeffer, B. A., Reynolds, N., Ferriby, H., Salls, W., Smith, D., Johnston, J. M., & Myer, M. (2024). Forecasting freshwater cyanobacterial harmful algal blooms for Sentinel-3 satellite resolved U.S. lakes and reservoirs. *Journal of Environmental Management*, *349*, 119518. DOI: 10.1016/j.jenvman.2023.119518 PMID: 37944321

Shahi, T. B., Xu, C.-Y., Neupane, A., & Guo, W. (2023). Recent Advances in Crop Disease Detection Using UAV and Deep Learning Techniques. *Remote Sensing (Basel)*, *15*(9), 2450. DOI: 10.3390/rs15092450

Sheffield, J., Wood, E. F., Pan, M., Beck, H., Coccia, G., Serrat-Capdevila, A., & Verbist, K. (2018). Satellite Remote Sensing for Water Resources Management: Potential for Supporting Sustainable Development in Data-Poor Regions. *Water Resources Research*, *54*(12), 9724–9758. DOI: 10.1029/2017WR022437

Shen, Y., Ren, Z., Fan, J., Xiao, J., Zhang, Y., & Liu, X. (2025). Fine-Scale Risk Mapping for Dengue Vector Using Spatial Downscaling in Intra-Urban Areas of Guangzhou, China. *Insects*, *16*(7), 661. DOI: 10.3390/insects16070661 PMID: 40725294

Shi, Y., Liu, X., Kok, S.-Y., Rajarethinam, J., Liang, S., Yap, G., Chong, C.-S., Lee, K.-S., Tan, S. S. Y., Chin, C. K. Y., Lo, A., Kong, W., Ng, L. C., & Cook, A. R. (2016). Three-Month Real-Time Dengue Forecast Models: An Early Warning System for Outbreak Alerts and Policy Decision Support in Singapore. *Environmental Health Perspectives*, *124*(9), 1369–1375. DOI: 10.1289/ehp.1509981 PMID: 26662617

Skendžić, S., Zovko, M., Živković, I. P., Lešić, V., & Lemić, D. (2021). The Impact of Climate Change on Agricultural Insect Pests. *Insects*, *12*(5), 440. DOI: 10.3390/insects12050440 PMID: 34066138

Sophia, Y., Roxy, M. K., Murtugudde, R., Karipot, A., Sapkota, A., Dasgupta, P., Baliwant, K., Saunik, S., Tiwari, A., Chattopadhyay, R., & Phalkey, R. K. (2025). Dengue dynamics, predictions, and future increase under changing monsoon climate in India. *Scientific Reports*, *15*(1), 1637. Advance online publication. DOI: 10.1038/s41598-025-85437-w PMID: 39837878

Southall, E., Brett, T. S., Tildesley, M. J., & Dyson, L. (2021). Early warning signals of infectious disease transitions: A review. *Journal of the Royal Society, Interface*, *18*(182), 20210555. Advance online publication. DOI: 10.1098/rsif.2021.0555 PMID: 34583561

Subedi, B., Poudel, A., & Aryal, S. (2023). The impact of climate change on insect pest biology and ecology: Implications for pest management strategies, crop production, and food security. *Journal of Agriculture and Food Research*, *14*, 100733. DOI: 10.1016/j.jafr.2023.100733

Sun, J., Xu, L., Huang, C., & Ng, E. Y. K. (2025). Climate change and health: The role of artificial intelligence in predictive surgical treatment. *Innovation and Emerging Technologies*, *12*, 2550004. Advance online publication. DOI: 10.1142/S2737599425500045

Sutherst, R. W. (2004). Global Change and Human Vulnerability to Vector-Borne Diseases. *Clinical Microbiology Reviews*, *17*(1), 136–173. DOI: 10.1128/CMR.17.1.136-173.2004 PMID: 14726459

Tesla, B., Demakovsky, L. R., Mordecai, E. A., Ryan, S. J., Bonds, M. H., Ngonghala, C. N., Brindley, M. A., & Murdock, C. C. (2018). Temperature drives Zika virus transmission: evidence from empirical and mathematical models. Proceedings of the Royal Society B: Biological Sciences, 285(1884), 20180795. https://doi.org/ DOI: 10.1098/rspb.2018.0795

Tiggeloven, T., Pfeiffer, S., Matanó, A., van den Homberg, M., Thalheimer, L., Reichstein, M., & Torresan, S. (2025). The role of artificial intelligence for early warning systems: Status, applicability, guardrails, and ways forward. *iScience*, *28*(11), 113689. DOI: 10.1016/j.isci.2025.113689 PMID: 41280665

Van der Weyde, L. K., Tobler, M. W., Gielen, M. C., Cozzi, G., Weise, F. J., Adams, T., Bauer, D., Bennitt, E., Bowles, M., Brassine, A., Broekhuis, F., Chase, M., Collins, K., Finerty, G. E., Golabek, K., Hartley, R., Henley, S., Isden, J., Keeping, D., & Flyman, M. V. (2021). Collaboration for conservation: Assessing countrywide carnivore occupancy dynamics from sparse data. *Diversity & Distributions*, *28*(5), 917–929. DOI: 10.1111/ddi.13386

Villanueva-Miranda, I., Xiao, G., & Xie, Y. (2025). Artificial intelligence in early warning systems for infectious disease surveillance: A systematic review. *Frontiers in Public Health*, *13*, 1609615. Advance online publication. DOI: 10.3389/fpubh.2025.1609615 PMID: 40626156

Visave, J. (2025). Transparency in AI for emergency management: Building trust and accountability. *AI and Ethics*, *5*(4), 3967–3980. DOI: 10.1007/s43681-025-00692-x

Wang, X., Xu, Y., & Xi, X. (2025). Spatiotemporal Risk Assessment of H5 Avian Influenza in China: An Interpretable Machine Learning Approach to Uncover Multi-Scale Drivers. *Animals (Basel)*, *15*(16), 2447. DOI: 10.3390/ani15162447 PMID: 40867775

Waqas, M., & Humphries, U. W. (2024). A critical review of RNN and LSTM variants in hydrological time series predictions. *MethodsX*, *13*, 102946. DOI: 10.1016/j.mex.2024.102946 PMID: 39324077

Womersley, F. C., Sousa, L. L., Humphries, N. E., Abrantes, K., Araujo, G., Bach, S. S., Barnett, A., Berumen, M. L., Lion, S. B., Braun, C. D., Clingham, E., Cochran, J. E. M., de la Parra, R., Diamant, S., Dove, A. D. M., Duarte, C. M., Dudgeon, C. L., Erdmann, M. V., Espinoza, E., & Sims, D. W. (2024). Climate-driven global redistribution of an ocean giant predicts increased threat from shipping. *Nature Climate Change*, *14*(12), 1282–1291. DOI: 10.1038/s41558-024-02129-5 PMID: 39650805

Xing, Y., Liu, X., & Wang, X. (2026). Integrating UAVs, satellite remote sensing, and machine learning in precision agriculture: Pathways to sustainable food production, resource efficiency, and scalable innovation. *Frontiers in Agronomy*, *7*, 1670380. Advance online publication. DOI: 10.3389/fagro.2025.1670380

Yu, H. (2025). Climate change unveils hidden microbial dangers. *Environmental Science and Ecotechnology*, *24*, 100544. DOI: 10.1016/j.ese.2025.100544 PMID: 40115027

Zeghina, A., Leborgne, A., Le Ber, F., & Vacavant, A. (2024). Deep learning on spatiotemporal graphs: A systematic review, methodological landscape, and research opportunities. *Neurocomputing*, *594*, 127861. DOI: 10.1016/j.neucom.2024.127861

Zeng, D., Cao, Z., & Neill, D. B. (2021). Artificial intelligence–enabled public health surveillance—from local detection to global epidemic monitoring and control. In *Artificial Intelligence in Medicine* (pp. 437–453). Elsevier., DOI: 10.1016/B978-0-12-821259-2.00022-3

Zhong, X. (2025). Adaptive Multi-Scale Deep Learning Framework for Intelligent Environmental Anomaly Detection and Predictive Analytics in Smart Cities. *Journal of Computer Science and Frontier Technologies, 1*(3), 65. DOI: 10.63313/JCSFT.9024

Zhou, Z., Jin, J., Yong, B., Huang, W., Yu, L., Yang, P., & Sun, D. (2025). Enhanced NDVI prediction accuracy in complex geographic regions by integrating machine learning and climate data—A case study of Southwest basin. *International Journal of Applied Earth Observation and Geoinformation, 139*, 104498. DOI: 10.1016/j.jag.2025.104498

Chapter 11
AI-Enabled Wastewater and Environmental Sensor Networks for Early Bio Surveillance:
Opportunities, Limitations, and Equity Implications

Furqan Mazhar
Department of Pharmacy, Faculty of Health and Pharmaceutical Sciences, University of Agriculture, Faisalabad, Pakistan

Eman Fatima
Department of Pharmacy, Faculty of Health and Pharmaceutical Sciences, University of Agriculture, Faisalabad, Pakistan

Abdul Ahad Sajid
http://orcid.org/0009-0008-4458-9925
Department Chemistry, Quaid e Azam University, Islamabad, Pakistan

Esha Hamdani
Department of Pharmacy, Faculty of Health and Pharmaceutical Sciences, University of Agriculture, Faisalabad, Pakistan

Hassnain Ahmad Naqvi
University of Engineering and Technology, Lahore, Pakistan

Rubab Amjad
Department of Pharmacy, Faculty of Health and Pharmaceutical Sciences, University of Agriculture, Faisalabad, Pakistan

Abu Sulman
http://orcid.org/0009-0008-7194-5219
University of Agriculture, Faisalabad, Pakistan

Sabira Sultana
Department of Eastern Medicine, Faculty of Medical Sciences, Government College University, Faisalabad, Pakistan

DOI: 10.4018/979-8-3373-9033-8.ch011

Adnan Afzal
http://orcid.org/0009-0001-4359-5695
Government College University, Faisalabad, Pakistan

Afsheen Mansoor
http://orcid.org/0000-0002-1666-2987
Shaheed Zulfiqar Ali Bhutto Medical University, Islamabad, Pakistan

ABSTRACT

Early bio surveillance is being reformed by the intersection of artificial intelligence (AI), wastewater-based epidemiology (WBE), and distributed environmental sensor networks. Through machine learning, predictive analytics, and anomaly detection based on real-time environmental sampling, AI-enabled systems will provide the opportunity to identify the outbreak of an infectious disease, the trend of antimicrobial resistance, a chemical exposure, and an emerging threat of biological attacks before it manifests itself clinically. This chapter reviews the technological principles of wastewater and environmental monitoring systems that operate on AI, sensor design, edge computing, data fusion, cloud-based analytics, and digital dashboards to support the decision-making process of the general population. It discusses the potential to improve the sensitivity, specificity, and timeliness of outbreak detection using the advanced modeling methods, including deep learning, spatiotemporal forecasting, and interpreting genomic signals.

1. INTRODUCTION

Surveillance refers to the ongoing organized assembling, study, explanation, and distribution of data (Kman & Bachmann, 2012). Biosurveillance offers details that enhance judgements about reducing the consequences of epidemics and biological attacks (Willis & Moore, 2014). Surveillance is considered the key tool for detecting public health events that affect all the countries, notably infectious diseases that are new. Not only is the use of surveillance useful for bioterror attacks, but also the data produced by surveillance systems is also useful in the detection and reaction to new outbreaks. These outbreaks are not concerned with conventional biological weapons, but their public health importance can be similarly serious. (Kman & Bachmann, 2012)

The four functions of basic surveillance include:

- Finding cases of disease in particular groups and documenting the data.
- Evaluating and verifying documented sickness details to discover epidemics.
- Prompting quick and suitable measures at the community or area level to allow suitable countrywide disease management and control of epidemics.

- Producing outbreak health data to support future handling of public health and community health initiatives and plans. (Kman & Bachmann, 2012; Castillo-Salgado, 2010)

i. Limitations of Traditional Bio Surveillance Systems

The current biosurveillance systems face major problems in ability of systems to share and use data, slowness, and data coordination. Different system designs and privately owned infrastructures often block smooth exchange of information between scientific organizations, medical service providers, and government health departments. Moreover, irregular information formats lower the dependability of immediate information exchange, while lag in data updates hinders prompt action during medical crises. (Amanna, 2025). These obstacles emphasize the need for AI methodologies capable of managing data sharing across different systems through flexible guiding structures. The unavailability of common standards for combining complex data types ranging from genetic codes and medical visual record to activity patterns and environmental data further limits prompt action. Advanced structures must therefore use adaptive rules and shared training models that honors data protection while maintaining evaluation accuracy. Carrying out such complex functions needs flexible designs that can review historical data and predict outbreak patterns and medical solutions immediately (Amanna, 2025; Falkner, 2025).

ii. Role of AI in Enhancing Biosurveillance

AI is an element of technology that includes creation and betterment of softwares to carry out different activities with better accuracy similar to human reasoning. Some of these activities include finding solutions, making choices and adapting. In the area of biosurveillance, AI can enhance the identification, monitoring and forecasting of hard-to-treat infections. (Ali et al., 2023; Ayesiga et al., 2025). AI programs, like machine learning (ML), deep learning and natural language processing can be applied in identifying patterns by studying massive collection of information regarding microorganisms. It can update responsible groups about the trends and abnormalities in trends, allowing the quick discovery of more epidemics. (Ayesiga et al., 2025).

AI based methods have been applied to check and forecast water quality measures. Such methods can analyze information from sensors and distant sensing systems, providing immediate evaluations that are key factors for handling water systems and making sure availability of cleaner water. In environmental monitoring domains, particularly water systems surveillance, AI methods are also successfully employed using variety of learning algorithm systems. For example, the radar-based rainfall predication was suggested by some ML models. Also, a hydrological regimen surveillance system was recently developed to solve measurement precision problems and enhance productivity. (Amani et al., 2020; Chen et al., 2020)

2. WASTEWATER AND ENVIRONMENTAL BIOSURVEILLANCE SYSTEMS

Wastewater-based Epidemiology (WBE)

Wastewater-based epidemiology (WBE) is considered a valuable approach to assist in medical policies by producing quick, comprehensive, and non-intrusive community-wide health checks, and has strongly developed and improved as a research field recently. The employment of public wastewater as a basis of medical insights is done by studying human waste biomarkers collected from communities in wastewater that show different perspectives on human wellness, activity patterns, contact with risks, and daily actions. (Bowes, 2024; Daughton, 2018; Daughton, 2001). Employed as an additional information source besides existing approaches, these measurable and factual evaluations of populations can be employed to make fact-driven choices, provide new ideas for accurate and productive use of resources, and make better modifications in plans. In the beginning of the 21st century, WBE was broadly acknowledged as a primary method for surveillance of patterns at community level. During COVID-19 worldwide outbreak, SARS-CoV-2 surveillance in wastewater to assist quick and directed medical actions provided strong approval of the field that continues undergoing change as health goals change, especially in case of infectious diseases. (Bowes, 2024

In WBE, wastewater is checked for the existence of substances, breakdown products or microbes on the basis of the fact that any substance taken in or consumed by the human body will be excreted in urine or feces. The amounts of waste in wastewater indicate how much the population using the sewerage is facing the threat. (Vitale et al., 2021).

The most established use of WBE is the study of a variety of illegal drugs used by a community. The possible uses of WBE are growing at a rapid rate. WBE is being used to evaluate use of drugs like nicotine, alcohol, and caffeine; medicinal consumption; contact with pollutants and toxins like bugs-killers, plastic additives, and fungal toxins; and in monitoring of disease-causing microbes like influenza, poliovirus, and hepatitis A, and for antimicrobial resistance (AMR). Interest is also developing in surveillance of internal bio-indicators of diseases like cancer, diabetes, and cardiovascular diseases, or other indicators of general well-being-like quality of diet or imbalance of antioxidants. (Eaton et al., 2022; Daughton, 2012)

Role of AI in WBE

The application of modern computer-based systems like ML, layered neural training, large-scale data analysis, and AI are vital for the success of a WBE. The idea of the IoT (Internet of Things) has a primary role in linking and sharing information among advanced tools and systems, making them smarter, more trackable, adjustable, and easy to handle. By harnessing the IoT, WBE sensors can be smoothly combined with modern industrial automation, advanced data analysis and digital healthcare, providing immediate connection, information exchange, analysis, and efficiency enhancement. This combination provides production of refined information, which is useful for making smart choices regarding biosurveillance. (Gonçalves, 2023).

Goal of AI integration in Biosurveillance

The current state of AI and IoT investigation is mainly focused on how these tools can assist in long-term progress and shared benefits. The achievement of strong ecological balance by the application of AI has helped investigators to deal with on-going ecological problems concerned with biosurveillance and pollution management, thus protecting threatened species, fighting climate crisis, and organizing waste handling. Biosurveillance measures the amount of toxins in an organism that is causing physical and chemical, molecular, and hereditary modifications in cells. Biomonitoring preserves the quality of the surroundings (Figure 1). IoT and AI have great advantages like self-operation, quickness, fairness, self-adjustment, the ability to work in extreme surroundings, and prompt and precise predictions. (Hait et al., 2024).

i. AI Enhanced Biosurveillance in Agriculture

Advanced agriculture faces closely related problems of enhancing output requirements and increasing crop pests. Bio-surveillance, using tools such as insect traps, detection procedures, and Un-manned Aerial Vehicles (UAVs), raises moral concerns vital for long-term cooperation. AI incorporation, using Geographic Information Systems (GIS), Global Positioning System (GPS), and Remote Sensing (RS), proved to be useful. GIS helps with mapping, minimizing costs, and lowering ecological effects. The role of AI in the management of pests is heterogeneous, and more attention is paid to timely detection, data forecasting, and precision in the agricultural sector. Although there are barriers to the use of AI in agriculture, there is a need to keep on innovating so as to deliver harmless food and enhanced bio surveillance. It reduces production cost and harmful environmental impacts.

ii. Accuracy in Agriculture

Self-driving Machinery: AI-powered machines with a tool accuracy improvement can apply pesticides or other chemicals to the parts where it is required only. It reduces the use of chemicals and their impact on the environment.

Optimization: AI can help to optimize the utilization of chemicals by alternating the rate based on real-time data such as the severity of the pest invasion.

iii. Support in Decision-making:

Data Combination: AI is able to unite informational sources of various types, and it entails weather prediction, records of field condition and insect monitoring records which can help farmers in making a comprehensive decision related to handling the pests in a way that is more efficient.

iv. Advanced Traps:

IoT: AI can be linked with advanced traps and lures that use IoT tools to lure and trap targeted insects. This information can be studied to get concept of pest activity trends, assisting in the establishment of more successful management strategies.

Figure 1. Goal of AI integration in bio surveillance

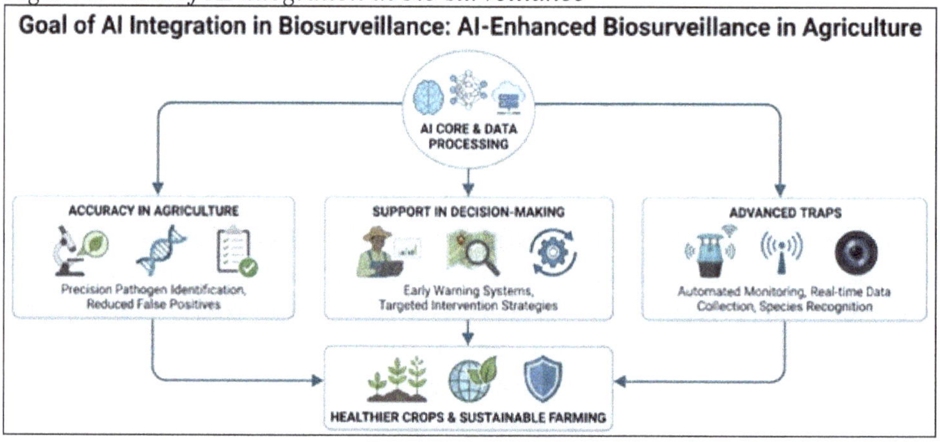

v. Biocontrol:

Evolutionary Computing Methods: AI can be employed successfully to refine the deployment of beneficial organisms or the application of biological insecticides. Genetic algorithms can help determine the most beneficial methods to deploy useful organisms to regulate pest number. (Yadav et al., 2023)

3. ARCHITECTURE OF AN AI-ENABLED SENSOR NETWORK

Biosurvelliance system have become more important for worldwide health protection, detecting infectious agents and observing ecological factors that influence the pattern of disease spread. (Asorose, E., *et al.*, 2025). Personalized medicine utilizes these systems by employing predictive computations to customize risk evaluation and treatment plans, often applying an artificial intelligence knowledge framework and predictive tools. (Ibitoye, J. S., *et al.* (2025). The increasing need for immediate health surveillance and treatment development highlights the need for AI-powered system integration. Epidemic forecasting tools, for instance, increasingly utilize AI learning workflows that integrate genetic, environmental, climatic, and behavioral datasets for advanced alerts. (Lin, H. W., *et al.* (2023).

3.1. Sensor Technologies and Development Strategies

Wastewater contains various pollutants that have been generated and released in harmful ways from household, industrial, commercial, and agricultural sources. (Ejeian *et al.*, 2018) The content and level of chemicals differ significantly and can be grouped into three main types:

1. Microbes
2. Mineral compounds
3. Bio- pollutants

Biosensors for the Detection of Organic Materials

Bio-pollutants can originate naturally or from human activity, and a wide range of pollutants can be found in factories, farms, and domestic wastewater. Organic substances in industrial discharge consist of different hydrocarbons, aromatic substances, chlorinated compounds, diphenylmethane and surfactants. Long-lasting organic contaminants, for example, polybrominated diphenyl ethers (PBDEs), polybrominated diphenyl (PBBs) and polychlorinated biphenyls (PCBs) exist in contaminated water. (Ejeian *et al.*, 2018)

Biosensors for the Detection of Heavy Metals (HMs)

Toxic metal pollution in wastewater is frequently linked to agricultural wastewater, chemical industry operations, mining operations, cosmetic formulations, nuclear by-products, metallurgical operations, coating processes and industrial wastewater. Metabolic processes in living beings rely on specific concentration limits of essential heavy metals such as Co, Cu, Fe, Mn, Ni, Se, and Zn. While non-essential heavy metals such as Pb, Cr, Cd, Hg, As, and Sb are toxic even at low concentrations and pose serious risks to human health and the environment. (Ejeian *et al.*, 2018)

Biosensors for the Detection of Microorganisms

The wide range of water-borne infectious diseases results in mortality and disease burden worldwide. Wastewater offers a favorable environment for various microbial species, including bacteria, viruses, protozoa, algae, fungi and yeasts. They are primarily discharged into wastewater from domestic activities and food processing industries, which are marked by high variability in their microbial composition. The level of pathogenic, nonpathogenic, and saprophytic microorganisms shows considerable variation in water waste samples. (Ejeian *et al.*, 2018).

3.2. IoT Integration and Data Pipelines

Public health and sanitation are essential elements of sustainable human development and national development, which come from a safe and pollution-free environment. Thus, continuous monitoring is necessary to guarantee that the population of any nation can maintain good health. Environmental survellience consist of strategic disaster planning and management, regulating various pollutants, and effectively handling the issues that arise due to harmful environmental conditions. Environmental surveillance deals with aquatic contamination, air pollution, harmful radiation exposure, climate variations, seismic activities, etc. The origin of pollution is contributed to by multiple factors, some of which are anthropogenic and others due to environmental factors, and the function of environmental monitoring is to accurately address the difficulties so that the environment is preserved for societal and global well-being. (Sayed et al., 2019).

IoT devices are integrated into wireless sensor networks for efficient waste handling, vehicle identification, thermal regulation and pollution monitoring and reduction. Therefore, advanced techniques of environment monitoring are known as smart environment monitoring systems, due to the use of IoT, AI and wireless sensing. (Bhoomika et al**.,** 2016).Evaluation of fire affected regions using multi-band spectral data captured through space based imaging techniques and remote sensing(Cicala et al., 2018), smart healthcare monitoring system, IoT based environmental sensing(Gaglio et al., 2014), an automated marine sensing system using multimodal sensing networks(Zhang et al., 2015), and many other SME methods are documented in recent studies.

3.3. Edge Computing Biosurveillance

Edge-based computing is extensively applied in detecting infectious diseases. A smart edge monitoring system employs edge computing to detect communicable diseases. (Adhikari & Munusamy, 2021). It is a remote surveillance system that is superior in performance and reliability to any conventional sensing system. (Aazam et al., 2021). The advanced edge platform assists doctors, health agencies, and medical institutions to access patient related information. This system is primarily applied to detect contact networks of the infected individuals within the population. This framework identifies infected individuals and helps to track their movement from the external environment. Edge computing saves patient-related information in a secure manner. (Zhou et al., 2021). Contagious diseases can be transmitted between individuals by contact with infected objects or by direct contact. The primary cause of infectious diseases is viral and bacterial agents of zoonotic or human origin. (Sengupta et al., 2020). Infectious disease assessment is the primary responsibility in the medical sector to provide safe living conditions. (Yang et al., 2021). The evaluation process helps to detect infected individuals in society by tracking individuals through a monitoring framework. Without effective surveillance mechanisms, infectious diseases rapidly propagate in the environment and lead to serious consequences for the global population. (Ojagh et al., 2021; Hosseini et al., 2020). Edge computing is an information technology model that brings processing nearer to the user. Edge computing is widely used to deliver improved services to the customer through network-based technologies. (Yang & Wang, 2020). To enhance performance as compared to traditional approaches and to increase prediction precision, AI-driven learning models are often used to extract complex spatial features. They are also commonly applied to detect and analyses biomedical data (Kumar et al., 2024).

4. ARTIFICIAL INTELLIGENCE METHODS FOR EARLY DETECTION

Developments in emerging digital technologies are rapidly increasing. Digital health system provides multiple possibilities to minimize human errors, enhance patient outcomes, and record data continuously. AI-based techniques, including machine learning and deep learning algorithms, are extensively applied in forecasting and detection of multiple disorders, especially those whose diagnosis is dependent on medical imaging or signal processing. (Uysal & Ozturk, 2020; Woldaregay et al., 2019). AI also helps to recognize population groups or geographic regions where illness or risky activities are common. (Bhatt & Pal, 2019; Yıldırım et al., 2018). Machine learning models have demonstrated remarkable performance in clinical image processing due to sophisticated algorithms that enable the automatic identification of enhanced features.ML is built on learning principles and can be classified into three main classes (Figure 2). Label-based learning (class labelling, trend modelling, and composition), unlabeled learning (relationship discovery, data grouping, and variable compression), and reward-based learning. (Wu et al., 2021).

Figure 2. Artificial intelligence methods for early detection

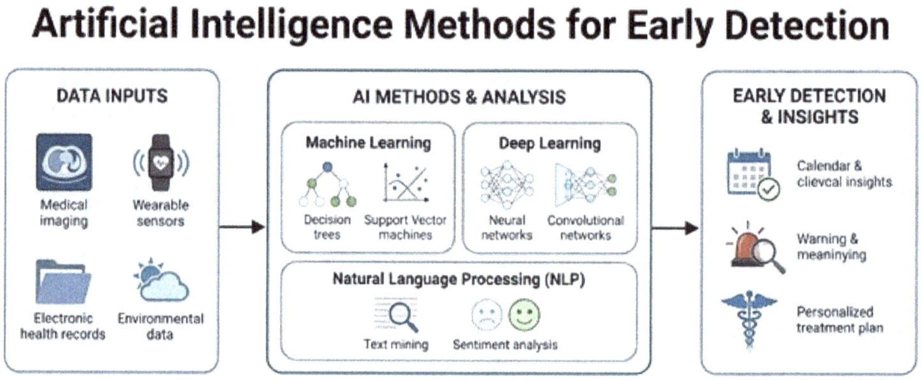

4.1. Machine Learning and Deep Learning Approaches

Multiple calculations and processes on the input database are carried out via ML models. Data preparation is the initial and critical stage to minimize prediction errors, increase processing efficiency, and enhance data reliability. Once processing is completed, imported features are identified and applied according to the chosen machine or deep learning model for image-based categorization. Attribute selection lowers feature space and enhance model performance (Dhal & Azad, 2022).

Machine learning allows systems to carry out clinical tasks.ML has a popular subset in medical image analysis known as deep learning. DL is an approach for developing ML models in which basic principles are arranged progressively to form a multi-layered neural architecture. In simpler terms, DL is an advanced form of machine learning for handling big data. (Xi et al., 2020; Rizk et al., 2019; Bhatt et al., 2021).

4.2. Artificial Intelligence Techniques in Disease Diagnosis and Prediction

Medical imaging modalities are one of the highly utilized instruments for clinical patient data. Still, it depends on manual evaluation and is prone to growing resource limitations. AI-based medical imaging diagnosis, especially using deep learning, has successfully addressed human mistakes due to limited expertise or errors. Identification of illness and contaminants of disease transmission needs ongoing monitoring and analysis of information. Immediate intervention using reliable information has a considerable societal and economic influence on the global population. (Minaee et al., 2020)

The application of AI in medicine has improved data collection and computational processing of valuable data, at a high level of use, including robotic surgery. (Kumar & Singla, 2021).

AI refers to the capability of machines to model human learning through visual and pattern analysis in difficult situations.AI in healthcare has revolutionised data collection, processing, and application to clinical decision making. (Tengnah et al., 2019).

4.3. The Effective Role of AI Technologies in Identifying and Predicting Human Disorders and Diseases

AI technologies are increasingly used in medicine to succeed and obtain a precise understanding of severe medical conditions. (Zitnik et al., 2019; Burgos & Colliot, 2020).

Since AI effectively processes visual data in the healthcare sector, it is more widely applied in illness detection and forecasting. (Moen et al., 2019). Machine learning models and large-scale data sets from electronic health records and wearables are key instruments to apply AI techniques effectively in the healthcare system to enhance medical diagnosis Classification of conditions, clinical decision making, assistive device effectiveness, and selection of optimal therapies, and ultimately supporting healthier and longer lives. AI is used to improve clinical evaluation and diagnosis quickly. (Chen et al., 2019;)

5. EQUITY IMPLICATIONS AND GLOBAL HEALTH CONSIDERATION

In global health, equity is making sure that all groups, no matter where they live, how much money they have, or what resources they have, have the same access to the advantages of any form of surveillance system that may keep an eye on health risks and environmental exposures. (August E et al., 2022). Global health prioritizes enhancing health and achieving health equity for all individuals globally. (koplan et al., 2009)

There are big gaps in equity all over the place, and we'll show you two examples from the present. Early attempts to stop COVID, one of the worst pandemics in history, showed huge differences in vaccine coverage between rich and poor countries. Vaccine apartheid meant that poor nations had fewer than 1% coverage at the start of 2021. On the other hand, rich countries used their neocolonial bargaining power, global policy might, and financial resources to get more than twice the number of doses they required to protect their citizens. (Harman S et al., 2021)

Another instance concerns inequities in access to surgical care. More than five billion people around the world don't have quick access to safe surgery and anaesthesia. Only 6% of the 313 million surgical procedures are done in low- and middle-income countries, where these procedures are most needed (Meara et al., 2015). Also, low- and middle-income countries, which make up 48% of the world's population, only have 20% of the world's surgical workforce, or 19% of all surgeons, 15% of anaesthesiologists, and 29% of obstetricians. Once more, even though this difference in outcomes has been noted, there hasn't been enough emphasis on partnerships and processes. (August E et al., 2022).

5.1. AI enabled Disease surveillance

In a world that is connected, infectious diseases like COVID-19 are a threat to everyone's health. In the last 20 years, resistant infectious diseases like severe acute respiratory syndrome (SARS), Middle East respiratory syndrome (MERS), H1N1 influenza (swine flu), Ebola virus, Zika virus, and now COVID-19 have been affecting global health defences and growing rapidly because of more people traveling, living in cities, climate change, and environmental damage. (Borda et al., 2022) Concurrently, this incident in global human health underscores the potential of artificial intelligence.

5.2. Disease surveillance

Disease surveillance intelligence encompasses efforts aimed at the rapid discovery of potential health risks, followed by their verification, evaluation, and research to facilitate public health control recommendations (Paquet et al., 2006).

"Epidemic intelligence" is a very important part of disease monitoring that is becoming more important because of the rise of big data (Kostkova et al., 2021; Kostkova, 2018).

Epidemic intelligence consists of two elements:

1. a part that is based on indicators
2. an event-based part (World Health Organization, 2016)

AI-enabled disease surveillance gathers and analyses huge amounts of unstructured, real-time data to help with epidemiological responses. The investigation of the prerequisites for a worldwide data-driven and pandemic intelligence surveillance system is established in the literature (Kostkova, 2013).

Wastewater-based and environmental biosurveillance

Wastewater-based and environmental biosurveillance systems produce high-frequency, population-level signals that can act as precursors to infectious disease outbreaks, antimicrobial resistance (AMR), and environmental health hazards. To make these signals part of normal public health work, we need standard ways to understand the data, alert systems that use thresholds, and explicit ways for technical experts and policymakers to talk to each other. Research has underscored the necessity of integrating AI-enabled biosurveillance into a One Health surveillance framework to guarantee synchronized responses across human, animal, and environmental health domains (Lawal et al., 2025). Policy translation also depends on institutional trust in AI-generated insights. Before taking action on biosurveillance alerts, public health officials typically need clear models, confusion about how to measure things, and validation against epidemiological ground truth. Systems based on AI providing explainable results and risk evaluation based on scenarios have a greater likelihood of being deployed in the policy sphere, particularly in cases of a public health emergency with greater accountability and public monitoring (Ali, 2024).

Policy Integration and Future Directions

The success of AI-driven wastewater and environmental sensor networks to detect the early signs of bio surveillance depends on whether they are integrated into the policy and decision-making of the population. Even though advancement in artificial intelligence and environmental sensing has significantly enhanced the early warning of infectious diseases and environmental health hazards, the transformation of complex bio surveillance information into meaningful public health actions remains a major challenge. Past studies suggest that wastewater-based environmental biomonitoring and epidemiology often operate concomitantly with, but not within, formal surveillance and governance institutions, therefore, limiting their policy effectiveness in times of population health crises (Lawal et al., 2025; Ali, 2024). Researchers argue that bio surveillance systems should be directly coupled with policy processes, emergency operations, and One Health governance systems to ensure timely and synchronized interventions on the part of the population (Singh et al., 2025).

6. OPPORTUNITIES AND PUBLIC HEALTH APPLICATIONS

Public health preparedness and response can be improved by AI-enabled environmental biosurveillance systems, particularly those harmonizing wastewater and environmental monitoring (Figure 3) (Singh et al., 2024).

Environmental biosurveillance can help early outbreak detection, monitoring of antimicrobial resistance, identification of environmental hazards, and evidence-based public health decision-making when combined with artificial intelligence and advanced analytics (Hendriksen et al., 2024; Patel et al., 2022).

6.1. Public Health Utility & Wastewater Monitoring

By collecting samples from sewage systems, wastewater surveillance has historically been used to track waterborne or fecal-orally transmitted microbes (Xagoraraki & O'Brien, 2020). Wastewater testing provides an early sign for Infectious diseases by isolating pathogens, including those from asymptomatic people (Singer et al., 2023).

6.1.1. Wastewater Surveillance: Monitoring Key Pathogens of Concern

Gastrointestinal pathogens: Campylobacter spp. are a primary cause of human gastroenteritis and diarrhea. Typhoid fever and paratyphoid fever are primarily caused by Salmonella enterica serovar Typhi and Para typhi, respectively. Giardia and Cryptosporidium, two protozoan parasites, are important waterborne pathogens and enteric pathogens of public health issues (Singh et al., 2024).

Respiratory pathogens: The focus on wastewater as a monitoring tool to provide early disease identification in the community has increased since it emerged in 2020 (Singh et al., 2024).

6.1.2. Implementation of Wastewater Monitoring

Keegan et al. analyzed the occurrence of PMMoV, Mpox virus, and SARS-CoV-2 in US community wastewater (Brighton et al., 2024). (Ahmed et al., 2020) provided proof of SARS-CoV-2 in Australian wastewater, providing awareness into community prevalence and assisting public health initiatives.

6.2. Pandemic Readiness and Early Outbreak Detection

6.2.1. The Issue With Conventional Surveillance

The majority of these approaches rely on collected information, such as hospitalization, prescription, and incidence and mortality rates (Gracia-Lor et al., 2018). The extent of the distribution is unknown because not everyone has access to infectious disease testing, making it hard to properly to oversee possible patients and asymptomatic patients (Qiu, 2020).

6.2.2. The Necessity of Early Warning and Readiness

Enhancing the supervision of infectious diseases, attaining early warning, and preventing infectious disease pandemics are crucial elements of public health (Sims & Kasprzyk-Hordern, 2020).

6.2.3. WBE's Function in Early Outbreak Identification

Wastewater-based epidemiology (WBE) can provide comprehensive and unbiased real-time assessments of the quality of both the public and environmental health. This data can also be tracked in real time in order to detect the deviation of the overall pattern as quickly as possible (Mao et al., 2020).

6.2.4. The Benefit of Pandemic Preparedness

When compared to other conventional surveillance systems, WBE can provide real-time analysis, and the timely reaction of the public services, making sure that the practice of possible health measures can be implemented before the infectious diseases, and outbreaks spread (Mao et al., 2020).

Regarding the COVID-19 pandemic, many researchers are trying to use WBE to identify SARS-CoV-2 in wastewater in order to examine possible virus carriers and providc alcrt of COVID-19 outbreaks in the community (Mao et al., 2020).

Figure 3. Opportunities and public health application

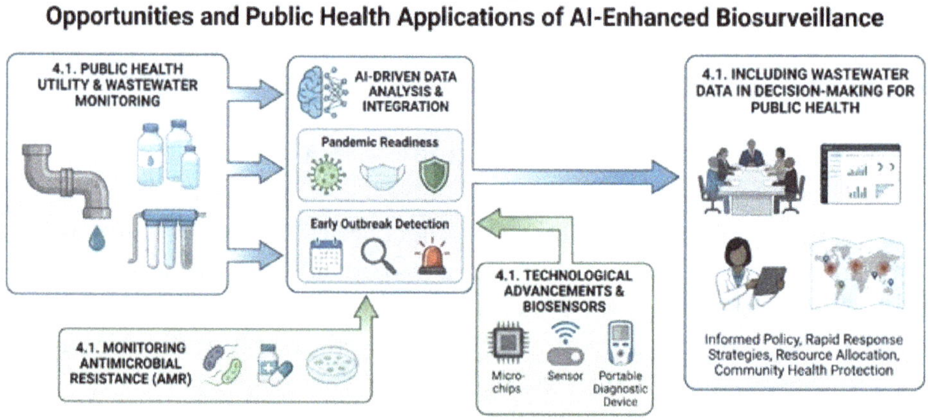

6.3. Monitoring Antimicrobial Resistance (AMR)

A critical threat to public health is antibiotic resistance. Many ARBs and ARGs can be detected in the feces or on the skin of healthy humans, but population estimates of antibiotic resistance that are independent of the healthcare system are lacking (Foxman et al., 2024). Since ARBs and ARGs are released from feces, urine, nasal mucus, skin, and sputum at all stages of infection and can be found in wastewater, wastewater surveillance for antibiotic resistance is feasible (Tiwari et al., 2024).

Because carbapenems are among the few antibiotics that are effective against ESBL bacteria, and because ESBL destroys beta-lactam antibiotics, extended-spectrum beta-lactamase (ESBL)-producing Enterobacteriaceae and carbapenem-resistant Enterobacteriaceae (CRE) are especially concerning (U.S. Department of Health and Human Services [HHS], 2019). The possibility of using wastewater to identify and measure ARGs and ARBs as a basis for public health action has highlighted as these substances have been found in wastewater all over the world (Hendriksen et al., 2019).

6.4. Technological Advancements & Biosensors

The rising possibility of more frequent disease epidemics demands the creation of policies that enable early detection (Balcázar, 2025).

The current COVID -19 crisis underscores the importance of early detection methods to monitor disease outbreaks. These methods also have the ability to determine the quantity and distribution of asymptomatic and pre-symptomatic individuals (Jiménez Rodríguez et al., 2022).

6.4.1. Conventional Epidemiology's Limitations

Most low-middle-income nations (LMICs) and low-capacity cities do not have the means and test technology significant to implement most of the atypical epidemiology measures (Kuster & Overgaard, 2020).

6.4.2. Early Outbreak Detection Using Wastewater-Based Epidemiology

With the help of a vast scale disease surveillance, wastewater-based epidemiology (WBE) is capable of preventing disease outbreaks as well as monitoring the presence of different infectious agents within wastewater (Xagoraraki & O'Brien, 2020). WBE is very useful in estimating the prevalence of the disease besides reporting confirmed active cases since it also considers both pre-symptomatic and asymptomatic patients (Jiménez Rodríguez et al., 2022).

6.4.3. Using Biosensors in WBE

The detection of human pathogenic viruses is of great importance to predict the occurrence of flare-ups. In-situ detection can be made in limited resources areas due to numerous benefits of biosensors, such as low cost, sensitivity, specificity, portability, and real-time information (Jiménez Rodríguez et al., 2022).

6.5. Including Wastewater Data in Decision-Making for Public Health

107 different pathogen species have been found in wastewater surveillance analyses between 1970 and 2024, including 41 bacteria, 57 viruses, 8 protozoa, and 1 fungus. A single pathogen (median = 1, range = 1 to 11) was the focus of most of surveillance studies. The taxonomy proposed by Schoch et al. (2020) and Lefkowitz et al. (2018) was used to categorize pathogen species. The most studied pathogen in wastewater during the COVID-19 pandemic was Beta coronavirus Pandemicum (SARS-CoV-2). Only 84 of the 974 included studies reported at least one public health intervention.

Studies of wastewater surveillance conducted by or in collaboration with public health institutes continue to be strongly linked to a higher probability of their application in public health initiatives (OR = 3.11, 95% CI = 1.17–9.28). As testing frequency increases, the odds of public health action remain notably higher (OR = 1.02, 95% CI = 1.01–1.04) (van der Drift et al., 2025).

Additional controls include community awareness, immunization, surveillance, and quarantine measures (van der Drift et al., 2025).

7. OPERATIONAL, ETHICAL, AND TECHNICAL RESTRICTIONS

7.1. Data Uncertainty and Quality

7.1.1. High-Profile Mistakes and Takeaways

Real-world examples highlight the importance of data quality and the grave repercussions of poor governance in a number of industries, including public administration, technology, healthcare, finance, and aerospace (Guillen-Aguinaga et al., 2025).

E.g.: NASA's Mars Climate Orbiter

It was destroyed due to a $125 million mismatch between imperial and metric units (Guillen-Aguinaga et al., 2025).

Lesson: In high-stakes fields, standardization, validation, and transparent governance procedures are essential.

7.1.2. Achievements: Data Quality as a Strategic Resource

Good governance and data of high quality are a competitive strength as evidenced by successful companies. Strict data standards provide reliability, and Netflix uses them in its personalized recommendation system (Yang, 2024).

Multicenter clinical databases also show that patient safety may be secured with semi-automated data quality assessments (Kahn et al., 2012).

7.1.3. Data Quality Assurance: Techniques and Best Practices

A key component of efficient data collection is the use of standardized methods to reduce errors (Pykes, 2024). Anomaly detection, uniform definitions across systems, routine maintenance and ongoing monitoring protect data integrity (Ehrlinger & Wöß, 2022).

7.1.4. The AI Era's Data Quality

Data must now be suited for algorithmic processing in addition to human interpretation (Cinar et al., 2025). For regulatory compliance, high-quality data must guarantee provenance, lineage, interpretability, and metadata (Cheong, 2024).

7.1.5. Synthesis: Towards Integrated Governance of Data Quality in the AI Era

AI can enhance data quality at scale, organizations must also make sure that AI systems are trained, validated, and tracked using high-quality, ethically sourced data (Asokan et al., 2025).

7.2. Issues With Model Bias and Validation

The study of creating Artificial Moral Agents (AMAs), robots or artificially intelligent computers that act morally or as though morally, is known as machine ethics or machine morality (Anderson & Anderson, 2011).

This raises the question of how robots would learn about the world and inherit morality, pro-survival attitudes, inconsistency, and insensitivity etc. like humans ("Ethics of Artificial Intelligence," n.d.).

Machine ethics is framed by some researchers as a component of the larger AI control or value alignment problem, which is the challenge of making sure that increasingly powerful systems pursue goals that are consistent with human values and oversight. Stuart Russell suggests that rather than optimizing a fixed, fully defined goal, beneficial systems should be built to strive to realize human preferences; remain unsure of what those preferences are; and learn about them from human behavior and feedback (Russell, 2019).

Challenges

AI's utility function may permit numerous dangerous situations that comply with a particular ethical framework but defy common sense (Yudkowsky, 2011).

Developers should set preliminary goals, since AIs may not share human inclinations. Principles must develop, and human limitation helps to avoid misuse via harmful agents ("Ethics of Artificial Intelligence," n.d.).

7.3. Infrastructure Limitations and Sensor Upkeep

Missing data, uncertainty, faults (outliers, bias, constant values, stuck-at-zeros and noise) are some of the errors in sensor data, which contribute to sensor data quality problems. These errors may affect the reliability of the collected measurements and may be induced by noise, sensor drifts, sensor failure, and calibration issues (Khan et al., 2020). Calibration-based methods are often used to correct the problems of bias, drifts, noise, and sensor failure.

To identify the model of sensor behavior typicality, (Wang et al., 2015) presented procedures of online blind detection and automatic calibration of sensor drifts through Principal Component Analysis. The value of sensor drift is then approximated with the help of the Kalman filter and to derive a better approximation to the actual sensor drift value the sensor reading is subtracted with the drift value.

The PCA-based detection, isolation, and reconstruction method to identify bias, drifts, noise, constant and stuck-at-zero errors was introduced (Uren et al., 2016). The method assumes that not all the sensors will fail simultaneously and employs a group of redundant measurements to validate sensor measurements (Khan et al., 2020).

7.4. Issues With Consent, Privacy, and Governance

Even though this use of a public health surveillance would raise a number of ethical and legal concerns, the proposed application of wastewater screening as a means to detect SARS-CoV-2 viral RNA has the potential to greatly enhance our technical capacity to identify, track, localize, and measure the transmission of COVID-19 (Gable et al., 2020).

Nevertheless, in some cases, wastewater surveillance can cause privacy concerns and raise the concerns of the violation of the Fourth Amendment, which prohibits unwarranted searches and seizures (Gable et al., 2020).

7.4.1. Problems utilizing wastewater information

Further legal and ethical concerns arise if wastewater data is used for other public health intensions (Fairchild et al., 2007).

Proof of SARS-CoV-2 in wastewater may be used by public health authorities to pursue various intervention strategies. Fourth Amendment issues could arise from conditional screening requirements based on SARS-CoV-2 detection in wastewater. Legal concerns rely on how samples are kept, by whom, and for what purposes. Government representatives must refrain from using their restrictive powers in a discriminatory manner (Gable et al., 2020).

CONCLUSION

With AI-controlled systems of wastewater and environmental detectors, early bio surveillance is going to be done in a way that is game-changing. They allow new avenues of securing health among the people even before such issues occur. Advanced sensing technologies, machine learning, and predictive analytics allow these systems to be able to detect new health issues more quickly than more traditional methods of monitoring. They will also be able to assist in preparedness to an outbreak and direct narrow-targeted treatment. They are more useful in identifying infectious diseases; they are used in checking antibiotic resistance, tracking environmental threats and other biosecurity measures. Nevertheless, such benefits can be reaped only to the fullest provided the critical technical, ethical, operational, and equity-related issues are addressed. Provided that these problems are not resolved, they may cause systems to be less trustworthy and undermine the confidence of the population. Such problems are low quality of the data, uncertainty in models, infrastructure challenges, and lapses in governance. Moreover, existing inequalities in global health are also likely to be exacerbated by discrepancies in technical access and institutional capability unless improved focus is placed on inclusive deployment techniques. AI-based bio surveillance must become part of the public health policy in the future since it is founded on the principles of openness, fairness, and collaboration. In order to ensure that new technologies are actually used to enhance the health of the population and not to cause harm or isolation, we must invest into interdisciplinary research, international standards, and democratic governance. Finally, the potential of AI-enabled wastewater and environmental bio surveillance is enormous, but its future dependability is pegged on the idea of smart policy, continuous international collaboration, and what can be considered as a commitment to ethical and equitable innovation.

REFERENCES

Aazam, M., Zeadally, S., & Flushing, E. F. (2021). Task offloading in edge computing. *Computer Networks*, *191*, 108019. DOI: 10.1016/j.comnet.2021.108019

Adhikari, M., & Munusamy, A. (2021). iCovidCare framework. *Internet of Things : Engineering Cyber Physical Human Systems*, *14*, 100385. DOI: 10.1016/j.iot.2021.100385 PMID: 38620813

Ahmed, W., Angel, N., Edson, J., Bibby, K., Bivins, A., O'Brien, J. W., Choi, P. M., Kitajima, M., Simpson, S. L., Li, J., Tscharke, B., Verhagen, R., Smith, W. J. M., Zaugg, J., Dierens, L., Hugenholtz, P., Thomas, K. V., & Mueller, J. F. (2020). First confirmed detection of SARS-CoV-2 in untreated wastewater in Australia: A proof of concept for the wastewater surveillance of COVID-19 in the community. *The Science of the Total Environment*, *728*, 138764. Advance online publication. DOI: 10.1016/j.scitotenv.2020.138764 PMID: 32387778

Ali, H. (2024). AI for pandemic preparedness and infectious disease surveillance: Predicting outbreaks, modeling transmission, and optimizing public health interventions. *International Journal of Research Publication and Reviews*, *5*(8), 4605–4619.

Ali, T., Ahmed, S., & Aslam, M. (2023). Artificial intelligence for antimicrobial resistance prediction: Challenges and opportunities towards practical implementation. *Antibiotics (Basel, Switzerland)*, *12*(3), 523. Advance online publication. DOI: 10.3390/antibiotics12030523 PMID: 36978390

Amani, M., Ghorbanian, A., Ahmadi, S. A., Kakooei, M., Moghimi, A., Mirmazloumi, S. M., & Brisco, B.. (2020). Google Earth Engine cloud computing platform for remote sensing big data applications: A comprehensive review. *IEEE Journal of Selected Topics in Applied Earth Observations and Remote Sensing*, *13*, 5326–5350. DOI: 10.1109/JSTARS.2020.3021052

Amanna, A. (2025). Deploying next-generation artificial intelligence ecosystems for real-time biosurveillance, precision health analytics and dynamic intervention planning in life science research. *Magna Scientia Advanced Biology and Pharmacy*, *16*(1), 38–54. DOI: 10.30574/msabp.2025.16.1.0066

Anderson, M., & Anderson, S. L. (Eds.). (2011). *Machine ethics*. Cambridge University Press. DOI: 10.1017/CBO9780511978036

Asokan, D. R., Smith, C. M., & Huq, F. A. (2025). Digitalisation as a catalyst for supplier diversity, equity and inclusion. *International Journal of Operations & Production Management*, •••, 1–26.

August, E., Tadesse, L., O'Neill, M. S., Eisenberg, J. N. S., Wong, R., Kolars, J. C., & Bekele, A. (2022). What is global health equity? A proposed definition. *Annals of Global Health, 88*(1), 50. Advance online publication. DOI: 10.5334/aogh.3754 PMID: 35860038

Ayesiga, I., Yeboah, M. O., Okoro, L. N., Edet, E. N., Gmanyami, J. M., Ovye, A., Atimango, L., Gadzama, B. N., Kembabazi, E., & Atwau, P. (2025). Artificial intelligence-enhanced biosurveillance for antimicrobial resistance in sub-Saharan Africa. *International Health, 17*(5), 795–803. DOI: 10.1093/inthealth/ihae081 PMID: 39545538

Balcázar, J. L. (2025). Wastewater-based epidemiology as a complementary tool for antimicrobial resistance surveillance: Overcoming barriers to integration. *BioEssays : News and Reviews in Molecular, Cellular and Developmental Biology, 47*(8), e70034. Advance online publication. DOI: 10.1002/bies.70034 PMID: 40522150

Bhatt, C., Kumar, I., Vijayakumar, V., Singh, K. U., & Kumar, A. (2021). The state of the art of deep learning models in medical science and their challenges. *Multimedia Systems, 27*(4), 599–613. DOI: 10.1007/s00530-020-00694-1

. Bhatt, V. K., & Pal, V. K. (2019). *Thyroid disease diagnosis using artificial neural networks.*

Borda, A., Molnar, A., Neesham, C., & Kostkova, P. (2022). Ethical issues in AI-enabled disease surveillance: Perspectives from global health. *Applied Sciences (Basel, Switzerland), 12*(8), 3890. Advance online publication. DOI: 10.3390/app12083890

Bowes, D. A. (2024). Towards a precision model for environmental public health: Wastewater-based epidemiology to assess population-level exposures and related diseases. *Current Epidemiology Reports, 11*(3), 131–139. DOI: 10.1007/s40471-024-00350-6

Brighton, K., Fisch, S., Wang, H., Vigil, K., & Aw, T. G. (2024). Targeted community wastewater surveillance for SARS-CoV-2 and Mpox virus during a festival mass-gathering event. *The Science of the Total Environment, 906*, 167443. Advance online publication. DOI: 10.1016/j.scitotenv.2023.167443 PMID: 37793442

Burgos, N., & Colliot, O. (2020). Brain disease prediction. *Current Opinion in Neurology, 33*(4), 439. DOI: 10.1097/WCO.0000000000000838 PMID: 32657885

Castillo-Salgado, C. (2010). Trends and directions of global public health surveillance. *Epidemiologic Reviews, 32*(1), 93–109. DOI: 10.1093/epirev/mxq008 PMID: 20534776

Chen, J., Remulla, D., Nguyen, J. H., Dua, A., Liu, Y., Dasgupta, P., & Hung, A. J. (2019). Current status of artificial intelligence applications in urology and their potential to influence clinical practice. *BJU International, 124*(4), 567–577. DOI: 10.1111/bju.14852 PMID: 31219658

Cheong, B. C. (2024). Transparency and accountability in AI systems: Safeguarding wellbeing in the age of algorithmic decision-making. *Frontiers in Human Dynamics, 6*, Article 1421273.

Cinar, R. F., Yuksek, G., Lale, T., Ekinci, S., Izci, D., & Ma'arif, A. (2025). SHAP-based framework for temporal detection of sensor drift in gas sensor arrays. *Journal of Robotics and Control, 6*(5), 2592–2601. DOI: 10.18196/jrc.v6i5.28048

Daughton, C. G. (2001). Illicit drugs in municipal sewage. In Daughton, C. G., & Jones-Lepp, T. L. (Eds.), *Pharmaceuticals and personal care products in the environment* (pp. 348–364). American Chemical Society. DOI: 10.1021/bk-2001-0791.ch020

Daughton, C. G. (2012). Using biomarkers in sewage to monitor community-wide human health: Isoprostanes as conceptual prototype. *The Science of the Total Environment, 424*, 16–38. DOI: 10.1016/j.scitotenv.2012.02.038 PMID: 22425170

Daughton, C. G. (2018). Monitoring wastewater for assessing community health: Sewage chemical-information mining (SCIM). *The Science of the Total Environment, 619–620*, 748–764. DOI: 10.1016/j.scitotenv.2017.11.102 PMID: 29161600

Dhal, P., & Azad, C. (2022). Feature selection survey. *Applied Intelligence, 52*(4), 4543. DOI: 10.1007/s10489-021-02550-9

Eaton, C. J., Coxon, S., Pattis, I., Chappell, A., Hewitt, J., & Gilpin, B. J. (2022). A framework for public health authorities to evaluate health determinants for wastewater-based epidemiology. *Environmental Health Perspectives, 130*(12), 125001. Advance online publication. DOI: 10.1289/EHP11115 PMID: 36520537

Ehrlinger, L., & Wöß, W. (2022). A survey of data quality measurement and monitoring tools. *Frontiers in Big Data, 5*, 850611. DOI: 10.3389/fdata.2022.850611 PMID: 35434611

Ejeian, F., Etedali, P., Mansouri-Tehrani, H. A., Soozanipour, A., Low, Z. X., Asadnia, M., & Razmjou, A.. (2018). Biosensors for wastewater monitoring: A review. *Biosensors & Bioelectronics, 118*, 66–79. DOI: 10.1016/j.bios.2018.07.019 PMID: 30056302

Ethics of artificial intelligence. (n.d.). In *Wikipedia*. Retrieved February 9, 2026, from https://en.wikipedia.org/wiki/Ethics_of_artificial_intelligence

Fairchild, A. L., Gable, L., Gostin, L. O., Bayer, R., Sweeney, P., & Janssen, R. S. (2007). Public goods, private data: HIV and the history, ethics, and uses of identifiable public health information. *Public Health Reports (Washington, D.C.)*, *122*(1, suppl), 7–15. DOI: 10.1177/00333549071220S103 PMID: 17354522

Falkner, S. (2025). Next-gen healthcare: Powered by artificial intelligence and smart biosensors. *Archives of Advanced Artificial Intelligence, Data Science, and Machine Learning*, *1*(1), 1–8.

Foxman, B., Salzman, E., Gesierich, C., Gardner, S., Ammerman, M., Eisenberg, M., & Wigginton, K. (2024). Wastewater surveillance of antibiotic-resistant bacteria for public health action: Potential and challenges. *American Journal of Epidemiology*, *194*(5), 1192–1199. DOI: 10.1093/aje/kwae419 PMID: 39475072

Gable, L., Ram, N., & Ram, J. L. (2020). Legal and ethical implications of wastewater monitoring of SARS-CoV-2 for COVID-19 surveillance. *Journal of Law and the Biosciences*, *7*(1), lsaa039. Advance online publication. DOI: 10.1093/jlb/lsaa039 PMID: 32793373

Gonçalves, J. (2023). The role of smart technologies in wastewater-based epidemiology. *Journal of Environmental Exposure Assessment*, *2*(4), 18. Advance online publication. DOI: 10.20517/jeea.2023.16

Gracia-Lor, E., Rousis, N. I., Hernández, F., Zuccato, E., & Castiglioni, S. (2018). Wastewater-based epidemiology as a novel biomonitoring tool to evaluate human exposure to pollutants. *Environmental Science & Technology*, *52*(18), 10224–10226. DOI: 10.1021/acs.est.8b01403 PMID: 30146868

Guillen Aguinaga, M., Aguinaga Ontoso, E., Guillen Aguinaga, L., Guillén Grima, F., & Aguinaga Ontoso, I. (2025). Data quality in the age of AI: A review of governance, ethics, and the FAIR principles. *Data*, *10*(12), 201. Advance online publication. DOI: 10.3390/data10120201

Hait, M., Sahu, P., Biswas, S., & Izah, S. C. (2024). Sustainable application of artificial intelligence in biomonitoring for environmental sustainability: Challenges and prospects. In *Biomonitoring of Pollutants in the Global South* (pp. 747–778). Springer., DOI: 10.1007/978-981-97-1658-6_22

Harman, S., Erfani, P., Goronga, T., Hickel, J., Morse, M., & Richardson, E. T. (2021). Global vaccine equity demands reparative justice—Not charity. *BMJ Global Health*, *6*(6), e006504. DOI: 10.1136/bmjgh-2021-006504 PMID: 34919057

Hendriksen, R. S., Munk, P., Njage, P., Van Bunnik, B., McNally, L., Lukjancenko, O., Röder, T., Nieuwenhuijse, D., Pedersen, S. K., Kjeldgaard, J., Kaas, R. S., Clausen, P. T. L. C., Vogt, J. K., Leekitcharoenphon, P., van de Schans, M. G. M., Zuidema, T., de Roda Husman, A. M., Rasmussen, S., Petersen, B., & Aarestrup, F. M. (2019). Global monitoring of antimicrobial resistance based on metagenomics analyses of urban sewage. *Nature Communications*, *10*(1), 1124. DOI: 10.1038/s41467-019-08853-3 PMID: 30850636

Hosseini, M. P.. (2020). Multimodal EEG and fMRI analysis. *Artificial Intelligence in Medicine*, *104*, 101813. DOI: 10.1016/j.artmed.2020.101813 PMID: 32498996

Kahn, M. G., Raebel, M. A., Glanz, J. M., Riedlinger, K., & Steiner, J. F. (2012). A pragmatic framework for single-site and multisite data quality assessment in electronic health record-based clinical research. *Medical Care*, *50*(Suppl.), S21–S29. DOI: 10.1097/MLR.0b013e318257dd67 PMID: 22692254

Khan, A., Zubair, M., & Ahmad, M. (2020). Data quality challenges in sensor networks: Issues, methods, and future directions. *Journal of Big Data*, *7*(1), 85. Advance online publication. DOI: 10.1186/s40537-026-01413-9 PMID: 32547903

Kman, N. E., & Bachmann, D. J. (2012). Biosurveillance: A review and update. *Advances in Preventive Medicine*, *2012*, 301408. Advance online publication. DOI: 10.1155/2012/301408 PMID: 22242207

Koplan, J. P., Bond, T. C., Merson, M. H., Reddy, K. S., Rodriguez, M. H., Sewankambo, N. K., & Wasserheit, J. N. (2009). Towards a common definition of global health. *Lancet*, *373*(9679), 1993–1995. DOI: 10.1016/S0140-6736(09)60332-9 PMID: 19493564

Kostkova, P. (2013). A roadmap to integrated digital public health surveillance: The vision and the challenges. In *Proceedings of the 22nd International World Wide Web Conference (WWW '13)* (pp. 687–694). DOI: 10.1145/2487788.2488024

Kostkova, P. (2018). Disease surveillance data sharing for public health: The next ethical frontiers. *Life Sciences, Society and Policy*, *14*(1), 16. DOI: 10.1186/s40504-018-0078-x PMID: 29971516

Kostkova, P., Saigí-Rubió, F., Eguia, H., Borbolla, D., Verschuuren, M., Hamilton, C., Azzopardi-Muscat, N., & Novillo-Ortiz, D. (2021). Data and digital solutions to support surveillance strategies in the context of the COVID-19 pandemic. *Frontiers in Digital Health*, *3*, 707902. DOI: 10.3389/fdgth.2021.707902 PMID: 34713179

. Kuster, A. C., & Overgaard, H. J. (2020). *A novel comprehensive metric to assess COVID-19 testing outcomes: Effects of geography, government, and policy response* [Preprint].

Lawal, O. P., Opara, I. J., Ayo-Ige, A., Eboh, N. A., Cos-Ibe, U., Forson, K. A. A. M., & Forson, K.. (2025). Artificial intelligence-integrated biosensors for antimicrobial resistance detection and surveillance: A review and future perspectives for global biosecurity. *Cureus*, *17*(11). Advance online publication. DOI: 10.7759/cureus.98098 PMID: 41477370

Mao, K., Zhang, K., Du, W., Ali, W., Feng, X., & Zhang, H. (2020). The potential of wastewater-based epidemiology as surveillance and early warning of infectious disease outbreaks. *Current Opinion in Environmental Science & Health*, *17*, 1–7. DOI: 10.1016/j.coesh.2020.04.006 PMID: 32395676

Meara, J. G., Leather, A. J., Hagander, L., Alkire, B. C., Alonso, N., Ameh, E. A., Bickler, S. W., Conteh, L., Dare, A. J., Davies, J., Mérisier, E. D., El-Halabi, S., Farmer, P. E., Gawande, A., Gillies, R., Greenberg, S. L. M., Grimes, C. E., Gruen, R. L., Ismail, E. A., & Yip, W. (2015). Global Surgery 2030: Evidence and solutions for achieving health, welfare, and economic development. *Lancet*, *386*(9993), 569–624. DOI: 10.1016/S0140-6736(15)60160-X PMID: 25924834

Minaee, S., Kafieh, R., Sonka, M., Yazdani, S., & Soufi, G. J. (2020). Deep-COVID: Predicting COVID-19 from chest X-ray images using deep transfer learning. *Medical Image Analysis*, *65*, 101794. DOI: 10.1016/j.media.2020.101794 PMID: 32781377

Moen, E., Bannon, D., Kudo, T., Graf, W., Covert, M., & Van Valen, D. (2019). Deep learning for cellular image analysis. *Nature Methods*, *16*(12), 1233–1246. DOI: 10.1038/s41592-019-0403-1 PMID: 31133758

Ojagh, S., Cauteruccio, F., Terracina, G., & Liang, S. H. (2021). Enhanced air quality prediction by edge-based spatiotemporal data preprocessing. *Computers & Electrical Engineering*, *96*, 107572. DOI: 10.1016/j.compeleceng.2021.107572

Paquet, C., Coulombier, D., Kaiser, R., & Ciotti, M. (2006). Epidemic intelligence: A new framework for strengthening disease surveillance in Europe. *Eurosurveillance*, *11*(12), 212–214. DOI: 10.2807/esm.11.12.00665-en PMID: 17370970

Patel, M., Emerich, D., & Singh, S. (2022). Review of antimicrobial resistance surveillance using wastewater-based approaches. *Environmental Monitoring and Assessment*.

Pykes, K. (2024). *10 signs of bad data: How to spot poor quality data*. DataCamp. https://www.datacamp.com/blog/10-signs-bad-data-quality

Qiu, J. (2020). Covert coronavirus infections could be seeding new outbreaks. *Nature*. Advance online publication. DOI: 10.1038/d41586-020-00822-x PMID: 32203376

. Rizk, Y., Hajj, N., Mitri, N., & Awad, M. (2019). Deep belief networks and cortical algorithms: A comparative study for supervised classification. Applied computing and informatics, 15(2), 81-93.

Russell, S. J. (2019). *Human compatible: Artificial intelligence and the problem of control*. Viking.

Schoch, C. L., Ciufo, S., Domrachev, M., Hotton, C. L., Kannan, S., Khovanskaya, R., Leipe, D., McVeigh, R., O'Neill, K., Robbertse, B., Sharma, S., Soussov, V., Sullivan, J. P., Sun, L., Turner, S., & Karsch-Mizrachi, I. (2020). NCBI taxonomy: A comprehensive update on curation, resources, and tools. *Database : The Journal of Biological Databases and Curation*, *2020*, baaa062. Advance online publication. DOI: 10.1093/database/baaa062 PMID: 32761142

Sengupta, A., Seal, A., Panigrahy, C., Krejcar, O., & Yazidi, A. (2020). Edge information based image fusion metrics using fractional order differentiation and sigmoidal functions. *IEEE Access : Practical Innovations, Open Solutions*, *8*, 88385–88398. DOI: 10.1109/ACCESS.2020.2993607

Sims, N., & Kasprzyk-Hordern, B. (2020). Future perspectives of wastewater-based epidemiology: Monitoring infectious disease spread and resistance to the community level. *Environment International*, *139*, 105689. DOI: 10.1016/j.envint.2020.105689 PMID: 32283358

. Singer, AC, Thompson, JR, Filho, CRM, Street, R, Li, X, Castiglioni, S, et al. A world of wastewater-based epidemiology. Nat Water. (2023) 1:408–15. Doi: Full Text | Google ScholarDOI: 10.1038/s44221-023-00083-8Crossref

Singh, S., Ahmed, A. I., Almansoori, S., Alameri, S., Adlan, A., Odivilas, G., Chattaway, M. A., Salem, S. B., Brudecki, G., & Elamin, W. (2024). A narrative review of wastewater surveillance: Pathogens of concern, applications, detection methods, and challenges. *Frontiers in Public Health*, *12*, 1445961. Advance online publication. DOI: 10.3389/fpubh.2024.1445961 PMID: 39139672

Tengnah, M. A. J., Sooklall, R., & Nagowah, S. D. (2019). A predictive model for hypertension diagnosis using machine learning techniques. In *Telemedicine technologies* (pp. 139–152). Academic Press. DOI: 10.1016/B978-0-12-816948-3.00009-X

Tiwari, A., Krolicka, A., Tran, T. T., Räisänen, K., Ásmundsdóttir, Á. M., Wikmark, O. G., Lood, R., & Pitkänen, T. (2024). Antibiotic resistance monitoring in wastewater in the Nordic countries: A systematic review. *Environmental Research, 246*, 118052. DOI: 10.1016/j.envres.2023.118052 PMID: 38163547

Uren, K. R., Van Schoor, G., Du Rand, C. P., & Botha, A. (2016). An integrated approach to sensor FDI and signal reconstruction in HTGRs–Part II: Case studies. *Annals of Nuclear Energy*, *87*, 739–749. DOI: 10.1016/j.anucene.2015.06.013

Uysal, G., & Ozturk, M. (2020). Alzheimer's diagnosis using ML. *Journal of Neuroscience Methods*, *33*, 108669. DOI: 10.1016/j.jneumeth.2020.108669 PMID: 32126274

Van der Drift, A. M. R., Welling, A., Arntzen, V., Nagelkerke, E., van der Beek, R. F. H. J., & de Roda Husman, A. M. (2025). Wastewater surveillance studies on pathogens and their use in public health decision making: A scoping review. *The Science of the Total Environment*, *993*, 179982. Advance online publication. DOI: 10.1016/j.scitotenv.2025.179982 PMID: 40592216

Vitale, D., Suárez-Varela, M. M., & Picó, Y. (2021). Wastewater-based epidemiology, a tool to bridge biomarkers of exposure, contaminants, and human health. [doi.org in Bing]. *Current Opinion in Environmental Science & Health, 20*, 100229. DOI: 10.1016/j.coesh.2021.100229

Wang, Y., Yang, A., Li, Z., Wang, P., & Yang, H. (2015, April). Blind drift calibration of sensor networks using signal space projection and Kalman filter. In *2015 IEEE tenth international conference on intelligent sensors, sensor networks and information processing (ISSNIP)* (pp. 1-6). IEEE.

Willis, H. H., & Moore, M. (2014). Improving the value of analysis for biosurveillance. *Decision Analysis, 11*(1), 63–81. DOI: 10.1287/deca.2013.0283

Woldaregay, A. Z., Årsand, E., Walderhaug, S., Albers, D., Mamykina, L., Botsis, T., & Hartvigsen, G. (2019). Data-driven modeling and prediction of blood glucose dynamics: Machine learning applications in type 1 diabetes. *Artificial Intelligence in Medicine*, *98*, 109–134. DOI: 10.1016/j.artmed.2019.07.007 PMID: 31383477

World Health Organization. Disease Outbreaks. 2016. Available online: https://www.emro.who.int/health-topics/disease-outbreaks/index.html (accessed on 12 July 2016).

Wu, X., Liu, X., & Zhou, Y. (2021, October). Review of unsupervised learning techniques. In *Proceedings of 2021 Chinese Intelligent Systems Conference:* Volume II (pp. 576-590). Singapore: Springer Singapore.

Xagoraraki, I., & O'Brien, E. (2020). Wastewater-based epidemiology for early detection of viral outbreaks. In *Women in water quality: Investigations by prominent female engineers* (pp. 75–97). Springer International Publishing.

Xi, X., Meng, X., Qin, Z., Nie, X., Yin, Y., & Chen, X. (2020). IA-net: Informative attention convolutional neural network for choroidal neovascularization segmentation in OCT images. *Biomedical Optics Express*, *11*(11), 6122–6136. DOI: 10.1364/BOE.400816 PMID: 33282479

Yadav, A., Yadav, S., Shahi, N., Yadav, P. K., & Kumar, A. (2023). Uses of artificial intelligence in agricultural pest management. In *Smart Agricultural Technologies for Sustainable Crop Production*. Springer., DOI: 10.1007/978-981-19-XXXX-X_13

Yang, F., & Wang, M. (2020). Big data evaluation review. *Frontiers of Engineering Management*, *7*, 27–46. DOI: 10.1007/s42524-020-0092-6

Yang, Y. (2024). Applications and challenges of big data in market analytics. Transactions on Economics. *Business and Management Research*, *9*, 450–458.

Yıldırım, Ö., Pławiak, P., Tan, R. S., & Acharya, U. R. (2018). Arrhythmia detection using deep convolutional neural network with long duration ECG signals. *Computers in Biology and Medicine*, *102*, 411–420. DOI: 10.1016/j.compbiomed.2018.09.009 PMID: 30245122

Yudkowsky, E. (2011). Complex value systems in friendly AI. In International Conference on Artificial General Intelligence (pp. 388–393). Berlin, Heidelberg: Springer Berlin Heidelberg.

Zhou, J. R., You, Z. H., Cheng, L., & Ji, B. Y. (2021). Prediction of lncRNA-disease associations via an embedding learning HOPE in heterogeneous information networks. *Molecular Therapy. Nucleic Acids*, *23*, 277–285. DOI: 10.1016/j.omtn.2020.10.040 PMID: 33425486

Zitnik, M., Nguyen, F., Wang, B., Leskovec, J., Goldenberg, A., & Hoffman, M. M. (2019). Machine learning for integrating data in biology and medicine: Principles, practice, and opportunities. *Information Fusion*, *50*, 71–91. DOI: 10.1016/j.inffus.2018.09.012 PMID: 30467459

Chapter 12
Biosurveillance of Food Systems:
Leveraging Digital Technology for Contamination Detection and Supply Chain Resilience

Sylvester Izah
http://orcid.org/0000-0001-5526-006X
Bayelsa Medical University, Nigeria

Kingsley Dunga
Kingsley Ozumba Mbadiwe University, Ogboko, Nigeria

Ayebapreye Kari
Bayelsa Medical University, Nigeria

Esther Ugo Alum
http://orcid.org/0000-0003-4105-8615
Kampala International University, Uganda

Morufu Raimi
http://orcid.org/0000-0001-5042-6729
Federal University, Otuoke, Nigeria

Chandrashekhar Uppin
Hensard University, Nigeria

ABSTRACT

Global food systems are increasingly vulnerable to contamination, emerging pathogens, and complex supply chain disruptions that threaten public and environmental health. This chapter examines how digital technologies enhance biosurveillance in food systems to enable rapid contamination detection and strengthen supply chain resilience. Integrated digital approaches demonstrate improved early detection of microbial and chemical hazards across farm-to-fork pathways. Real-time data from IoT sensors and AI analytics support proactive risk identification and faster response actions. Blockchain-based traceability enhances transparency, accountability, and trust within food supply networks. Big data modeling enables prediction of contamination hotspots and supports informed decision-making. Combined digital

DOI: 10.4018/979-8-3373-9033-8.ch012

platforms improve coordination among food safety, environmental, and public health stakeholders. Addressing data governance, infrastructure gaps, and algorithmic bias remains essential for effective implementation.

1. INTRODUCTION

The demand for strong food safety measures has intensified amid escalating global challenges related to contamination, foodborne diseases, and disruptions within food supply chains (Ogwu et al., 2024a; Izah et al., 2024; Yarkwan et al., 2024). Growing concern surrounding food safety incidents has attracted heightened public attention and regulatory oversight, particularly as reported cases of contamination and foodborne illnesses continue to rise (Oriekhoe et al., 2024). Conventional supply chain systems frequently show limitations in transparency, traceability, and response speed, increasing the likelihood of large-scale public health consequences. These limitations highlight the need for innovative approaches that improve detection capabilities and responsiveness during food safety incidents (Sri Vigna Hema & Manickavasagan, 2024).

Food safety challenges generate consequences that extend beyond immediate illness, influencing public health systems, nutritional outcomes, and environmental sustainability (Iyiola et al., 2025a; Odubo et al., 2024; Izah et al., 2022). Foodborne diseases affect millions of individuals globally, with an estimated 600 million illness cases reported annually, resulting in considerable health and economic burdens (Sri Vigna Hema & Manickavasagan, 2024). Weak food safety controls threaten food security and compromise nutritional standards, particularly in vulnerable populations. Recent global disruptions, including the COVID-19 pandemic, exposed weaknesses in food supply chains and revealed how rapidly food system instability can affect availability and safety (Luo et al., 2022). These developments intensified the focus on resilient systems capable of managing both routine risks and unexpected shocks.

Digital technologies have gained prominence as food systems adapt to these evolving challenges. Blockchain technology and the Internet of Things (IoT) have gained attention for improving traceability, transparency, and accountability across food supply chains (Chopra & Pathrotkar, 2024; Ogidi & Izah, 2024). IoT-enabled sensors support continuous monitoring of environmental and safety parameters such as temperature, humidity, and contamination indicators, allowing rapid identification of anomalies (Abass et al., 2024). Real-time data streams improve situational awareness and support timely decision-making during potential contamination events. These technological advances support a shift toward proactive biosurveillance, reducing delays between detection and intervention.

Blockchain technology contributes further value by providing decentralized and tamper-resistant records of food production, processing, and distribution activities. Such systems allow stakeholders to trace food products across multiple supply chain stages with greater accuracy and reliability (Oriekhoe et al., 2024). Enhanced traceability improves consumer confidence while also supporting faster, more targeted recall processes during contamination events (Mukherjee et al., 2021). When combined with IoT technologies, blockchain systems strengthen coordination among producers, distributors, regulators, and retailers. These integrated digital solutions reflect a changing approach to food safety governance and operational oversight.

Despite promising benefits, digital transformation within food systems introduces several challenges that warrant careful evaluation. High implementation costs and unequal access to digital infrastructure create barriers, particularly in low-resource settings. Technical issues related to interoperability, standardization, and data governance remain persistent concerns (Jin & Xie, 2023). Expanding reliance on digital platforms also increases exposure to cyber risks, which could compromise sensitive food safety information. Ethical considerations related to data ownership and privacy further complicate adoption processes. Addressing these challenges remains essential for achieving sustainable and equitable improvements in food safety management (Nugroho et al., 2023).

This chapter focuses on the role of digital technologies in strengthening biosurveillance within food systems across the entire supply chain. Emphasis is placed on real-time monitoring, data integration, and advanced analytics for detecting biological and chemical contamination. Attention is given to collaborative frameworks, One Health integration, and emerging digital platforms that support proactive food safety management. The chapter highlights improved early detection, faster response to food safety threats, and stronger supply chain resilience as key benefits. These advances contribute to enhanced public health protection and more sustainable, trustworthy global food systems.

2. CONCEPTUAL FRAMEWORK OF BIOSURVEILLANCE IN FOOD SYSTEMS

Figure 1 illustrates the conceptual framework of biosurveillance in food systems, highlighting core definitions, contrasting traditional and technology-enhanced surveillance approaches, outlining a multi-tiered farm-to-fork structure, and demonstrating integration between public health and environmental monitoring. The framework presents biosurveillance as an interconnected process that links hazard detection, supply chain oversight, and cross-sector data integration to strengthen food safety outcomes. It visually represents how coordinated surveillance across production, processing, distribution, and consumption stages supports early risk identification and informed decision-making. This framework provides a foundation for understanding how modern biosurveillance systems operate within increasingly complex global food systems.

Figure 1. Possible conceptual framework of biosurveillance in food systems

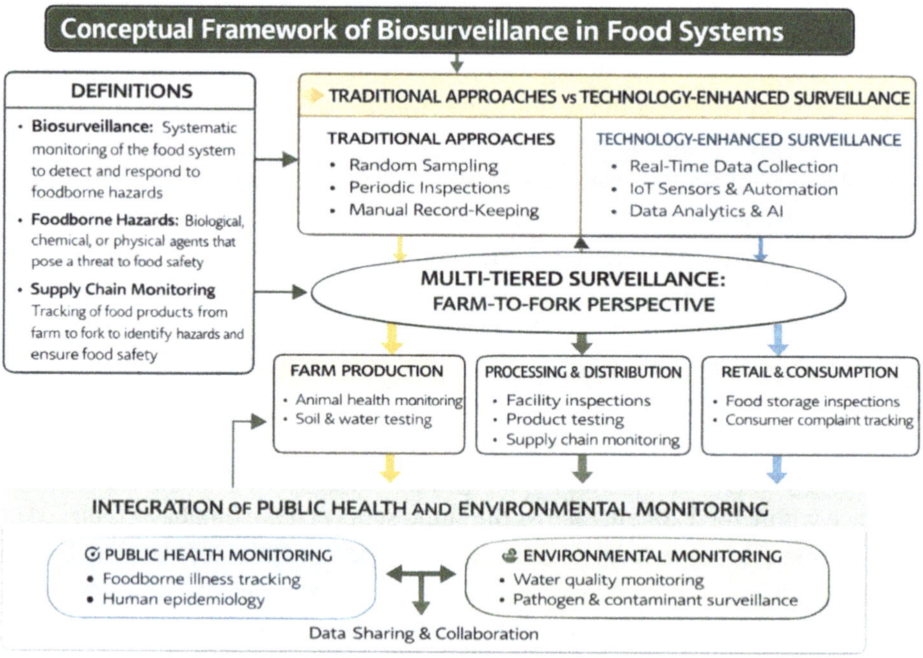

Typically, biosurveillance in food systems refers to the systematic monitoring, analysis, and interpretation of data related to health threats arising from foodborne pathogens and contaminants across the food production and supply continuum. This process addresses foodborne hazards, defined as biological, chemical, or physical agents capable of contaminating food and causing adverse health effects (McNamara et al., 2007). Effective biosurveillance relies on coordinated data collection from agricultural, environmental, and public health domains, enabling a broad view of risk factors influencing food safety (Primavilla et al., 2021). Such integration supports timely identification of emerging threats and improves the capacity of food systems to respond to potential contamination events.

Foodborne hazards originate from multiple sources within the food system, including agricultural practices, animal health, inadequate sanitation, and failures during food processing or distribution (Ogwu et al., 2024b; Sahoo et al., 2022). Pathogens such as *Salmonella* and *Escherichia coli* can move through interconnected stages of the food chain, resulting in widespread exposure and significant public health consequences (Phillips et al., 2022). Continuous identification and monitoring of these hazards remain central to biosurveillance efforts, as they support early detection and reduce the likelihood of large-scale outbreaks. Clear definitions and standardized classification of hazards improve consistency in surveillance activities across jurisdictions.

Supply chain monitoring represents a key component of biosurveillance, focusing on tracking food products from origin through consumption. This process involves documenting production conditions, transportation environments, storage parameters, and handling practices to verify compliance with food safety standards (Keawdounglek et al., 2025). Enhanced monitoring systems increase visibility across supply chain stages, allowing stakeholders to identify deviations that may signal emerging risks. Rapid access to reliable data supports faster response actions, reducing the duration and impact of contamination events.

Traditional food safety surveillance approaches have historically depended on periodic inspections, sample-based laboratory testing, and retrospective outbreak investigations. While these methods contributed to foundational food safety practices, they often lacked real-time data and integration across supply chain stages (Gabriël et al., 2022). Surveillance activities tended to focus on responding after illnesses occurred rather than identifying risks before exposure. As food systems expanded and globalized, limitations associated with delayed detection and fragmented data became more apparent, particularly given rising incidences of foodborne disease (Yin et al., 2025).

Technology-enhanced surveillance offers substantial improvements through continuous data collection and automated analysis. Digital tools such as the Internet of Things and blockchain systems support ongoing monitoring of environmental conditions and product integrity throughout the supply chain (Tesson et al., 2020). IoT sensors provide real-time information on temperature, humidity, and contamination indicators, while blockchain platforms create transparent records of food movement and transactions (Badul et al., 2021). These technologies enable earlier identification of unsafe conditions and support preventive interventions that reduce food safety risks.

A multi-tiered surveillance structure based on a farm-to-fork perspective strengthens biosurveillance by addressing risks at every stage of food production and distribution. This approach emphasizes continuous oversight beginning with primary production and extending through processing, transport, retail, and consumption (Nastasijević et al., 2025). Monitoring activities conducted at each tier support identification of critical control points where targeted actions can limit hazard transmission (Humboldt-Dachroeden & Mantovani, 2021). Coordinated surveillance across these tiers enhances consistency and effectiveness in food safety management.

Farm-to-fork surveillance enables evaluation of contamination pathways and facilitates collaboration among producers, processors, distributors, and regulators. Systems designed to monitor animal health, processing hygiene, and retail conditions generate complementary data that strengthen risk assessment. Integrated data sharing across tiers improves situational awareness and supports coordinated responses during food safety incidents. This structure contributes to maintaining product quality and consumer confidence across the food system.

Integration of public health and environmental monitoring remains essential for effective biosurveillance. Environmental factors such as water quality, soil conditions, and wildlife interactions influence the presence and spread of foodborne hazards (Van Boxstael et al., 2013). Environmental contamination events, including agricultural runoff, can introduce pathogens into food production systems, creating risks that extend beyond individual facilities (Hyde et al., 2016). Linking environmental data with food safety surveillance improves understanding of upstream risk drivers.

The One Health framework supports integration by promoting collaboration among public health, veterinary, and environmental sectors. Shared data systems and coordinated surveillance activities improve assessment of risks affecting human, animal, and ecosystem health (Humboldt-Dachroeden & Mantovani, 2021). Cross-sector collaboration enables regulators and stakeholders to develop informed policies and interventions addressing food safety challenges across the entire system (García-Díez et al., 2023). This integrated approach strengthens biosurveillance capacity and supports safer, more resilient food systems.

3. DIGITAL TECHNOLOGIES FOR CONTAMINATION DETECTION

Figure 2 illustrates the role of digital technologies in contamination detection, highlighting how IoT-enabled sensors, rapid diagnostics supported by artificial intelligence, blockchain-based traceability, and big data analytics collectively strengthen food safety surveillance. The figure presents these technologies as complementary components that operate across multiple stages of the food supply chain, supporting early detection, transparency, and predictive risk management. Together, these digital tools enable more timely identification of hazards and support informed interventions before contamination events escalate.

Figure 2. The role of digital technologies in contamination detection

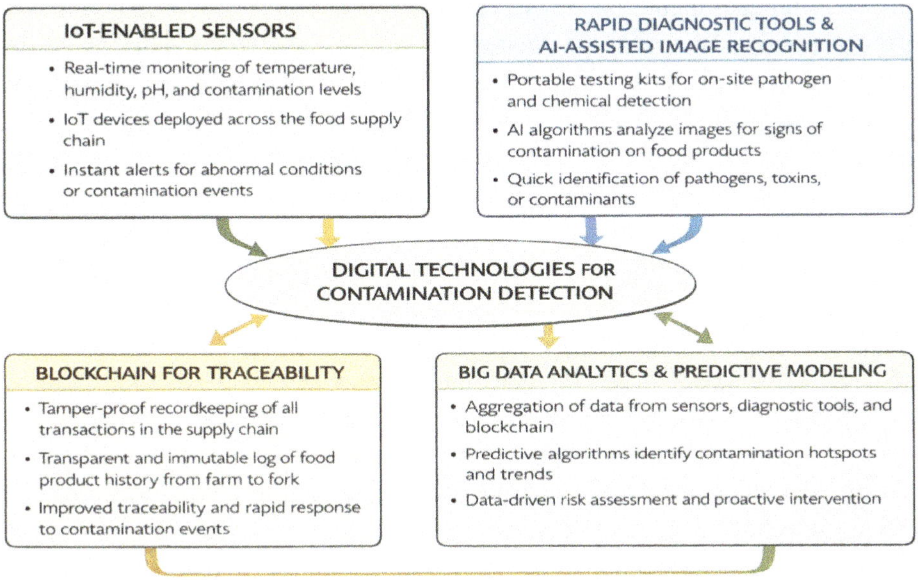

Typically, the adoption of Internet of Things (IoT) technology has significantly transformed contamination detection within food systems, particularly through the deployment of sensors designed for continuous pathogen and chemical monitoring. IoT-enabled sensors detect foodborne pathogens such as *Salmonella*, *Escherichia coli*, and *Listeria*, along with chemical contaminants including pesticide residues and heavy metals, at critical points across the supply chain (Achieng & Ogundaini, 2024; Tsiouris et al., 2024). Continuous data collection supports compliance with food safety standards and reduces the likelihood that contaminated products reach consumers (Igwama et al., 2024). This capability improves situational awareness and strengthens preventive food safety management.

IoT sensors also monitor environmental conditions that influence microbial growth and food quality, including temperature, humidity, and storage conditions. These parameters play a central role in controlling contamination risks during production, transport, and storage (Arzilli et al., 2024). Real-time access to environmental data allows stakeholders to identify deviations from safe thresholds and initiate corrective actions without delay (Nikita et al., 2023). As sensor technologies advance, improvements in sensitivity and reliability continue to expand their role in protecting public health and maintaining product integrity (Venkatraman et al., 2023).

Advances in rapid diagnostic tools further strengthen contamination detection by enabling faster and more accurate identification of hazards. These tools increasingly rely on artificial intelligence and machine learning to analyze visual and biological data from food samples (Ajala et al., 2024; Damiati, 2020). AI-assisted image recognition techniques identify pathogens through pattern analysis, reducing reliance on time-intensive laboratory procedures (Tsiouris et al., 2024). This capability supports timely decision-making in sectors requiring rapid processing, including meat production and fresh produce handling (Idahor et al., 2025).

Rapid diagnostic technologies also enhance traceability by enabling swift identification of contaminated batches, supporting faster recalls and limiting exposure during outbreaks (Alahakoon et al., 2023). Integration of diagnostic outputs with historical contamination data and environmental monitoring improves predictive accuracy and strengthens risk assessment processes (Aryffin et al., 2025). These digital approaches streamline food safety operations while addressing persistent challenges associated with delayed detection and fragmented information systems.

Blockchain technology contributes to contamination detection by strengthening traceability and ensuring integrity of supply chain records. Decentralized, tamper-resistant ledgers document food transactions from production through consumption, allowing stakeholders to trace product histories with high reliability (Achieng & Ogundaini, 2024; Isiaka et al., 2024). Enhanced traceability supports rapid identification of contamination sources and facilitates targeted recalls, reducing public health impacts (Suartana & Putra, 2022). Transparent recordkeeping improves coordination among supply chain actors and regulatory authorities.

The immutability of blockchain records promotes accountability by ensuring that recorded transactions remain verifiable and resistant to unauthorized modification. This feature strengthens consumer confidence by increasing assurance regarding product authenticity and safety (Lan et al., 2022). Blockchain platforms also support regulatory compliance by providing auditable records of handling practices and safety checks, reinforcing oversight mechanisms within food systems (El Idrissi et al., 2021). These attributes position blockchain as a critical component of modern contamination detection frameworks.

Big data analytics further enhance contamination detection through advanced pattern recognition and predictive modeling. Large datasets generated by IoT sensors, diagnostic tools, and blockchain systems provide insights into trends associated with contamination risks (Damiati, 2020; Wan et al., 2024). Analytical models identify contamination hotspots by correlating environmental conditions, production practices, and historical incident data. Predictive modeling supports early intervention strategies that reduce the likelihood of widespread contamination events.

Integration of diverse data sources strengthens analytical accuracy and supports comprehensive risk assessments. Environmental variables, operational records, and public health data collectively inform targeted responses and efficient resource allocation. This approach enables focused interventions at high-risk points within the supply chain, improving prevention while reducing operational costs (Yatsyna

& Kudinov, 2023). As data processing capabilities expand, big data analytics continue to support adaptive food safety strategies aligned with evolving production and distribution environments.

4. ENHANCING SUPPLY CHAIN RESILIENCE THROUGH DIGITAL PLATFORMS

Figure 3 illustrates how digital platforms enhance supply chain resilience through real-time monitoring, data-driven risk assessment, predictive logistics, and integrated datasets spanning food safety, environmental, and public health domains. The figure demonstrates how these digital components operate collectively to improve visibility, responsiveness, and coordination across food supply chains. It highlights the role of continuous data flows and analytical tools in supporting early detection of contamination events and reducing system-wide vulnerabilities.

Figure 3. Overview of digital platforms that could enhance supply chain resilience

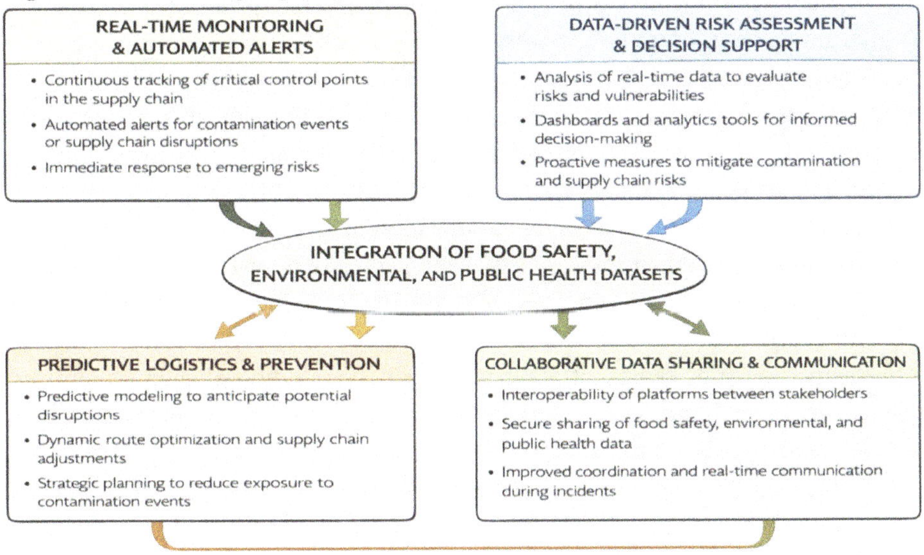

Typically, digital platforms have significantly strengthened supply chain resilience through the deployment of real-time monitoring systems supported by IoT-enabled sensors and advanced data analytics. These systems continuously track critical parameters such as temperature, humidity, and contaminant concentrations across production, storage, and distribution stages (Deng & Feng, 2020; Oriekhoe et al.,

2024). Automated alerts are generated when deviations from safe thresholds occur, allowing rapid response actions such as pausing distribution or initiating product recalls. Timely alerts reduce the likelihood of contaminated products reaching consumers and limit the scale of foodborne illness outbreaks (Xue-ting, 2023).

Continuous data collection also improves operational efficiency within food safety management. Immediate access to sensor data supports faster decision-making during transportation and storage processes. For instance, alerts triggered by temperature deviations in refrigerated transport enable corrective actions before product quality deteriorates (Liu et al., 2022). Early interventions reduce spoilage, limit economic losses, and strengthen consumer trust in food safety systems (Zhang et al., 2023). Real-time monitoring therefore contributes to both safety assurance and supply chain stability.

Digital platforms further support resilience through data-driven risk assessment and decision support mechanisms. Aggregation of diverse datasets, including historical contamination records, environmental data, and operational metrics, enables more detailed risk analyses (Chen et al., 2025). These analyses help identify vulnerable points within supply chains where contamination or disruption risks are elevated. Insights generated from integrated data allow stakeholders to prioritize interventions and allocate resources more effectively.

Machine learning algorithms enhance decision support by identifying patterns and trends associated with contamination risks. Predictive insights derived from real-time and historical data support preventive strategies that reduce exposure before incidents occur (Oriekhoe et al., 2024). For example, suppliers can adjust sourcing decisions or processing controls for ingredients linked to higher contamination likelihoods based on environmental and pathogen prevalence data (Xue-ting, 2023). Such data-informed strategies strengthen preparedness and reduce reliance on reactive responses.

Predictive logistics represent another critical contribution of digital platforms to supply chain resilience. Advanced analytics transform logistics management by anticipating disruptions rather than responding after failures occur. Big data models forecast demand fluctuations, transportation delays, and operational bottlenecks, supporting optimized planning across supply networks (Chen et al., 2021; Nisa'Habib Elias et al., 2024). AI-driven analysis of logistics data reveals patterns indicating where disruptions are likely, allowing early adjustments to mitigate their effects (Amrutha & Gomathy, 2024).

Predictive logistics applications include modifying delivery routes in anticipation of delays, adjusting inventory levels to accommodate demand variability, and reallocating resources to reduce congestion. These actions lower exposure to food safety risks associated with prolonged storage or transport interruptions. Financial losses linked to wastage and inefficiencies are also reduced through improved forecasting accuracy (Wang, 2023). Enhanced logistical reliability supports continuity and adaptability within increasingly complex global food supply chains.

Integration of food safety, environmental, and public health datasets further strengthens resilience by supporting a holistic understanding of food system risks. Unified digital platforms that connect these datasets provide insights into interactions among environmental conditions, production practices, and health outcomes (Li et al., 2022; Hamad & Bhattacharya, 2025). Integrated data environments improve risk assessments and encourage collaboration among agricultural producers, environmental agencies, and public health authorities.

Cross-referencing agricultural sensor data with public health surveillance information enables early identification of emerging foodborne outbreaks (Kurniawan et al., 2023; Hasan et al., 2023). Shared data platforms enhance transparency, traceability, and accountability across supply chains while supporting coordinated responses to contamination events. Improved communication among stakeholders reduces fragmentation and strengthens trust throughout the food system. Integrated datasets therefore support cohesive responses to food safety challenges and promote resilient supply chains aligned with public health priorities (Zhang et al., 2022; Feng et al., 2020).

5. APPLIED DIGITAL SURVEILLANCE FOR FOOD SAFETY AND SUPPLY CHAIN RESILIENCE

Figure 4 illustrates key case studies and practical applications of digital technologies in food safety, focusing on microbial pathogen detection, biosurveillance in aquaculture, chemical contamination warning systems, and AI-based monitoring of complex supply networks. It highlights how these applications translate digital innovations into operational food safety practices across diverse production environments. It demonstrates how real-world implementations support early detection, informed decision-making, and resilience across modern food systems.

Figure 4. Overview of practical applications of digital technologies in food safety

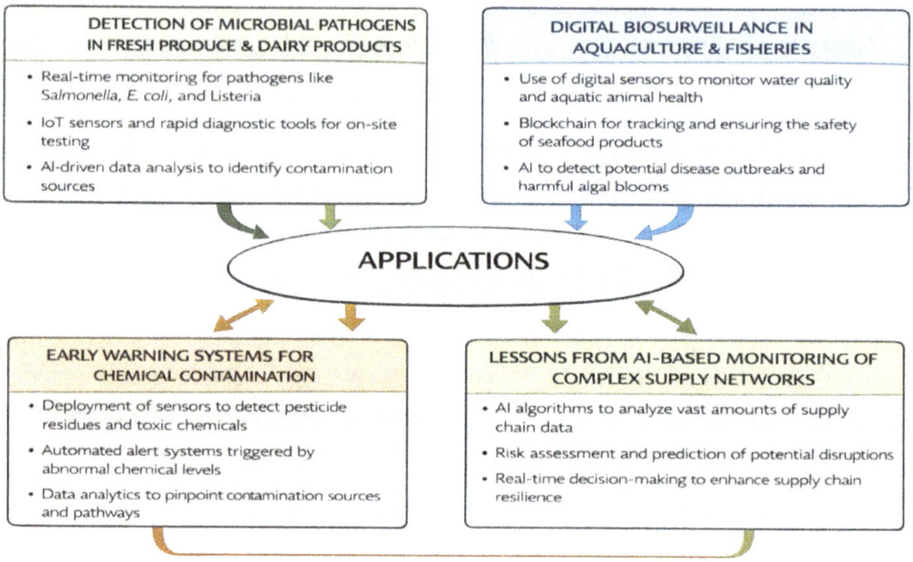

Typically, digital technologies have substantially improved the detection of microbial pathogens in fresh produce and dairy products through continuous monitoring and advanced data analytics. IoT-enabled sensors allow constant assessment of environmental and product conditions that influence pathogen growth, including temperature and humidity (Odimarha et al., 2024; Zhang, 2024). Automated alerts notify stakeholders when deviations from safety thresholds occur, enabling timely corrective actions that reduce the likelihood of foodborne illness outbreaks. In dairy processing environments, sensor-based monitoring supports immediate identification of contamination risks, strengthening microbial control before products enter distribution channels (Kakade & Vitalkar, 2024).

Machine learning techniques further enhance pathogen detection by analyzing complex datasets derived from microbial studies and environmental observations. Pattern recognition algorithms identify conditions associated with heightened contamination risk and link them to specific pathogens (Joel et al., 2024). These analytical capabilities support proactive food safety management by guiding targeted interventions and improving understanding of contamination dynamics. Predictive insights generated from data-driven models contribute to reducing reliance on reactive responses and improving consistency in food safety practices.

Digital biosurveillance systems have also demonstrated strong value within aquaculture and fisheries (Iyiola et al., 2025b; Izah et al., 2025a). These systems integrate IoT devices, wearable sensors, and environmental monitoring tools to track fish health and detect early signs of disease or contamination (Abhulimen & Ejike, 2024; Atadoga et al., 2024). Continuous monitoring of water quality parameters such as pH, dissolved oxygen, and harmful algal presence supports timely identification of conditions that threaten aquatic species and consumer safety. Early detection reduces the likelihood of disease spread within aquatic populations and limits downstream food safety risks.

AI-driven analytics enhance aquaculture biosurveillance by enabling prediction of disease outbreaks based on historical data and real-time environmental conditions (Zhang, 2024). Predictive models support informed management decisions, including adjustments to feeding regimes, stocking densities, and water treatment strategies. These digital tools contribute to safer and more sustainable aquaculture practices while addressing broader challenges related to environmental degradation and resource management (Adekola & Dada, 2024).

Early warning systems for chemical contamination represent another critical application of digital technologies in food safety. These systems integrate IoT sensors, data analytics platforms, and machine learning models to monitor pesticide residues, heavy metals, and other hazardous substances across food supply chains (Pan et al., 2024). Centralized data integration from environmental monitoring systems and agricultural input records allows rapid identification of chemical levels exceeding safety thresholds (Anitha & Patil, 2018; Yang, 2024). Automated alerts facilitate prompt intervention, limiting consumer exposure to harmful contaminants.

Predictive models embedded within early warning systems analyze historical patterns and real-time data to forecast contamination risks before incidents escalate (Al Khatib et al., 2024). Regulatory agencies benefit from timely alerts that support swift review processes and distribution controls. These systems strengthen accountability across the food industry and reinforce transparency, contributing to increased stakeholder confidence in food safety governance (Waller & Fawcett, 2013).

AI-based monitoring of complex supply networks has generated valuable insights into improving food safety resilience (Izah et al., 2025b; Iyiola et al., 2025b). AI platforms analyze large volumes of data across interconnected supply chain nodes, including production sites, processing facilities, logistics networks, and retail outlets (Mahi, 2024). Predictive models identify emerging disruptions and vulnerabilities, supporting faster and more adaptive responses to food safety threats.

AI algorithms also reveal correlations among variables influencing food safety, such as supplier performance, environmental conditions, and logistical constraints (Abhulimen & Ejike, 2024; Abaku et al., 2024). These insights support optimized inventory management and prioritization of high-risk products, reducing contamination incidence. Lessons from AI integration emphasize the value of data sharing, interdisciplinary collaboration, and system adaptability in strengthening food supply chains against unpredictable challenges (Oyewole et al., 2024; Ogbewele et al., 2024).

6. CHALLENGES AND LIMITATIONS OF DIGITAL BIOSURVEILLANCE

Figure 5 illustrates the major challenges and limitations associated with digital biosurveillance, emphasizing data governance concerns, infrastructural constraints, algorithmic limitations, and integration barriers across stakeholders and agencies. It presents these challenges as interconnected factors that influence the effectiveness, equity, and reliability of digital biosurveillance systems. It highlights how technical, ethical, and organizational issues can limit the full realization of digital innovations in food safety and public health surveillance.

Figure 5. Major challenges and limitations associated with digital biosurveillance

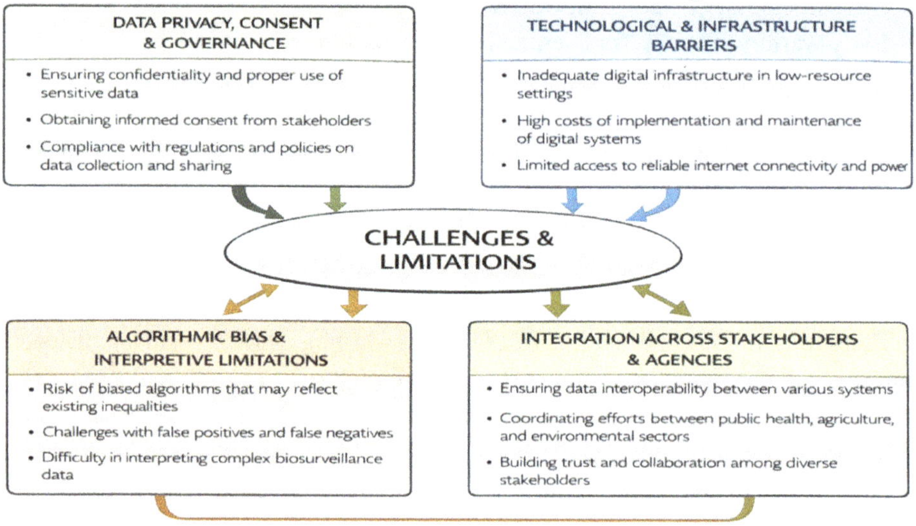

As digital biosurveillance systems expand, concerns related to data privacy, consent, and governance have become increasingly prominent. These systems often rely on the collection, storage, and sharing of sensitive health and production data, raising ethical and legal questions surrounding informed consent and data ownership (Bhartiya & Mehrotra, 2015). In many regulatory environments, comprehensive data protection frameworks remain underdeveloped, creating vulnerabilities related to confidentiality breaches and misuse of information. Risks increase further when multiple datasets are integrated, as broader data access points heighten exposure to unauthorized use (Mazimwe et al., 2019).

Trust plays a central role in determining stakeholder participation in biosurveillance initiatives. Uncertainty regarding how data are collected, processed, and shared can discourage engagement from producers, workers, and communities (Binamungu & Kimaro, 2025). Clear governance structures that define data responsibilities, consent mechanisms, and access rights support transparency and accountability. Well-defined policies regarding ethical data use strengthen confidence in digital biosurveillance systems and improve long-term participation across sectors.

Technological and infrastructure barriers present substantial challenges, particularly in low-resource settings. Many regions lack reliable electricity, broadband connectivity, and digital infrastructure required to support real-time monitoring systems (Gupta et al., 2023). These limitations restrict deployment of advanced tools such as IoT sensors, cloud-based analytics, and automated reporting platforms. Inconsistent infrastructure reduces data continuity and undermines the effectiveness of surveillance systems that depend on continuous information flows.

Limited availability of skilled personnel further constrains implementation. Operation and maintenance of advanced digital platforms require technical expertise that may be scarce in resource-limited contexts (Rahman & Szabó, 2023). As a result, reliance on manual or legacy surveillance methods persists, contributing to delays, incomplete datasets, and reduced responsiveness. The digital divide reinforces inequities in food safety and public health protection, leaving vulnerable populations with limited access to modern surveillance benefits. Addressing these challenges requires sustained investment in infrastructure development, workforce training, and technology adaptation tailored to local conditions.

Algorithmic bias and interpretive limitations introduce additional complexities in digital biosurveillance. Artificial intelligence and machine learning systems rely on historical datasets that may reflect existing structural biases (Amabie et al., 2023; Iyiola et al., 2025c), leading to skewed predictions or uneven performance across populations (Chatfield & Alanazi, 2015). Models trained on limited or non-representative data can produce outputs that fail to generalize across diverse production systems or geographic contexts. Such biases risk distorting risk assessments and influencing decision-making in ways that exacerbate disparities.

False positives also pose significant challenges for biosurveillance credibility. Incorrect alerts indicating contamination or disease outbreaks can trigger unnecessary interventions, disrupt supply chains, and generate public concern (Calvetti et al., 2024). Repeated inaccuracies may erode trust among stakeholders and reduce confidence in surveillance outputs. Transparent algorithm design, routine performance evaluation, and contextual validation are essential practices for minimizing misinterpretation and improving reliability across varied environments.

Effective digital biosurveillance depends heavily on coordination among multiple stakeholders, including public health agencies, regulatory authorities, industry actors, and technology providers. Integration challenges arise when organizations operate within isolated data systems, follow differing standards, or maintain incompatible reporting structures (Palojoki & Vuokko, 2023). Fragmentation limits information sharing and reduces the ability to develop unified situational awareness during food safety incidents.

Cross-agency collaboration introduces additional governance complexities related to data ownership, decision authority, and accountability (Kim & Tak, 2019). Disagreements over responsibilities and control can slow response efforts and hinder system interoperability. Establishing shared standards, interoperable platforms, and harmonized data-sharing agreements supports smoother collaboration and improves system-wide effectiveness. Consensus-driven governance models and clearly defined operational protocols enhance coordination and contribute to more efficient biosurveillance initiatives (Pubellier, 2005).

Together, these challenges illustrate that digital biosurveillance extends beyond technical deployment. Ethical governance, equitable infrastructure development, algorithmic accountability, and institutional collaboration collectively shape system performance. Addressing these limitations remains essential for advancing reliable, inclusive, and effective digital biosurveillance across food systems.

7. POLICY, ETHICAL, AND PUBLIC HEALTH IMPLICATIONS

Table 1 summarizes the key policy, ethical, and public health implications associated with the adoption of digital biosurveillance systems, highlighting the relationships among regulatory development, ethical safeguards, and food security outcomes. It illustrates how aligned governance structures, ethical oversight, and transparent implementation can enhance early risk detection, strengthen public trust, and support resilient food systems. It also emphasizes the importance of interoperability, fairness, and accountability in ensuring that digital biosurveillance contributes positively to public health protection.

Table 1. Policy, ethical, and public health implications of digital biosurveillance

Policy & Regulation	Ethical Considerations	Public Health & Food Security Implications
Development of regulatory frameworks that support the adoption and standardization of digital biosurveillance systems	Ensuring data privacy, informed consent, and responsible data governance in automated monitoring systems	Enhanced early detection and prevention of foodborne and environmental health risks
Alignment of national and international food safety, environmental, and public health regulations	Addressing ethical concerns related to surveillance, data ownership, and potential misuse of collected data	Improved public trust through transparent monitoring and communication mechanisms
Establishment of data-sharing policies and interoperability standards across agencies and sectors	Mitigating risks of algorithmic bias and ensuring fairness and accountability in AI-driven systems	Strengthened food security through timely risk management and resilient supply chains
Policy recommendations for integrating digital tools with existing health and environmental governance structures	Balancing technological efficiency with human oversight and ethical decision-making	Increased confidence in food systems, supporting long-term public health protection

Typically, the rapid advancement of digital biosurveillance technologies has intensified the need for regulatory frameworks that support their responsible adoption and standardization. Governments and health authorities play a central role in defining policies that clarify stakeholder responsibilities, data governance structures, and operational boundaries for biosurveillance systems (Wittmer et al., 2025). Effective regulation includes provisions for informed consent, data privacy protection, and secure data-sharing mechanisms across institutions, ensuring alignment with national and international health and privacy laws (Göktaş & Grzybowski, 2025). Experiences from wastewater-based surveillance during the COVID-19 pandemic demonstrated both the value of digital monitoring and the necessity of governance structures that support evidence-based public health decision-making (Wittmer et al., 2025). Regulatory clarity supports innovation while safeguarding ethical and legal standards.

Ethical considerations represent a critical dimension of automated monitoring and data sharing within digital biosurveillance systems. Automation of health and environmental data collection raises concerns related to privacy, consent, and algorithmic bias, particularly when artificial intelligence systems influence decision-making processes (Göktaş & Grzybowski, 2025). Bias embedded within training datasets can result in uneven outcomes or misinformed interventions, affecting fairness and equity across populations (Pūraitė et al., 2020). Ethical frameworks that prioritize transparency, accountability, and explainability in algorithmic systems support responsible implementation. Clearly defined consent mechanisms remain essential, ensuring individuals understand how data are collected, processed, and shared. Dynamic consent models offer flexible approaches that allow individuals to maintain control over their data across its lifecycle, reinforcing ethical participation and data stewardship.

Public trust and transparency strongly influence the effectiveness of digital biosurveillance initiatives. Confidence in surveillance systems depends on clear communication regarding data collection practices, system objectives, and safeguards against misuse (Mastrogiovanni, 2025). Limited transparency or perceived opacity can undermine acceptance and participation, weakening surveillance effectiveness. Ethical responsibility extends to demonstrating integrity in data handling and reinforcing accountability through transparent governance mechanisms (Bhatnagar, 2025). Transparent biosurveillance systems also support food security by enabling real-time insights that inform production, distribution, and risk management decisions. Open data-sharing practices among stakeholders improve responsiveness to contamination events and strengthen coordination across supply chains (Tabassum et al., 2025).

Harmonization of digital biosurveillance tools with existing health and environmental policies remains essential for coherent system performance. Collaboration among policymakers, public health authorities, environmental agencies, and technology developers supports the development of interoperable frameworks that enable efficient data exchange and coordinated responses (Sharma & Maharshi, 2024; Keerthi & Jayan, 2025). Establishing shared data standards and interoperable protocols allows diverse digital platforms to function cohesively, enhancing situational awareness across sectors. Integrated governance approaches support responses to complex challenges such as zoonotic disease emergence and environmental hazards affecting food safety (Zhorzholiani, 2025). Continuous dialogue among stakeholders ensures that policies adapt to technological innovation while addressing evolving ethical and public health considerations.

8. FUTURE DIRECTIONS AND RESEARCH OPPORTUNITIES

Figure 6 illustrates future directions and research opportunities in digital biosurveillance, emphasizing technological innovation, cross-sector collaboration, system scalability, and strategic planning for resilient food systems. It presents these dimensions as interconnected pathways that guide the evolution of biosurveillance toward more adaptive, predictive, and integrated food safety frameworks. It also highlights how advances in sensing technologies, data analytics, and governance structures collectively support long-term improvements in food system resilience.

Figure 6. Future directions and research opportunities in digital biosurveillance

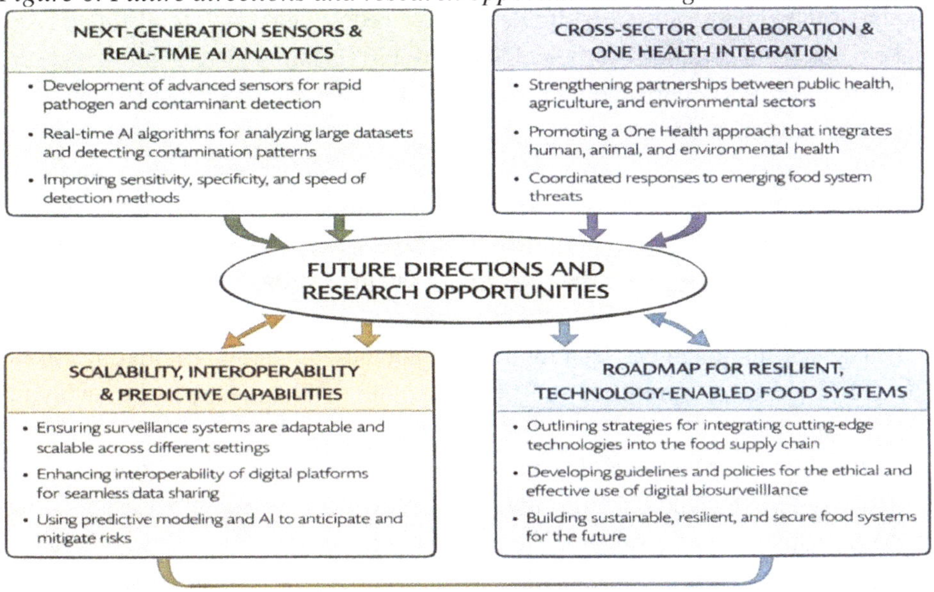

Advances in next-generation sensors and real-time artificial intelligence analytics represent a central research direction for digital biosurveillance. Emerging sensor technologies, including nanosensors and advanced biosensors, enable monitoring of environmental and food safety parameters with improved precision and sensitivity (Mastrogiovanni, 2025). When combined with AI-driven analytics, these sensors generate high-resolution data streams that support early identification of contamination risks and rapid intervention across food supply chains (Andrianto et al., 2025). Enhanced sensing capabilities improve detection accuracy for biological and chemical hazards, strengthening preventive food safety management.

Real-time AI analytics further enhance biosurveillance by enabling continuous interpretation of large and complex datasets. Advanced algorithms identify emerging patterns and trends within surveillance data, supporting anticipatory responses rather than delayed reactions (Sinha, 2025). These analytical capabilities support risk forecasting and situational awareness, contributing to more adaptive food safety strategies. Continued research into sensor miniaturization, durability, and AI model optimization remains essential for expanding the effectiveness of next-generation biosurveillance technologies.

Cross-sector collaboration and One Health integration represent another important area for future development. Digital biosurveillance systems benefit from coordinated data sharing among public health authorities, agricultural sectors, environmental agencies, and local communities (Khan & Ullah, 2024). The One

Health approach recognizes the interconnected relationships among human, animal, and environmental health, providing a framework for addressing food safety risks that emerge from complex ecological interactions (Haetami, 2025). Integrated surveillance improves understanding of zoonotic disease pathways and environmental drivers of contamination.

Future research should prioritize development of collaborative data platforms that support interoperability and shared governance across sectors. Integrated frameworks enable aggregation of diverse datasets, supporting more accurate risk assessments and coordinated intervention strategies (Chauhan & Dutta, 2025). Strengthening cross-sector partnerships improves early detection of emerging threats and enhances alignment between food safety policies and public health objectives. These collaborative approaches contribute to more resilient responses to evolving food system risks.

Scalability and interoperability remain critical considerations as digital biosurveillance systems expand across regions and contexts. Effective deployment requires platforms capable of adapting to differences in infrastructure capacity, governance environments, and resource availability (Almagharbeh et al., 2025). Scalable architectures that utilize cloud computing and decentralized data-sharing models support broader adoption and improve access to biosurveillance tools, particularly in low-resource settings (Yang et al., 2025). Research focused on scalable system design supports equitable implementation across diverse food systems.

Interoperability enables seamless data exchange among surveillance platforms, enhancing analytical depth and predictive accuracy. Adoption of shared data standards and communication protocols facilitates integration of environmental, production, and health data streams (Chauhan & Dutta, 2025; Kamara, 2025). Interoperable systems strengthen predictive modeling capabilities, supporting identification of contamination hotspots and emerging risks (Izah 2026). Continued development of flexible, standards-based platforms remains essential for improving system performance and usability.

Development of a structured roadmap for resilient, technology-enabled food systems represents a strategic priority for advancing digital biosurveillance. Such a roadmap outlines objectives, performance metrics, and phased implementation strategies that align technological innovation with food safety and public health goals (Al-Omari, 2026). Incorporating stakeholder perspectives and local contexts supports equitable system design and effective adoption across communities.

Ethical governance considerations remain integral to roadmap development. Issues related to data privacy, algorithmic bias, transparency, and accountability influence acceptance and long-term sustainability of biosurveillance initiatives (Ahmed et al., 2025). Frameworks that encourage stakeholder collaboration, ethical oversight, and public engagement strengthen trust and support shared responsibility for food safety outcomes (Wirata et al., 2025). Together, these future directions guide the evolution of digital biosurveillance toward more resilient, inclusive, and adaptive food systems.

9. CONCLUSION

Digital biosurveillance has become a critical response to interconnected food safety, public health, and environmental challenges within global food systems. Conceptual frameworks emphasize real-time monitoring, farm-to-fork surveillance, and integration of public health and environmental data to manage foodborne hazards effectively. Technology-enhanced approaches advance beyond traditional methods through IoT sensors, AI diagnostics, blockchain traceability, and big data analytics. Practical applications across fresh produce, dairy, aquaculture, and chemical monitoring demonstrate improved early warnings, transparency, and system resilience.

Digital platforms further strengthen supply chain resilience through continuous monitoring, data-driven decision support, predictive logistics, and integrated multi-sector datasets. Persistent challenges related to data privacy, infrastructure limitations, algorithmic bias, and stakeholder coordination require strong governance and collaborative frameworks. Future progress depends on next-generation sensors, real-time AI analytics, One Health collaboration, and interoperable systems. A strategic and ethical roadmap supports technology-enabled food systems that protect public health, enhance sustainability, and maintain trust across global food networks.

REFERENCES

Abaku, E. A., Edunjobi, T. E., & Odimarha, A. C. (2024). Theoretical approaches to AI in supply chain optimization: Pathways to efficiency and resilience. *International Journal of Science and Technology Research Archive, 6*(1), 092–107.

Abass, T., Eruaga, M. A., Itua, E. O., & Bature, J. T. (2024). Advancing food safety through IoT: Real-time monitoring and control systems. *International Medical Science Research Journal, 4*(3), 276–283. DOI: 10.51594/imsrj.v4i3.919

Abhulimen, A. O., & Ejike, O. G.Angela Omozele AbhulimenOnyinye Gift Ejike. (2024). Solving supply chain management issues with AI and big data analytics for future operational efficiency. *Computer Science & IT Research Journal, 5*(8), 1780–1805. DOI: 10.51594/csitrj.v5i8.1396

Achieng, M. S., & Ogundaini, O. O. (2024). Big data analytics for integrated infectious disease surveillance in Sub-Saharan Africa. *South African Journal of Information Management, 26*(1), 1668. DOI: 10.4102/SAJIM.v26i1.1668

Ahmed, M. M., Okesanya, O. J., Oweidat, M., Othman, Z. K., Musa, S. S., & Lucero-Prisno, D. E.III. (2025). The ethics of data mining in healthcare: Challenges, frameworks, and future directions. *BioData Mining, 18*(1), 47. DOI: 10.1186/s13040-025-00461-w PMID: 40646553

Ajala, O. A., Arinze, C. A., Ofodile, O. C., Okoye, C. C., & Daraojimba, O. D.Olakunle Abayomi AjalaChuka Anthony ArinzeOnyeka Chrisanctus OfodileChinwe Chinazo OkoyeObinna Donald Daraojimba. (2024). Reviewing advancements in privacy-enhancing technologies for big data analytics in an era of increased surveillance. *World Journal of Advanced Engineering Technology and Sciences, 11*(1), 294–300. DOI: 10.30574/wjaets.2024.11.1.0060

Al Khatib, I., Alasheh, S., & Shamayleh, A. (2024). The drivers of complexity in inventory management within the healthcare industry: A systematic review. [IJSSMET]. *International Journal of Service Science, Management, Engineering, and Technology, 15*(1), 1–26. DOI: 10.4018/IJSSMET.347332

Al-Omari, A. A. (2026). Future trends in educational leadership: Tech advances, evolving paradigms, and personalized learning. In Roache, D. (Ed.), *Advancing safety, service, and performance in organizations with impactful leadership* (pp. 25–50). IGI Global Scientific Publishing., DOI: 10.4018/979-8-3373-5686-0.ch002

Alahakoon, D., Nawaratne, R., Xu, Y., De Silva, D., Sivarajah, U., & Gupta, B. (2023). Self-building artificial intelligence and machine learning to empower big data analytics in smart cities. *Information Systems Frontiers*, *25*(1), 221–240. DOI: 10.1007/s10796-020-10056-x

Almagharbeh, W. T., Alharrasi, M., Rony, M. K. K., Kabir, S., Ahmed, S. K., & Alrazeeni, D. M. (2025). Ethical and institutional readiness for artificial intelligence in nursing: An umbrella review. *International Nursing Review*, *72*(4), e70111. DOI: 10.1111/inr.70111 PMID: 41045018

Amabie, T., Izah, S. C., Ogwu, M. C., & Hait, M. (2023). Harmonizing tradition and technology: The synergy of artificial intelligence in traditional medicine. In *Herbal medicine phytochemistry: Applications and trends* (pp. 1–23). Springer International Publishing.

Amrutha, B. K., & Gomathy, B. (2024). Research survey on role of blockchain technology in agriculture. *International Journal of Scientific Research in Engineering and Management*, *8*(6), 1–5.

Andrianto, T., Tangit, T. M., & Minh, N. C. (2025). Adoption of artificial intelligence (AI) technology in enhancing tourist experience: A conceptual model. *Journal of Tourism. Hospitality and Travel Management*, *3*(1), 53–66. DOI: 10.58229/jthtm.v3i1.302

Anitha, P., & Patil, M. M. (2018). A review on data analytics for supply chain management: A case study. *International Journal of Information Engineering and Electronic Business*, *14*(5), 30.

Aryffin, H. A. K., Sahbudin, M. A. B., Pitchay, S. A., Abhalim, A. H., & Sahbudin, I. (2025). Technological trends in epidemic intelligence for infectious disease surveillance: A systematic literature review. *PeerJ. Computer Science*, *11*, e2874. DOI: 10.7717/peerj-cs.2874 PMID: 40567698

Arzilli, G., De Vita, E., Pasquale, M., Carloni, L. M., Pellegrini, M., Di Giacomo, M., Esposito, E., Porretta, A. D., & Rizzo, C. (2024). Innovative techniques for infection control and surveillance in hospital settings and long-term care facilities: A scoping review. *Antibiotics (Basel, Switzerland)*, *13*(1), 77. DOI: 10.3390/antibiotics13010077 PMID: 38247635

Atadoga, A., Obi, O. C., Osasona, F., Onwusinkwue, S., Daraojimba, A. I., & Dawodu, S. O. (2024). AI in supply chain optimization: A comparative review of USA and African trends. *International Journal of Science and Research Archive*, *11*(1), 896–903. DOI: 10.30574/ijsra.2024.11.1.0156

Badul, S., Abia, A. L., Amoako, D. G., Perrett, K., Bester, L. A., & Essack, S. Y. (2021). From the farms to the dining table: The distribution and molecular characteristics of antibiotic-resistant *Enterococcus* spp. in intensive pig farming in South Africa. *Microorganisms*, *9*(5), 882. DOI: 10.3390/microorganisms9050882 PMID: 33918989

Bhartiya, S., & Mehrotra, D. (2015). Applying CHAID algorithm to investigate critical attributes of secured interoperable health data exchange. *International Journal of Electronic Healthcare*, *8*(1), 25–50. DOI: 10.1504/IJEH.2015.071638 PMID: 26559072

. Bhatnagar, S. (2025). *Digital trust certification: Building resilient, ethical, and citizen-centric digital system.*

Binamungu, L. P., & Kimaro, H. C. (2025). Problematizing the determinants of success of interoperability of information systems through the lens of stakeholder theory and enterprise interoperability framework. *The Electronic Journal on Information Systems in Developing Countries*, *91*(4), e70020. DOI: 10.1002/isd2.70020

Calvetti, D., de Mattos Nascimento, D. L., de Araújo Filho, F. N. M., Abreu, R. H. V., & Papadopoulos, N. A. (2024). Linked data to streaming data sensors sources for direct access. *Brazilian Journal of Operations & Production Management*, *21*(1). Advance online publication. DOI: 10.14488/BJOPM.1993.2024

Chatfield, A. T., & AlAnazi, J. (2015). Collaborative governance matters to e-government interoperability: An analysis of citizen-centric integrated interoperable e-government implementation in Saudi Arabia. [IJPADA]. *International Journal of Public Administration in the Digital Age*, *2*(3), 24–44. DOI: 10.4018/ijpada.2015070102

Chauhan, S., & Dutta, A. (2025). ARTIFICIAL INTELLIGENCE IN STUDENT PRIVACY AND DATA SECURITY. *International Journal of Advanced Research in Computer Science*, *16*(3).

Chen, F., Zhao, C., Yang, X., Luo, N., & Sun, C. (2025). A lightweight accountable parallel blockchain architecture based on redactable blockchain for agri-food traceability. *Foods*, *14*(4), 623. DOI: 10.3390/foods14040623 PMID: 40002066

Chen, H., Chen, Z., Lin, F., & Zhuang, P. (2021). Effective management for blockchain-based agri-food supply chains using deep reinforcement learning. *IEEE Access : Practical Innovations, Open Solutions*, *9*, 36008–36018. DOI: 10.1109/ACCESS.2021.3062410

Chopra, T., & Pathrotkar, N. (2024). Implementation of a food traceability system utilizing the Internet of Things and blockchain technology to enhance safety measures within the food supply chain. In *BIO Web of Conferences* (Vol. 82, p. 05016). EDP Sciences. DOI: 10.1051/bioconf/20248205016

Damiati, S. A. (2020). Digital pharmaceutical sciences. *AAPS PharmSciTech, 21*(6), 206. DOI: 10.1208/s12249-020-01747-4 PMID: 32715351

Deng, M., & Feng, P. (2020). A food traceability system based on blockchain and radio frequency identification technologies. *Journal of Computer and Communications, 8*(9), 17–27. DOI: 10.4236/jcc.2020.89002

El Idrissi, A. H., Larfaoui, F., Dhingra, M., Johnson, A., Pinto, J., & Sumption, K. (2021). Digital technologies and implications for Veterinary Services. *Revue Scientifique et Technique (International Office of Epizootics), 40*(2), 455–468. DOI: 10.20506/rst.40.2.3237 PMID: 34542102

Feng, H., Wang, X., Duan, Y., Zhang, J., & Zhang, X. (2020). Applying blockchain technology to improve agri-food traceability: A review of development methods, benefits and challenges. *Journal of Cleaner Production, 260*, 121031. DOI: 10.1016/j.jclepro.2020.121031

Gabriël, S., Dorny, P., Saelens, G., & Dermauw, V. (2022). Foodborne parasites and their complex life cycles challenging food safety in different food chains. *Foods, 12*(1), 142. DOI: 10.3390/foods12010142 PMID: 36613359

García-Díez, J., Saraiva, S., Moura, D., Grispoldi, L., Cenci-Goga, B. T., & Saraiva, C. (2023). The importance of the slaughterhouse in surveilling animal and public health: A systematic review. *Veterinary Sciences, 10*(2), 167. DOI: 10.3390/vetsci10020167 PMID: 36851472

Goktas, P., & Grzybowski, A. (2025). Shaping the future of healthcare: Ethical clinical challenges and pathways to trustworthy AI. *Journal of Clinical Medicine, 14*(5), 1605. DOI: 10.3390/jcm14051605 PMID: 40095575

Gupta, K. K., Saha, S., Sahoo, S. K., & Goswami, S. S. (2023). Revolutionizing healthcare industry 5.0: Exploring the potential of blockchain technology for medical applications. *J. Technol. Innov. Energy, 2*(3), 76–93. DOI: 10.56556/jtie.v2i3.610

Haetami, H. (2025). AI-driven educational transformation in Indonesia: From learning personalization to institutional management. *AL-ISHLAH: Jurnal Pendidikan, 17*(2), 1819–1832. DOI: 10.35445/alishlah.v17i2.7448

Hamad, A., & Bhattacharya, T. (2025). *Examining the potential and hurdles of blockchain technology in food systems: An overview*. Food Safety and Health. DOI: 10.1002/fsh3.70017

Hasan, I., Habib, M., & Mohamed, Z. (2023). Blockchain database and iot: A technology driven agri-food supply chain. *International Supply Chain Technology Journal*, *9*(3), 40–45. DOI: 10.20545/isctj.v09.i03.01

Humboldt-Dachroeden, S., & Mantovani, A. (2021). Assessing environmental factors within the one health approach. *Medicina (Kaunas, Lithuania)*, *57*(3), 240. DOI: 10.3390/medicina57030240 PMID: 33807528

Hyde, R., Hoflund, A. B., & Pautz, M. (2016). One HACCP, two approaches: Experiences with and perceptions of the hazard analysis and critical control point food safety management systems in the United States and the EU. *Administration & Society*, *48*(8), 962–987. DOI: 10.1177/0095399714548266

Idahor, C. O., Esomu, E. J. O., Ogbonna, N., Momoh, Z., Ogbeide, O. A., Ikhu-Omoregbe, O., Adigwe, A., Erhabor, O. M., Osaghae, O., Orons, N., & Ikhu-Omoregbe, O. A. (2025). Infectious disease surveillance in the era of big data and AI: Opportunities and pitfalls. *Cureus*, *17*(10). Advance online publication. DOI: 10.7759/cureus.93929 PMID: 41200593

Igwama, G. T., Nwankwo, E. I., Emeihe, E. V., & Ajegbile, M. D.Geneva Tamunobarafiri IgwamaEjike Innocent NwankwoEbube Victor EmeiheMojeed Dayo Ajegbile. (2024). AI and big data analytics for enhancing public health surveillance in rural communities. *International Journal of Applied Research in Social Sciences*, *6*(8), 1797–1823. DOI: 10.51594/ijarss.v6i8.1427

Isiaka, A. B., Anakwenze, V. N., Ilodinso, C. R., Anaukwu, C. G., Ezeokoli, C. M., Noi, S. M., Agboola, G. O., & Adonu, R. (2024). Harnessing artificial intelligence for early detection and management of infectious disease outbreaks. *International Journal of Innovative Research and Development*, *13*(2), 52–65.

Iyiola, A. O., Asiedu, B., Ogwu, M. C., & Izah, S. C. (2025a). Sustainability through AI in fisheries. In Izah, S. C., Ogwu, M. C., & Iyiola, A. O. (Eds.), *Artificial intelligence in fisheries: Transformative potentials and challenges*. CRC Press. DOI: 10.1201/9781003506287-5

Iyiola, A. O., Izah, S. C., & Ogwu, M. C. (2025b). Fish stock assessment using AI algorithms. In Izah, S. C., Ogwu, M. C., & Iyiola, A. O. (Eds.), *Artificial intelligence in fisheries: Transformative potentials and challenges*. CRC Press. DOI: 10.1201/9781003506287-4

Izah, S. C. (2026). Risk assessment frameworks and predictive modeling for food safety decision-making. In Ogwu, M., & Izah, S. (Eds.), *Mitigation strategies and risk management for food contaminants* (pp. 85–110). IGI Global Scientific Publishing., DOI: 10.4018/979-8-3373-7052-1.ch003

Izah, S. C., Amabie, T., Ogwu, M. C., & Iyiola, A. O. (2025b). Artificial intelligence for monitoring aquatic pollutants: Safeguarding ecosystem health and sustainable fisheries development. In Izah, S. C., Ogwu, M. C., & Iyiola, A. O. (Eds.), *Artificial intelligence in fisheries: Transformative potentials and challenges*. CRC Press. DOI: 10.1201/9781003506287-6

Izah, S. C., Etebu, E. N., Aigberua, A. O., Odubo, T. C., & Iniamagha, I. (2022). A meta-analysis of microbial contaminants in selected ready-to-eat foods in Bayelsa State, Nigeria: Public health implications and risk-reduction strategies. *Hygiene and Environmental Health Advances*, *4*, 100017. DOI: 10.1016/j.heha.2022.100017

Izah, S. C., Odubo, T. C., Ogwu, M. C., & Hait, M. (2024). Aspects of microorganisms in the food industry. In Ogwu, M. C., Izah, S. C., & Ntuli, N. R. (Eds.), *Food safety and quality in the global south* (pp. 399–425). Springer., DOI: 10.1007/978-981-97-2428-4_13

Izah, S. C., Ogwu, M. C., & Iyiola, A. (2025a). *Artificial intelligence in fisheries: Transformative potentials and challenges*. CRC Press. DOI: 10.1201/9781003506287

Jin, H., & Xie, Y. (2023). A review of research on the impact of digitalization on agricultural supply chain security. *Digitalization and Management Innovation*, 378–385.

Joel, O. S., Oyewole, A. T., Odunaiya, O. G., & Soyombo, O. T.Olorunyomi Stephen JoelAdedoyin Tolulope OyewoleOlusegun Gbenga OdunaiyaOluwatobi Timothy Soyombo. (2024). Leveraging artificial intelligence for enhanced supply chain optimization: A comprehensive review of current practices and future potentials. *International Journal of Management & Entrepreneurship Research*, *6*(3), 707–721. DOI: 10.51594/ijmer.v6i3.882

Kakade, S. L., & Vitalkar, S. (2024). Inventory management in pharmaceutical industry. *International Journal of Scientific Research in Engineering and Management*, *8*(4), 1–5.

Kamara, R. D. (2025). Bridging the gap: Opportunities, challenges and strategies for AI deployment in public service delivery. *Public Administration and Regional Development*, (28), 334–356. DOI: 10.34132/pard2025.28.02

Keawdounglek, V., Chaiyakul, S., & Aunkham, A. (2025). The development of the "Laab Nuer Model" for food safety management in handling traditional Lanna cuisine in Thailand. *PLoS One*, *20*(9), e0331933. DOI: 10.1371/journal.pone.0331933 PMID: 41004446

Keerthi, M., & Jayan, A. (2025). AI in global governance: Unleashing potential, overcoming challenges, and addressing ethical concerns. *MET SRUJAN*, *1*(1), 55–65. DOI: 10.34047/SJ/20251105

Khan, I. A., & Ullah, A. (2024). The role of artificial intelligence in enhancing social governance: A framework for ethical implementation and policy development in Pakistan. *Journal of Management & Social Science*, *1*(4), 274–289.

Kim, A. J., & Tak, S. (2019). Implementation system of a biosurveillance system in the Republic of Korea and its legal ramifications. *Health Security*, *17*(6), 462–467. DOI: 10.1089/hs.2019.0071 PMID: 31800333

Kurniawan, M., Suparno, S., & Vanany, I. (2023, December). Implementation blockchain technology and the internet of things in halal supply chain traceability and food safety: A systematic literature review. In *Proceeding International Conference of Technology on Community and Environmental Development* (Vol. 1, No. 2). DOI: 10.18196/ictced.v1i2.75

Lan, H. Y., Ubina, N. A., Cheng, S. C., Lin, S. S., & Huang, C. T. (2022). Digital twin architecture evaluation for intelligent fish farm management using modified analytic hierarchy process. *Applied Sciences (Basel, Switzerland)*, *13*(1), 141. DOI: 10.3390/app13010141

Li, Y., Zhang, X., Zhao, Z., Xu, J., Jiang, Z., Yu, J., & Cui, X. (2022). Research on grain food blockchain traceability information management model based on master–slave multichain. *Computational Intelligence and Neuroscience*, *2022*(1), 7498025. DOI: 10.1155/2022/7498025 PMID: 36605726

Liu, P., Ren, S., Wang, J., Yuan, S., Nian, Y., & Li, Y. (2022). A blockchain consensus optimization-based algorithm for food traceability. *Mobile Information Systems*, *2022*(1), 1529983. DOI: 10.1155/2022/1529983

Luo, J., Leng, S., & Bai, Y. (2022). Food supply chain safety research trends from 1997 to 2020: A bibliometric analysis. *Frontiers in Public Health*, *9*, 742980. DOI: 10.3389/fpubh.2021.742980 PMID: 35186862

Mahi, R. (2024). Optimizing supply chain efficiency in the manufacturing sector through AI-powered analytics. *International Journal of Management Information Systems and Data Science*, *1*(1), 41–50.

Mastrogiovanni, S. (2025). AI-driven data governance for smart cities: Balancing privacy, efficiency, and public trust. *South Florida Journal of Development, 6*(9), e5810–e5810. Mazimwe, A., Hammouda, I., & Gidudu, A. (2019). An empirical evaluation of data interoperability: A case of the disaster management sector in Uganda. *ISPRS International Journal of Geo-Information, 8*(11), 484.

McNamara, P. E., Miller, G. Y., Liu, X., & Barber, D. A. (2007). A farm-to-fork stochastic simulation model of pork-borne salmonellosis in humans: Lessons for risk ranking. *Agribusiness, 23*(2), 157–172. DOI: 10.1002/agr.20115

Mukherjee, U., Dutta, S., & Bandyopadhyay, S. K. (2021). Assembling blockchain and IoT for smart food-supply chain. *Asian Journal of Advanced Agricultural Research, 16*, 49–58. DOI: 10.9734/ajaar/2021/v16i330177

Nastasijevic, I., Kundacina, I., Jaric, S., Pavlovic, Z., Radovic, M., & Radonic, V. (2025). Recent advances in biosensor technologies for meat production chain. *Foods, 14*(5), 744. DOI: 10.3390/foods14050744 PMID: 40077447

Nikita, M., Kashyap, L., Singh, B. P., & Tyagi, D. (2023). AI enabled advance surveillance system. *International Research Journal of Modernization in Engineering Technology and Science, 5*, 6056–6061.

Nisa'Habib Elias, S. K., Usman, S., & Chuprat, S. (2024). Discovering the global landscape of agri-food and blockchain: A bibliometric review. *International Journal of Advanced Computer Science and Applications, 15*(4).

Nugroho, G., Tedjakusuma, F., Lo, D., Romulo, A., Pamungkas, D. H., & Kinardi, S. A. (2023). Review of the application of digital transformation in food industry. *Journal of Current Science and Technology, 13*(3), 774–790. DOI: 10.59796/jcst.V13N3.2023.1285

Odimarha, A. C., Ayodeji, S. A., & Abaku, E. A.Agnes Clare OdimarhaSodrudeen Abolore AyodejiEmmanuel Adeyemi Abaku. (2024). Machine learning's influence on supply chain and logistics optimization in the oil and gas sector: A comprehensive analysis. *Computer Science & IT Research Journal, 5*(3), 725–740. DOI: 10.51594/csitrj.v5i3.976

Odubo, T. C., Izah, S. C., & Ogwu, M. C. (2024). Mycological safety in tiger nut (*Cyperus esculentus* L.) drinks: Cultivation, contaminants, and compliance strategies. In *Food safety and quality in the global south* (pp. 463–487). Springer Nature Singapore. DOI: 10.1007/978-981-97-2428-4_15

Ogidi, O. I., & Izah, S. C. (2024). Emerging technologies in the global south food industry: Prospects and challenges. In *Food safety and quality in the global south* (pp. 197–231).

Ogwu, M. C., Izah, S. C., & Ntuli, N. R. (2024a). *Food safety and quality in the global south*. Springer. DOI: 10.1007/978-981-97-2428-4

Ogwu, M. C., Izah, S. C., & Ntuli, N. R. (2024b). Food security complexities in the global south. In Ogwu, M. C., Izah, S. C., & Ntuli, N. R. (Eds.), *Food safety and quality in the global south* (pp. 3–33). Springer., DOI: 10.1007/978-981-97-2428-4_1

Oriekhoe, O. I., Ilugbusi, B. S., & Adisa, O. (2024). Ensuring global food safety: Integrating blockchain technology into food supply chains. *Engineering Science & Technology Journal*, *5*(3), 811–820. DOI: 10.51594/estj.v5i3.905

Oyewole, A. T., Okoye, C. C., Ofodile, O. C., & Ejairu, E. (2024). Reviewing predictive analytics in supply chain management: Applications and benefits. *World Journal of Advanced Research and Reviews*, *21*(3), 568–574. DOI: 10.30574/wjarr.2024.21.3.0673

Palojoki, S., & Vuokko, R. (2023). Semantic interoperability of EHRs: Review of approaches for enhancing the goals of European Health Data Space.

Pan, Y., Wang, X., & Ye, Q. (2024). Enhancing supply chain management through artificial intelligence: A case study of JD Logistics. *Advances in Economics. Management and Political Sciences*, *109*(1), 116–121. DOI: 10.54254/2754-1169/109/2024BJ0127

Phillips, C., Chapman, B., Agunos, A., Carson, C. A., Parmley, E. J., Reid-Smith, R. J., Smith, B. A., & Murphy, C. P. (2022). A scoping review of factors potentially linked with antimicrobial-resistant bacteria from turkeys (iAM. AMR Project). *Epidemiology and Infection*, *150*, e153. DOI: 10.1017/S0950268822001224 PMID: 35843720

Primavilla, S., Roila, R., Zicavo, A., Ortenzi, R., Branciari, R., Shtylla Kika, T., Valiani, A., & Ranucci, D. (2021). *Salmonella* spp. in pigs slaughtered in small and medium-sized abattoirs in central Italy: Preliminary results on occurrence and control strategies. *Applied Sciences (Basel, Switzerland)*, *11*(16), 7600. DOI: 10.3390/app11167600

Pubellier, C. (2005). Lithuanian geographic information infrastructure@ spatial data sharing with Lithuania. *Geodezija ir Kartografija*, *31*(2), 47–53.

Pūraitė, A., Zuzevičiūtė, V., Bereikienė, D., Skrypko, T., & Shmorgun, L. (2020). Algorithmic governance in public sector: Is digitization a key to effective management. *Independent Journal of Management & Production*, *11*(9), 2149–2170. DOI: 10.14807/ijmp.v11i9.1400

Rahman, M. M., & Szabó, G. (2023). Assessing the status of national spatial data infrastructure (NSDI) of Bangladesh. *ISPRS International Journal of Geo-Information*, *12*(6), 236. DOI: 10.3390/ijgi12060236

Sahoo, M., Panigrahi, C., & Aradwad, P. (2022). Management strategies emphasizing advanced food processing approaches to mitigate food borne zoonotic pathogens in food system. *Food Frontiers*, *3*(4), 641–665. DOI: 10.1002/fft2.153

Sharma, K., & Maharshi, N. (2024). Social impacts of AI-powered online reviews: An ethical consideration. *Journal of Informatics Education and Research*, *4*(3).

Sinha, R. K.Ritesh Kumar Sinha. (2025). Data engineering as social infrastructure: Building platforms for equitable global development. *World Journal of Advanced Research and Reviews*, *26*(3), 1128–1135. DOI: 10.30574/wjarr.2025.26.3.2172

Sri Vigna Hema, V., & Manickavasagan, A. (2024). Blockchain implementation for food safety in supply chain: A review. *Comprehensive Reviews in Food Science and Food Safety*, *23*(5), e70002. DOI: 10.1111/1541-4337.70002 PMID: 39217509

Suartana, M., & Putra, R. E. (2022, December). Software-defined networking (SDN) traffic analysis using big data analytic approach. In *International Joint Conference on Science and Engineering 2022 (IJCSE 2022)* (pp. 243–250). Atlantis Press. DOI: 10.2991/978-94-6463-100-5_25

Tabassum, A., Ghaznavi, I., Abd-Alrazaq, A., & Qadir, J. (2025). Exploring the application of AI and extended reality technologies in metaverse-driven mental health solutions: Scoping review. *Journal of Medical Internet Research*, *27*, e72400. DOI: 10.2196/72400 PMID: 40829151

Tesson, V., Federighi, M., Cummins, E., de Oliveira Mota, J., Guillou, S., & Boué, G. (2020). A systematic review of beef meat quantitative microbial risk assessment models. *International Journal of Environmental Research and Public Health*, *17*(3), 688. DOI: 10.3390/ijerph17030688 PMID: 31973083

Tsiouris, V., Mantzios, T., Kiskinis, K., Papadopoulos, G. A., Giannenas, I., Marín, C., Carrasquer, F., Karvelis, P., Filippitzi, M. E., & Fortomaris, P. (2024). Backyard flock sampling and artificial intelligence: A dual strategy for early detection of avian influenza and Newcastle disease. *German Journal of Veterinary Research*, *4*(4), 73–80. DOI: 10.51585/gjvr.2024.4.0111

Van Boxstael, S., Habib, I., Jacxsens, L., De Vocht, M., Baert, L., Van de Perre, E., Rajkovic, A., Lopez-Galvez, F., Sampers, I., Spanoghe, P., De Meulenaer, B., & Uyttendaele, M. (2013). Food safety issues in fresh produce: Bacterial pathogens, viruses and pesticide residues indicated as major concerns by stakeholders in the fresh produce chain. *Food Control, 32*(1), 190–197. DOI: 10.1016/j.foodcont.2012.11.038

Venkatraman, S., Parvin, S., Mansoor, W., & Gawanmeh, A. (2023). Big data analytics and internet of things for personalised healthcare: Opportunities and challenges. *International Journal of Electrical and Computer Engineering, 13*(4).

Waller, M. A., & Fawcett, S. E. (2013). Click here for a data scientist: Big data, predictive analytics, and theory development in the era of a maker movement supply chain. *Journal of Business Logistics, 34*(4), 249–252. DOI: 10.1111/jbl.12024

Wan, Y., Peng, Z., Wu, F., Gao, J., & Li, M. (2024). Understanding public concerned ethic issues and ethical acceptance of AI surveillance technology: Analysis from social media data. In *Electronics, Communications and Networks* (pp. 717–722). IOS Press.

Wang, M. (2023). Design and implementation of the food supply chain traceability system based on the blockchain. *Highlights in Science. Engineering and Technology, 32*, 14–23. DOI: 10.54097/hset.v32i.4931

Wirata, G., Gunawan, M. S., Judijanto, L., Sarmento, A. D. A., & Permono, A. R. (2025). The role of e-government innovation in enhancing public service performance and strengthening transparent governance. *International Journal of Accounting and Economics Studies, 12*(6), 710–717. DOI: 10.14419/zxpzp606

Wittmer, J., Prouse, C., & Arefin, M. R. (2025). Digitalizing sewage: The politics of producing, sharing, and operationalizing data from wastewater-based surveillance. *Environment and Planning C: Politics and Space*, 23996544241313454.

Xueting, H. (2023). Blockchain-enabled smart packaging: Enhancing food traceability and consumer confidence in the Chinese food industry. *The Frontiers of Society. Science and Technology, 5*(12), 1–7.

Yang, C. (2024). Innovation in cross-border supply chain inventory management driven by big data. *Advances in Economics. Management and Political Sciences, 76*(1), 66–73.

Yang, H., Xu, J., & Li, S. (2025). Mapping research trends and hotspots in sports ethics: A visual bibliometric analysis using CiteSpace. *SAGE Open, 15*(4), 21582440251404196. DOI: 10.1177/21582440251404196

Yarkwan, B., Isaac, T. O., Okopi, A., & Izah, S. C. (2024). Evidence of the toxic potentials of agrochemicals on human health and biodiversity: Carcinogens and mutagens. In *Food safety and quality in the global south* (pp. 331–359). Springer Nature Singapore. DOI: 10.1007/978-981-97-2428-4_11

Yatsyna, Y., & Kudinov, I. (2023). Directions for the implementation of innovative analytical and statistical technologies as a tool for corruption counteraction. *Laisvalaikio Tyrimai*, *2*(22), 17–29. DOI: 10.33607/elt.v2i22.1440

Yin, B., Tan, G., Muhammad, R., Liu, J., & Bi, J. (2025). AI-powered innovations in food safety from farm to fork. *Foods*, *14*(11), 1973. DOI: 10.3390/foods14111973 PMID: 40509500

Zhang, D. (2024). AI integration in supply chain and operations management: Enhancing efficiency and resilience. *Applied and Computational Engineering*, *90*(1), 8–13. DOI: 10.54254/2755-2721/90/2024MELB0060

Zhang, X., Li, Y., Peng, X., Zhao, Z., Han, J., & Xu, J. (2022). Information traceability model for the grain and oil food supply chain based on trusted identification and trusted blockchain. *International Journal of Environmental Research and Public Health*, *19*(11), 6594. DOI: 10.3390/ijerph19116594 PMID: 35682180

Zhang, Y., Wu, X., Ge, H., Jiang, Y., Sun, Z., Ji, X., Jia, Z., & Cui, G. (2023). A blockchain-based traceability model for grain and oil food supply chain. *Foods*, *12*(17), 3235. DOI: 10.3390/foods12173235 PMID: 37685168

Zhorzholiani, T. (2025). The ethical and societal implications of AI in economic policy and public administration. In *AI use in social sciences* (pp. 125–156). IGI Global Scientific Publishing. DOI: 10.4018/979-8-3373-2612-2.ch005

Chapter 13

Plant-Based RNA Interference Powered by AI for Agricultural Safety

Sabyasachi Pramanik
http://orcid.org/0000-0002-9431-8751
Haldia Institute of Technology, India

ABSTRACT

Crop protection, gene control, and agricultural biotechnology might all be revolutionized by combining artificial intelligence (AI) with plant RNA interference (RNAi) techniques. A natural biological process called RNA interference (RNAi) inhibits gene expression, providing potential means of improving agricultural characteristics, boosting disease resistance, and managing pests. This article examines how AI is refining gene silencing processes, speeding up target gene identification, and enhancing RNAi delivery methods in order to optimize RNA interference technologies. By anticipating efficient RNA sequences, reducing off-target effects, and permitting cross-species applications, AI-driven methods improve RNAi design. By providing environmentally benign substitutes for conventional chemical pesticides and boosting crop tolerance to environmental stresses, the use of AI in RNAi has great potential for sustainable agriculture.

DOI: 10.4018/979-8-3373-9033-8.ch013

INTRODUCTION

An Overview of the Biological Function of Plant RNA Interference (RNAi)

A conserved chemical process called RNA interference (RNAi) is essential for post-transcriptional gene silencing in a variety of species, including plants. It includes short non-coding RNAs that target certain mRNA molecules for translational suppression or destruction in order to control gene expression. The review emphasizes many kinds of regulatory non-coding small RNAs, including siRNAs, miRNAs, and phasiRNAs, each of which adds to the complexity of RNAi-mediated gene expression control and has a unique function in gene regulation. Many biological activities in plants depend on RNAi. It facilitates the control of genes related to stress impressions and metabolic processes, allowing plants to adjust to changes in their surroundings. By specifically targeting viral RNA for destruction, RNAi also acts as a defensive mechanism against viral infections, shielding the plant from illness. Additionally, by regulating genes that affect growth patterns and organ formation, it contributes to the regulation of plant architecture and development. The natural mechanism of RNA interference (RNAi), which manages gene expression in eukaryotic cells, was originally identified in the late 1990s. This finding created new opportunities for biotechnology and genetic research, especially in the field of agriculture. The ability of RNAi to silence genes in different species was the main focus of early research on the technology. Scientists showed that tiny RNA molecules may successfully suppress gene expression, resulting in a loss of function akin to that caused by genetic defects. The use of RNAi in agricultural contexts was made possible by this groundbreaking breakthrough. In "proof of concept" trials, researchers used RNA interference (RNAi) to target critical genes in agricultural pests including insects, worms, and fungus. Significant mortality was achieved when these pests were exposed to different RNAi delivery techniques, including as feeding, soaking, and microinjection. This achievement prompted further research on the possibilities of RNAi in crop protection. After these early achievements, scientists started looking into in vivo RNAi, which suggests that plants may generate RNAi signals on their own. By producing deadly RNAi signals within the plants, this strategy sought to manage insect populations by successfully exploiting the plants as a defensive mechanism against predators. RNAi technology has been incorporated into agricultural biotechnology throughout time, resulting in the creation of genetically engineered crops capable of expressing RNAi constructs. These crops increase agricultural output and sustainability by targeting certain pests or illnesses. With continuous research aimed at increasing its effectiveness and broadening its uses, RNA interference (RNAi) is now acknowledged as a potent instrument in agricultural biotechnology. The method has the potential to reap crops which are better resistant to diseases and pests, which would eventually lead to more effective agricultural systems. In conclusion, the development of RNA interference (RNAi) in agricultural biotechnology has advanced from early discoveries to real-world uses, showcasing its potential to completely transform crop yield and pest control.

Difficulties in Creating and Implementing Successful RNAi Tactics

Developing and implementing successful RNA interference (RNAi) tactics requires overcoming a number of important obstacles. The efficient transport of RNA interference (RNAi) molecules, like microRNA (miRNA) or short interfering RNA (siRNA), to target cells is one of the main challenges. Following delivery, these molecules encounter a number of obstacles, such as quick glomerular excretion and reticuloendothelial system (RES) identification, which causes their early removal from the circulation. The breakdown of RNA interference treatments by extracellular nucleases is another difficulty. The stability and effectiveness of RNAi molecules are greatly diminished by this degradation, making it challenging to provide the intended gene silencing effect. Moreover, obtaining efficient cellular and tissue penetration continues to be a significant challenge. RNAi molecules' therapeutic potential is limited since they often have difficulty penetrating cellular membranes and reaching their designated intracellular targets. Innovative delivery methods that can successfully encapsulate RNAi treatments are also desperately needed. Numerous tactics, such as carrier-mediated methods, chemical alterations, and bioconjugation with particular ligands, have been investigated. However, creating a system that effectively targets certain tissues is still quite difficult. Furthermore, it's critical to precisely target the appropriate genes. Mistargeting may lower the treatment's overall efficacy and have unforeseen repercussions.

Lastly, the clinical use of RNAi therapies is also hampered by safety and regulatory issues. The implementation of these medicines may be delayed by the complicated process of ensuring that they are safe for human use and adhere to regulatory norms. In conclusion, transport hurdles, nuclease degradation, cellular penetration issues, the need for novel delivery mechanisms, precise gene targeting, and regulatory considerations are the main obstacles to developing and implementing successful RNA interference tactics. For RNAi to be successfully used in clinical settings, these issues must be resolved. AI is becoming an innovative technology in biology. Significant advances in research and applications have resulted from the growing integration of artificial intelligence (AI) (Ahale et al., 2024) into numerous biological subfields. Scientists' approaches to complicated biological systems and issues are changing as a result of this confluence. AI's capacity to evaluate massive datasets is among its most noteworthy effects on biology. Previously difficult-to-manage data sets, such genetic sequences and clinical data, may now be handled and interpreted by researchers thanks to artificial intelligence algorithms.

At the front of this change are machine learning algorithms (Dharshini et al., 2025). They speed up the drug development process, find new biomarkers, and reveal hidden patterns in biological data. This capacity is essential for creating novel treatments and improving our knowledge of illnesses. AI-driven technologies are also improving the customization of illness diagnosis and treatment. AI improves patient outcomes by forecasting the course of diseases and customizing treatment regimens; in today's healthcare system, a customized approach is becoming more and more crucial. Significant advancements in genomics research are being fuelled by AI. It improves our knowledge of gene regulation, makes it possible to identify genetic variations linked to illnesses more precisely, and helps predict protein structures. These developments enable focused therapies by offering crucial insights into disease processes. Notwithstanding the advantages, ethical concerns about privacy, prejudice, and responsibility are brought up by the use of AI in biology. In order to guarantee responsible AI research and use in biological sciences, it is imperative that scientists, legislators, and regulatory agencies work together to define

Ethical Standards

Looking forward, the intersection of biology and artificial intelligence is set to have a big impact on research. Researchers may open up new knowledge gaps, transform healthcare, and address urgent global issues by embracing this partnership, ushering in a new age of innovation and discovery.

LITERATURE REVIEW

Key approaches for integrating multimodal omics data, such as concatenation-based, transformation-based, and network-based (Ayyaswamy, 2026) techniques, are examined in the literature review of (Chaudhary et al., 2025). It talks on the successes and difficulties in the sector and emphasizes the growing relevance of massive language models in AI-driven integration. The study emphasizes the need of further research to improve precision medicine by fusing phenotypic insights, omics data, and AI platforms. The application of deep learning techniques for integrating several omics data types, such as proteomics, metabolomics, and genomes, in precision medicine is the main topic of (Karki et al., 2025) literature review. It draws attention to difficulties including the difficulty of combining various datasets and the need for sophisticated computing methods. In order to enhance disease knowledge and patient outcomes, the study highlights the significance of combining omics data () with phenotypic information. It also recommends that future research will need multidisciplinary cooperation to address these hurdles. Omics and phenotypic data integration in precision medicine is covered in (Ullah et al., 2025)'s literature review. It emphasizes how technological advancements and electronic health records (EHR) have made it easier for scientific research to move toward big data. In order to customize diagnosis and therapy, the study highlights the significance of integrating metabolomics, proteomics, and genomes with medical data. The study outlines future research paths for enhancing data harmonization and analytical frameworks in precision medicine, while also addressing issues with data integration and the need for sophisticated machine learning approaches. The PhenMap (Cosme et al., 2025) machine-learning technique for cancer subtype categorization is introduced in the literature review in Nyamundanda et al. (2020), addressing difficulties in identifying clinically significant tumour subtypes. In order to find functional subtypes, the program combines phenotypic and omics data using a modified factor analysis approach. PhenMap offers potential therapeutic implications in personalized cancer therapy by identifying new subgroups with varying responses to therapies, which have been validated using datasets related to breast cancer. In order to improve knowledge of cancer heterogeneity and its clinical significance, the research also investigates continuous subtypes.

Goals of the Research

This chapter aims to investigate how RNA interference (RNAi) technology and artificial intelligence (AI) may be combined in the area of plant biology. The chapter's goal is to demonstrate how AI may improve the planning, development, and use of RNA interference (RNAi) techniques for gene control and crop protection. It talks about how AI-driven (Rithik & Naren, 2024) technologies may speed up target gene identification, enhance RNA interference delivery methods, and improve gene silencing processes.

Scope of Research

1. AI Integration with RNAi Technology: How AI may be used to RNAi target prediction, RNAi construct optimization, and plant delivery system design.
2. Crop Protection: Examining how AI might enhance plant defences, disease resistance, and pest management strategies based on RNA interference.

3. Gene Regulation: Analyzing how artificial intelligence (AI) may improve plant gene regulation using RNA interference (RNAi), which would help manipulate growth patterns and metabolic pathways for increased agricultural output.
4. The use of AI-enabled RNAi technologies presents a number of obstacles, such as distribution, targeting specificity, and scalability. The future of these technologies in sustainable agriculture is also discussed.

THE BASICS OF PLANT RNA INTERFERENCE

The tiny RNA molecules target certain messenger RNA (mRNA) (Gurawalia et al., 2024) molecules for translational repression or destruction, which starts the gene silencing process. An overview of the siRNA, miRNA, and gene silencing mechanisms involved in RNA interference is provided below:

1. The pathway of small interfering RNA (siRNA)
    - Initiation: Long double-stranded RNA (dsRNA), which may come from a variety of sources including viral infections or transgenic expression in plants, is what sets off RNA interference. An enzyme known as Dicer breaks down the dsRNA into brief segments of around 21–23 nucleotides. We refer to these little segments as siRNAs.
2. Pathway of MicroRNA (miRNA)
    - Start: miRNAs are naturally occurring single-stranded RNA molecules that are produced by the transcription of miRNA genes found in the genome. Long primary miRNA (pri-miRNA) transcripts are the earliest form of these precursor miRNAs (pre-miRNAs).

COMPREHENDING THE RNA INTERFERENCE MECHANISM

One essential biological mechanism that controls gene expression and silences genes is RNA interference (RNAi). Double-stranded RNA (dsRNA), which may be produced intentionally or result from viral infections, is what starts it. The type III endoribonuclease enzyme (Haider et al., 2026) Dicer breaks down this dsRNA into tiny RNA pieces that are between 19 and 21 base pairs long. Both strands of these small interfering RNAs (siRNAs) exhibit distinctive dinucleotide overhangs.

One siRNA strand—called the guide strand—is kept in RISC, while the other strand is thrown away. By directing RISC to complementary mRNA targets, the guide strand effectively silences the gene by causing mRNA breakdown.

Apart from siRNAs, another family of short RNA molecules involved in gene regulation are microRNAs (miRNAs). Dicer processes miRNAs, which are produced from longer primary transcripts, similarly to siRNAs.

In plants, endogenous versus exogenous RNAi

1. RNAi that is endogenous

The term "endogenous RNAi" describes RNA interference systems that are activated by RNA molecules that come from the genome of the plant. These RNA molecules are crucial for controlling regular cellular functions and preserving the genetic integrity of the plant.

- Triggering RNA Sources: In endogenous RNA interference, the plants own genes are usually the source of the triggering RNA. This may consist of:
    - The tiny, non-coding RNAs known as microRNAs (miRNAs) are produced by endogenous genes. By attaching to target mRNAs and either destroying or blocking their translation, miRNAs control the expression of genes.
    - Phased small interfering RNAs, or phasiRNAs, are made from long, complementary RNAs that form a number of secondary small RNA molecules that influence gene expression and aid in the defence and growth of plants.
    - Endogenous siRNAs: These may originate from inverted repeats, transposonse, or repeated regions in the plant's DNA. Their key functions include controlling transposable element expression and preserving genomic stability.

2. RNAi exogenously

RNA interference caused by foreign or external RNA molecules put into the plant is referred to as exogenous RNA interference (RNAi). This mechanism has grown to be a potent tool in crop protection and genetic engineering, primarily being exploited for defence against viruses and transgenes.

- Sources of Triggering RNA: RNA molecules that are brought into the plant from outside sources rather than being synthesized by the plant itself are usually what cause exogenous RNA interference. These may consist of:
    - Viral RNA: The plant's RNAi machinery can identify and silence exogenous RNA molecules from the virus in response to viral infections.

- Artificial RNA: In genetic engineering, some genes or transgenes (Rahman et al., 2026) may be silenced by introducing synthetic siRNAs or exogenous dsRNA into plants.

- Mechanism: Exogenous RNA interference involves either introducing foreign dsRNA into the plant by gene delivery methods like direct injection or Agrobacterium-mediated transformation, or it may be applied externally using procedures like soaking or spraying. The plant's Dicer enzyme transforms the injected dsRNA into siRNAs, which then interact with the RISC complex to silence the target mRNA.

KNOWING THE DIFFERENCES BETWEEN ENDOGENOUS AND EXOGENOUS RNAI IN PLANTS

The term "endogenous RNA interference" (RNAi) describes RNAi processes that start within the plant. In this mechanism, plants use their own genes to create hairpin double-stranded RNAs (hpRNAs) and double-stranded RNAs (dsRNAs).

Conversely, exogenous RNA interference (RNAi) entails introducing foreign dsRNAs or hpRNAs into the plant system. Target genes in pathogens are suppressed as a result of the plant absorbing and processing these foreign RNA molecules similarly to endogenous dsRNAs. Through direct targeting of their virulence genes, this technique has been investigated as a biotechnological means of improving plant resistance against a variety of diseases, including viruses and fungi. Without requiring the plant to be genetically modified, exogenous RNA interference (RNAi) techniques like Host-Induced Gene Silencing (HIGS) (Kathirvel et al., 2025) use the plant's own RNAi machinery to fight against infections.

Endogenous and Exogenous RNAi Comparison

- Source: Exogenous RNAi is derived from external sources (such as viruses or artificial introduction), while endogenous RNAi is produced within the plant.
- Mechanism: Both routes produce siRNAs, which induce gene silencing by targeting mRNAs for translation suppression or destruction, using comparable processing mechanisms.
- Application: Exogenous RNAi is a targeted strategy used in agricultural biotechnology to increase plant resistance against certain infections, while endogenous RNAi acts as a natural defensive system, shielding the plant from pathogens and controlling gene expression.

In summary, exogenous RNAi provides novel ways to increase plant resilience using biotechnological techniques without genetic change, while endogenous RNAi is a natural mechanism that shields plants from a variety of dangers. Both have important uses in agriculture and are essential to plant defence systems.

Techniques of Delivery (e.g., Topical Sprays, Nano-Particles, Transgenic Expression)

One of the most important parts of using RNA interference (RNAi) for crop protection and gene control is getting RNAi molecules to plants. RNAi constructs may be introduced into plants via a variety of techniques, each with unique benefits and drawbacks. The main delivery techniques for plant RNA interference applications are listed below:

1. Expression of Transgenics
   - Description: To express RNAi constructs, plants are genetically modified using this technique. The plants are modified to generate hairpin RNA (hpRNA) or double-stranded RNA (dsRNA) molecules, which then initiate the RNA interference process in the plant cells.

These structures are designed to target certain genes, either those of pests and diseases or indigenous plant genes.

2. Sprays for Topics
   - Description: Using this technique, RNAi molecules (such as siRNA or dsRNA) are sprayed directly onto plant surfaces. RNAi reactions are brought on by the plant cells absorbing the sprayed RNA molecules.
   - Method: To improve the absorption of RNA molecules by plant tissues, buffers or surfactants may be included in the formulation of topical sprays. Depending on the pest or target route, the RNA molecules may be administered to the leaves, stems, or roots.
3. Nanoparticles
   - Description: The use of nanoparticles as RNAi molecule delivery systems is being investigated more and more. These nanoparticles provide a more regulated and efficient delivery system by encasing RNA molecules and facilitating their entrance into plant cells.
   - Method: Nanoparticles, such as liposomes, silica nanoparticles, or chitosan-based nanoparticles, are loaded with RNA molecules (such as siRNA or dsRNA). After that, these particles are sprayed on plants or absorbed via their roots. The RNAi molecules are shielded from dete-

rioration by the nanoparticles, which also improve their stability and plant cell absorption.

AN OVERVIEW OF THE DELIVERY METHODS

There are distinct benefits and difficulties associated with each of these delivery modalities. The target plant species, the intended result, and whether genetic modification is appropriate in a particular situation all influence the technique selection.

Investigating Biologics Delivery Techniques

High-concentration protein medications may be delivered precisely to the site of action by topical application of biologics. By reducing bloodstream dilution, this technique improves therapy efficacy. Nevertheless, despite its promise, only a small number of topical medications have been approved by the FDA, suggesting that further research is necessary in this field. This emphasizes the necessity for more study to broaden the use of topical biologic treatments.

The ease and patient choice of oral administration systems have drawn attention. Only two oral biologics have received FDA approval since 2015, however a recently introduced oral biologic is much less expensive—about 3% less than the typical injectable biologic. A streamlined regulatory approval procedure that does away with costly fermentation and purification procedures is mostly to blame for this cost decrease. When it comes to patient convenience and cost-effectiveness, oral biologics provide a compelling substitute for injections.

Although it isn't covered in detail in the study, transgenic expression is the technique of genetically changing organisms to make biologics. This strategy may be useful for economically generating biologics like proteins. This is not covered in detail in the study, but it is nevertheless a crucial technique for producing biologics on a big scale, particularly for therapies that call for high-volume manufacture.

A possible technique to improve the delivery of nucleic acid medications is the use of nanoparticles. These nanoparticles are a promising target for further study and development as they may increase the stability and bioavailability of biologics. They are especially crucial for delivering biologics like RNA-based medicines because of their capacity to enhance cellular absorption and shield biologics from destruction.

The analysis highlights a number of cutting-edge delivery methods, including chewing gum, patches, and sprays that are now undergoing clinical studies. These techniques aim to increase the efficacy and accessibility of biologics, particularly in low-income areas where infrastructural constraints may cause standard delivery methods to fail. These alternate administration methods may improve patient compliance and increase the accessibility of biologic treatments.

The study highlights the need for further investigation into other delivery systems, especially ones that may get around the problems with the biologic delivery systems in use today. With an emphasis on enhancing stability, bioavailability, and patient comfort, this involves investigating novel formulations and technologies that might make the administration of biologics simpler and more efficient.

There lies a lot of room for innovation in delivery systems, even if topical administration and oral delivery are promising approaches for biologics. The distribution, efficacy, and accessibility of biologic therapies might all be greatly improved by using nanoparticles (Jangra & Tiwari, 2025) and other cutting-edge techniques. To overcome the drawbacks of current delivery methods and provide fresh approaches that enhance patient outcomes and broaden the use of biologics, further research is required.

Current Uses Include Support for Gene Editing, Virus Resistance, and Pest Management

In contemporary biotechnology, RNA interference (RNAi) has several uses, particularly in agriculture. Viral resistance, gene editing support, and pest management are some of its main applications. By using RNA interference (RNAi) to selectively silence certain genes, these applications provide a potent tool for enhancing crop resilience, managing pests, and assisting with genetic alteration procedures.

RNA interference has been used in pest management to target genes in insects and other pests that damage crops. RNA interference (RNAi) may interfere with these pests' growth, reproduction, or metabolism by creating RNA molecules that precisely target vital genes in them. This can eventually result in the pests' death or decreased population. For instance, RNA interference (RNAi) has been effectively used to cotton and corn to shield them against insect pests including bollworm and corn rootworm. Compared to conventional chemical pesticides, this method is less likely to cause resistance and may target pests without harming non-target species, making it ecologically beneficial.

An inventive method of defending plants against viral infections is provided by RNA interference (RNAi) in viral resistance. When a virus infects a plant, the RNAi machinery may identify and break down the infection's RNA, stopping it from spreading throughout the plant. This strategy has been investigated to fight plant viruses in crops including potatoes, papaya, and tobacco, greatly lowering crop losses and increasing productivity. The development of virus-resistant crop types and the replacement of conventional chemical treatments are both possible with RNAi-based viral resistance.

ARTIFICIAL INTELLIGENCE'S FUNCTION IN RNAI DESIGN

RNA interference (RNAi) technology design and optimization are becoming more dependent on artificial intelligence (AI) (Ayyaswamy & Vianny, 2025). Data-driven (Fortune et al., 2025) RNA sequence optimization, which includes finding efficient RNA sequences for gene silencing and reducing off-target effects, is one of the main areas where AI (Naren & Kaliappan, 2025) makes a substantial contribution. AI-driven methods are revolutionizing the design of RNA interference molecules, improving their accuracy, effectiveness, and suitability for a variety of species and illnesses.

RNA Sequence Optimization Based on Data

In order to ensure that RNA sequences for RNA interference (RNAi) applications are both efficient in silencing target genes and minimize unwanted interactions with non-target genes, AI models are being used to improve their design.

- Target Prediction: AI systems, especially machine learning-based ones, are able to identify possible RNAi target locations by analyzing massive genomic data sets. To determine the most effective target locations for RNA interference, these models are trained using enormous volumes of RNA sequencing data and established gene silencing results. AI can determine which candidate sequences are most likely to successfully silence the target gene by looking at variables including sequence specificity, RNA secondary structure, and sequence context.
- Off-Target Minimization: AI-based models take care of this by anticipating and reducing these off-target interactions using computational methods. In order to create RNAi molecules with more selectivity, machine learning models are taught to identify patterns in RNA sequences that might result in off-target binding. AI can also mimic how RNA interference molecules interact with an organism's whole transcriptome, seeing possible off-targets and recommending sequence changes to prevent them.

AI for Predicting Secondary Structures

AI is essential for RNA secondary structure prediction and stability modelling, especially deep learning methods. Knowledge RNA function, particularly for RNA interference (RNAi) applications, requires a knowledge of RNA folding and structure.

Deep Learning for Stability Modelling and RNA Folding

- RNA Folding: RNA molecules fold into certain three-dimensional shapes that are essential to their operation. The way RNA interacts with other molecules, including as proteins and RNAi-related factors, is determined by its secondary structure, which consists of base pairing, loops, and stems. The algorithms used in traditional RNA secondary structure prediction techniques were derived from heuristic principles and thermodynamic models. But when it comes to large-scale datasets or complicated RNA sequences with non-canonical structures, these methods often failed.

RNA folding predictions have been greatly improved by deep learning techniques, particularly neural networks and convolutional neural networks (CNNs) (Bhattacharya, D. et al. 2025). Even in situations where standard models fail, deep learning algorithms may learn intricate patterns and reliably predict the secondary structures of RNA sequences by training on substantial datasets of known RNA sequences and their related structures. Complex sequence-structure connections that are essential to the stability and function of RNA may be taken into account by these models.

- Stability Modelling: The strength and makeup of RNA molecules base pairs, loops, and other structural elements influence its stability. For RNAi applications to be successful, stability is essential since the RNA molecules must retain their structure in order to effectively silence genes. By examining enormous sequence databases and identifying the critical elements affecting stability, deep learning algorithms are able to predict RNA stability. By forecasting the thermodynamic stability of RNA molecules, these models may provide light on how they are expected to behave in biological systems.

The RNNs (Madhavakumar D., et al. 2025) and LSTM (Vianny, M. et al. 2025) AI models are ideal for predicting RNA stability because they can identify long-range relationships in RNA structures and sequences. By using these models, scientists may forecast which RNA sequences will be most stable in physiological settings, enhancing the design of RNA interference molecules for increased efficacy.

- AI Integration with Conventional Models: Although deep learning (Kathirvel et al., 2025) has sophisticated features, it is often used with conventional thermodynamic models to provide forecasts that are more accurate. Predictions based on databases of known RNA structures may be improved in accuracy by using AI models. By using the fundamental understanding of thermodynamic principles and the powers of artificial intelligence's pattern recognition, this blend of deep learning and classical methods enables the best of both worlds.

RNA FOLDING AND STABILITY MODELLING USING DEEP LEARNING

Habib et al. (2025) presents a unique method for RNA secondary structure prediction termed "Structural Stability-Aware Deep Learning" that blends domain-specific expertise with cutting-edge machine learning methods. Prediction accuracy is greatly increased using this strategy, particularly when there is a lack of high-quality data.

The creation of the Energy Neural Network (Energy NN), which makes use of GCNs (Kathirvel et al. 2025) and CNNs, is a crucial part of this strategy. Energy NN provides crucial information on RNA stability and folding by estimating the free energy of RNA based on its sequence and secondary structure. This approach improves current machine learning frameworks without changing their design by including an energy loss element that is computed from both actual and anticipated structures, enabling more reliable and accurate predictions.

Incorporating this link into the machine learning model, the research also highlights the relationship between stability and free energy. The precision, generalizability, and interpretability of RNA secondary structure predictions are greatly enhanced by this integration. The model consistently outperformed earlier models like E2EFold and UFold after being thoroughly tested on well-known datasets. Metrics like F1 score, accuracy, and recall improved with the improved framework, demonstrating its efficacy in RNA folding predictions.

This approach's wide application is one of its advantages. The approach is appropriate for a variety of RNA research fields as it is not limited to certain RNA types or machine learning techniques. This wide range of applications represents a major breakthrough in the generalization of machine learning algorithms for the prediction of RNA secondary structures.

In order to better incorporate domain knowledge into machine learning frameworks, the paper identifies future research topics. These developments may result in even more precise and trustworthy predictions of RNA structure, creating new avenues for RNA-based treatments and uses.

According to (Kumar & Khurana, 2025), the use of deep learning methods in RNA folding and stability modelling is a noteworthy development in the area. This method improves RNA structure predictions' precision and interpretability, setting the stage for more successful RNA-based applications in biotechnology and medicine.

Combining Bioinformatics Pipelines

Gene sequence analysis and annotation have been transformed by the incorporation of Artificial Intelligence (AI) methods, including as neural networks (Naren & Gunasekaran, 2025) and Natural Language Processing (NLP) (Salem & Eid, 2025), into bioinformatics workflows. The enormous volumes of data produced in genomics research can now be processed, analyzed, and interpreted much more easily because to these sophisticated computational techniques, which also provide more precise and effective tools for comprehending intricate biological data.

Gene Sequence Analysis Using Neural Networks

Gene sequence analysis now requires the use of neural networks, especially deep learning models. These models may identify complicated patterns in nucleotide sequences that are often too difficult to find using conventional techniques. Neural networks are used in gene sequence analysis to uncover regulatory regions, forecast gene function, and find possible mutations associated with illnesses. Convolutional neural networks (CNNs), for instance, may be used to identify local sequence patterns in genomic sequences that indicate gene activity or interactions with other molecular components. Their capacity to learn from high-dimensional data makes it possible to identify patterns that were previously missed, which is essential for expanding our knowledge of genomics.

Gene Sequence Annotation Using NLP

Gene sequence annotation is increasingly being done using Natural Language Processing (NLP) approaches. When seen as "language," gene sequences may be examined similarly to how text is examined by NLP algorithms. In sequence annotation, where the objective is to provide meaningful labels to portions of DNA, RNA, or protein sequences based on predetermined criteria, this comparison is very helpful.

Large genomic datasets may be automatically annotated by researchers using NLP models like transformers or recurrent neural networks. To anticipate the activities of unannotated genes or to find biological features like promoters, enhancers, or binding sites for regulatory proteins, these models may be trained using annotated datasets that are already available. Gene ontology (GO) term assignment, in which sequences are connected to particular biological processes, cellular components, or molecular activities, may also be accomplished using NLP-based technologies.

Problems and Prospects for the Future

Even though AI-assisted siRNA design has a lot of promise, there are still issues with getting RNAi molecules to the target pathogens. For instance, it is more challenging to deliver siRNAs to soil-borne pathogens like Fusarium than to foliar diseases, necessitating the creation of more effective delivery methods. Furthermore, as some infections may develop ways to evade RNAi silencing, further study is required to determine the long-term efficacy of RNAi-based resistance.

An innovative strategy for plant pathogen resistance is AI-assisted siRNA design, mainly during treating viral infections and illnesses such as Fusarium wilt. Researchers are developing more efficient, accurate and long-lasting plant protection strategies by using AI's capacity to forecast ideal siRNA sequences, opening the door to healthier crops and more robust agricultural systems.\

AI-ASSISTED PLANT PATHOGEN RESISTANCE: AN ANALYSIS OF FUSARIUM WILT

The possibility of AI-assisted RNA interference (RNAi) technology for creating resistance to Fusarium wilt, a serious disease that impacts many crops, especially tomatoes. This work emphasizes how RNA interference (RNAi) targets certain genes in the pathogen to increase plant resistance.

Key genes in Fol, including as those encoding β-1, 3-glucanosyltransferase (GAS1) and peroxisomal biogenesis factor (PEX6), were silenced by the researchers using RNAi technology. The capacity of the fungal pathogen to infect and produce illness is greatly influenced by these genes.

In order to do this, the researchers used Agrobacterium-mediated transformation to introduce the RNAi constructs into transgenic tomato plants. By ensuring the silencing constructs' steady incorporation into the plant genome, this technique enabled the plants to consistently generate certain small interfering RNAs (siRNAs) that target the Fol genes.

By confirming that the RNAi constructs were functional and successfully silenced the targeted genes in the transgenic plants, stem-loop RT-PCR analysis validated the effectiveness of the RNAi technique. The resistance to Fusarium wilt was significantly increased in these transgenic tomato lines. Compared to tomato plants that were not changed, the plants that were infected with Fol showed delayed disease signs and had a greater survival rate. This suggested that the pathogen may be effectively resisted by RNAi-mediated silencing of the targeted genes.

In conclusion, the research shows how well AI-assisted RNAi technology works for developing defenses against plant diseases such as tomato Fusarium wilt. This strategy, which targets genes unique to pathogens, offers a potential way to increase crop health, resilience, and production in addition to offering insightful information about gene activity.

Enhancement of Crop Traits

RNAi has evolved as a revolutionary technique for improving some crop characteristics, including postponing ripening, lowering anti-nutritional factors, boosting disease resistance, and raising yield. More accurate targeting, increased effectiveness, and stronger regulation design was attainable by the incorporation of AI into RNA interference (RNAi) applications, guaranteeing that these trait enhancements are long-lasting and ecologically benign. Here, we examine how AI helps optimize RNAi for different agricultural trait improvements.

Delaying Ripening With RNAi

Unwanted side effects have been a problem for traditional techniques like genetically altering fruit ripening processes. However, RNAi enables more focused regulation of genes linked to ripening.

By evaluating plant genomic data to find important genes involved in the ripening process, such as those coding for enzymes like polygalacturonase or ethylene biosynthetic enzymes, AI creates a major part in targeting gene selection. In order to postpone ripening without yielding to the fruit's overall quality or nutritive content, machine learning algorithms assist in predicting which genes should be silenced. AI is used to improve the RNA sequences for effective silencing when possible targets have been discovered.

By mimicking the way the intended RNA molecules would interact with the target gene and its mRNA, AI also increases the effectiveness of RNA interference. This prediction lessens off-target effects that could impact other significant characteristics while also ensuring that the RNA molecules are successful in silencing the gene. Additionally, AI can help optimize delivery systems, such topical sprays or transgenic plants, to guarantee that RNAi molecules efficiently reach the intended regions.

AI may assist in evaluating the possible advantages and disadvantages of RNAi-based crop alterations in terms of regulatory design, making sure that the modifications adhere to national and international agricultural laws. AI-powered models are able to assess the trait's long-term durability across generations and forecast the environmental effects of delayed ripening.

Reducing Anti-Nutritional Factors With RNAi

Reducing anti-nutritional elements that might harm human health or the quality of animal feed such as lectins, protease inhibitors and alkaloids, is another crucial area of agricultural trait improvement by RNA interference. Certain grains, legumes, and soybeans are examples of crops that often include anti-nutritional elements. By precisely silencing the genes that produce these chemicals, RNA interference (RNAi) technology enhances the nutritional profile of these crops.

By examining the genomes of crops to find the genes that produce anti-nutritional substances, AI aids in target selection. Potential targets for silencing may be swiftly identified by deep learning algorithms that have been trained on vast amounts of data from genetic databases. Based on their relative significance in generating toxic chemicals and their impact on crop development overall, AI may also rank which genes should be targeted first.

By modelling the potential effects of gene silencing on the plant's larger metabolic pathways, artificial intelligence (AI) may increase effectiveness. This makes it possible for the RNAi molecules to be strong enough to lessen anti-nutritional elements without impairing other advantageous characteristics like yield or growth. In order to increase RNA molecules' stability and plant absorption, AI models may also optimize their design.

AI aids in regulation design by forecasting the environmental effect and safety of silencing these genes. AI simulations can evaluate these risks and guarantee that the trait modification complies with regulatory standards established by agencies like the USDA or EFSA (European Food Safety Authority) because silencing anti-nutritional factors may affect non-target genes or result in unintended consequences.

THE CONTRIBUTION OF AI TO THE ADVANCEMENT OF RNAI TECHNOLOGY FOR CROP TRAIT IMPROVEMENT

By offering the resources required for effective target gene discovery, improving RNA sequence design, forecasting effectiveness, and guaranteeing regulatory compliance, AI improves RNAi applications in both postponing ripening and lowering anti-nutritional factors. These AI-powered methods shorten the trial-and-error stage that is usually involved in the creation of agricultural traits, resulting in quicker development schedules and more accurate outcomes.

In order to ensure that the genetic tweaks enhance crops in ways that are advantageous for farmers, consumers, and the environment, researchers may use AI to design tailored RNAi constructs that target certain features with minimal off-target impacts. Furthermore, AI may assist in the creation of delivery systems, ensuring that the RNA molecules are efficiently produced and distributed to the appropriate tissues, whether via transgenic plants or other techniques.

With precise control over processes like ripening and anti-nutritional factor reduction, AI-driven RNA interference (RNAi) technology is a potent tool for improving agricultural attributes. AI increases the efficiency of RNAi in agricultural trait modification by maximizing target selection, enhancing efficacy, and guaranteeing regulatory compliance. These developments have the potential to produce more robust, nutrient-dense, and sustainable crops, supporting sustainable agricultural methods and global food security.

PLATFORMS, DATASETS, AND TOOLS

The design, development, and use of RNA interference (RNAi) in plant and animal research are supported by a number of specific AI tools and platforms. By improving RNAi's accuracy, efficacy, and efficiency, these methods enable researchers to more precisely target certain genes while reducing off-target effects. Some popular AI tools for RNAi design are listed below:

siDirect

A technique called siDirect was created to forecast siRNA sequences that would effectively interact with RNA. The program analyzes a target gene sequence using sophisticated algorithms to find the best siRNA candidates. Considering elements like secondary structure, sequence accessibility, and the possibility of off-target effects, siDirect is an effective technique for creating siRNAs that effectively quiet target genes in both plants and animals. Researchers may choose the optimal RNA sequences for efficient gene silencing by using the tool's target-specific predictions.

- Features:
  - o Accurately predicts siRNA sequences.
  - o Reduces consequences that are not intended.
  - o Offers statistical comments on each siRNA's efficacy.
- Use: o In genetic research, siRNA molecules are designed.
  - o Facilitates effective gene silencing in medicinal applications based on RNA interference.

dsCheck

A program called dsCheck assesses the secondary structure of double-stranded RNA (dsRNA) molecules. It is particularly helpful in RNA interference (RNAi) applications where the stability and efficacy of the RNA molecule are greatly influenced by its shape. dsCheck may assist in identifying possible areas of instability that might compromise the effectiveness of gene silencing by mimicking the structure of dsRNA. In order to guarantee that the RNAi molecules are stable and functional after they are injected into the target organism, this technology use machine learning methods to predict the thermodynamic characteristics of dsRNA.

- Characteristics: o Examines dsRNA's secondary structure.
  - o Finds potential unstable areas that can lower RNA interference effectiveness.

Assists in maximizing the stability and silencing potential of RNA molecules

- Use: o Guarantees the stability of dsRNA molecules in RNA interference studies.
  - o Assists in creating more resilient RNA interference molecules for research on plants and animals.

sRNATarget

A program called sRNATarget is intended to identify miRNA and siRNA target locations in the genomes of plants and animals. sRNATarget examines RNA sequences to find areas where short RNAs (such as siRNAs or microRNAs) may bind and effectively silence genes using statistical models and machine learning. Researchers may use this tool to determine the optimal target areas for RNAi-based treatments, which has wide-ranging implications in functional genomics and gene regulatory investigations.

- Characteristics: o Forecasts small RNA molecule target sites (siRNAs and miRNAs).

Target accessibility and effectiveness ratings are provided.

High binding specificity miRNA and siRNA candidates are selected with its assistance.

- Use: o Crucial for creating RNA molecules capable of efficiently silencing certain genes.

By discovering important gene targets, it aids in the development of RNAi-based treatments.

ADDITIONAL DATASETS AND PLATFORMS

Researchers in the area of RNAi have access to a number of other platforms and datasets in addition to these tools:

1. RNAiDB: This database offers thorough details on RNAi-targeted genes, the effectiveness of silencing, and associated biological facts.
2. Integrated RNAi Platforms: A few of these platforms provide integrated solutions that include experimental procedures, RNA molecule optimization, and gene targeting. These systems have the ability to automate the design of RNA interference and provide real-time feedback on how well siRNA or miRNA suppresses gene expression.

The way scientists create and refine RNA-based treatments is being completely transformed by the incorporation of AI tools like as siDirect, dsCheck, and sRNA-Target into RNAi research. The creation of more accurate and successful RNAi treatments is made possible by these tools, which provide crucial assistance in target gene identification, sequence design, and effectiveness prediction. Platforms like RNAiDB and TargetScan also make a contribution by providing extensive datasets that facilitate functional analysis and target discovery. Gene regulation, agricultural trait improvement, and medicinal applications will evolve more quickly as RNAi research becomes more effective due to the ongoing development of AI technology and databases.

NEW AI PLATFORMS COMBINING PHENOTYPIC OUTPUTS WITH OMICS DATA

More and more, new AI platforms are being created to use these linked information to more effectively forecast, optimize, and control a range of biological processes.

1. AI-powered Integration of Omics Data

High-dimensional (Sudha & Kathirvel, 2025) biological data from methods such as mass spectrometry and next-generation sequencing (NGS) is referred to as omics data. Although these datasets provide a thorough understanding of the molecular composition of an organism, their intricacy may be too great for conventional analytical techniques to handle. AI systems that combine phenotypic results with omics data are becoming more potent instruments to close this gap.

AI may be used to connect observable phenotypic parameters like growth rates, disease resistance, or agricultural yields with molecular profiles (transcriptomic, proteomic, and genomic). AI can determine the important genes, metabolites, or protein interactions that control phenotypic variation by examining these connections. To forecast and enhance crop performance, for instance, AI models may be built on plant omics data and phenotypic characteristics like drought resistance, disease tolerance, or nutrient absorption.

2. AI Examples and Platforms
    - DeepPhenotype: This artificial intelligence platform analyzes phenotypic data and connects it to underlying genetic variation. Deep learning models are used by the platform to automatically analyze high-throughput phenotypic data (such as photos of plant growth and disease symptoms) and correlate them with genetic data. This allows for the identification of genetic loci linked to desirable traits and offers insights into gene-environment interactions. Precision breeding and genetic modification of animals and crops may benefit from this platform.
    - Gene-phenotype association tools (e.g., GWAS and deep learning): To associate genetic polymorphisms with certain phenotypic features, genome-wide association studies (GWAS) and deep learning methods are being included into AI systems. Large datasets produced by GWAS and other omics technologies may be handled well by AI models, allowing for more precise predictions of complex features like crop production, disease resistance, or nutritional content. Large datasets may be sorted through by AI algorithms, which can also identify subtle genetic-phenotypic relationships that conventional approaches would overlook.
    - PhenoBox: The platform may assist guide choices on crop varieties that perform well in certain conditions and enhance breeding techniques by merging this phenotypic data with omics data. PhenoBox predicts how genetic changes (or particular environmental factors) will affect these phenotypic traits using AI-driven models. AI and CRISPR-based platforms: New AI platforms are also being combined with CRISPR-Cas9 gene-editing technology to optimize gene-editing strategy design. These systems discover certain genes or regulatory elements that may be al-

tered for desired qualities by combining phenotypic results with omics data (e.g., whole-genome sequencing). AI integration aids in prioritizing target genes according to their anticipated influence on phenotypic qualities like stress tolerance, yield enhancement, or disease resistance.

The following important RNAi datasets are accessible to the general public and may be used to train AI models:

1. The RNAiDB (RNAi Database)

A comprehensive database, RNAiDB lists information on RNA interference (RNAi) in a variety of species. It contains information on RNAi-related pathways, targeted genes, silencing effectiveness, and experimental settings. RNAiDB is a useful tool for training models that predict RNAi performance in various plant, animal, and human systems as researchers use it to determine how RNAi affects gene expression and phenotypic results.

- Content: RNAi pathways, target genes, and the effectiveness of gene silencing.
- Use in AI Models: By learning from previous RNAi studies and gene-target data, AI models may use RNAiDB data to forecast RNAi efficiency, off-target effects, and possible therapeutic results.

2. The Database for RNA Interference Screening (RNAi-SCREEN)

Large libraries of siRNAs are utilized to silence genes in human cells using RNAi-SCREEN, a platform for high-throughput RNAi screening that offers thorough phenotypic annotations. This dataset is particularly helpful for training models that seek to improve RNAi design and detect any off-target effects.

- Content: Experimental settings, phenotypic responses (such as cell viability and morphological changes), and gene silencing data.
- Application in AI Models: RNAi-SCREEN data may be utilized to train AI models that anticipate phenotypic consequences, optimize RNAi sequences, and evaluate the efficacy of silencing.

3. Datasets for siRNA Design and Prediction Tools

Curated information on siRNA design, sequence efficiency, and target predictions may be found in a number of publicly accessible datasets from RNAi design programs including siDirect, RNAiDesign, and sRNATarget. These files provide data on the silencing efficacy of siRNA sequences that target certain genes in plants and animals.

- Content: target genes, silencing effectiveness, off-target effects, and siRNA sequences.
- Use in AI Models: These datasets may be used to train AI models that forecast siRNA effectiveness, reduce off-target effects, and improve RNAi construct design for a range of applications.

4. The GDC or Genomic Data Commons

A plethora of information about RNAi research connected to cancer may be found at the GDC, which is run by the National Cancer Institute. Gene expression levels, silencing effectiveness, RNA-seq data, and related phenotypic data are all included in this collection. It provides useful information on changes in gene expression after RNAi therapies, despite its primary emphasis on cancer.

- Content: RNA-seq data, post-RNAi gene expression data, and RNAi targets associated with cancer.
- Use in AI Models: This dataset may be used to train AI models that identify important genes involved in the course of cancer and forecast the results of gene silencing for cancer.

FUTURE DIRECTIONS

With promising prospects for improving crop protection, disease control, and overall agricultural output, artificial intelligence (AI) has the potential to further RNA interference (RNAi) technology. New and creative methods are being developed as biotechnology and AI advance. Some of the major future paths that will influence the development of AI-driven RNAi technologies and their uses in medicine and agriculture are listed below.

Generative AI's Potential for Creating Innovative RNA Structures

The generation of novel RNA constructs, where AI models can independently produce new RNA sequences that are tailored for silencing certain genes, is a promising use of generative AI. By learning from enormous databases of current RNA sequences and their corresponding silencing effectiveness, generative AI might significantly expedite the trial-and-error process that is generally necessary for traditional RNAi design techniques.

Generative models, such variation autoencoders (VAEs) or Generative Adversarial Networks (GANs) (Gaikwad et al., 2025), may produce unique RNA structures with the required properties, including low off-target effects, high specificity, and stability. These models might search a much wider range of sequences and find RNA sequences that are resistant to environmental changes and very good at silencing target genes. Generative AI (Vanidhasri & Kathirvel, 2025) has the potential to greatly speed up the development of RNAi-based solutions in gene therapy and agriculture by producing RNA libraries customized for certain crops, illnesses, or infections.

CONCLUSION

An innovative development in agricultural biotechnology is the combination of artificial intelligence (AI) with plant RNA interference (RNAi) techniques. Artificial Intelligence is changing crop protection, disease resistance, and gene control by improving RNAi technology. AI models are enhancing the design of RNAi constructs, guaranteeing improved targeting, and reducing off-target effects by using AI's skills in data-driven RNA sequence optimization. The effectiveness and efficiency of RNAi applications in crops might be greatly increased by AI's capacity to forecast RNAi efficacy, improve delivery methods, and create highly precise RNA sequences, leading to more ecologically friendly and sustainable solutions. AI and RNAi simultaneously have vast possibility to solve some of the most important issues facing agriculture today, such as improving crop attributes, managing pests, and developing virus resistance. In order to ensure that agricultural systems can fulfil the needs of a rising global population while reducing environmental damage, artificial intelligence (AI) may help forecast RNAi targets, optimize genetic constructs, and design innovative delivery mechanisms that can respond to evolving threats. Additionally, this strategy has the potential to produce crops which are resilient to pests, diseases, and climate change, which would increase food security and promote sustainable farming methods. However, multidisciplinary cooperation between plant biologists, AI researchers, and agritech entrepreneurs is required to fully exploit the promise of AI-enabled RNAi. In order to overcome the present constraints, such as data scarcity, species variety, and regulatory obstacles, collaboration across these sectors will be crucial. Together, these professionals can improve RNAi structures, hone AI models, and guarantee the scalability and safety of these technologies. Facilitating the implementation of RNAi-based solutions and guaranteeing their responsible and broad usage in agriculture would also require promoting partnerships between academic institutions, business (Naren & Saravanan, 2025), and regulatory agencies. To sum up, combining AI with RNAi technology is a significant step in transforming agricultural biotechnology. RNAi-based crop improvement and protection technologies will be crucial in finding the control of sustainable agriculture in years to come, provided that AI continues to progress and that specialists from other disciplines work together more closely. This innovative partnership between artificial intelligence and plant biology will open the door to more intelligent and productive agricultural methods, which will eventually strengthen and maintain the world's food supply.

REFERENCES

Ahale, S., Rakhonde, G., Bhateja, S., Kuppuraj, J., & Mishra, S. (2024). Disease and pest control through advance technology. In K. Pandey, N. L. Kushwaha, C. B. Pande, & K. G. Singh (Eds.), *Artificial intelligence and smart agriculture* (Advances in Geographical and Environmental Sciences). Springer. https://doi.org/DOI: 10.1007/978-981-97-0341-8_21

Ayyaswamy, K. (2026). AI technologies for improving education systems. In Malahlela, M., & Sadiki, M. (Eds.), *Breaking structural inequality and enhancing social justice in higher education* (pp. 65–92). IGI Global Scientific Publishing., DOI: 10.4018/979-8-3373-3236-9.ch003

Ayyaswamy, K. N., & Vianny, M. (2025). Strategies for combating criminal use and abuse of artificial intelligence. In *Avoiding ad fraud and supporting brand safety: Programmatic advertising solutions*. IGI Global., DOI: 10.4018/979-8-3693-7041-4.ch009

Bhattacharya, D., Das, A., Satyaprakash, K., & Singh, M. K. (2025). Advanced biotechnological tools and techniques for muscle food identification. In Nimmanapalli, R., & Kumar Singh, M. (Eds.), *Trends in animal-based foods*. Springer., DOI: 10.1007/978-3-032-00575-5_9

Chaudhary, A., Chaudhary, P., Singh, S., Bhandari, G., & Tripathi, V. (2025). Advances in contemporary tools for detecting and diagnosing plant pathogens. In J.-T. Chen, M. Khan, A. Parveen, & J. K. Patra (Eds.), *Molecular and biotechnological tools for plant disease management* (Interdisciplinary Biotechnological Advances). Springer. https://doi.org/DOI. 10.1007/978-981-97-7510-1_9

Cosme, F., Vilela, A., & Gonçalves, B. (2025). Biotechnology in quality analysis and sensory evaluation. In Gaikwad, P. S., & Choudhary, P. (Eds.), *Biotechnological innovations in food processing*. Springer., DOI: 10.1007/978-3-031-96526-5_14

Dharshini, L. C. P., Salim, P., Sharanya, K., Palanisamy, B., Banerjee, S., & Mandal, A. K. A. (2025). RNAi: A technology for enhancing commercially valuable traits in industrial crops. In N. Kumar (Ed.), *Industrial crops improvement* (Sustainable Landscape Planning and Natural Resources Management). Springer. https://doi.org/DOI: 10.1007/978-3-031-75937-6_7

Fortune, A., Aime, A., Raymond, D., & Kumar, S. (2025). Nanotechnology in medicine: A double-edged sword for health outcomes. *Health Nanotechnology*, *1*(1), 9. DOI: 10.1186/s44301-025-00008-2

Gaikwad, P. S., Choudhary, P., Shaikh, K., Dange, D., & Haria, R. B. (2025). Introduction to biotechnology in food processing. In Gaikwad, P. S., & Choudhary, P. (Eds.), *Biotechnological innovations in food processing*. Springer., DOI: 10.1007/978-3-031-96526-5_1

Gurawalia, H., Shinde, P. B., & Sharma, K. K. (2024). Embracing Industry 4.0 ingredients in synthetic biology for the development of a carbon-neutral economy. In Chandel, A. K. (Ed.), *Biorefinery and Industry 4.0: Empowering sustainability (Green Energy and Technology)*. Springer., DOI: 10.1007/978-3-031-51601-6_8

. Habib, M., Singh, S., Jan, S., Jan, K., & Bashir, K. (2025). The future of the future foods: Understandings from the past towards SDG-2. npj Science of Food, 9(1), Article 138.

Haider, K., Abbas, D., Ullah, F., Ijaz, M., Ikram, M., Kabir, K., Khan, K. A., & Galian, J. (2026). Climate-driven insect pest outbreaks and food security risks: A review of adaptive strategies for resilient agriculture. *Journal of Pest Science*, *99*(1), 12. DOI: 10.1007/s10340-025-01985-0

Jangra, A., & Tiwari, S. (2025). Empowering nutrition: Biofortification to combat hidden hunger. In Pathak, H., Lakra, W. S., Gopalakrishnan, A., & Bansal, K. C. (Eds.), *Advances in agri-food systems*. Springer., DOI: 10.1007/978-981-96-0763-1_17

. Karki, R., Karki, K., Bashyal, A., Parajuli, R., Rai, D., Adhikari, A., ... & Ojha, P. (2025). Biobased Fortification: Current Challenges and Perspectives. *Sustainable Food Fortification: Biobased Approaches and Strategies*, 1-52.

Kathirvel, N., Gobinath, V. M., & Maheswaran, C. P. (2025). Effective leadership strategies for enhancing student success in higher education. In *New horizons in leadership: Inclusive explorations in health, technology, and education* (pp. 61–91). IGI Global., DOI: 10.4018/979-8-3693-6437-6.ch003

Kumar, N., & Khurana, S. P. (2025). An application of biopesticides in control of pest and crop protection: An eco-friendly management. In *Microbial biocontrol techniques: Importance in ensuring food security* (pp. 95–124). Springer Nature Singapore.

Madhavakumar, D., Nelson, A., Naren, A. R., & Prathap, S. (2025). An application of Pythagoras theorem: From Heron's formula to derive the foci of an ellipse. In *Educational AI humanoid computing devices for cyber nomads* (pp. 73–98). IGI Global., DOI: 10.4018/979-8-3693-8985-0.ch004

Naren, A., & Gunasekaran, K. (2025). Future cyber physical system security. In *Impact of digitalization on communication dynamics*. IGI Global., DOI: 10.4018/979-8-3693-3579-6.ch010

Naren, A., & Kaliappan, P. (2025). Mathematical approach: AI-based feature selection in machine learning. In *Serverless computing concepts in contemporary legal systems*. IGI Global., DOI: 10.4018/979-8-3693-7205-0.ch012

Naren, A., & Saravanan, N. (2025). Entrepreneurship and business culture in rural area. In *Digital disruption in hospitality: Sustainable hospitality: A global perspective on people, planet, and profit*. Emerald Publishing., DOI: 10.1108/978-1-83662-360-120251009

Rahman, M. U., Shah, J. A., Khan, M. N., Bilal, H., Zhu, D., Du, Z., & Mu, D. S. (2026). Innovative approaches to combat antimicrobial resistance: A review of emerging therapies and technologies. *Probiotics and Antimicrobial Proteins*, *18*(3), 3597–3614. DOI: 10.1007/s12602-025-10676-2 PMID: 41026406

Rithik, K., & Naren, A. K. (2024). The IoT of robotics: The frontier of automation. In *Artificial intelligence-enabled digital twin for smart manufacturing*. Scrivener Publishing/Wiley., DOI: 10.1002/9781394303601.ch13

Salem, K. F. M., & Eid, N. A. (2025). Modern breeding strategies and tools in bread wheat (*Triticum aestivum* L.). In K. F. M. Salem, J. M. Al-Khayri, & S. M. Jain (Eds.), *Breeding and biotechnology of grass and bast fiber crops* (Advances in Plant Breeding Strategies, Vol. 12). Springer. https://doi.org/DOI: 10.1007/978-3-032-00403-1_8

. Sudha, & Kathirvel, A. (2025). Reliable path selection in AODV: A QoS-based strategy for MANET. *Utilitas Mathematica, 122*(1), 780–794.

Ullah, F., Guru-Pirasanna-Pandi, G., Murtaza, G., Sarangi, S., Gul, H., Li, X., Chavarín-Gómez, L. E., Ramírez-Romero, R., Guedes, R. N. C., Desneux, N., & Lu, Y. (2025). Evolving strategies in agroecosystem pest control: Transitioning from chemical to green management. *Journal of Pest Science*, *98*(4), 2307–2324. DOI: 10.1007/s10340-025-01939-6

Vanidhasri, R., & Kathirvel, A. (2025, August). Improving EEG Signal Classification in Brain-Computer Interfaces via Grey Wolf Optimizer-based Deep Learning. In *2025 3rd International Conference on Sustainable Computing and Data Communication Systems (ICSCDS)* (pp. 1312-1318). IEEE.

Vianny, M., Ayyaswamy, K., Rithik, G., Naren, K., & Anthony Raj, A. (2025). Private AI: An overview. In *AI-driven breakthroughs in antimicrobial resistance*. IGI Global., DOI: 10.4018/979-8-3693-7550-1

Chapter 14

Antimicrobial Resistance Surveillance Using Artificial Intelligence and Big Data

Matthew Chidozie Ogwu
http://orcid.org/0000-0001-6054-1667
Appalachian State University, USA

Olugbemiga Ojo Aliu
http://orcid.org/0000-0003-4942-7454
Auchi Polytechnic, Auchi, Nigeria

ABSTRACT

This chapter examines how artificial intelligence (AI) and big data can enhance AMR surveillance by linking microbiology, genomics, metagenomics, electronic health records, antimicrobial use data, wastewater signals, animal health records, food-chain data, geospatial indicators, and public-health intelligence into a more anticipatory One Health surveillance architecture. It argues that AI should not be treated as a replacement for microbiology but as an interpretive layer that increases the speed, scale, and analytic depth of surveillance. The chapter reviews major data sources, machine-learning methods, genomic and metagenomic applications, predictive analytics, wastewater-based epidemiology, clinical decision support, stewardship integration, governance requirements, and implementation challenges in low-resource settings. It also emphasizes the risks of biased data, poor interoperability, weak external validation, algorithmic opacity, privacy breaches, false alarms, and inequitable surveillance capacity.

DOI: 10.4018/979-8-3373-9033-8.ch014

1.0. INTRODUCTION

Antimicrobial resistance is not a single pathogen problem. It is an evolving ecological, clinical, agricultural, environmental, and governance challenge in which microbial populations adapt under selective pressure created by antimicrobial use, infection prevention failures, environmental contamination, agricultural production systems, travel, trade, and health-system inequalities. The global burden analysis by Laxminarayan et al. (2013) emphasized the need for global AMR solutions, and Murray et al. (2022) estimated that bacterial AMR was associated with millions of deaths in 2019, reinforcing the need to treat AMR as a major public-health and biosurveillance priority rather than a narrow hospital microbiology issue. The World Health Organization (WHO, 2015) similarly placed surveillance and evidence generation at the center of the Global Action Plan on Antimicrobial Resistance, while the Global Antimicrobial Resistance and Use Surveillance System (GLASS) was developed to strengthen standardized reporting across countries (WHO, 2022). AMR surveillance is difficult because resistance does not spread through one pathway. A resistant strain may emerge in a hospital, circulate silently in the community, be selected through agricultural antimicrobial use, move through food systems, enter wastewater, persist in environmental reservoirs, or travel across borders with humans, animals, and commodities. This cross-sectoral framing aligns with wider One Health scholarship emphasizing that biological risks emerge through interactions among food systems, environmental contamination, agro-aqua production, public health, and ecological change (Izah & Ogwu, 2025; Izah et al., 2024; Ogwu et al., 2024). Resistance genes may be carried on plasmids and other mobile genetic elements, allowing horizontal transfer across bacterial lineages and ecological compartments (Davies & Davies, 2010; Martínez, 2008). Surveillance must therefore monitor organisms, genes, drugs, people, animals, environments, practices, and institutions. This makes AMR one of the clearest examples of why biosurveillance requires a One Health architecture.

Artificial intelligence and big data are increasingly proposed as solutions to AMR surveillance gaps. Machine learning can classify resistance phenotypes from genomic features, detect unusual clusters, forecast resistance trends, prioritize laboratory testing, extract relevant information from clinical notes, identify high-risk patients or facilities, and integrate heterogeneous signals across sectors (Ali et al., 2023; Ren et al., 2022; Rušić et al., 2024). Big data infrastructures can connect laboratory information systems, electronic health records, pharmacy records, genomic databases, veterinary surveillance, food-chain monitoring, and wastewater metagenomics. Yet the central argument of this chapter is deliberately cautious: AI is not a substitute for microbiology, epidemiology, or stewardship. Its value lies in augmenting verified surveillance systems by improving integration, interpretation, timeliness, and decision support. Despite substantial advances in artificial intelligence, genomic surveillance, and antimicrobial stewardship, existing literature remains fragmented. Most studies focus on isolated technological applications, individual pathogens, or healthcare settings, with limited integration across clinical, veterinary, food, and environmental surveillance systems. Furthermore, relatively few conceptual studies synthesize AI, big data, genomics, environmental surveillance, and One Health governance into a unified biosurveillance framework. This chapter addresses this gap by proposing an integrated conceptual architecture that connects these domains within a practical decision-support framework for antimicrobial resistance surveillance.

The chapter aims to develops an advanced framework for AMR surveillance using AI and big data. It begins with the foundations of AMR surveillance, moves through the data ecosystem and analytical methods, evaluates genomic and metagenomic applications, and then examines early warning systems, One Health integration, stewardship, governance, and implementation barriers. Throughout the chapter, the emphasis is on confirmed evidence, practical decision-making, and responsible use of AI in biosurveillance.

2.0. CONCEPTUAL FOUNDATIONS OF AMR SURVEILLANCE

AMR surveillance is the systematic collection, analysis, interpretation, and dissemination of data on antimicrobial resistance and antimicrobial use to guide public-health action, clinical practice, infection prevention, stewardship, and policy. At its core, surveillance connects microbiological evidence to decisions. It should answer practical questions: which pathogens are becoming resistant, to which drugs, in which populations, in which places, through which routes, and with what consequences for treatment, transmission, and prevention?

Phenotypic antimicrobial susceptibility testing remains the operational backbone of AMR surveillance. Minimum inhibitory concentrations, disk diffusion results, and standardized interpretive criteria provide clinically meaningful information about whether a pathogen is susceptible, intermediate, or resistant to a specific antimicrobial. Phenotypic data are essential because genotype does not always predict phenotype perfectly. Resistance expression can depend on gene regulation, gene copy number, permeability, efflux, inoculum effects, species background, testing method, and local interpretive standards. Genotypic surveillance, especially whole-genome sequencing, adds major value by identifying resistance genes, mutations, plasmids, outbreak relationships, and transmission pathways, but it must be interpreted alongside phenotypic and epidemiological evidence (Ellington et al., 2017; Quainoo et al., 2017).

AMR surveillance systems can be passive, active, sentinel, laboratory based, case based, syndromic, event based, or environmental. Passive systems depend on routinely generated laboratory reports and are often easier to sustain but vulnerable to under-detection and biased sampling. Active systems collect data according to predefined protocols and can produce higher quality evidence but require greater resources. Sentinel systems monitor selected sites and are useful where national coverage is not feasible. Case-based systems connect microbiological findings with patient outcomes, exposure history, and treatment data, providing richer information for action (Mo et al., 2023). Wastewater and environmental surveillance complement clinical systems by capturing population-level resistome signals, including from people who are not sampled clinically (Tiwari et al., 2022).

The One Health perspective expands the surveillance unit from the patient to the interconnected system. Human medicine, veterinary medicine, livestock production, aquaculture, food processing, wastewater, soil, surface water, wildlife, and antimicrobial manufacturing can all influence resistance emergence and spread (Thakur & Gray, 2019). Integrated One Health surveillance therefore requires data interoperability across sectors that have different sampling traditions, regulatory mandates, laboratory standards, incentives, and data-governance cultures. This is precisely where big data systems and AI can contribute, provided they are built on standardized, trusted, and ethically governed data streams (Aenishaenslin et al., 2021; Delpy et al., 2024).

3.0. DATA SOURCES FOR AI-ENABLED AMR SURVEILLANCE

AI-enabled AMR surveillance depends on the quality, coverage, and interoperability of underlying data. The most important data streams include clinical microbiology results, electronic health records, pharmacy and antimicrobial consumption data, infection prevention records, genomic and metagenomic sequences, wastewater measurements, veterinary and food-chain data, environmental indicators, and contextual information such as mobility, climate, land use, and socioeconomic conditions. Each source provides a partial view of the resistance landscape. The purpose of big data integration is not to merge everything indiscriminately, but to connect complementary signals in ways that improve detection, attribution, forecasting, and intervention.

Clinical microbiology laboratories produce organism identification, specimen source, susceptibility results, minimum inhibitory concentrations, resistance phenotypes, and sometimes molecular test results. These data are close to clinical action, but they are biased toward patients who seek care, are sampled, and have access to diagnostic services. Electronic health records add demographics, diagnoses, comorbidities, treatments, hospitalization history, device use, travel history, and outcomes. Pharmacy data provide antimicrobial exposure, prescription patterns, duration, route, and stewardship signals. Together, laboratory, EHR, and pharmacy data can support facility-level risk prediction, empirical therapy guidance, outbreak detection, and stewardship evaluation (Kelly et al., 2019; Pennisi et al., 2025).

Genomic data provide higher resolution. Whole-genome sequencing can identify resistance determinants, strain relatedness, plasmid content, virulence factors, and transmission clusters. It is particularly valuable for pathogens such as carbapenem-resistant Enterobacterales, methicillin-resistant Staphylococcus aureus, vancomycin-resistant Enterococcus, multidrug-resistant tuberculosis, Salmonella, and Neisseria gonorrhoeae. Metagenomic data from wastewater, soil, food systems, and animal environments can characterize resistomes without requiring culture of every organism. However, metagenomics detects genes and sequence fragments, not necessarily viable pathogens or clinically expressed resistance. Its outputs require careful normalization, quality control, and interpretation (Duarte et al., 2020; Hendriksen et al., 2019; Tiwari et al., 2022). Food-chain surveillance is particularly important in poultry systems, where antibiotic residues may create direct public health risks and indicate broader antimicrobial-use practices that shape resistance selection pressure (Izah et al., 2025).

Veterinary, livestock, aquaculture, and food-chain data are essential because antimicrobial use and resistance selection occur beyond human healthcare. Surveillance may include animal clinical isolates, slaughterhouse sampling, retail meat testing, farm antimicrobial use, aquaculture treatments, manure management, and foodborne pathogen monitoring. Environmental data include wastewater treatment plants, hospital effluents, surface water, sediments, agricultural runoff, and antimicrobial residues. These data help identify reservoirs, selection pressures, and transmission interfaces that would remain invisible in hospital-only surveillance (Singh et al., 2021; Thakur & Gray, 2019). The effectiveness of AI-enabled AMR surveillance depends fundamentally on the quality, completeness, and interoperability of the underlying datasets. Missing metadata, inconsistent laboratory standards, duplicate records, reporting delays, selection bias, and heterogeneous data formats can substantially reduce predictive performance and introduce systematic errors into AI models. Surveillance systems must therefore prioritize rigorous data governance, standardized metadata collection, quality assurance procedures, and continuous validation before analytical models are deployed. Investment in data quality often provides greater improvements in surveillance performance than increasing algorithmic complexity.

Digital event-based data can complement formal systems by detecting early signals from reports, publications, media, outbreak notices, and public-health alerts. Natural language processing can extract pathogen-drug-location relationships from unstructured documents, but such signals must be verified. In AMR surveillance, false alarms can distort policy and clinical behavior, while missed signals can delay response. AI systems should therefore treat informal data as triage inputs rather than definitive evidence. Figure 1 illustrates a simplified conceptual framework of an AI-enabled One Health AMR surveillance platform that integrates diverse, multi-sectoral data streams to support evidence-based public health decision-making. The platform synthesizes information from human health, animal health, genomics and metagenomics, antimicrobial stewardship, food systems, wastewater and environmental surveillance, and mobility, climate, and socioeconomic datasets. By fusing these complementary data sources through artificial intelligence, the system enables integrated early warning, source attribution, predictive forecasting, and decision support for AMR surveillance. This systems-based architecture highlights the importance of cross-sectoral data integration in improving the detection, monitoring, and management of antimicrobial resistance while strengthening preparedness and response across the One Health continuum.AI adds value by integrating verified human, animal, food, wastewater, environmental, genomic, antimicrobial-use, and contextual data into actionable surveillance intelligence (Figure 1).

Figure 1. Conceptual framework for an AI-enabled one health antimicrobial resistance surveillance platform

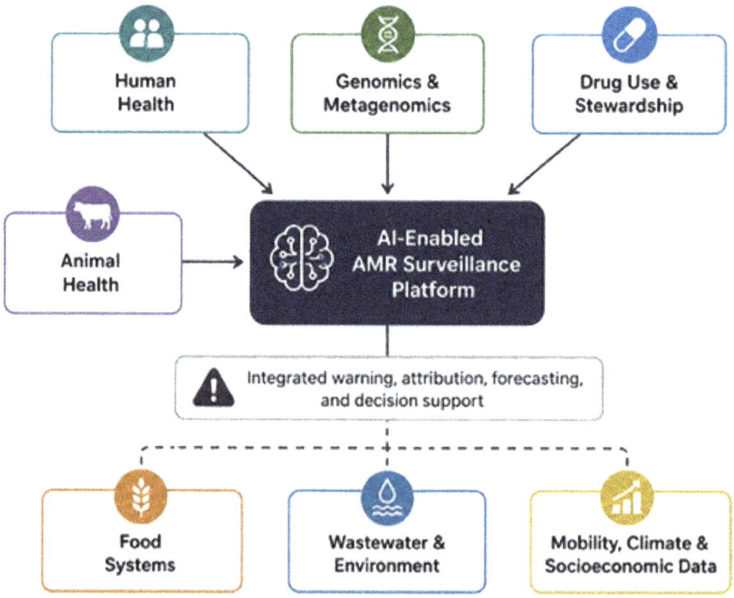

4.0. BIG DATA INFRASTRUCTURE: FROM FRAGMENTED RECORDS TO SURVEILLANCE INTELLIGENCE

Big data in AMR surveillance is not defined only by volume. It is defined by heterogeneity, velocity, linkage, provenance, and decision value. A national AMR platform may need to connect microbiology laboratories, hospital systems, outpatient clinics, pharmacies, veterinary laboratories, agricultural agencies, wastewater utilities, genomic sequencing centers, and public-health authorities. The technical challenge is substantial, but the governance challenge is greater. Data must be standardized enough for comparison, secure enough to protect privacy, and actionable enough to change decisions.

Interoperability begins with common data elements. At minimum, AMR surveillance records should capture organism identification, specimen type, date, location, susceptibility method, interpretive standard, antimicrobial tested, patient or host category, duplicate status, and relevant metadata. For genomic data, metadata should include sequencing platform, assembly quality, reference database, resistance-calling pipeline, coverage thresholds, sample source, and epidemiological context. For wastewater and environmental data, metadata should include sampling method, catchment population, flow conditions, treatment stage, extraction method, sequencing depth, normalization approach, and potential point sources. Without such metadata, AI models may learn artifacts rather than resistance biology.

Data harmonization also requires de-duplication and denominator logic. Counting every isolate can overrepresent patients with prolonged infections or intensive testing. Counting only the first isolate per patient may improve comparability but can miss within-patient resistance evolution. Resistance proportions can be misleading without information on testing practices, sampling criteria, clinical syndrome, and population denominator. AI models trained on inconsistent denominators will produce unstable outputs. Surveillance architecture must therefore encode analytic rules rather than leaving them implicit.

Cloud platforms and federated systems offer different advantages. Centralized data lakes can support rapid integration and large-scale analytics but raise concerns about privacy, sovereignty, and governance. Federated learning can allow models to learn across institutions without moving raw patient-level data, which is attractive for privacy-preserving AMR analytics (Rieke et al., 2020). However, federated systems still require harmonized variables, shared model specifications, bias auditing, and governance agreements. The technical promise does not remove the need for institutional trust.

The ultimate output of big data infrastructure should be surveillance intelligence, not simply dashboards. Dashboards can show resistance trends, maps, and alerts, but intelligence requires interpretation: whether a signal is real, what is driving it, which populations are at risk, what interventions are feasible, and how success will be measured. AI should support this interpretation by quantifying uncertainty, ranking hypotheses, identifying data gaps, and linking signals to response protocols. Effective AMR surveillance depends on integrating heterogeneous data sources that capture the biological, clinical, environmental, and societal dimensions of resistance emergence and spread (Table 1). Each data domain contributes unique insights into AMR dynamics, ranging from pathogen detection and antimicrobial consumption to environmental reservoirs and contextual drivers. However, individual datasets also possess inherent limitations related to sampling bias, data quality, interoperability, privacy, and representativeness. Table 1 summarizes the principal surveillance domains, representative data types, their contributions to integrated AMR surveillance, and the major challenges that must be addressed to develop robust, AI-enabled One Health surveillance systems.

Table 1. Major data domains for AI-enabled AMR surveillance.

Data domain	Examples	Surveillance value	Major limitation
Clinical microbiology	Organism identification, AST results, MICs, specimen source	Defines clinically relevant resistance and supports treatment and outbreak detection	Biased toward tested patients and variable laboratory access
Electronic health records	Diagnosis, comorbidities, admissions, devices, outcomes	Adds patient risk factors and outcome context	Privacy concerns, missingness, coding heterogeneity
Pharmacy and antimicrobial use	Prescriptions, consumption, duration, stewardship interventions	Links selection pressure to resistance trends	May lack indication and adherence data
Genomics and metagenomics	WGS, plasmids, ARGs, wastewater resistomes	Detects mechanisms, transmission, and environmental reservoirs	Genotype-to-phenotype uncertainty and database bias
Animal and food systems	Veterinary isolates, farm use, slaughterhouse and retail food testing	Supports One Health source attribution and regulation	Uneven sampling and sectoral data silos
Contextual data	Geography, mobility, sanitation, climate, land use	Improves risk mapping and forecasting	Risk of ecological fallacy and confounding

The development of interoperable AMR surveillance systems should be guided by the FAIR data principles, which emphasize that data should be Findable, Accessible, Interoperable, and Reusable. FAIR-compliant datasets facilitate reproducibility, improve collaboration across institutions, enhance machine learning performance, and support long-term sustainability of surveillance programs. Adherence to internationally recognized metadata standards also promotes data sharing across human, animal, environmental, and food surveillance systems while preserving appropriate governance and privacy protections.

5.0. AI AND MACHINE-LEARNING METHODS FOR AMR SURVEILLANCE

Machine learning can support AMR surveillance at several levels: classification, prediction, anomaly detection, clustering, natural language extraction, network analysis, geospatial risk mapping, and decision support. Classical algorithms such as logistic regression, random forests, support vector machines, gradient boosting, and regularized regression remain useful because they are interpretable and perform well on structured data. Deep learning can model complex sequence, image, text, and temporal data but requires larger datasets and careful validation. In AMR surveillance, the most advanced method is not automatically the best method. Fit-for-purpose modeling is more important than algorithmic novelty.

Genomic resistance prediction is one of the most active areas. Models may use known resistance genes, point mutations, k-mers, pan-genome features, plasmid markers, or combinations of genomic signals to predict phenotypic resistance. Ren et al. (2022) reviewed machine-learning approaches for predicting AMR from whole-genome sequencing, while organism-specific studies have applied machine learning to Escherichia coli, Pseudomonas aeruginosa, Helicobacter pylori, Actinobacillus pleuropneumoniae, and other pathogens (Liu et al., 2020; Moradigaravand et al., 2018; Noman et al., 2023; Yu et al., 2024). These methods can accelerate interpretation, but their performance varies by pathogen, drug, resistance mechanism, sample size, population structure, and the quality of phenotypic labels.

Anomaly detection is especially relevant to biosurveillance. A system may flag an unusual rise in carbapenem resistance, a new resistance gene in wastewater, a cluster of genetically related isolates across hospitals, or an unexpected resistance phenotype in a pathogen-drug pair. Such alerts require calibration because low thresholds produce alarm fatigue, while high thresholds delay detection. AI systems should combine statistical anomaly detection with microbiological plausibility, known outbreak definitions, laboratory confirmation, and public-health review. Natural language processing can extract AMR-relevant information from clinical notes, discharge summaries, laboratory comments, outbreak reports, stewardship notes, and scientific literature. NLP may identify prior colonization, travel, device exposure, antimicrobial intolerance, infection syndrome, or informal outbreak descriptions. Yet clinical language is variable, context dependent, and prone to ambiguity. A note stating that a patient had 'no history of MRSA' must not be interpreted as MRSA colonization. NLP pipelines require careful annotation, validation, and human review before operational use.

Network analysis can identify transmission relationships among patients, facilities, farms, wastewater catchments, laboratories, and geographic regions. When combined with genomic data, network models can help distinguish clonal spread from repeated emergence or plasmid-mediated dissemination. Geospatial AI can map AMR hotspots and link resistance to antibiotic use, sanitation, healthcare access, livestock density, environmental contamination, climate, and mobility. Such analyses are valuable for prioritization, but they can also stigmatize communities if communicated without context. Surveillance maps should be accompanied by uncertainty, denominator information, and structural explanations rather than implying blame. Surveillance and Decision Support responsible pipeline links data ingestion, harmonization, feature engineering, modeling, validation, action, and feedback (Figure 2). Figure 2 presents a simplified end-to-end workflow illustrating the analytical pipeline underpinning an AI-enabled AMR surveillance system. The process begins with the integration of heterogeneous data sources—including antimicrobial susceptibility testing (AST), whole-genome sequencing (WGS), environmental DNA (eDNA), and electronic health records (EHRs)—followed by data quality assurance, harmonization, and feature engineering to generate robust analytical inputs (Figure 2). Predictive models are subsequently developed and rigorously validated using external datasets, uncertainty quantification, and bias assessment before informing public health actions such as outbreak alerts, antimicrobial stewardship interventions, infection prevention and control (IPC), and policy development. Importantly, a continuous feedback loop allows surveillance outcomes and intervention effectiveness to refine subsequent model development, supporting adaptive learning and continuous improvement of AI-driven AMR surveillance

Figure 2. AI-AMR surveillance pipeline - end-to-end workflow for AI-enabled antimicrobial resistance

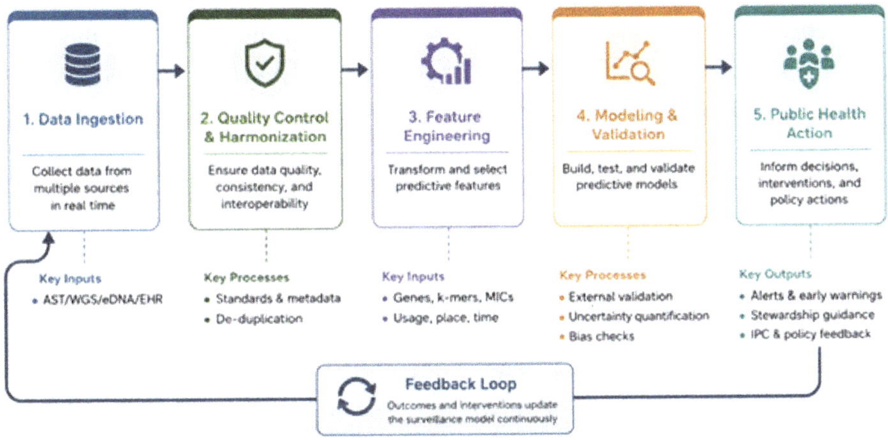

6.0. GENOMIC AND METAGENOMIC SURVEILLANCE

Whole-genome sequencing has transformed pathogen surveillance by providing high-resolution information on strain relatedness, resistance determinants, virulence factors, plasmids, and transmission pathways. For AMR, WGS can identify known genes such as blaCTX-M, blaKPC, blaNDM, blaOXA variants, mecA, vanA, mcr genes, and mutations associated with fluoroquinolone, rifampicin, macrolide, or tuberculosis drug resistance. It can also distinguish closely related isolates during outbreak investigations and help determine whether resistance is spreading through clonal expansion, plasmid transfer, or repeated independent acquisition (Ellington et al., 2017; Quainoo et al., 2017).

The genotype-to-phenotype relationship remains a central challenge. Some resistance mechanisms are well characterized, such as mecA-mediated methicillin resistance in Staphylococcus aureus or specific mutations in Mycobacterium tuberculosis. Others are more complex, involving combinations of genes, regulatory changes, efflux pumps, porin loss, enzyme expression, or species-specific background effects. A gene detected in a genome may not be expressed, may be truncated, or may not produce clinically relevant resistance. Conversely, resistance may occur through novel mechanisms absent from reference databases. Therefore, genomic surveillance should be tied to phenotypic validation, curated databases, and periodic reassessment of interpretive rules.

Metagenomic surveillance expands the lens from isolates to communities. Wastewater, animal facilities, soils, sediments, and food-processing environments contain mixed microbial communities with diverse resistance genes. Metagenomics can detect ARGs across populations and environments without culturing individual organisms. Hendriksen et al. (2019) demonstrated the potential of global sewage surveillance for mapping AMR genes, and systematic reviews have highlighted wastewater surveillance as a complementary tool for AMR monitoring (Tiwari et al., 2022). Hospital wastewater may be particularly informative because it reflects antimicrobial use, high-risk patients, and healthcare-associated pathogens, although it may also contain strong point-source biases.

Metagenomic AMR surveillance must address several technical issues: sampling representativeness, DNA extraction bias, sequencing depth, read classification, reference database completeness, gene-length normalization, host assignment, and distinguishing gene abundance from viable transmission risk. A high abundance of a resistance gene in wastewater does not automatically mean clinical risk is increasing. It may reflect shifts in microbial biomass, hospital discharge, industrial contamination, antibiotic residues, or sampling conditions. AI can help integrate metagenomic signals with clinical and environmental data, but the interpretation must remain microbiologically grounded. Future AMR surveillance will increasingly benefit from integrating multiple layers of biological information beyond genomic sequencing alone. Transcriptomics provides insights into gene expression, proteomics characterizes functional protein activity, metabolomics captures biochemical responses, while pangenomic analyses improve understanding of strain diversity and accessory genomes. Integrating these complementary datasets through AI may improve resistance prediction, identify emerging resistance mechanisms, and strengthen understanding of microbial adaptation within complex ecological systems.

Long-read sequencing and hybrid assembly are increasingly important because many resistance genes occur on plasmids and other mobile genetic elements. Short reads may detect genes but fail to reconstruct their genomic context. Long reads can help determine whether resistance genes are linked to plasmids, virulence factors, or host-specific markers. This matters for surveillance because the same resistance gene can have different risk implications depending on whether it is chromosomal, plasmid-borne, associated with a high-risk clone, or present in a low-abundance environmental organism. Table 2 summarizes major analytical approaches used in AI-enhanced antimicrobial resistance biosurveillance, linking each method to typical input data, operational applications, and key implementation cautions. Table 2 emphasizes that AMR analytics range from rule-based interpretation of known resistance genes and mutations to machine learning, deep learning, anomaly detection, natural language processing, and geospatial modelling. Whole-genome sequencing and machine learning are increasingly used to predict resistance phenotypes and support surveillance, but their reliability depends on data quality, sampling design, phenotype labels, model validation, and biological interpretability (Liu et al., 2020). AMR surveillance also requires a One Health perspective because resistance emerges and circulates across human, animal, agricultural, and environmental systems, making integrated genomic, clinical, ecological, and spatial data essential for effective monitoring and intervention (Kahn, 2017).

Table 2. AI methods used in AMR surveillance and their operational cautions.

Analytical approach	Typical input	Common AMR application	Operational caution
Rule-based resistance calling	Known ARGs and mutations	Rapid interpretation of WGS using curated databases	Misses novel or context-dependent mechanisms
Classical machine learning	Genes, k-mers, MIC labels, metadata	Prediction of resistance phenotypes	Can learn sampling structure rather than causal biology
Deep learning	Sequences, spectra, images, text	Complex feature extraction and pattern recognition	Requires large datasets and careful validation
Anomaly detection	Time series, facility trends, gene abundance	Early warning for emerging resistance clusters	Needs calibrated thresholds to avoid alert fatigue
NLP	Clinical notes, reports, literature	Extraction of exposure, syndrome, and outbreak information	Ambiguity and documentation bias
Geospatial modeling	Resistance, location, usage, environment	Hotspot mapping and prioritization	Requires denominators and careful communication

7.0. PREDICTIVE ANALYTICS AND EARLY WARNING SYSTEMS

AMR surveillance becomes more powerful when it moves from retrospective description to anticipatory warning. Predictive analytics can forecast resistance proportions, identify facilities at risk of outbreaks, estimate future treatment failure risk, detect emergence of rare phenotypes, and prioritize laboratory or infection-control resources. Forecasting can be conducted at multiple levels: patient, ward, hospital, region, pathogen, drug, or environmental catchment. The goal is not to predict the future perfectly, but to provide decision-makers with earlier, more structured warnings. Environmental AMR surveillance should also be linked to broader microbial water-quality monitoring, since contaminated water systems can function as exposure pathways for pathogens and resistance-associated microbial communities, particularly in under-resourced regions where sanitation and treatment infrastructure may be limited (Izah & Ogwu, 2025).

Clinical risk prediction models may estimate the probability that a patient has an infection caused by a resistant organism based on prior microbiology, recent antibiotic exposure, hospitalization, residence in long-term care, travel, devices, comorbidities, and local resistance prevalence. Such models can support empirical therapy decisions, but they must be externally validated and integrated carefully into antimicrobial stewardship. Overly broad use of risk scores could increase unnecessary use of last-line antibiotics, which would worsen selection pressure. Therefore, decision-support models should be evaluated not only for predictive accuracy but for clinical outcomes, antibiotic consumption, resistance trends, and equity. The agro-aqua food production chain is a particularly important AMR surveillance interface because chemical contaminants, microbial hazards, aquaculture practices, food-processing environments, and public health outcomes intersect within the One Health triad (Ogwu et al., 2024).

Population-level forecasting uses time series of resistance, antimicrobial consumption, hospitalization, mobility, wastewater signals, and environmental conditions. Forecasting models may include autoregressive methods, Bayesian hierarchical models, gradient boosting, recurrent neural networks, or hybrid mechanistic-statistical approaches. In surveillance, interpretability and robustness often matter more than marginal improvements in accuracy. A transparent model that identifies rising resistance and gives credible uncertainty intervals may be more useful than a black-box model with slightly better retrospective performance. Early warning systems require defined triggers. A trigger may be a statistically unusual increase in resistance, detection of a critical gene, identification of a novel organism-drug combination, sudden wastewater emergence of a carbapenemase, or genomic evidence of cross-facility transmission. Each trigger should be linked to a response pathway: confirmatory testing, isolate referral, sequencing, infection-control review, stewardship audit, environmental investigation, public communication, or policy action. Without response pathways, AI alerts become noise.

Predictive analytics should also incorporate uncertainty. Model outputs should communicate confidence, data completeness, population coverage, and known limitations. Public-health users need to know whether an alert reflects a strong signal across multiple data streams or a weak signal from sparse data. AI systems should be designed to increase situational awareness, not to hide uncertainty behind a single risk score.

8.0. ONE HEALTH APPLICATIONS ACROSS HUMAN, ANIMAL, FOOD, AND ENVIRONMENTAL SYSTEMS

AMR is a One Health problem because antimicrobial selection and resistance transmission occur across humans, animals, food systems, and environments. AI-enabled surveillance can help integrate these domains, but integration must be purposeful. The objective is not simply to place all data into one platform; it is to understand cross-sector transmission, shared selection pressures, and intervention points. In human healthcare, AI-AMR surveillance can support hospital antibiograms, outbreak detection, empirical therapy guidance, infection prevention, screening strategies, and stewardship audits. Facility-level models can identify wards with rising resistance, detect unusual organism-drug patterns, or link resistance trends to antimicrobial prescribing and device-associated infection rates. Genomic surveillance can identify transmission clusters that traditional epidemiology misses. However, hospital data are biased toward severe illness and may not represent community resistance.

In animal health and agriculture, AI systems can integrate antimicrobial use, veterinary diagnoses, farm management, animal movement, slaughterhouse testing, foodborne pathogen isolates, and environmental runoff. Such systems can identify high-risk production practices, track resistant Salmonella, Campylobacter, Escherichia coli, or Enterococcus, and support more targeted stewardship in livestock and aquaculture. They must also avoid simplistic narratives that blame farmers without considering market pressures, veterinary access, animal welfare, biosecurity, and regulatory contexts. Surveillance should also inform prevention-oriented research on antimicrobial alternatives and complementary control strategies, including natural antimicrobial products whose effectiveness and safety require rigorous evaluation before public health application (Ogwu & Izah, 2025a).

Food-chain surveillance links animal reservoirs and human exposure. Retail meat testing, genomic comparison of foodborne isolates, outbreak investigations, and supply-chain data can identify routes by which resistant pathogens reach consumers. AI can support source attribution by combining genomic, temporal, geographic, and trade data. However, source attribution is probabilistic and should not be overstated when sampling is uneven across sectors. In agricultural biosurveillance, pathogen monitoring increasingly intersects with disease-prevention technologies, including nanotechnology-based strategies for controlling fungal crop pathogens and reducing losses that may otherwise increase chemical and antimicrobial intervention pressures (Ogwu & Izah, 2025b). Environmental biosurveillance also benefits from quantitative risk-assessment traditions developed for pollutant monitoring, where statistical interpretation, exposure context, spatial variability, and uncertainty are central to translating environmental measurements into management decisions (Ogwu et al., 2025a).

Environmental surveillance captures the broader ecology of AMR. Wastewater, river systems, soils, sediments, wildlife, and antimicrobial manufacturing discharge can all contain resistance genes and resistant organisms. Wastewater-based surveillance is attractive because it is population-level, relatively passive, and can include underserved populations that are not well represented in clinical data (Tiwari et al., 2022). Yet environmental resistomes are complex. Surveillance must distinguish background environmental resistance, human-associated resistance, clinically critical genes, and viable transmission pathways. A mature One Health AMR platform would connect these sectors through shared metadata, harmonized laboratory standards, genomic comparability, and decision-relevant indicators. Canada's AMRNet and other data-driven One Health initiatives illustrate the movement toward integrated surveillance, although implementation remains uneven (Rudnick et al., 2022). The major barrier is not only technology. It is the alignment of mandates, data governance, funding, workforce capacity, and accountability across sectors.

9.0. AMR STEWARDSHIP AND POLICY DECISION SUPPORT

Surveillance is only valuable if it informs action. In AMR, the most important actions include antimicrobial stewardship, infection prevention and control, vaccination, sanitation, diagnostic access, regulation of antimicrobial use, environmental controls, and public-health response. AI can support these actions by translating surveillance data into risk stratification, prescribing guidance, targeted alerts, resource prioritization, and evaluation of intervention impact. Antimicrobial stewardship seeks to optimize antimicrobial use to improve patient outcomes while reducing unnecessary selection pressure. AI can support stewardship by identifying inappropriate prescribing, predicting likely resistance, recommending narrower therapy when susceptibility data become available, detecting prolonged courses, and prioritizing cases for pharmacist or infectious-disease review (Pennisi et al., 2025). However, stewardship algorithms must be carefully governed. If models are trained on historically excessive broad-spectrum prescribing, they may reproduce poor practice. If they prioritize only individual treatment risk, they may ignore population-level resistance consequences.

Policy decision support operates at a broader level. National action plans need information on priority pathogens, geographic hotspots, antimicrobial consumption, laboratory capacity, resistance trends, and intervention gaps. AI dashboards can help decision-makers compare regions, identify under-sampled populations, prioritize surveillance investments, and evaluate whether stewardship or infection-control policies are changing trends. However, a dashboard is not a policy. Data must be translated into governance mechanisms, budgets, regulations, procurement systems, and accountability. Successful deployment of AI-supported antimicrobial stewardship requires more than accurate prediction models. Healthcare institutions frequently encounter challenges related to clinician acceptance, workflow integration, alert fatigue, limited computational infrastructure, inadequate digital literacy, regulatory uncertainty, and financial constraints. Consequently, implementation strategies should incorporate stakeholder engagement, user-centered system design, continuous model evaluation, and organizational change management to maximize adoption and long-term sustainability.

AI can also support scenario analysis. For example, models can estimate how resistance trends might change under reduced fluoroquinolone use, improved hand hygiene, expanded vaccination, wastewater intervention, livestock antimicrobial restrictions, or increased diagnostic testing. Such models should be treated as decision aids rather than forecasts of certainty. Their value lies in clarifying assumptions, trade-offs, and plausible consequences. The most important principle is that AMR surveillance should close the loop between detection and action. A system that detects resistance but cannot trigger stewardship, infection control, laboratory confirmation, or policy response is incomplete. AI can improve the speed of detection, but institutions determine whether detection becomes prevention.

10.0. ETHICS, GOVERNANCE, AND RESPONSIBLE AI

AI-enabled AMR surveillance raises ethical, legal, and governance challenges that must be addressed at the design stage. The data used in AMR surveillance may include patient records, genomic sequences, facility identifiers, prescribing practices, farm information, wastewater catchment data, and geospatial indicators. These data can reveal sensitive information about individuals, communities, hospitals, farms, companies, and municipalities. Responsible governance must balance public-health benefit with privacy, fairness, transparency, and accountability. Algorithmic bias is a major risk. AMR data are not randomly sampled. They overrepresent people with access to healthcare, severe infections, hospital exposure, or diagnostic testing. They underrepresent populations with limited laboratory access, rural communities, informal settlements, and low-resource regions. Machine-learning models trained on biased data may perform poorly for underrepresented populations or falsely identify risk based on healthcare access rather than biology. Recent work has also emphasized that bacterial population structure and biased sampling can confound genomic AMR prediction, meaning that models may learn lineage associations rather than causal resistance mechanisms (Yu et al., 2025).

Explainability matters because surveillance decisions have consequences. A model that flags a hospital, farm, community, or region as high risk should provide interpretable reasons, uncertainty estimates, and data-quality indicators. Explainability does not mean every model must be simple, but it does mean that users should understand the evidence behind alerts. This is especially important when AI outputs affect antibiotic prescribing, infection-control measures, regulatory inspections, or public communication. Privacy-preserving analytics should be considered where patient-level or facility-level data are sensitive. De-identification, aggregation, access controls, federated learning, secure enclaves, differential privacy, and data-use agreements can reduce risk. Yet technical privacy measures are not enough. Communities and institutions should know how data will be used, who benefits, and how harms will be prevented. In environmental surveillance, wastewater catchments can raise community privacy concerns even when individual data are not collected.

Responsible AI governance should require model documentation, external validation, bias assessment, version control, audit trails, human oversight, and post-deployment monitoring. Model cards or similar documentation can describe intended use, training data, validation settings, known limitations, and inappropriate uses. AMR models should be periodically re-evaluated because resistance mechanisms, clinical practice, sequencing methods, and population structures change over time. Figure 3 presents a conceptual framework outlining the six foundational principles underpinning responsible AI-enabled antimicrobial resistance (AMR) surveillance within a One Health context. At the center is a responsible AI surveillance framework supported by scientific validity, explainability and accountability, actionability and stewardship, sustainability and interoperability, privacy and data governance, and equity and capacity building. These interconnected principles ensure that AI systems are scientifically robust, transparent, ethically governed, interoperable across sectors, and capable of generating equitable and actionable insights for antimicrobial stewardship and public health decision-making. The framework emphasizes that successful implementation of AI in AMR surveillance requires balancing technological innovation with ethical governance, stakeholder trust, data protection, and long-term sustainability to maximize public health impact while minimizing unintended risks.

Figure 3. Governance principles for responsible AI-AMR surveillance. Trustworthy systems require scientific validity, equity, privacy, explainability, actionability, sustainability, and interoperability

Responsible AI in AMR surveillance should be guided by internationally recognized principles including transparency, explainability, accountability, fairness, privacy protection, robustness, reproducibility, and human oversight. Continuous auditing of model performance, algorithmic bias, and decision outcomes should accompany routine deployment to ensure that AI systems remain scientifically reliable, socially acceptable, and ethically responsible throughout their operational lifecycle.

11.0 IMPLEMENTATION IN LOW-RESOURCE AND DATA-LIMITED SETTINGS

The promise of AI-enabled AMR surveillance will remain incomplete if it benefits only high-income settings with mature laboratory networks and digital infrastructure. Many regions with high AMR burden face limited microbiology capacity, inconsistent susceptibility testing, shortages of trained staff, weak information systems, fragmented reporting, limited sequencing access, and inadequate funding. These gaps are not peripheral; they shape the global evidence base. If AI models are built mainly from high-income datasets, they may fail where surveillance is most needed.

Implementation should begin with surveillance fundamentals: reliable specimen collection, quality-assured microbiology, standardized susceptibility testing, basic data management, trained personnel, and feedback to clinicians and public-health authorities. AI cannot compensate for unreliable data. In low-resource settings, high-impact AI may initially focus on practical tasks such as automated data cleaning, de-duplication, missingness detection, antibiogram generation, laboratory quality dashboards, outbreak triage, and prioritization of samples for confirmatory testing or sequencing.

Digital tools should be designed for context. Systems requiring continuous broadband, expensive cloud subscriptions, or highly specialized bioinformatics support may fail after pilot funding ends. Sustainable platforms should support offline functionality where needed, open standards, local ownership, modular deployment, and training. Integrative digital tools for AMR data management in One Health settings demonstrate that informatics can strengthen surveillance when adapted to local workflows (Yadav et al., 2023). Equity also requires genomic capacity building. Sequencing provides major surveillance value, but it requires sample logistics, equipment, reagents, bioinformatics, quality control, data storage, and interpretation. Regional sequencing hubs, shared pipelines, cloud-based analysis, and training networks can help, but they should not extract data without building local capacity. Equitable AMR surveillance must include fair authorship, data sovereignty, benefit sharing, and support for national action plans.

The most practical model is tiered surveillance. Basic microbiology and standardized reporting should cover as many sites as possible. Sentinel sites can collect richer case-based data. Reference laboratories can conduct confirmatory testing and sequencing. Wastewater and environmental sampling can provide population-level signals. AI can integrate these tiers while explicitly representing uncertainty and coverage gaps. Table 3 presents a tiered implementation pathway for AI-enhanced antimicrobial resistance biosurveillance, moving from routine laboratory reporting to integrated One Health intelligence platforms. Table 3 emphasizes that AI adoption should be staged according to surveillance maturity, data quality, workforce capacity, and governance readiness. At the basic level, AI can support cleaning, de-duplication, and automated antibiogram generation, but these tools are only meaningful when laboratory antimicrobial susceptibility testing is quality assured. At more advanced levels, AI can assist with risk stratification, trend detection, genomic cluster identification, resistome monitoring, and cross-sector dashboarding. However, AMR biosurveillance cannot be built on algorithms alone; it requires standardized sampling, trained personnel, bioinformatics capacity, contextual metadata, interoperable systems, and governance agreements that allow human, animal, food, and environmental data to be meaningfully linked. This staged approach is consistent with the One Health view that AMR emerges across interconnected clinical, agricultural, environmental, and ecological systems and therefore requires integrated surveillance rather than isolated data streams (Kahn, 2017). Long-term sustainability also depends on appropriate financing mechanisms. International partnerships involving organizations such as the World Health Organization, Africa CDC, the Fleming Fund, the World Bank, and regional public health agencies can support laboratory strengthening, workforce development, genomic sequencing capacity, digital infrastructure, and institutional governance. Sustainable financing models should prioritize local ownership while avoiding excessive dependence on externally funded pilot projects.

Table 3. Tiered implementation model for AI-enabled AMR surveillance

Implementation tier	Core activity	AI contribution	Minimum enabling condition
Basic surveillance	Routine AST and standardized reporting	Automated cleaning, de-duplication, antibiograms	Quality-assured laboratory testing
Sentinel surveillance	Richer case-based AMR data	Risk stratification and trend detection	Stable sentinel sites and trained personnel
Reference laboratory	Confirmatory testing and sequencing	Cluster detection and genomic interpretation	Bioinformatics and quality control
Environmental surveillance	Wastewater and ecosystem monitoring	Resistome trend analysis and early warning	Sampling protocols and contextual metadata
Integrated One Health platform	Human-animal-food-environment linkage	Cross-sector analytics and dashboards	Governance agreements and interoperability

12.0. FUTURE DIRECTIONS

The future of AMR surveillance will depend on integrated, adaptive, and explainable systems. The strongest systems will combine phenotypic susceptibility testing, genomic sequencing, metagenomic environmental monitoring, antimicrobial use data, clinical outcomes, animal health data, food-chain surveillance, and contextual indicators. AI will help connect these streams, but the scientific foundation must remain microbiological validity and epidemiological interpretation. Real-time or near-real-time genomic surveillance will become more common as sequencing costs decline and automated pipelines improve. Long-read sequencing may improve plasmid tracking and mobile-element reconstruction. Metagenomic wastewater surveillance may become a routine complement to clinical surveillance, especially for early detection of high-risk genes and community-level trends. Federated learning may allow institutions to collaborate without sharing raw patient data. Explainable AI methods may improve trust and operational use.

Another major direction is causal surveillance. Many current models identify correlations between resistance and predictors such as antibiotic use, location, or prior hospitalization. Public-health action requires stronger causal inference: which interventions are likely to reduce resistance, under what conditions, and with what unintended effects? Future AI-AMR systems should incorporate causal diagrams, quasi-experimental designs, intervention evaluation, and feedback from stewardship programs.

Climate and environmental change should also be integrated. Temperature, flooding, sanitation disruption, migration, agricultural change, and land-use transformation may influence AMR emergence and spread. Forecasting future AMR risks will require linking microbial data with environmental and socioeconomic drivers. However, such models must avoid speculative overreach. They should identify plausible risk pathways and guide surveillance priorities rather than claim deterministic prediction. AMR surveillance must become more participatory. Public trust is essential when surveillance involves hospitals, farms, wastewater, and communities. Stakeholders should be involved in defining surveillance questions, acceptable data uses, communication strategies, and intervention priorities. AI-enabled AMR surveillance will succeed only if it is scientifically valid, socially legitimate, and operationally useful.

13.0. OPERATIONAL FRAMEWORK FOR AI-ENABLED AMR BIOSURVEILLANCE

A practical AI-AMR surveillance system should be designed as an operational workflow rather than as a collection of disconnected models. The workflow begins with problem definition. Public-health agencies, hospitals, veterinary authorities, environmental regulators, and community stakeholders must specify the decision that surveillance is expected to support. A model built to guide empirical therapy requires different data, validation, and risk tolerance from a model designed to detect regional emergence of carbapenemases in wastewater. A model intended to prioritize sequencing of clinical isolates requires different governance from a model that informs national restrictions on antimicrobial use. Clear problem definition prevents a common failure in digital surveillance: building technically impressive systems that do not answer operational questions.

The next stage is surveillance design. This involves selecting pathogens, drugs, populations, sectors, locations, sampling frequency, and laboratory methods. Priority pathogen-drug combinations should reflect clinical severity, transmissibility, treatment importance, local epidemiology, and international reporting needs. In human health, these may include carbapenem-resistant Enterobacterales, methicillin-resistant Staphylococcus aureus, vancomycin-resistant Enterococcus, multidrug-resistant Acinetobacter baumannii, drug-resistant Neisseria gonorrhoeae, and multidrug-resistant Mycobacterium tuberculosis. In food and animal systems, surveillance may prioritize Salmonella, Campylobacter, Escherichia coli, Enterococcus, and sector-specific pathogens. In wastewater, gene targets such as carbapenemases, extended-spectrum beta-lactamases, colistin resistance markers, and vancomycin resistance genes may be monitored alongside metagenomic resistome profiles.

Data engineering then converts surveillance observations into analyzable evidence. This includes data cleaning, organism-name normalization, antimicrobial-name standardization, duplicate handling, temporal alignment, geocoding, metadata validation, missingness assessment, and linkage across laboratory, pharmacy, clinical, environmental, and genomic datasets. Many AI failures begin at this stage because models are trained on data that appear large but are not comparable. For example, resistance rates may change because a laboratory changed susceptibility breakpoints, a hospital began screening high-risk patients, a region added new sentinel sites, or a wastewater catchment changed sampling design. AI systems must therefore retain provenance information so that models can distinguish biological signals from surveillance artifacts.

Model development should proceed from simple to complex. Baseline descriptive analytics, statistical process control, and transparent regression models should be established before deploying more complex machine-learning systems. This allows users to understand expected patterns and detect when advanced models add real value. For genomic prediction, models should be evaluated against curated phenotypic labels and tested across lineages, laboratories, regions, and time periods. For wastewater surveillance, models should be evaluated against sampling replicates, flow variation, population normalization, and clinical trends. For clinical risk prediction, models should be evaluated not only for discrimination and calibration but also for clinical utility, prescribing outcomes, and equity across patient groups.

Implementation requires human-in-the-loop decision pathways. AMR surveillance models should not automatically trigger high-consequence interventions without expert review. Alerts should be triaged by microbiologists, epidemiologists, infection prevention teams, stewardship specialists, veterinarians, environmental scientists, or public-health officers depending on the signal. A wastewater detection of a carbapenemase gene may trigger repeat sampling, source investigation, comparison with hospital isolates, and risk communication. A genomic cluster across hospitals may trigger infection-control review and isolate referral. A rise in fluoroquinolone resistance may trigger stewardship audit and guideline review. In each case, AI should accelerate evidence synthesis while preserving professional accountability.

Evaluation must close the loop. A surveillance system should be judged by whether it improves detection, shortens response time, reduces inappropriate antimicrobial use, prevents transmission, improves treatment, expands equitable coverage, or strengthens policy. Model accuracy alone is insufficient. A highly accurate model that does not change decisions has limited public-health value. Conversely, a modestly accurate early warning tool may be valuable if it triggers timely confirmatory testing and prevents spread. Evaluation should therefore include process metrics, epidemiological outcomes, stewardship outcomes, equity indicators, and user trust.

14.0. RISK CONTROLS, QUALITY ASSURANCE, AND FAILURE MODES

AI-enabled AMR surveillance must be designed with explicit risk controls because errors can affect treatment, public confidence, institutional reputation, and policy. The most common failure mode is biased sampling. Clinical isolates do not represent the general population; they represent people who were tested. Wastewater samples do not represent every individual equally; they represent catchment systems shaped by infrastructure, flow, industrial discharge, hospital inputs, and sampling design. Genomic databases overrepresent well-funded laboratories, outbreak investigations, and high-priority pathogens. Models trained on these data may therefore learn surveillance practices rather than resistance biology. Bias assessment should be routine, not optional. Another important failure mode is poor external validity. Models may perform well within the dataset where they were developed but fail when applied to new hospitals, regions, lineages, testing platforms, or later time periods. AMR evolves; laboratory methods change; prescribing practices shift; and reference databases expand. For this reason, surveillance models should be externally validated across geography, time, population structure, and laboratory context. Post-deployment monitoring should also track model drift, calibration decay, changes in feature importance, and disagreement between predicted and observed resistance.

A related challenge is weak genotype-to-phenotype interpretation. Resistance genes are important, but their clinical meaning depends on organism background, gene expression, gene integrity, regulatory context, plasmid copy number, permeability, efflux, and testing conditions. Some genotype-based systems may overcall resistance by treating gene presence as equivalent to clinical resistance, while others may undercall resistance by missing novel or context-dependent mechanisms. The safest approach is layered interpretation: high-confidence markers can support rapid classification, uncertain markers should trigger additional review, and discordant genotype–phenotype results should be investigated rather than forced into a single label. A further limitation is action without capacity. Detecting AMR earlier has limited value if laboratories cannot confirm the signal, clinicians cannot change practice, infection-control teams cannot respond, or policy agencies cannot intervene. Surveillance modernization should therefore be linked to capacity building, including laboratory strengthening, workforce development, bioinformatics training, stewardship programs, diagnostic supply chains, governance agreements, and sustainable financing. AI may reveal gaps faster than institutions can solve them; without investment, this can produce frustration rather than impact.

Communication is also a major vulnerability. AMR signals are probabilistic and context dependent, yet public communication often demands simple messages. A poorly communicated map of resistant organisms can stigmatize a hospital, farm, municipality, or country. Conversely, excessive caution may delay necessary action. Communication should distinguish detection from confirmed outbreak, gene abundance from clinical risk, association from causation, and local signal from generalizable trend. Trustworthy communication is part of biosurveillance, not an afterthought. This broader surveillance logic is consistent with recent edited works emphasizing detection, monitoring, and risk management of food contaminants, biomonitoring of pollutants, sustainable food safety, and One Health implications of agrochemical exposure (Ogwu & Izah, 2026a, 2026b; Izah et al., 2024; Ogwu et al., 2023). Although this chapter provides a comprehensive conceptual synthesis of AI-enabled antimicrobial resistance surveillance, it does not present primary empirical data or experimentally validate the proposed framework. Implementation requirements may vary considerably across countries, healthcare systems, and surveillance infrastructures. Furthermore, the rapidly evolving nature of artificial intelligence means that surveillance technologies, regulatory frameworks, and analytical methodologies will continue to change. Future empirical research should evaluate the operational performance of the proposed framework across diverse One Health settings and resource contexts.

Quality assurance should therefore be embedded across the system. Laboratory quality assurance should include standardized AST methods, reference strains, external quality assessment, and confirmatory testing. Genomic quality assurance should include read quality, assembly metrics, contamination checks, database versioning, and reproducible pipelines. Data quality assurance should include metadata completeness, missingness reporting, de-duplication rules, denominator definitions, and audit trails. Model quality assurance should include transparent documentation, validation, fairness analysis, uncertainty reporting, and planned retirement or recalibration when performance degrades. These controls make AI-AMR surveillance credible, defensible, and useful.

15.0. CONCLUSION

AI and big data can transform AMR surveillance from delayed reporting into anticipatory biosurveillance, but only if they are embedded within strong laboratory, epidemiological, and governance systems. AMR emerges across interconnected human, animal, food, wastewater, and environmental domains. No single dataset can capture this complexity. The value of AI lies in integrating heterogeneous evidence, detecting patterns earlier, quantifying uncertainty, supporting stewardship, and linking surveillance to action. The central caution is equally important. AI cannot replace phenotypic testing, genomic validation, microbiological expertise, infection prevention, antimicrobial stewardship, or public-health governance. Poor data will produce poor models. Biased sampling will produce biased predictions. Black-box systems without accountability may undermine trust. Surveillance systems that detect signals without response capacity will not reduce AMR burden. A high-impact AI-AMR surveillance framework should therefore be One Health in scope, evidence-based in design, transparent in analytics, equitable in implementation, and adaptive in operation. It should integrate clinical, genomic, environmental, veterinary, food-chain, and antimicrobial-use data while protecting privacy and data sovereignty. It should make uncertainty visible, support human decision-makers, and evaluate whether interventions reduce resistance and improve outcomes. In this sense, the future of AMR biosurveillance is not simply more data or more advanced algorithms. It is the disciplined integration of verified evidence, responsible AI, and coordinated public-health action. The future of antimicrobial resistance biosurveillance will depend not only on increasingly sophisticated artificial intelligence algorithms but also on trusted governance, interoperable data ecosystems, multidisciplinary collaboration, and equitable implementation across the One Health continuum. AI should be viewed as an enabling technology that strengthens scientific interpretation rather than replacing microbiological expertise, epidemiological investigation, or public health decision-making.

REFERENCES

Aenishaenslin, C., Häsler, B., Ravel, A., Parmley, E. J., Mediouni, S., Bennani, H., Stärk, K. D. C., & Buckeridge, D. L. (2021). Evaluating the integration of One Health in surveillance systems for antimicrobial use and resistance: A conceptual framework. *Frontiers in Veterinary Science*, *8*, 611931. DOI: 10.3389/fvets.2021.611931 PMID: 33842569

Ali, T., Ahmed, S., & Aslam, M. (2023). Artificial intelligence for antimicrobial resistance prediction: Challenges and opportunities towards practical implementation. *Antibiotics (Basel, Switzerland)*, *12*(3), 523. DOI: 10.3390/antibiotics12030523 PMID: 36978390

Davies, J., & Davies, D. (2010). Origins and evolution of antibiotic resistance. *Microbiology and Molecular Biology Reviews : MMBR*, *74*(3), 417–433. DOI: 10.1128/MMBR.00016-10 PMID: 20805405

Delpy, L., Astbury, C. C., Aenishaenslin, C., Ruckert, A., Penney, T. L., Wiktorowicz, M., Ciss, M., Benko, R., & Bordier, M. (2024). Integrated surveillance systems for antibiotic resistance in a One Health context: A scoping review. *BMC Public Health*, *24*(1), 1717. DOI: 10.1186/s12889-024-19158-6 PMID: 38937706

Duarte, A. S. R., Stärk, K. D., Munk, P., Leekitcharoenphon, P., Bossers, A., Luiken, R., Sarrazin, S., Lukjancenko, O., Pamp, S. J., Bortolaia, V., Nissen, J. N., Kirstahler, P., Van Gompel, L., Poulsen, C. S., Kaas, R. S., Hellmér, M., Hansen, R. B., Gomez, V. M., & Hald, T. (2020). Addressing learning needs on the use of metagenomics in antimicrobial resistance surveillance. *Frontiers in Public Health*, *8*, 38. DOI: 10.3389/fpubh.2020.00038 PMID: 32158739

Ellington, M. J., Ekelund, O., Aarestrup, F. M., Canton, R., Doumith, M., Giske, C., Grundman, H., Hasman, H., Holden, M. T. G., Hopkins, K. L., Iredell, J., Kahlmeter, G., Köser, C. U., MacGowan, A., Mevius, D., Mulvey, M., Naas, T., Peto, T., Rolain, J.-M., & Woodford, N. (2017). The role of whole genome sequencing in antimicrobial susceptibility testing of bacteria: Report from the EUCAST Subcommittee. *Clinical Microbiology and Infection : The Official Publication of the European Society of Clinical Microbiology and Infectious Diseases*, *23*(1), 2–22. DOI: 10.1016/j.cmi.2016.11.012 PMID: 27890457

Hendriksen, R. S., Munk, P., Njage, P., van Bunnik, B., McNally, L., Lukjancenko, O., Röder, T., Nieuwenhuijse, D., Pedersen, S. K., Kjeldgaard, J., Kaas, R. S., Clausen, P. T. L. C., Vogt, J. K., Leekitcharoenphon, P., van de Schans, M. G. M., Zuidema, T., de Roda Husman, A. M., Rasmussen, S., Petersen, B., & Aarestrup, F. M. (2019). Global monitoring of antimicrobial resistance based on metagenomics analyses of urban sewage. *Nature Communications*, *10*(1), 1124. DOI: 10.1038/s41467-019-08853-3 PMID: 30850636

Izah, S. C., Nurmahanova, A., Ogwu, M. C., Toktarbay, Z., Umirbayeva, Z., Ussen, K., Koibasova, L., Nazarbekova, S., Tynybekov, B., & Guo, Z. (2025). Public health risks associated with antibiotic residues in poultry food products. *Journal of Agriculture and Food Research*, *21*, 101815. DOI: 10.1016/j.jafr.2025.101815

Izah, S. C., & Ogwu, M. C. (2025). Modeling solutions for microbial water contamination in the Global South for public health protection. *Frontiers in Microbiology*, *16*, 1504829. DOI: 10.3389/fmicb.2025.1504829 PMID: 40241726

Izah, S. C., Ogwu, M. C., & Shahsavani, A. (2024). *Air pollutants in the context of One Health: Fundamental, sources, and impacts*. Springer Cham., DOI: 10.1007/978-3-031-74165-4

Kelly, C. J., Karthikesalingam, A., Suleyman, M., Corrado, G., & King, D. (2019). Key challenges for delivering clinical impact with artificial intelligence. *BMC Medicine*, *17*(1), 195. DOI: 10.1186/s12916-019-1426-2 PMID: 31665002

Laxminarayan, R., Duse, A., Wattal, C., Zaidi, A. K. M., Wertheim, H. F. L., Sumpradit, N., Vlieghe, E., Hara, G. L., Gould, I. M., Goossens, H., Greko, C., So, A. D., Bigdeli, M., Tomson, G., Woodhouse, W., Ombaka, E., Peralta, A. Q., Qamar, F. N., Mir, F., & Cars, O. (2013). Antibiotic resistance: The need for global solutions. *The Lancet. Infectious Diseases*, *13*(12), 1057–1098. DOI: 10.1016/S1473-3099(13)70318-9 PMID: 24252483

Liu, Z., Deng, D., Lu, H., Sun, J., Lv, L., Li, S., Peng, G., Ma, X., Li, J., Li, Z., Rong, T., & Wang, G. (2020). Evaluation of machine learning models for predicting antimicrobial resistance of Actinobacillus pleuropneumoniae from whole genome sequences. *Frontiers in Microbiology*, *11*, 48. DOI: 10.3389/fmicb.2020.00048 PMID: 32117101

Martínez, J. L. (2008). Antibiotics and antibiotic resistance genes in natural environments. *Science*, *321*(5887), 365–367. DOI: 10.1126/science.1159483 PMID: 18635792

Mo, Y., Ding, Y., Cao, Y., Hopkins, J., Ashley, E. A., Waithira, N., Wannapinij, P., Lee, S. J., Paterson, D. L., Doorn, H. R., & Turner, P. (2023). ACORN (A Clinically-Oriented Antimicrobial Resistance Surveillance Network) II: Protocol for case based antimicrobial resistance surveillance. *Wellcome Open Research*, *8*, 179. DOI: 10.12688/wellcomeopenres.19210.1 PMID: 37854055

Moradigaravand, D., Palm, M., Farewell, A., Mustonen, V., Warringer, J., & Parts, L. (2018). Prediction of antibiotic resistance in Escherichia coli from large-scale pan-genome data. *PLoS Computational Biology*, *14*(12), e1006258. DOI: 10.1371/journal.pcbi.1006258 PMID: 30550564

Murray, C. J. L., Ikuta, K. S., Sharara, F., Swetschinski, L., Robles Aguilar, G., Gray, A., Han, C., Bisignano, C., Rao, P., Wool, E., Johnson, S. C., Browne, A. J., Chipeta, M. G., Fell, F., Hackett, S., Haines-Woodhouse, G., Hamadani, B. H. K., Kumaran, E. A. P., McManigal, B., & Naghavi, M. (2022). Global burden of bacterial antimicrobial resistance in 2019: A systematic analysis. *Lancet*, *399*(10325), 629–655. DOI: 10.1016/S0140-6736(21)02724-0 PMID: 35065702

Noman, S. M., Zeeshan, M., Arshad, J., Deressa Amentie, M., Shafiq, M., Yuan, Y., Zeng, M., Li, X., Xie, Q., & Jiao, X. (2023). Machine learning techniques for antimicrobial resistance prediction of Pseudomonas aeruginosa from whole genome sequence data. *Computational Intelligence and Neuroscience*, *2023*(1), 5236168. DOI: 10.1155/2023/5236168 PMID: 36909968

Ogwu, M. C., & Izah, S. C. (2025a). Honey as a natural antimicrobial. *Antibiotics (Basel, Switzerland)*, *14*(3), 255. DOI: 10.3390/antibiotics14030255 PMID: 40149066

Ogwu, M. C., & Izah, S. C. (2025b). Nanotechnology for fungal pathogen control in crops: Innovations, public health impacts, and disease prevention. *Frontiers in Fungal Biology*, *6*, 1653214. DOI: 10.3389/ffunb.2025.1653214 PMID: 40959466

Ogwu, M. C., & Izah, S. C. (Eds.). (2026a). *Detection, identification, and monitoring of food contaminants*. IGI Global Scientific Publishing., DOI: 10.4018/979-8-3373-3982-5

Ogwu, M. C., & Izah, S. C. (Eds.). (2026b). *Mitigation strategies and risk management for food contaminants*. IGI Global Scientific Publishing., DOI: 10.4018/979-8-3373-7052-1

Ogwu, M. C., Izah, S. C., Aigberua, A. O., & Ngun, C. T. (2024). Editorial: Detection, risk analysis, and monitoring of chemical contaminants from agro-aqua food production and processing: Implications on the One Health triad. *Frontiers in Sustainable Food Systems*, *8*, 1501930. DOI: 10.3389/fsufs.2024.1501930

Ogwu, M. C., Izah, S. C., & Ntuli, N. R. (Eds.). (2024). *Food safety and quality in the Global South*. Springer Nature., DOI: 10.1007/978-981-97-2428-4

Pennisi, F., Pinto, A., Ricciardi, G. E., Signorelli, C., & Gianfredi, V. (2025). The role of artificial intelligence and machine learning models in antimicrobial stewardship in public health: A narrative review. *Antibiotics (Basel, Switzerland)*, *14*(2), 134. DOI: 10.3390/antibiotics14020134 PMID: 40001378

Quainoo, S., Coolen, J. P. M., van Hijum, S. A. F. T., Huynen, M. A., Melchers, W. J. G., van Schaik, W., & Wertheim, H. F. L. (2017). Whole-genome sequencing of bacterial pathogens: The future of nosocomial outbreak analysis. *Clinical Microbiology Reviews*, *30*(4), 1015–1063. DOI: 10.1128/CMR.00016-17 PMID: 28855266

Ren, Y., Chakraborty, T., Doijad, S., Falgenhauer, L., Falgenhauer, J., Goesmann, A., Hauschild, A.-C., Schwengers, O., & Heider, D. (2022). Prediction of antimicrobial resistance based on whole-genome sequencing and machine learning. *Bioinformatics (Oxford, England)*, *38*(2), 325–334. DOI: 10.1093/bioinformatics/btab681 PMID: 34613360

Rieke, N., Hancox, J., Li, W., Milletarì, F., Roth, H. R., Albarqouni, S., Bakas, S., Galtier, M. N., Landman, B. A., Maier-Hein, K., Ourselin, S., Sheller, M., Summers, R. M., Trask, A., Xu, D., Baust, M., & Cardoso, M. J. (2020). The future of digital health with federated learning. *NPJ Digital Medicine*, *3*(1), 119. DOI: 10.1038/s41746-020-00323-1 PMID: 33015372

Rudnick, W., Mukhi, S. N., Reid-Smith, R. J., German, G. J., Nichani, A., & Mulvey, M. R. (2022). Overview of Canada's Antimicrobial Resistance Network (AMRNet): A data-driven One Health approach to antimicrobial resistance surveillance. *Canada Communicable Disease Report = Relevé des Maladies Transmissibles au Canada*, *48*(11-12), 522. PMID: 38173468

Singh, K. S., Anand, S., Dholpuria, S., Sharma, J. K., Blankenfeldt, W., & Shouche, Y. S. (2021). Antimicrobial resistance dynamics and the One Health strategy: A review. *Environmental Chemistry Letters*, *19*(4), 2995–3007. DOI: 10.1007/s10311-021-01238-3

Thakur, S., & Gray, G. C. (2019). The mandate for a global One Health approach to antimicrobial resistance surveillance. *The American Journal of Tropical Medicine and Hygiene*, *100*(2), 227–228. DOI: 10.4269/ajtmh.18-0973 PMID: 30608047

Tiwari, A., Kurittu, P., Al-Mustapha, A. I., Heljanko, V., Johansson, V., Thakali, O., Mishra, S. K., Lehto, K.-M., Lipponen, A., Oikarinen, S., Pitkänen, T., & Heikinheimo, A. (2022). Wastewater surveillance of antibiotic-resistant bacterial pathogens: A systematic review. *Frontiers in Microbiology*, *13*, 977106. DOI: 10.3389/fmicb.2022.977106 PMID: 36590429

World Health Organization. (2015). *Global action plan on antimicrobial resistance*. World Health Organization.

World Health Organization. (2022). *Global antimicrobial resistance and use surveillance system (GLASS) report: 2022*. World Health Organization.

Yadav, S. K., Shrestha, L., Acharya, J., Gompo, T. R., Chapagain, S., & Jha, R. (2023). Integrative digital tools to strengthen data management for antimicrobial resistance surveillance in the "One Health" domain in Nepal. *Tropical Medicine and Infectious Disease*, *8*(6), 291. DOI: 10.3390/tropicalmed8060291 PMID: 37368709

Yu, J., Jia, Y., Yu, Q., Lin, L., Li, C., Chen, B., Zhong, P., Lin, X., Li, H., Sun, Y., Zhong, X., He, Y., Huang, X., Lin, S., & Pan, Y. (2024). Deciphering complex antibiotic resistance patterns in Helicobacter pylori through whole genome sequencing and machine learning. *Frontiers in Cellular and Infection Microbiology*, *13*, 1306368. DOI: 10.3389/fcimb.2023.1306368 PMID: 38379956

Yu, Y., Wheeler, N. E., & Barquist, L. (2025). Biased sampling driven by bacterial population structure confounds machine learning prediction of antimicrobial resistance. *PLoS Biology*, *23*(12), e3003539. DOI: 10.1371/journal.pbio.3003539 PMID: 41401143

Compilation of References

Aazam, M., Zeadally, S., & Flushing, E. F. (2021). Task offloading in edge computing. *Computer Networks*, *191*, 108019. DOI: 10.1016/j.comnet.2021.108019

Abaku, E. A., Edunjobi, T. E., & Odimarha, A. C. (2024). Theoretical approaches to AI in supply chain optimization: Pathways to efficiency and resilience. *International Journal of Science and Technology Research Archive, 6*(1), 092–107.

Abass, T., Eruaga, M. A., Itua, E. O., & Bature, J. T. (2024). Advancing food safety through IoT: Real-time monitoring and control systems. *International Medical Science Research Journal*, *4*(3), 276–283. DOI: 10.51594/imsrj.v4i3.919

Abba, S., Crabol, Y., de Bouillé, J. G., Dinh, A., & Morquin, D. (2025). Artificial intelligence and infectious diseases: Scope and perspectives. *Infectious Diseases Now*, *55*(7), 105131. Advance online publication. DOI: 10.1016/j.idnow.2025.105131 PMID: 40759336

Abdeldayem, O., Dabbish, A., Habashy, M., Mostafa, M., Elhefnawy, M., Amin, L., Al-Sakkari, E., Ragab, A., & Rene, E. (2021). Viral outbreaks detection and surveillance using wastewater-based epidemiology, viral air sampling, and machine learning techniques: A comprehensive review and outlook. *The Science of the Total Environment*, *803*, 149834–149834. DOI: 10.1016/j.scitotenv.2021.149834 PMID: 34525746

Abejirinde, I.-O. O., Itanyi, I. U., Kornas, K., Zaim, R., Guha, S., Chui, V., Lipscombe, L., Rosella, L. C., & Shaw, J. (2025). Principles and practices of community engagement in AI for population health: Formative qualitative study of the AI for diabetes prediction and prevention project. *Journal of Participatory Medicine*, *17*, e69497. DOI: 10.2196/69497 PMID: 41037755

Abhulimen, A. O., & Ejike, O. G.Angela Omozele AbhulimenOnyinye Gift Ejike. (2024). Solving supply chain management issues with AI and big data analytics for future operational efficiency. *Computer Science & IT Research Journal*, *5*(8), 1780–1805. DOI: 10.51594/csitrj.v5i8.1396

Abueg, M., Hinch, R., Wu, N., Liu, L., Probert, W., Wu, A., Eastham, P., Shafi, Y., Rosencrantz, M., Dikovsky, M., Cheng, Z., Nurtay, A., Abeler-Dörner, L., Bonsall, D., McConnell, M. V., O'Banion, S., & Fraser, C. (2021). Modeling the effect of exposure notification and non-pharmaceutical interventions on COVID-19 transmission in Washington state. *npj. Digital Medicine*, *4*(1), 49. Advance online publication. DOI: 10.1038/s41746-021-00422-7 PMID: 33712693

Achieng, M. S., & Ogundaini, O. O. (2024). Big data analytics for integrated infectious disease surveillance in Sub-Saharan Africa. *South African Journal of Information Management*, *26*(1), 1668. DOI: 10.4102/SAJIM.v26i1.1668

Adav, S. S., & Ng, K. W. (2025). The multifaceted role of hair as a biospecimen: Recent advances in precision medicine and forensic science. *Experimental & Molecular Medicine*, *57*(10), 1–17. DOI: 10.1038/s12276-025-01548-4 PMID: 41145616

Adhikari, M., & Munusamy, A. (2021). iCovidCare framework. *Internet of Things : Engineering Cyber Physical Human Systems*, *14*, 100385. DOI: 10.1016/j.iot.2021.100385 PMID: 38620813

Adyapady, R. R., & Annappa, B. (2023). A comprehensive review of facial expression recognition techniques. *Multimedia Systems*, *29*(1), 73–103. DOI: 10.1007/s00530-022-00984-w

Aenishaenslin, C., Häsler, B., Ravel, A., Parmley, E. J., Mediouni, S., Bennani, H., Stärk, K. D. C., & Buckeridge, D. L. (2021). Evaluating the integration of One Health in surveillance systems for antimicrobial use and resistance: A conceptual framework. *Frontiers in Veterinary Science*, *8*, 611931. DOI: 10.3389/fvets.2021.611931 PMID: 33842569

Agarwal, S., LeFevre, A. E., Lee, J., L'Engle, K., Mehl, G., Sinha, C., & Labrique, A. (2016). Guidelines for reporting of health interventions using mobile phones: Mobile health (mHealth) evidence reporting and assessment (mERA) checklist. *BMJ (Clinical Research Ed.)*, *352*, i1174. DOI: 10.1136/bmj.i1174 PMID: 26988021

Aghaei, M., Khoshnamvand, N., Janjani, H., Dehghani, M., & Karri, R. (2024). Exposure to environmental pollutants: A mini-review on the application of the wastewater-based epidemiology approach. *Journal of Environmental Health Science & Engineering*, *22*(1), 65–74. DOI: 10.1007/s40201-024-00895-0 PMID: 38887772

Ahale, S., Rakhonde, G., Bhateja, S., Kuppuraj, J., & Mishra, S. (2024). Disease and pest control through advance technology. In K. Pandey, N. L. Kushwaha, C. B. Pande, & K. G. Singh (Eds.), *Artificial intelligence and smart agriculture* (Advances in Geographical and Environmental Sciences). Springer. https://doi.org/DOI: 10.1007/978-981-97-0341-8_21

Ahmed, M. M., Okesanya, O. J., Oweidat, M., Othman, Z. K., Musa, S. S., & Lucero-Prisno, D. E.III. (2025). The ethics of data mining in healthcare: Challenges, frameworks, and future directions. *BioData Mining*, *18*(1), 47. DOI: 10.1186/s13040-025-00461-w PMID: 40646553

Ahmed, W., Angel, N., Edson, J., Bibby, K., Bivins, A., O'Brien, J. W., Choi, P. M., Kitajima, M., Simpson, S. L., Li, J., Tscharke, B., Verhagen, R., Smith, W. J. M., Zaugg, J., Dierens, L., Hugenholtz, P., Thomas, K. V., & Mueller, J. F. (2020). First confirmed detection of SARS-CoV-2 in untreated wastewater in Australia: A proof of concept for the wastewater surveillance of COVID-19 in the community. *The Science of the Total Environment*, *728*, 138764. Advance online publication. DOI: 10.1016/j.scitotenv.2020.138764 PMID: 32387778

Ajala, O. A., Arinze, C. A., Ofodile, O. C., Okoye, C. C., & Daraojimba, O. D.Olakunle Abayomi AjalaChuka Anthony ArinzeOnyeka Chrisanctus OfodileChinwe Chinazo OkoyeObinna Donald Daraojimba. (2024). Reviewing advancements in privacy-enhancing technologies for big data analytics in an era of increased surveillance. *World Journal of Advanced Engineering Technology and Sciences*, *11*(1), 294–300. DOI: 10.30574/wjaets.2024.11.1.0060

Al Khatib, I., Alasheh, S., & Shamayleh, A. (2024). The drivers of complexity in inventory management within the healthcare industry: A systematic review. [IJSSMET]. *International Journal of Service Science, Management, Engineering, and Technology*, *15*(1), 1–26. DOI: 10.4018/IJSSMET.347332

Alahakoon, D., Nawaratne, R., Xu, Y., De Silva, D., Sivarajah, U., & Gupta, B. (2023). Self-building artificial intelligence and machine learning to empower big data analytics in smart cities. *Information Systems Frontiers*, *25*(1), 221–240. DOI: 10.1007/s10796-020-10056-x

AlBahrani, S., AlZahrani, S. J., Al-Maqati, T. N., Almehbash, A., Alshammari, A., Bujlai, R., Ba Taweel, S., Almasabi, F., AlAmari, A., & Al-Tawfiq, J. A. (2024). Dynamic patterns and predominance of respiratory pathogens post-COVID-19: Insights from a two-year analysis. *Journal of Epidemiology and Global Health*, *14*(2), 311–318. DOI: 10.1007/s44197-024-00213-9 PMID: 38589755

Albahri, A. S., Khaleel, Y. L., Habeeb, M. A., Ismael, R. D., Hameed, Q. A., Deveci, M., Homod, R. Z., Albahri, O. S., Alamoodi, A. H., & Alzubaidi, L. (2024). A systematic review of trustworthy artificial intelligence applications in natural disasters. *Computers & Electrical Engineering*, *118*, 109409. DOI: 10.1016/j.compeleceng.2024.109409

Alcock, B. P., Raphenya, A. R., Lau, T. T. Y., Tsang, K. K., Bouchard, M., Edalatmand, A., Huynh, W., Nguyen, A.-L. V., Cheng, A. A., Liu, S., Min, S. Y., Miroshnichenko, A., Tran, H.-K., Werfalli, R. E., Nasir, J. A., Oloni, M., Speicher, D. J., Florescu, A., Singh, B., & McArthur, A. G. (2020). CARD 2020: Antibiotic resistome surveillance with the Comprehensive Antibiotic Resistance Database. *Nucleic Acids Research*, *48*(D1), D517–D525. DOI: 10.1093/nar/gkz935 PMID: 31665441

Alhassan, J. A. K., & Wills, O. (2024). Public health surveillance through community health workers: A scoping review of evidence from 25 low-income and middle-income countries. *BMJ Open*, *14*(4), e079776. DOI: 10.1136/bmjopen-2023-079776 PMID: 38582533

Ali, A., Jayaraman, R., Azar, E., & Maalouf, M. (2024). A comparative analysis of machine learning and statistical methods for evaluating building performance: A systematic review and future benchmarking framework. *Building and Environment*, *252*, 111268. DOI: 10.1016/j.buildenv.2024.111268

Ali, A., Snášel, V., & Platoš, J. (2025). Health-FedNet: A privacy-preserving federated learning framework for scalable and secure healthcare analytics. *Results in Engineering*, *27*, 106484. DOI: 10.1016/j.rineng.2025.106484

Ali, H. (2024). AI for pandemic preparedness and infectious disease surveillance: Predicting outbreaks, modeling transmission, and optimizing public health interventions. *International Journal of Research Publication and Reviews*, *5*(8), 4605–4619.

Ali, T., Ahmed, S., & Aslam, M. (2023). Artificial intelligence for antimicrobial resistance prediction: Challenges and opportunities towards practical implementation. *Antibiotics (Basel, Switzerland)*, *12*(3), 523. Advance online publication. DOI: 10.3390/antibiotics12030523 PMID: 36978390

Ali, W., Zhang, H., Wang, Z., Chang, C., Javed, A., Ali, K., Du, W., Niazi, N., Mao, K., & Yang, Z. (2021). Occurrence of various viruses and recent evidence of SARS-CoV-2 in wastewater systems. *Journal of Hazardous Materials*, *414*, 125439–125439. DOI: 10.1016/j.jhazmat.2021.125439 PMID: 33684818

Ali, Y., Iqbal, S., Aatif, H. M., Naveed, K., Khan, A. A., Ijaz, M., Magsi, M. M., Ahmad, S., Syed, A. U. A., Magsi, M. A., Iqbal, R. K., Bukhari, N. A., Hatamleh, A. A., & Raza, A. (2023). Predicting Stripe Rust Severity in Wheat Using Meteorological Data with Environmental Response Modeling. *Journal of King Saud University. Science*, *35*(4), 102591. DOI: 10.1016/j.jksus.2023.102591

Almagharbeh, W. T., Alharrasi, M., Rony, M. K. K., Kabir, S., Ahmed, S. K., & Alrazeeni, D. M. (2025). Ethical and institutional readiness for artificial intelligence in nursing: An umbrella review. *International Nursing Review*, *72*(4), e70111. DOI: 10.1111/inr.70111 PMID: 41045018

Al-Omari, A. A. (2026). Future trends in educational leadership: Tech advances, evolving paradigms, and personalized learning. In Roache, D. (Ed.), *Advancing safety, service, and performance in organizations with impactful leadership* (pp. 25–50). IGI Global Scientific Publishing., DOI: 10.4018/979-8-3373-5686-0.ch002

Alsaedi, S., Ogasawara, M., Alarawi, M., Gao, X., & Gojobori, T. (2025). AI-powered precision medicine: Utilizing genetic risk factor optimization to revolutionize healthcare. *NAR Genomics and Bioinformatics*, *7*(2), lqaf038. Advance online publication. DOI: 10.1093/nargab/lqaf038 PMID: 40330081

Althouse, B. M., Scarpino, S. V., Meyers, L. A., Ayers, J. W., Bargsten, M., Baumbach, J., Brownstein, J. S., Castro, L., Clapham, H., Cummings, D. A. T., Del Valle, S., Eubank, S., Fairchild, G., Finelli, L., Generous, N., George, D., Harper, D. R., Hébert-Dufresne, L., Johansson, M. A., & Vespignani, A. (2015). Enhancing disease surveillance with novel data streams: Challenges and opportunities. *EPJ Data Science*, *4*(1), 17. DOI: 10.1140/epjds/s13688-015-0054-0 PMID: 27990325

Aly, M. (2025). Revolutionizing online education: Advanced facial expression recognition for real-time student progress tracking via deep learning model. *Multimedia Tools and Applications*, *84*(13), 12575–12614. DOI: 10.1007/s11042-024-19392-5

Amabie, T., Izah, S. C., Ogwu, M. C., & Hait, M. (2023). Harmonizing tradition and technology: The synergy of artificial intelligence in traditional medicine. In *Herbal medicine phytochemistry: Applications and trends* (pp. 1–23). Springer International Publishing.

Amani, M., Ghorbanian, A., Ahmadi, S. A., Kakooei, M., Moghimi, A., Mirmazloumi, S. M., & Brisco, B.. (2020). Google Earth Engine cloud computing platform for remote sensing big data applications: A comprehensive review. *IEEE Journal of Selected Topics in Applied Earth Observations and Remote Sensing*, *13*, 5326–5350. DOI: 10.1109/JSTARS.2020.3021052

Amanna, A. (2025). Deploying next-generation artificial intelligence ecosystems for real-time biosurveillance, precision health analytics and dynamic intervention planning in life science research. *Magna Scientia Advanced Biology and Pharmacy, 16*(1), 38–54. DOI: 10.30574/msabp.2025.16.1.0066

Amin, V., Bowes, D., & Halden, R. (2022). Systematic scoping review evaluating the potential of wastewater-based epidemiology for monitoring cardiovascular disease and cancer. *The Science of the Total Environment, 858*, 160103–160103. DOI: 10.1016/j.scitotenv.2022.160103 PMID: 36370774

Ammenwerth, E., Schnell-Inderst, P., Machan, C., & Siebert, U. (2008). The effect of electronic prescribing on medication errors and adverse drug events: A systematic review. *Journal of the American Medical Informatics Association : JAMIA, 15*(5), 585–600. DOI: 10.1197/jamia.M2667 PMID: 18579832

Amrutha, B. K., & Gomathy, B. (2024). Research survey on role of blockchain technology in agriculture. *International Journal of Scientific Research in Engineering and Management, 8*(6), 1–5.

Anderson, M., & Anderson, S. L. (Eds.). (2011). *Machine ethics*. Cambridge University Press. DOI: 10.1017/CBO9780511978036

Andrejevic, M. (2004). The work of watching one another: Lateral surveillance, risk, and governance. *Surveillance & Society, 2*(4), 479–497. DOI: 10.24908/ss.v2i4.3359

Andrianto, T., Tangit, T. M., & Minh, N. C. (2025). Adoption of artificial intelligence (AI) technology in enhancing tourist experience: A conceptual model. *Journal of Tourism. Hospitality and Travel Management, 3*(1), 53–66. DOI: 10.58229/jthtm.v3i1.302

Angelou, A., Pappa, A., Markov, P. V., Gewehr, S., Stilianakis, N. I., & Kioutsioukis, I. (2025). Early warning system of the seasonal west nile virus infection risk in humans in northern greece, 2020–2024. *Scientific Reports, 15*(1), 7129. Advance online publication. DOI: 10.1038/s41598-025-91996-9 PMID: 40021915

Anitha, P., & Patil, M. M. (2018). A review on data analytics for supply chain management: A case study. *International Journal of Information Engineering and Electronic Business, 14*(5), 30.

Anyamba, A., Chretien, J.-P., Small, J., Tucker, C. J., Formenty, P. B., Richardson, J. H., Britch, S. C., Schnabel, D. C., Erickson, R. L., & Linthicum, K. J. (2009). Prediction of a Rift Valley fever outbreak. *Proceedings of the National Academy of Sciences of the United States of America, 106*(3), 955–959. DOI: 10.1073/pnas.0806490106 PMID: 19144928

Aplakidou, E., Vergoulidis, N., Chasapi, M., Venetsianou, N., Kokoli, M., Panagiotopoulou, E., Iliopoulos, I., Karatzas, E., Pafilis, E., Georgakopoulos-Soares, I., Kyrpides, N., Pavlopoulos, G., & Baltoumas, F. (2024). Visualizing metagenomic and metatranscriptomic data: A comprehensive review. *Computational and Structural Biotechnology Journal, 23*, 2011–2033. DOI: 10.1016/j.csbj.2024.04.060 PMID: 38765606

Artzrouni, M. (2020). Are models useful? Reflections on simple epidemic projection models and the COVID-19 pandemic. *The Mathematical Intelligencer, 42*(3), 1–9. DOI: 10.1007/s00283-020-09997-7 PMID: 34934263

Aryffin, H., Sahbudin, M., Pitchay, S., Halim, A., & Sahbudin, I. (2025). Technological trends in epidemic intelligence for infectious disease surveillance: A systematic literature review. *PeerJ. Computer Science, 11*, e2874. Advance online publication. DOI: 10.7717/peerj-cs.2874 PMID: 40567698

Arzilli, G., De Vita, E., Pasquale, M., Carloni, L. M., Pellegrini, M., Di Giacomo, M., Esposito, E., Porretta, A. D., & Rizzo, C. (2024). Innovative techniques for infection control and surveillance in hospital settings and long-term care facilities: A scoping review. *Antibiotics (Basel, Switzerland), 13*(1), 77. DOI: 10.3390/antibiotics13010077 PMID: 38247635

Asan, O., Bayrak, A. E., & Choudhury, A. (2020). Artificial Intelligence and Human Trust in Healthcare: Focus on Clinicians. *Journal of Medical Internet Research, 22*(6), e15154. DOI: 10.2196/15154 PMID: 32558657

Asokan, D. R., Smith, C. M., & Huq, F. A. (2025). Digitalisation as a catalyst for supplier diversity, equity and inclusion. *International Journal of Operations & Production Management*, •••, 1–26.

Astobiza, M. A., Alonso, M., & Ortega Lozano, R. (2025). Trust and AI in healthcare: A systematic review. *Monash Bioethics Review*. Advance online publication. DOI: 10.1007/s40592-025-00272-z

Atadoga, A., Obi, O. C., Osasona, F., Onwusinkwue, S., Daraojimba, A. I., & Dawodu, S. O. (2024). AI in supply chain optimization: A comparative review of USA and African trends. *International Journal of Science and Research Archive, 11*(1), 896–903. DOI: 10.30574/ijsra.2024.11.1.0156

August, E., Tadesse, L., O'Neill, M. S., Eisenberg, J. N. S., Wong, R., Kolars, J. C., & Bekele, A. (2022). What is global health equity? A proposed definition. *Annals of Global Health, 88*(1), 50. Advance online publication. DOI: 10.5334/aogh.3754 PMID: 35860038

Aung, M. M., & Chang, Y. S. (2014). Traceability in a food supply chain: Safety and quality perspectives. *Food Control*, *39*, 172–184. DOI: 10.1016/j.foodcont.2013.11.007

Ayesiga, I., Yeboah, M. O., Okoro, L. N., Edet, E. N., Gmanyami, J. M., Ovye, A., Atimango, L., Gadzama, B. N., Kembabazi, E., & Atwau, P. (2025). Artificial intelligence-enhanced biosurveillance for antimicrobial resistance in sub-Saharan Africa. *International Health*, *17*(5), 795–803. DOI: 10.1093/inthealth/ihae081 PMID: 39545538

Ayyaswamy, K. (2026). AI technologies for improving education systems. In Malahlela, M., & Sadiki, M. (Eds.), *Breaking structural inequality and enhancing social justice in higher education* (pp. 65–92). IGI Global Scientific Publishing., DOI: 10.4018/979-8-3373-3236-9.ch003

Ayyaswamy, K. N., & Vianny, M. (2025). Strategies for combating criminal use and abuse of artificial intelligence. In *Avoiding ad fraud and supporting brand safety: Programmatic advertising solutions*. IGI Global., DOI: 10.4018/979-8-3693-7041-4.ch009

Badul, S., Abia, A. L., Amoako, D. G., Perrett, K., Bester, L. A., & Essack, S. Y. (2021). From the farms to the dining table: The distribution and molecular characteristics of antibiotic-resistant *Enterococcus* spp. in intensive pig farming in South Africa. *Microorganisms*, *9*(5), 882. DOI: 10.3390/microorganisms9050882 PMID: 33918989

Bagave, P., Westberg, M., Janssen, M., & Ding, A. Y. (2025). *Accountability framework for healthcare AI systems: Towards joint accountability in decision making* (Preprint). arXiv. https://doi.org//arXiv.2509.03286DOI: 10.48550

Balcázar, J. L. (2025). Wastewater-based epidemiology as a complementary tool for antimicrobial resistance surveillance: Overcoming barriers to integration. *BioEssays : News and Reviews in Molecular, Cellular and Developmental Biology*, *47*(8), e70034. Advance online publication. DOI: 10.1002/bies.70034 PMID: 40522150

Baltrusaitis, K., Brownstein, J. S., Scarpino, S. V., Bakota, E., Crawley, A. W., Conidi, G., Gunn, J. E., Gray, J., Zink, A., Santillana, M., & Smolinski, M. S. (2018). Comparison of crowd-sourced, electronic health records-based, and traditional healthcare-based influenza-tracking systems at multiple spatial resolutions in the United States. *BMC Infectious Diseases*, *18*(1), 403. DOI: 10.1186/s12879-018-3322-3 PMID: 30111305

Banerjee, S., & Van Der Heijden, M. (2022). Soil microbiomes and one health. *Nature Reviews. Microbiology*, *21*(1), 6–20. DOI: 10.1038/s41579-022-00779-w PMID: 35999468

Barboza, P., Vaillant, L., Le Strat, Y., Hartley, D. M., Nelson, N. P., Mawudeku, A., Madoff, L. C., Linge, J. P., Collier, N., Brownstein, J. S., Yangarber, R., Astagneau, P., & Valenciano, M. (2013). Factors influencing performance of Internet-based biosurveillance systems used in epidemic intelligence for early detection of infectious diseases outbreaks. *PLoS One*, *8*(3), e57252. DOI: 10.1371/journal.pone.0057252 PMID: 23472077

Bazzano, A. N., Mantsios, A., Mattei, N., Kosorok, M. R., & Culotta, A. (2025). AI can be a powerful social innovation for public health if community engagement is at the core. *Journal of Medical Internet Research*, *27*(1), e68198. Advance online publication. DOI: 10.2196/68198 PMID: 39841529

Bedford, J., Farrar, J., Ihekweazu, C., Kang, G., Koopmans, M., & Nkengasong, J. (2019). A new twenty-first century science for effective epidemic response. *Nature*, *575*(7781), 130–136. DOI: 10.1038/s41586-019-1717-y PMID: 31695207

Benedum, C. M., Seidahmed, O. M. E., Eltahir, E. A. B., & Markuzon, N. (2018). Statistical modeling of the effect of rainfall flushing on dengue transmission in Singapore. *PLoS Neglected Tropical Diseases*, *12*(12), e0006935. DOI: 10.1371/journal.pntd.0006935 PMID: 30521523

Bergquist, R., Zheng, J.-X., & Zhou, X.-N. (2024). Synergistic integration of climate change and zoonotic diseases by artificial intelligence: A holistic approach for sustainable solutions. *Science in One Health*, *3*, 100070. DOI: 10.1016/j.soh.2024.100070 PMID: 39077386

Bernardo, T. M., Rajik, A., Young, I., Robiadek, K., Pham, M. T., & Funk, J. A. (2013). Scoping review on search queries and social media for disease surveillance: A chronology of innovation. *Journal of Medical Internet Research*, *15*(7), e2740. Advance online publication. DOI: 10.2196/jmir.2740 PMID: 23896182

Bhardwaj, T., & Sumangali, K. (2025). An explainable federated blockchain framework with privacy-preserving AI optimization for securing healthcare data. *Scientific Reports*, *15*(1), 21799. Advance online publication. DOI: 10.1038/s41598-025-04083-4 PMID: 40595873

Bhartiya, S., & Mehrotra, D. (2015). Applying CHAID algorithm to investigate critical attributes of secured interoperable health data exchange. *International Journal of Electronic Healthcare*, *8*(1), 25–50. DOI: 10.1504/IJEH.2015.071638 PMID: 26559072

Bhattacharya, D., Das, A., Satyaprakash, K., & Singh, M. K. (2025). Advanced biotechnological tools and techniques for muscle food identification. In Nimmanapalli, R., & Kumar Singh, M. (Eds.), *Trends in animal-based foods*. Springer., DOI: 10.1007/978-3-032-00575-5_9

Bhatt, C., Kumar, I., Vijayakumar, V., Singh, K. U., & Kumar, A. (2021). The state of the art of deep learning models in medical science and their challenges. *Multimedia Systems*, *27*(4), 599–613. DOI: 10.1007/s00530-020-00694-1

Binamungu, L. P., & Kimaro, H. C. (2025). Problematizing the determinants of success of interoperability of information systems through the lens of stakeholder theory and enterprise interoperability framework. *The Electronic Journal on Information Systems in Developing Countries*, *91*(4), e70020. DOI: 10.1002/isd2.70020

Bisogni, C., Castiglione, A., Hossain, S., Narducci, F., & Umer, S. (2022). Impact of deep learning approaches on facial expression recognition in healthcare industries. *IEEE Transactions on Industrial Informatics*, *18*(8), 5619–5627. DOI: 10.1109/TII.2022.3141400

Bivins, A., Kaya, D., Bibby, K., Simpson, S., Bustin, S., Shanks, O., & Ahmed, W. (2021). Variability in RT-qPCR assay parameters indicates unreliable SARS-CoV-2 RNA quantification for wastewater surveillance. *Water Research*, *203*, 117516–117516. DOI: 10.1016/j.watres.2021.117516 PMID: 34412018

Bokulich, N. A., Lewis, Z. T., Boundy-Mills, K., & Mills, D. A. (2016). A new perspective on microbial landscapes within food production. *Current Opinion in Biotechnology*, *37*, 182–189. DOI: 10.1016/j.copbio.2015.12.008 PMID: 26773388

Boogaerts, T., Ahmed, F., Choi, P., Tscharke, B., O'Brien, J., De Loof, H., Gao, J., Thai, P., Thomas, K., Mueller, J., Hall, W., Covaci, A., & Van Nuijs, A. (2021). Current and future perspectives for wastewater-based epidemiology as a monitoring tool for pharmaceutical use. *The Science of the Total Environment*, *789*, 148047. DOI: 10.1016/j.scitotenv.2021.148047 PMID: 34323839

Boolchandani, M., D'Souza, A. W., & Dantas, G. (2019). Sequencing-based methods and resources to study antimicrobial resistance. *Nature Reviews. Genetics*, *20*(6), 356–370. DOI: 10.1038/s41576-019-0108-4 PMID: 30886350

Borda, A., Molnar, A., Neesham, C., & Kostkova, P. (2022). Ethical issues in AI-enabled disease surveillance: Perspectives from global health. *Applied Sciences (Basel, Switzerland)*, *12*(8), 3890. Advance online publication. DOI: 10.3390/app12083890

Bosona, T., & Gebresenbet, G. (2013). Food traceability as an integral part of logistics management in food and agricultural supply chain. *Food Control*, *33*(1), 32–48. DOI: 10.1016/j.foodcont.2013.02.004

Boudou, M., Garvey, P., ÓhAiseadha, C., O'Dwyer, J., Burke, D. T., & Hynds, P. (2025). Effects of the 2018 European summer heatwave on the incidence of sporadic bacterial gastroenteritis in a temperate maritime climate region (Republic of Ireland). *One Health*, *20*, 101105. DOI: 10.1016/j.onehlt.2025.101105 PMID: 40547209

Boutros, F., Struc, V., Fierrez, J., & Damer, N. (2023). Synthetic data for face recognition: Current state and future prospects. *Image and Vision Computing*, *135*, 104688. DOI: 10.1016/j.imavis.2023.104688

Bowes, D. A. (2024). Towards a precision model for environmental public health: Wastewater-based epidemiology to assess population-level exposures and related diseases. *Current Epidemiology Reports*, *11*(3), 131–139. DOI: 10.1007/s40471-024-00350-6

Bozzini, A., Huo, L., Brugnaro, S., Morgante, G., Persson, H. J., Finozzi, V., Battisti, A., & Faccoli, M. (2025). Multispectral drone images for the early detection of bark beetle infestations: Assessment over large forest areas in the Italian South-Eastern Alps. *Frontiers in Forests and Global Change*, *8*, 1532954. Advance online publication. DOI: 10.3389/ffgc.2025.1532954

Brighton, K., Fisch, S., Wang, H., Vigil, K., & Aw, T. G. (2024). Targeted community wastewater surveillance for SARS-CoV-2 and Mpox virus during a festival mass-gathering event. *The Science of the Total Environment*, *906*, 167443. Advance online publication. DOI: 10.1016/j.scitotenv.2023.167443 PMID: 37793442

Brownstein, J. S., Freifeld, C. C., & Madoff, L. C. (2009). Digital disease detection—Harnessing the web for public health surveillance. *The New England Journal of Medicine*, *360*(21), 2153–2157. DOI: 10.1056/NEJMp0900702 PMID: 19423867

Brownstein, J. S., Freifeld, C. C., Reis, B. Y., & Mandl, K. D. (2008). Surveillance sans frontières: Internet-based emerging infectious disease intelligence and the HealthMap project. *PLoS Medicine*, *5*(7), e151. DOI: 10.1371/journal.pmed.0050151 PMID: 18613747

Buckeridge, D. L. (2007). Outbreak detection through automated surveillance: A review of the determinants of detection. *Journal of Biomedical Informatics*, *40*(4), 370–379. DOI: 10.1016/j.jbi.2006.09.003 PMID: 17095301

Budd, J., Miller, B. S., Manning, E. M., Lampos, V., Zhuang, M., Edelstein, M., Rees, G., Emery, V. C., Stevens, M. M., Keegan, N., Short, M. J., Pillay, D., Manley, E., Cox, I. J., Heymann, D., Johnson, A. M., & McKendry, R. A. (2020). Digital technologies in the public-health response to COVID-19. *Nature Medicine*, *26*(8), 1183–1192. DOI: 10.1038/s41591-020-1011-4 PMID: 32770165

Burgos, N., & Colliot, O. (2020). Brain disease prediction. *Current Opinion in Neurology*, *33*(4), 439. DOI: 10.1097/WCO.0000000000000838 PMID: 32657885

Callahan, B. J., McMurdie, P. J., Rosen, M. J., Han, A. W., Johnson, A. J. A., & Holmes, S. P. (2016). DADA2: High-resolution sample inference from Illumina amplicon data. *Nature Methods*, *13*(7), 581–583. DOI: 10.1038/nmeth.3869 PMID: 27214047

Calvetti, D., de Mattos Nascimento, D. L., de Araújo Filho, F. N. M., Abreu, R. H. V., & Papadopoulos, N. A. (2024). Linked data to streaming data sensors sources for direct access. *Brazilian Journal of Operations & Production Management*, *21*(1). Advance online publication. DOI: 10.14488/BJOPM.1993.2024

Câmara, J., Perestrelo, R., Berenguer, C., Andrade, C., Gomes, T., Olayanju, B., Kabir, A., Rocha, C., Teixeira, J., & Pereira, J. (2022). Green Extraction Techniques as Advanced Sample Preparation Approaches in Biological, Food, and Environmental Matrices: A Review. *Molecules (Basel, Switzerland)*, *27*. Advance online publication. DOI: 10.3390/molecules27092953 PMID: 35566315

Caminade, C., Van Dijk, J., Baylis, M., & Williams, D. (2015). Modelling recent and future climatic suitability for fasciolosis in Europe. *Geospatial Health*, *9*(2), 301. DOI: 10.4081/gh.2015.352 PMID: 25826311

Campanella, P., Lovato, E., Marone, C., Fallacara, L., Mancuso, A., Ricciardi, W., & Specchia, M. L. (2016). The impact of electronic health records on healthcare quality: A systematic review and meta-analysis. *European Journal of Public Health*, *26*(1), 60–64. DOI: 10.1093/eurpub/ckv122 PMID: 26136462

Carlson, C. J., Albery, G. F., Merow, C., Trisos, C. H., Zipfel, C. M., Eskew, E. A., Olival, K. J., Ross, N., & Bansal, S. (2022). Climate change increases cross-species viral transmission risk. *Nature*, *607*(7919), 555–562. DOI: 10.1038/s41586-022-04788-w PMID: 35483403

Carneiro, R., Nika, M., Gil-Solsona, R., Diamanti, K., Thomaidis, N., Corominas, L., & Gago-Ferrero, P. (2024). A critical review of wastewater-based epidemiology as a tool to evaluate the unintentional human exposure to potentially harmful chemicals. *Analytical and Bioanalytical Chemistry*, *417*(3), 495–511. DOI: 10.1007/s00216-024-05596-z PMID: 39422714

Carrillo de Albornoz, S., Sia, K.-L., & Harris, A. (2022). The effectiveness of teleconsultations in primary care: Systematic review. *Family Practice*, *39*(1), 168–182. DOI: 10.1093/fampra/cmab077 PMID: 34278421

Carroll, S. R., Garba, I., Figueroa-Rodríguez, O. L., Holbrook, J., Lovett, R., Materechera, S., Parsons, M., Raseroka, K., Rodriguez-Lonebear, D., Rowe, R., Sara, R., Walker, J. D., Anderson, J., & Hudson, M. (2020). The CARE Principles for Indigenous Data Governance. *Data Science Journal*, *19*, 43. Advance online publication. DOI: 10.5334/dsj-2020-043

Casino, F., Kanakaris, V., Dasaklis, T. K., Moschuris, S., & Rachaniotis, N. P. (2021). Modeling food supply chain traceability based on blockchain technology. *IFAC-PapersOnLine*, *54*(1), 115–121. DOI: 10.1016/j.ifacol.2019.11.620

Castillo-Salgado, C. (2010). Trends and directions of global public health surveillance. *Epidemiologic Reviews*, *32*(1), 93–109. DOI: 10.1093/epirev/mxq008 PMID: 20534776

Catarro, G., Pelixo, R., Feijó, M., Rosado, T., Socorro, S., Araujo, A., & Gallardo, E. (2025). Analytical Approaches Using GC-MS for the Detection of Pollutants in Wastewater Towards Environmental and Human Health Benefits: A Comprehensive Review. *Chemosensors (Basel, Switzerland)*, *13*(7), 253. Advance online publication. DOI: 10.3390/chemosensors13070253

Chamberlain, S. D., Singh, I., Ariza, C., Daitch, A., & Dalziel, B. D. (2020). Real-time detection of COVID-19 epicenters within the United States using a network of smart thermometers (preprint). medRxiv. https://doi.org/DOI: 10.1101/2020.04.06.20039909

Charlebois, S., Sterling, B., Haratifar, S., & Naing, S. K. (2014). Comparison of global food traceability regulations and requirements. *Comprehensive Reviews in Food Science and Food Safety*, *13*(5), 1104–1123. DOI: 10.1111/1541-4337.12101

Charles-Smith, L. E., Reynolds, T. L., Cameron, M. A., Conway, M., Lau, E. H. Y., Olsen, J. M., Pavlin, J. A., Shigematsu, M., Streichert, L. C., Suda, K. J., & Corley, C. D. (2015). Using social media for actionable disease surveillance and outbreak management: A systematic literature review. *PLoS One*, *10*(10), e0139701. DOI: 10.1371/journal.pone.0139701 PMID: 26437454

Chatfield, A. T., & AlAnazi, J. (2015). Collaborative governance matters to e-government interoperability: An analysis of citizen-centric integrated interoperable e-government implementation in Saudi Arabia. [IJPADA]. *International Journal of Public Administration in the Digital Age*, *2*(3), 24–44. DOI: 10.4018/ijpada.2015070102

Chaudhary, A., Chaudhary, P., Singh, S., Bhandari, G., & Tripathi, V. (2025). Advances in contemporary tools for detecting and diagnosing plant pathogens. In J.-T. Chen, M. Khan, A. Parveen, & J. K. Patra (Eds.), *Molecular and biotechnological tools for plant disease management* (Interdisciplinary Biotechnological Advances). Springer. https://doi.org/DOI: 10.1007/978-981-97-7510-1_9

Chauhan, S., & Dutta, A. (2025). ARTIFICIAL INTELLIGENCE IN STUDENT PRIVACY AND DATA SECURITY. *International Journal of Advanced Research in Computer Science*, *16*(3).

Cheah, B. C., Vicente, C. R., & Chan, K. R. (2025). Machine learning and artificial intelligence for infectious disease surveillance, diagnosis, and prognosis. *Viruses*, *17*(7), 882. DOI: 10.3390/v17070882 PMID: 40733500

Chel, Y. H., & Poh, L. L. (2025). Brain tumor classification in MRI: Insights from LIME and Grad-CAM explainable AI techniques. *IEEE Access : Practical Innovations, Open Solutions*.

Chen, F., Wang, L., Hong, J., Jiang, J., & Zhou, L. (2024). Unmasking bias in AI: A systematic review of bias detection and mitigation strategies in electronic health record-based models. *Journal of the American Medical Informatics Association : JAMIA*, *31*(5), 1172–1183. DOI: 10.1093/jamia/ocae060 PMID: 38520723

Chen, F., Zhao, C., Yang, X., Luo, N., & Sun, C. (2025). A lightweight accountable parallel blockchain architecture based on redactable blockchain for agri-food traceability. *Foods*, *14*(4), 623. DOI: 10.3390/foods14040623 PMID: 40002066

Chen, H., Chen, Z., Lin, F., & Zhuang, P. (2021). Effective management for blockchain-based agri-food supply chains using deep reinforcement learning. *IEEE Access : Practical Innovations, Open Solutions*, *9*, 36008–36018. DOI: 10.1109/ACCESS.2021.3062410

Chen, J., Remulla, D., Nguyen, J. H., Dua, A., Liu, Y., Dasgupta, P., & Hung, A. J. (2019). Current status of artificial intelligence applications in urology and their potential to influence clinical practice. *BJU International*, *124*(4), 567–577. DOI: 10.1111/bju.14852 PMID: 31219658

Chen, X., Chughtai, A. A., & MacIntyre, C. R. (2023). Artificial intelligence in public health: The potential of epidemic early warning systems. *Infectious Disease Reports*, *15*(3), 222–236. DOI: 10.3390/idr15030024

Chen, X., & Moraga, P. (2025). Forecasting dengue across Brazil with LSTM neural networks and SHAP-driven lagged climate and spatial effects. *BMC Public Health, 25*(1), 973. Advance online publication. DOI: 10.1186/s12889-025-22106-7 PMID: 40075398

Chen, X., & Moraga, P. (2026). Dengue forecasting and outbreak detection in Brazil using LSTM: Integrating human mobility and climate factors. *Infectious Disease Modelling, 11*(1), 338–354. DOI: 10.1016/j.idm.2025.11.002 PMID: 41321683

Chen, X., Wang, Q., Cui, B., Chen, G., Xie, T., & Yang, W. (2023). Ecological time lags in biodiversity response to habitat changes. *Journal of Environmental Management, 346*, 118965. DOI: 10.1016/j.jenvman.2023.118965 PMID: 37741191

Cheong, B. C. (2024). Transparency and accountability in AI systems: Safeguarding wellbeing in the age of algorithmic decision-making. *Frontiers in Human Dynamics, 6*, Article 1421273.

Cheong, J. H., Jolly, E., Xie, T., Byrne, S., Kenney, M., & Chang, L. J. (2023). Py-feat: Python facial expression analysis toolbox. *Affective Science, 4*(4), 781–796. DOI: 10.1007/s42761-023-00191-4 PMID: 38156250

Chiu, C. Y., & Miller, S. A. (2019). Clinical metagenomics. *Nature Reviews. Genetics, 20*(6), 341–355. DOI: 10.1038/s41576-019-0113-7 PMID: 30918369

Chopra, T., & Pathrotkar, N. (2024). Implementation of a food traceability system utilizing the Internet of Things and blockchain technology to enhance safety measures within the food supply chain. In *BIO Web of Conferences* (Vol. 82, p. 05016). EDP Sciences. DOI: 10.1051/bioconf/20248205016

Ciannella, S., González-Fernández, C., & Gómez-Pastora, J. (2023). Recent progress on wastewater-based epidemiology for COVID-19 surveillance: A systematic review of analytical procedures and epidemiological modeling. *The Science of the Total Environment, 878*, 162953–162953. DOI: 10.1016/j.scitotenv.2023.162953 PMID: 36948304

Cinar, R. F., Yuksek, G., Lale, T., Ekinci, S., Izci, D., & Ma'arif, A. (2025). SHAP-based framework for temporal detection of sensor drift in gas sensor arrays. *Journal of Robotics and Control, 6*(5), 2592–2601. DOI: 10.18196/jrc.v6i5.28048

Ciuffreda, L., Rodríguez-Pérez, H., & Flores, C. (2021). Nanopore sequencing and its application to the study of microbial communities. *Computational and Structural Biotechnology Journal, 19*, 1497–1511. DOI: 10.1016/j.csbj.2021.02.020 PMID: 33815688

Cook, D. A., Levinson, A. J., Garside, S., Dupras, D. M., Erwin, P. J., & Montori, V. M. (2008). Internet-based learning in the health professions: A meta-analysis. *Journal of the American Medical Association*, *300*(10), 1181–1196. DOI: 10.1001/jama.300.10.1181 PMID: 18780847

Corburn, J., Vlahov, D., Mberu, B., Riley, L., Caiaffa, W. T., Rashid, S. F., Ko, A., Patel, S., Jukur, S., Martínez-Herrera, E., Jayasinghe, S., Agarwal, S., Nguendo-Yongsi, B., Weru, J., Ouma, S., Edmundo, K., Oni, T., & Ayad, H. (2020). Slum Health: Arresting COVID-19 and Improving Well-Being in Urban Informal Settlements. *Journal of Urban Health : Bulletin of the New York Academy of Medicine*, *97*(3), 348–357. DOI: 10.1007/s11524-020-00438-6 PMID: 32333243

Cornet, V. J., Cantin, N. E., Joyce, K. E., Leggat, W., Ainsworth, T. D., & Heron, S. F. (2025). Enhancing coral bleaching predictive tools through integrating sensitivity to heat exposure. *Biological Conservation*, *302*, 110958. DOI: 10.1016/j.biocon.2024.110958

Corrin, T., Rabeenthira, P., Young, K., Mathiyalagan, G., Baumeister, A., Pussegoda, K., & Waddell, L. (2024). A scoping review of human pathogens detected in untreated human wastewater and sludge. *Journal of Water and Health*, *22*(2), 436–449. DOI: 10.2166/wh.2024.326 PMID: 38421635

Cosme, F., Vilela, A., & Gonçalves, B. (2025). Biotechnology in quality analysis and sensory evaluation. In Gaikwad, P. S., & Choudhary, P. (Eds.), *Biotechnological innovations in food processing*. Springer., DOI: 10.1007/978-3-031-96526-5_14

Costa, C., Antonucci, F., Pallottino, F., Aguzzi, J., Sarriá, D., & Menesatti, P. (2013). A review on agri-food supply chain traceability by means of RFID technology. *Food and Bioprocess Technology*, *6*(2), 353–366. DOI: 10.1007/s11947-012-0958-7

Cramer, E. Y., Ray, E. L., Lopez, V. K., Bracher, J., Brennen, A., Rivadeneira, A. J. C., Gerding, A., Gneiting, T., House, K. H., Huang, Y., Jayawardena, D., Kanji, A. H., Khandelwal, A., Le, K., Mühlemann, A., Niemi, J., Stark, A., Shah, A., Wattanachit, N., & Reich, N. G. (2022). Evaluation of individual and ensemble probabilistic forecasts of COVID-19 mortality in the United States. *Proceedings of the National Academy of Sciences of the United States of America*, *119*(15), e2113561119. DOI: 10.1073/pnas.2113561119 PMID: 35394862

Crawford, A., & Serhal, E. (2020). Digital health equity and COVID-19: The innovation curve cannot reinforce the social gradient of health. *Journal of Medical Internet Research*, *22*(6), e19361. DOI: 10.2196/19361 PMID: 32452816

Creydt, M., & Fischer, M. (2019). Blockchain and more - Algorithm driven food traceability. *Food Control*, *105*, 45–51. DOI: 10.1016/j.foodcont.2019.05.019

Cuadros, D., Chen, X., Li, J., Omori, R., & Musuka, G. (2024). Advancing Public Health Surveillance: Integrating Modeling and GIS in the Wastewater-Based Epidemiology of Viruses, a Narrative Review. *Pathogens (Basel, Switzerland), 13*(8), 685. Advance online publication. DOI: 10.3390/pathogens13080685 PMID: 39204285

da Costa, J. M. F., Costa, A. C., Silveira, C. da S., & Gonçalves, S. T. N. Marcos Junior, A. D., & Cavalcanti, L. P. de G. (2026). Forecasting and Early Warning Systems for Dengue Outbreaks: Updated Narrative Review. Revista Da Sociedade Brasileira de Medicina Tropical, 59. https://doi.org/DOI: 10.1590/0037-8682-0429-2025

Dabbene, F., Gay, P., & Tortia, C. (2014). Traceability issues in food supply chain management: A review. *Biosystems Engineering, 120*, 65–80. DOI: 10.1016/j.biosystemseng.2013.09.006

Damiani, G. (2019). Data-driven decision making and stakeholders involvement in public health: A fascinating challenge for sustainable development. In 53 Days of Preventive Medicine – International Congress (Niš, Serbia). Public Health Institute Niš; Faculty of Medicine, University of Niš.

Damiati, S. A. (2020). Digital pharmaceutical sciences. *AAPS PharmSciTech, 21*(6), 206. DOI: 10.1208/s12249-020-01747-4 PMID: 32715351

Daughton, C. G. (2001). Illicit drugs in municipal sewage. In Daughton, C. G., & Jones-Lepp, T. L. (Eds.), *Pharmaceuticals and personal care products in the environment* (pp. 348–364). American Chemical Society. DOI: 10.1021/bk-2001-0791.ch020

Daughton, C. G. (2012). Using biomarkers in sewage to monitor community-wide human health: Isoprostanes as conceptual prototype. *The Science of the Total Environment, 424*, 16–38. DOI: 10.1016/j.scitotenv.2012.02.038 PMID: 22425170

Daughton, C. G. (2018). Monitoring wastewater for assessing community health: Sewage chemical-information mining (SCIM). *The Science of the Total Environment, 619–620*, 748–764. DOI: 10.1016/j.scitotenv.2017.11.102 PMID: 29161600

Davies, J., & Davies, D. (2010). Origins and evolution of antibiotic resistance. *Microbiology and Molecular Biology Reviews : MMBR, 74*(3), 417–433. DOI: 10.1128/MMBR.00016-10 PMID: 20805405

Davis, B., Brown, C., Gupta, S., Calarco, J., Liguori, K., Milligan, E., Harwood, V., Pruden, A., & Keenum, I. (2023). Recommendations for the use of metagenomics for routine monitoring of antibiotic resistance in wastewater and impacted aquatic environments. *Critical Reviews in Environmental Science and Technology, 53*(19), 1731–1756. DOI: 10.1080/10643389.2023.2181620 de Boer, J., & Harvey, J. (2020). Range-Expansion in Processionary Moths and Biological Control. *Insects, 11*(5), 267. DOI: 10.3390/insects11050267 PMID: 32353938

De Oliveira, A., De Melo Vieira, A., & Santos, J. (2023). Trends and challenges in analytical chemistry for multi-analysis of illicit drugs employing wastewater-based epidemiology. *Analytical and Bioanalytical Chemistry*, 1 - 10. https://doi.org/DOI: 10.1007/s00216-023-04644-4

Deiner, M. S., Consoli, S., & Brownstein, J. S. (2025). AI-driven epidemic intelligence: The future of outbreak detection and response. *Frontiers in Artificial Intelligence, 8*, 1645467. Advance online publication. DOI: 10.3389/frai.2025.1645467 PMID: 40810005

Delplanque, A., Linchant, J., Vincke, X., Lamprey, R., Théau, J., Vermeulen, C., Foucher, S., Ouattara, A., Kouadio, R., & Lejeune, P. (2024). Will artificial intelligence revolutionize aerial surveys? A first large-scale semi-automated survey of African wildlife using oblique imagery and deep learning. *Ecological Informatics, 82*, 102679. DOI: 10.1016/j.ecoinf.2024.102679

Delpy, L., Astbury, C. C., Aenishaenslin, C., Ruckert, A., Penney, T. L., Wiktorowicz, M., Ciss, M., Benko, R., & Bordier, M. (2024). Integrated surveillance systems for antibiotic resistance in a One Health context: A scoping review. *BMC Public Health, 24*(1), 1717. DOI: 10.1186/s12889-024-19158-6 PMID: 38937706

Deng, M., & Feng, P. (2020). A food traceability system based on blockchain and radio frequency identification technologies. *Journal of Computer and Communications, 8*(9), 17–27. DOI: 10.4236/jcc.2020.89002

Desai, A. N., Kraemer, M. U. G., Bhatia, S., Cori, A., Nouvellet, P., Herringer, M., Cohn, E. L., Carrion, M., Brownstein, J. S., Madoff, L. C., & Lassmann, B. (2019). Real-time epidemic forecasting: Challenges and opportunities. *Health Security, 17*(4), 268–275. DOI: 10.1089/hs.2019.0022 PMID: 31433279

Deutsch, C. A., Tewksbury, J. J., Tigchelaar, M., Battisti, D. S., Merrill, S. C., Huey, R. B., & Naylor, R. L. (2018). Increase in crop losses to insect pests in a warming climate. *Science, 361*(6405), 916–919. DOI: 10.1126/science.aat3466 PMID: 30166490

Dhal, P., & Azad, C. (2022). Feature selection survey. *Applied Intelligence, 52*(4), 4543. DOI: 10.1007/s10489-021-02550-9

Dharshini, L. C. P., Salim, P., Sharanya, K., Palanisamy, B., Banerjee, S., & Mandal, A. K. A. (2025). RNAi: A technology for enhancing commercially valuable traits in industrial crops. In N. Kumar (Ed.), *Industrial crops improvement* (Sustainable Landscape Planning and Natural Resources Management). Springer. https://doi.org/ DOI: 10.1007/978-3-031-75937-6_7

Di Tommaso, P., Chatzou, M., Floden, E. W., Barja, P. P., Palumbo, E., & Notredame, C. (2017). Nextflow enables reproducible computational workflows. *Nature Biotechnology*, *35*(4), 316–319. DOI: 10.1038/nbt.3820 PMID: 28398311

Dias, J. S. (2021). Nutritional quality and health benefits of vegetables: A review. *Food and Nutrition Sciences*, *3*(10), 1354–1374. DOI: 10.4236/fns.2012.310179

Djemiel, C., Maron, P., Terrat, S., Dequiedt, S., Cottin, A., & Ranjard, L. (2022). Inferring microbiota functions from taxonomic genes: A review. *GigaScience*, *11*, giab090. Advance online publication. DOI: 10.1093/gigascience/giab090 PMID: 35022702

Doerwald, F., Stalling, I., Recke, C., Busse, H., Shrestha, R., Rach, S., & Bammann, K. (2024). A rapid review of digital approaches for the participatory development of health-related interventions. *Frontiers in Public Health*, *12*, 1461422. DOI: 10.3389/fpubh.2024.1461422 PMID: 39678234

Domingo, J., Souza, M., Nadal, M., & Barbosa, F. (2025). A One Health approach to polybrominated diphenyl ethers (PBDEs): Integrating human, animal, and environmental health perspectives. *Chemosphere*, *386*, 144644. DOI: 10.1016/j.chemosphere.2025.144644 PMID: 40829294

Drew, D. A., Nguyen, L. H., Steves, C. J., Menni, C., Freydin, M., Varsavsky, T., Sudre, C. H., Cardoso, M. J., Ourselin, S., Wolf, J., Spector, T. D., Chan, A. T., Chan, A. T., Drew, D. A., Nguyen, L. H., Joshi, A. D., Guo, C.-G., Ma, W., Lo, C.-H., & Spector, T. (2020). Rapid implementation of mobile technology for real-time epidemiology of COVID-19. *Science*, *368*(6497), 1362–1367. DOI: 10.1126/science.abc0473 PMID: 32371477

Duarte, A. S. R., Stärk, K. D., Munk, P., Leekitcharoenphon, P., Bossers, A., Luiken, R., Sarrazin, S., Lukjancenko, O., Pamp, S. J., Bortolaia, V., Nissen, J. N., Kirstahler, P., Van Gompel, L., Poulsen, C. S., Kaas, R. S., Hellmér, M., Hansen, R. B., Gomez, V. M., & Hald, T. (2020). Addressing learning needs on the use of metagenomics in antimicrobial resistance surveillance. *Frontiers in Public Health*, *8*, 38. DOI: 10.3389/fpubh.2020.00038 PMID: 32158739

Dunleavy, G., Nikolaou, C. K., Nifakos, S., Atun, R., Law, G. C. Y., & Tudor Car, L. (2019). Mobile digital education for health professions: Systematic review and meta-analysis by the Digital Health Education Collaboration. *Journal of Medical Internet Research*, *21*(2), e12937. DOI: 10.2196/12937 PMID: 30747711

Eaton, C. J., Coxon, S., Pattis, I., Chappell, A., Hewitt, J., & Gilpin, B. J. (2022). A framework for public health authorities to evaluate health determinants for wastewater-based epidemiology. *Environmental Health Perspectives*, *130*(12), 125001. Advance online publication. DOI: 10.1289/EHP11115 PMID: 36520537

Eby, P., Peel, A. J., Hoegh, A., Madden, W., Giles, J. R., Hudson, P. J., & Plowright, R. K. (2022). Pathogen spillover driven by rapid changes in bat ecology. *Nature*, *613*(7943), 340–344. DOI: 10.1038/s41586-022-05506-2 PMID: 36384167

Eddamiri, S., Bassine, F. Z., Hakam, O., Kambiet, P. L. K., & Chehbouni, A. (2025). Applications of artificial intelligence for forecasting and managing extremes. In *Climate Change and Rainfall Extremes in Africa* (pp. 91–113). Elsevier., DOI: 10.1016/B978-0-443-28867-8.00005-8

Ehrlinger, L., & Wöß, W. (2022). A survey of data quality measurement and monitoring tools. *Frontiers in Big Data*, *5*, 850611. DOI: 10.3389/fdata.2022.850611 PMID: 35434611

Ejeian, F., Etedali, P., Mansouri-Tehrani, H. A., Soozanipour, A., Low, Z. X., Asadnia, M., & Razmjou, A.. (2018). Biosensors for wastewater monitoring: A review. *Biosensors & Bioelectronics*, *118*, 66–79. DOI: 10.1016/j.bios.2018.07.019 PMID: 30056302

El Idrissi, A. H., Larfaoui, F., Dhingra, M., Johnson, A., Pinto, J., & Sumption, K. (2021). Digital technologies and implications for Veterinary Services. *Revue Scientifique et Technique (International Office of Epizootics)*, *40*(2), 455–468. DOI: 10.20506/rst.40.2.3237 PMID: 34542102

Elendu, C., Amaechi, D. C., Elendu, T. C., Jingwa, K. A., Okoye, O. K., John Okah, M., Ladele, J. A., Farah, A. H., & Alimi, H. A. (2023). Ethical implications of AI and robotics in healthcare. *Revista de Medicina (São Paulo)*, *102*(50), e36671. DOI: 10.1097/MD.0000000000036671 PMID: 38115340

Ellahi, R. M., Wood, L. C., & Bekhit, A. E.-D. A. (2023). Blockchain-Based Frameworks for Food Traceability: A Systematic Review. *Foods*, *12*(16), 3026. DOI: 10.3390/foods12163026 PMID: 37628025

Ellington, M. J., Ekelund, O., Aarestrup, F. M., Canton, R., Doumith, M., Giske, C., Grundman, H., Hasman, H., Holden, M. T. G., Hopkins, K. L., Iredell, J., Kahlmeter, G., Köser, C. U., MacGowan, A., Mevius, D., Mulvey, M., Naas, T., Peto, T., Rolain, J.-M., & Woodford, N. (2017). The role of whole genome sequencing in antimicrobial susceptibility testing of bacteria: Report from the EUCAST Subcommittee. *Clinical Microbiology and Infection : The Official Publication of the European Society of Clinical Microbiology and Infectious Diseases*, *23*(1), 2–22. DOI: 10.1016/j.cmi.2016.11.012 PMID: 27890457

Ellis, D. I., Muhamadali, H., & Allen, D. P.. (2022). Fingerprinting food: Current technologies for the detection of food fraud and adulteration. *Trends in Food Science & Technology*, *119*, 122–136. DOI: 10.1016/j.tifs.2021.11.016

Ensor, K., Schedler, J., Sun, T., Schneider, R., Mulenga, A., Wu, J., Stadler, L., & Hopkins, L. (2024). Online trend estimation and detection of trend deviations in sub-sewershed time series of SARS-CoV-2 RNA measured in wastewater. *Scientific Reports*, *14*(1), 5575. Advance online publication. DOI: 10.1038/s41598-024-56175-2 PMID: 38448481

Ercolini, D. (2013). High-throughput sequencing and metagenomics: Moving forward in the culture-independent analysis of food microbial ecology. *Applied and Environmental Microbiology*, *79*(10), 3148–3155. DOI: 10.1128/AEM.00256-13 PMID: 23475615

Ethics of artificial intelligence. (n.d.). In *Wikipedia*. Retrieved February 9, 2026, from https://en.wikipedia.org/wiki/Ethics_of_artificial_intelligence

European Commission. (2022). Food safety: From the farm to the fork. Publications Office of the European Union. https://food.ec.europa.eu/horizontal-topics/farm-fork-strategy_en

European Parliament and Council of the European Union. (2016). Regulation (EU) 2016/679 on the protection of natural persons with regard to the processing of personal data and on the free movement of such data (General Data Protection Regulation). *Official Journal of the European Union, L, 119*, 1–88.

European Parliament and Council of the European Union. (2024). Regulation (EU) 2024/1689 laying down harmonised rules on artificial intelligence (Artificial Intelligence Act). *Official Journal of the European Union, L 2024/1689.*

Fagherazzi, G., Goetzinger, C., Rashid, M. A., Aguayo, G. A., & Huiart, L. (2020). Digital health strategies to fight COVID-19 worldwide: Challenges, recommendations, and a call for papers. *Journal of Medical Internet Research*, *22*(6), e19284. DOI: 10.2196/19284 PMID: 32501804

Fairchild, A. L., Gable, L., Gostin, L. O., Bayer, R., Sweeney, P., & Janssen, R. S. (2007). Public goods, private data: HIV and the history, ethics, and uses of identifiable public health information. *Public Health Reports (Washington, D.C.)*, *122*(1, suppl), 7–15. DOI: 10.1177/00333549071220S103 PMID: 17354522

Falkner, S. (2025). Next-gen healthcare: Powered by artificial intelligence and smart biosensors. *Archives of Advanced Artificial Intelligence, Data Science, and Machine Learning*, *1*(1), 1–8.

Fanelli, V., Mascio, I., Miazzi, M. M., Savoia, M. A., De Giovanni, C., & Montemurro, C. (2021). Molecular Approaches to Agri-Food Traceability and Authentication: An Updated Review. *Foods*, *10*(7), 1644. DOI: 10.3390/foods10071644 PMID: 34359514

Farkas, K., Williams, R., Hillary, L., Garcia-Delgado, A., Jameson, E., Kevill, J., Wade, M., Grimsley, J., & Jones, D. (2024). Harnessing the Power of Next-Generation Sequencing in Wastewater-Based Epidemiology and Global Disease Surveillance. *Food and Environmental Virology*, *17*(1), 5. Advance online publication. DOI: 10.1007/s12560-024-09616-0 PMID: 39614945

Farooq, Z., Rocklöv, J., Wallin, J., Abiri, N., Sewe, M. O., Sjödin, H., & Semenza, J. C. (2022). Artificial intelligence to predict West Nile virus outbreaks with eco-climatic drivers. *The Lancet Regional Health. Europe*, *17*, 100370. DOI: 10.1016/j.lanepe.2022.100370 PMID: 35373173

Faye, S. L. B., Nkweteyim, D., Sow, G. H. C., Diop, B., Diongue, F. B., Cisse, B., Dickson, N., Dia, N., Badu, K., Ayana, G., Ba, M. L., Mensah Gyening, R.-M. O., Ondua, M., Ndiaye, F., Diouf, M., King, R., Duclos, V., Musa, E., Goitom, M., & Kong, J. (2025). Reimagining artificial intelligence for zoonotic disease detection in Africa: A decolonial approach rooted in community engagement and local knowledge. *AI and Ethics*, *5*(5), 5327–5354. DOI: 10.1007/s43681-025-00779-5

Fdez-Sanromán, A., Bernárdez-Rodas, N., Rosales, E., Pazos, M., González-Romero, E., & Sanromán, M. (2025). Biosensor Technologies for Water Quality: Detection of Emerging Contaminants and Pathogens. *Biosensors (Basel)*, *15*(3), 189. Advance online publication. DOI: 10.3390/bios15030189 PMID: 40136986

Feng, H., Wang, X., Duan, Y., Zhang, J., & Zhang, X. (2020). Applying blockchain technology to improve agri-food traceability: A review of development methods, benefits and challenges. *Journal of Cleaner Production*, *260*, 121031. DOI: 10.1016/j.jclepro.2020.121031

Floridi, L., Cowls, J., Beltrametti, M., Chatila, R., Chazerand, P., Dignum, V., Luetge, C., Madelin, R., Pagallo, U., Rossi, F., Schafer, B., Valcke, P., & Vayena, E. (2018). AI4People—An ethical framework for a good AI society. *Minds and Machines*, *28*(4), 689–707. DOI: 10.1007/s11023-018-9482-5 PMID: 30930541

Folasole, A. (2023). Data analytics and predictive modelling approaches for identifying emerging zoonotic infectious diseases: Surveillance techniques, prediction accuracy, and public health implications. *International Journal of Engineering Technology Research & Management*, *7*(12), 292–300.

Fontaine, D., Vielzeuf, V., Genestier, P., Limeux, P., Santucci-Sivilotto, S., Mory, E., Darmon, N., Lanteri-Minet, M., Mokhtar, M., Laine, M., & Vistoli, D. (2022). Artificial intelligence to evaluate postoperative pain based on facial expression recognition. *European Journal of Pain : EJP*, *26*(6), 1282–1291. DOI: 10.1002/ejp.1948 PMID: 35352426

Fontes, C., Hohma, E., Corrigan, C. C., & Lütge, C. (2022). AI-powered public surveillance systems: Why we (might) need them and how we want them. *Technology in Society*, *71*, 102137. DOI: 10.1016/j.techsoc.2022.102137

Fortune, A., Aime, A., Raymond, D., & Kumar, S. (2025). Nanotechnology in medicine: A double-edged sword for health outcomes. *Health Nanotechnology*, *1*(1), 9. DOI: 10.1186/s44301-025-00008-2

Foxman, B., Salzman, E., Gesierich, C., Gardner, S., Ammerman, M., Eisenberg, M., & Wigginton, K. (2024). Wastewater surveillance of antibiotic-resistant bacteria for public health action: Potential and challenges. *American Journal of Epidemiology*, *194*(5), 1192–1199. DOI: 10.1093/aje/kwae419 PMID: 39475072

Francombe, J., Booth, A., & Buckley Woods, H. (2022). Digital innovations for public health functions: A scoping review of international evidence. *JMIR Public Health and Surveillance*, *8*(7), e34605. DOI: 10.2196/34605 PMID: 35605152

Franz, E., Gras, L. M., & Dallman, T. (2014). Significance of whole genome sequencing for food safety and public health. *Food Microbiology*, *40*, 102–108. DOI: 10.1016/j.fm.2013.12.003

Freifeld, C. C., Mandl, K. D., Reis, B. Y., & Brownstein, J. S. (2008). HealthMap: Global infectious disease monitoring through automated classification and visualization of Internet media reports. *Journal of the American Medical Informatics Association : JAMIA*, *15*(2), 150–157. DOI: 10.1197/jamia.M2544 PMID: 18096908

Frohlich, K. L., & Abel, T. (2014). Environmental justice and health practices: Understanding how health inequities arise at the local level. In *From Health Behaviours to Health Practices* (pp. 43–56). Palgrave Macmillan. DOI: 10.1002/9781118898345.ch5

Gable, L., Ram, N., & Ram, J. L. (2020). Legal and ethical implications of wastewater monitoring of SARS-CoV-2 for COVID-19 surveillance. *Journal of Law and the Biosciences*, *7*(1), lsaa039. Advance online publication. DOI: 10.1093/jlb/lsaa039 PMID: 32793373

Gabriël, S., Dorny, P., Saelens, G., & Dermauw, V. (2022). Foodborne parasites and their complex life cycles challenging food safety in different food chains. *Foods*, *12*(1), 142. DOI: 10.3390/foods12010142 PMID: 36613359

Galvez, J. F., Mejuto, J. C., & Simal-Gandara, J. (2018). Future challenges on the use of blockchain for food traceability analysis. *Trends in Analytical Chemistry*, *107*, 222–232. DOI: 10.1016/j.trac.2018.08.011

Gao, Q., Chen, L., & Huang, Z. (2026). Opportunities and challenges of artificial intelligence in public health: A systematic review on technological efficacy, ethical dilemmas, and governance pathways. *Frontiers in Public Health*, *13*, 1748797. DOI: 10.3389/fpubh.2025.1748797 PMID: 41613098

Gao, Y., Li, X., Fan, X., Zhao, J., & Zhang, Z. (2022). Wastewater treatment plants as reservoirs and sources for antibiotic resistance genes: A review on occurrence, transmission, and removal. *Journal of Water Process Engineering*, *46*, 102539. Advance online publication. DOI: 10.1016/j.jwpe.2021.102539

García-Díez, J., Saraiva, S., Moura, D., Grispoldi, L., Cenci-Goga, B. T., & Saraiva, C. (2023). The importance of the slaughterhouse in surveilling animal and public health: A systematic review. *Veterinary Sciences*, *10*(2), 167. DOI: 10.3390/vetsci10020167 PMID: 36851472

Garcia-Vozmediano, A., Romano, A., Begovoeva, M., Pitti, M., Crescio, E., Brenda, A., Di Roberto, M., Gioia, A., Giraldo, A., Massone, E., Lanzarini, M. N., Raggio, A., De Vita, E., Ru, G., & Maurella, C. (2025). Integrating statistical and machine-learning approaches for Salmonella enterica surveillance in northwestern Italy: A One Health data-driven framework. *Microorganisms*, *13*(12), 2773. DOI: 10.3390/microorganisms13122773 PMID: 41471975

Gardy, J. L., & Loman, N. J. (2018). Towards a genomics-informed, real-time, global pathogen surveillance system. *Nature Reviews. Genetics*, *19*(1), 9–20. DOI: 10.1038/nrg.2017.88 PMID: 29129921

Ge, H., Zhu, Z., Dai, Y., Wang, B., & Wu, X. (2022). Facial expression recognition based on deep learning. *Computer Methods and Programs in Biomedicine, 215*, 106621. DOI: 10.1016/j.cmpb.2022.106621 PMID: 35164903

Geneviève, L. D., Martani, A., Wangmo, T., Paolotti, D., Koppeschaar, C., Kjelsø, C., Guerrisi, C., Hirsch, M., Woolley-Meza, O., Lukowicz, P., Flahault, A., & Elger, B. S. (2019). Participatory Disease Surveillance Systems: Ethical Framework. *Journal of Medical Internet Research, 21*(5), e12273. DOI: 10.2196/12273 PMID: 31124466

George, A., Ecabert, C., Shahreza, H. O., Kotwal, K., & Marcel, S. (2024). Edgeface: Efficient face recognition model for edge devices. *IEEE Transactions on Biometrics, Behavior, and Identity Science, 6*(2), 158–168. DOI: 10.1109/TBIOM.2024.3352164

Ghafarian, F., Wieland, R., Lüttschwager, D., & Nendel, C. (2022). Application of extreme gradient boosting and Shapley Additive explanations to predict temperature regimes inside forests from standard open-field meteorological data. Environmental Modelling &amp. *Environmental Modelling & Software, 156*, 105466. DOI: 10.1016/j.envsoft.2022.105466

Ghaffarian, S., Taghikhah, F. R., & Maier, H. R. (2023). Explainable artificial intelligence in disaster risk management: Achievements and prospective futures. *International Journal of Disaster Risk Reduction, 98*, 104123. Advance online publication. DOI: 10.1016/j.ijdrr.2023.104123

Ghassemi, M., Oakden-Rayner, L., & Beam, A. L. (2021). The false hope of current approaches to explainable artificial intelligence in health care. *The Lancet. Digital Health, 3*(11), e745–e750. DOI: 10.1016/S2589-7500(21)00208-9 PMID: 34711379

Ginsberg, J., Mohebbi, M. H., Patel, R. S., Brammer, L., Smolinski, M. S., & Brilliant, L. (2009). Detecting influenza epidemics using search engine query data. *Nature, 457*(7232), 1012–1014. DOI: 10.1038/nature07634 PMID: 19020500

Gire, S. K., Goba, A., Andersen, K. G., Sealfon, R. S. G., Park, D. J., Kanneh, L., Jalloh, S., Momoh, M., Fullah, M., Dudas, G., Wohl, S., Moses, L. M., Yozwiak, N. L., Winnicki, S., Matranga, C. B., Malboeuf, C. M., Qu, J., Gladden, A. D., Schaffner, S. F., & Sabeti, P. C. (2014). Genomic surveillance elucidates Ebola virus origin and transmission during the 2014 outbreak. *Science, 345*(6202), 1369–1372. DOI: 10.1126/science.1259657 PMID: 25214632

Giri, P. A., & Gupta, P. (2024). Transforming disease surveillance through artificial intelligence. *Indian Journal of Community Medicine : Official Publication of Indian Association of Preventive & Social Medicine, 49*(5), 663–664. DOI: 10.4103/ijcm.ijcm_459_24 PMID: 39421515

Gogoi, G. J., Das, S., & Chakraborty, A. (2025). Bayesian Neural Network based analysis of supervised control and data acquisition of Horizontal Axis Wind Turbine for fault detection. *Wind Energy and Engineering Research*, *4*, 100016. DOI: 10.1016/j.weer.2025.100016

Goktas, P., & Grzybowski, A. (2025). Shaping the future of healthcare: Ethical clinical challenges and pathways to trustworthy AI. *Journal of Clinical Medicine*, *14*(5), 1605. DOI: 10.3390/jcm14051605 PMID: 40095575

Golan, E., Krissoff, B., Kuchler, F., Calvin, L., Nelson, K., & Price, G. (2004). Traceability in the U.S. food supply: Economic theory and industry studies. United States Department of Agriculture Economic Research Service. Agricultural Economic Report No. 830. DOI: 10.22004/AG.ECON.33939

Gomes de Sousa, J., Marques, A. D., & Silva, N. (2025). Harnessing artificial intelligence for enhanced public health surveillance: A narrative review. *Frontiers in Public Health*, *13*, 1601151. Advance online publication. DOI: 10.3389/fpubh.2025.1601151 PMID: 40809756

Gonçalves, J. (2023). The role of smart technologies in wastewater-based epidemiology. *Journal of Environmental Exposure Assessment*, *2*(4), 18. Advance online publication. DOI: 10.20517/jeea.2023.16

Gracia-Lor, E., Rousis, N. I., Hernández, F., Zuccato, E., & Castiglioni, S. (2018). Wastewater-based epidemiology as a novel biomonitoring tool to evaluate human exposure to pollutants. *Environmental Science & Technology*, *52*(18), 10224–10226. DOI: 10.1021/acs.est.8b01403 PMID: 30146868

Grassly, N., Shaw, A., & Owusu, M. (2024). Global wastewater surveillance for pathogens with pandemic potential: Opportunities and challenges. *The Lancet Microbe*, *100939*. Advance online publication. DOI: 10.1016/j.lanmic.2024.07.002 PMID: 39222653

Greenhalgh, T., Wherton, J., Papoutsi, C., Lynch, J., Hughes, G., Hinder, S., Fahy, N., Procter, R., & Shaw, S. (2017). Beyond adoption: A new framework for theorizing and evaluating nonadoption, abandonment, and challenges to the scale-up, spread, and sustainability of health and care technologies. *Journal of Medical Internet Research*, *19*(11), e367. DOI: 10.2196/jmir.8775 PMID: 29092808

Guillen Aguinaga, M., Aguinaga Ontoso, E., Guillen Aguinaga, L., Guillén Grima, F., & Aguinaga Ontoso, I. (2025). Data quality in the age of AI: A review of governance, ethics, and the FAIR principles. *Data*, *10*(12), 201. Advance online publication. DOI: 10.3390/data10120201

Gu, J., Peng, Y., Lu, H., Chang, X., & Chen, G. (2022). A novel fault diagnosis method of rotating machinery via VMD, CWT and improved CNN. *Measurement : Journal of the International Measurement Confederation*, *200*, 111635. DOI: 10.1016/j.measurement.2022.111635

Gupta, K. K., Saha, S., Sahoo, S. K., & Goswami, S. S. (2023). Revolutionizing healthcare industry 5.0: Exploring the potential of blockchain technology for medical applications. *J. Technol. Innov. Energy*, *2*(3), 76–93. DOI: 10.56556/jtie.v2i3.610

Gupta, S., Modgil, S., Lee, C.-K., & Sivarajah, U. (2023). The future is yesterday: Use of AI-driven facial recognition to enhance value in the travel and tourism industry. *Information Systems Frontiers*, *25*(3), 1179–1195. DOI: 10.1007/s10796-022-10271-8 PMID: 35529102

Gurawalia, H., Shinde, P. B., & Sharma, K. K. (2024). Embracing Industry 4.0 ingredients in synthetic biology for the development of a carbon-neutral economy. In Chandel, A. K. (Ed.), *Biorefinery and Industry 4.0: Empowering sustainability (Green Energy and Technology)*. Springer., DOI: 10.1007/978-3-031-51601-6_8

Guruswamy, S., Pojić, M., Subramanian, J., Mastilović, J., Sarang, S., Subbana-gounder, A., Stojanović, G., & Jeoti, V. (2022). Toward better food security using concepts from Industry 5.0. *Sensors (Basel)*, *22*(21), 8377. DOI: 10.3390/s22218377 PMID: 36366073

Hadfield, J., Megill, C., Bell, S. M., Huddleston, J., Potter, B., Callender, C., Sa-gulenko, P., Bedford, T., & Neher, R. A. (2018). Nextstrain: Real-time tracking of pathogen evolution. *Bioinformatics (Oxford, England)*, *34*(23), 4121–4123. DOI: 10.1093/bioinformatics/bty407 PMID: 29790939

Haetami, H. (2025). AI-driven educational transformation in Indonesia: From learn-ing personalization to institutional management. *AL-ISHLAH: Jurnal Pendidikan*, *17*(2), 1819–1832. DOI: 10.35445/alishlah.v17i2.7448

Haider, H. (2009). Community-based approaches to peacebuilding in conflict-affected and fragile contexts.

Haider, K., Abbas, D., Ullah, F., Ijaz, M., Ikram, M., Kabir, K., Khan, K. A., & Galian, J. (2026). Climate-driven insect pest outbreaks and food security risks: A review of adaptive strategies for resilient agriculture. *Journal of Pest Science*, *99*(1), 12. DOI: 10.1007/s10340-025-01985-0

Hait, M., Sahu, P., Biswas, S., & Izah, S. C. (2024). Sustainable application of artificial intelligence in biomonitoring for environmental sustainability: Challenges and prospects. In *Biomonitoring of Pollutants in the Global South* (pp. 747–778). Springer., DOI: 10.1007/978-981-97-1658-6_22

Hamad, A., & Bhattacharya, T. (2025). *Examining the potential and hurdles of blockchain technology in food systems: An overview*. Food Safety and Health. DOI: 10.1002/fsh3.70017

Hammouche, R., Attia, A., Akhrouf, S., & Akhtar, Z. (2022). Gabor filter bank with deep autoencoder based face recognition system. *Expert Systems with Applications*, *197*, 116743. DOI: 10.1016/j.eswa.2022.116743

Hapicha, E. M. H. (2025). Integrating predictive analytics and community engagement in Kenya: A framework for technology-driven wildlife conservation in private conservancies. National Security: A Journal of the National Defence University–Kenya, 3(1), 50–67.

Haque, S., Mengersen, K., Barr, I., Wang, L., Yang, W., Vardoulakis, S., Bambrick, H., & Hu, W. (2024). Towards development of functional climate-driven early warning systems for climate-sensitive infectious diseases: Statistical models and recommendations. *Environmental Research*, *249*, 118568. DOI: 10.1016/j.envres.2024.118568 PMID: 38417659

Harman, S., Erfani, P., Goronga, T., Hickel, J., Morse, M., & Richardson, E. T. (2021). Global vaccine equity demands reparative justice—Not charity. *BMJ Global Health*, *6*(6), e006504. DOI: 10.1136/bmjgh-2021-006504 PMID: 34919057

Hasan, I., Habib, M., & Mohamed, Z. (2023). Blockchain database and iot: A technology driven agri-food supply chain. *International Supply Chain Technology Journal*, *9*(3), 40–45. DOI: 10.20545/isctj.v09.i03.01

Hassoun, A., Aït-Kaddour, A., Abu-Mahfouz, A. M., Rathod, N. B., Bader, F., Barba, F. J., Biancolillo, A., Cropotova, J., Galanakis, C. M., Jambrak, A. R., Lorenzo, J. M., Luz, C., Mañes, J., Moltó, J. C., Nuñez, J. M. M., Ozogul, F., & Regenstein, J. M. (2022). The fourth industrial revolution in the food industry—Part I: Industry 4.0 technologies. *Critical Reviews in Food Science and Nutrition*, •••, 1–17. DOI: 10.1080/10408398.2022.2034735 PMID: 35114860

Health Level Seven International. (2019). *FHIR Release 4 (Version 4.0.1)*. https://hl7.org/fhir/R4/

Hedges, L. V. (2011). Comment on 'Multivariate meta-analysis: Potential and promise'. *Statistics in Medicine*, *30*(20), 2499. DOI: 10.1002/sim.4227 PMID: 25522447

Hendriksen, R. S., Munk, P., Njage, P., van Bunnik, B., McNally, L., Lukjancenko, O., Röder, T., Nieuwenhuijse, D., Pedersen, S. K., Kjeldgaard, J., Kaas, R. S., Clausen, P. T. L. C., Vogt, J. K., Leekitcharoenphon, P., van de Schans, M. G. M., Zuidema, T., de Roda Husman, A. M., Rasmussen, S., Petersen, B., & Aarestrup, F. M. (2019). Global monitoring of antimicrobial resistance based on metagenomics analyses of urban sewage. *Nature Communications*, *10*(1), 1124. Advance online publication. DOI: 10.1038/s41467-019-08853-3 PMID: 30850636

Hernández, F., Fabregat-Safont, D., Campos-Mañas, M., & Quintana, J. (2023). Efficient Validation Strategies in Environmental Analytical Chemistry: A Focus on Organic Micropollutants in Water Samples. *Annual Review of Analytical Chemistry*, *16*(1), 401–428. Advance online publication. DOI: 10.1146/annurev-anchem-091222-112115 PMID: 37068748

Himeidan, Y. (2016). Rift Valley fever: Current challenges and future prospects. *Research and Reports in Tropical Medicine*, *1*, 1. Advance online publication. DOI: 10.2147/RRTM.S63520 PMID: 30050334

Hobbs, J. E. (2020). Food supply chains during the COVID-19 pandemic. *Canadian Journal of Agricultural Economics*, *68*(2), 171–176. DOI: 10.1111/cjag.12237

Holton, E., Sims, N., Jagadeesan, K., Standerwick, R., & Kasprzyk-Hordern, B. (2022). Quantifying community-wide antimicrobials usage via wastewater-based epidemiology. *Journal of Hazardous Materials*, *436*, 129001. DOI: 10.1016/j.jhazmat.2022.129001 PMID: 35594673

Hosseini, M. P.. (2020). Multimodal EEG and fMRI analysis. *Artificial Intelligence in Medicine*, *104*, 101813. DOI: 10.1016/j.artmed.2020.101813 PMID: 32498996

Huang, C.-H., Lin, C.-Y., Yang, C.-Y., Chan, T.-C., Chiang, P.-H., & Chen, Y.-H. (2021). Relationship between the Incidence of Dengue Virus Transmission in Traditional Market and Climatic Conditions in Kaohsiung City. *The Canadian Journal of Infectious Diseases & Medical Microbiology*, *2021*, 1–10. DOI: 10.1155/2021/9916642 PMID: 34422144

Huang, C., Lin, C., Nguyen, M., Hussain, A., Bui, X., & Ngo, H. (2023). A review of biosensor for environmental monitoring: Principle, application, and corresponding achievement of sustainable development goals. *Bioengineered*, *14*(1), 58–80. DOI: 10.1080/21655979.2022.2095089 PMID: 37377408

Huang, Y., Xu, C., Ji, M., Xiang, W., & He, D. (2020). Medical service demand forecasting using a hybrid model based on ARIMA and self-adaptive filtering method. *BMC Medical Informatics and Decision Making*, *20*(1), 237. Advance online publication. DOI: 10.1186/s12911-020-01256-1 PMID: 32950059

Huizer, M., Ter Laak, T., De Voogt, P., & Van Wezel, A. (2021). Wastewater-based epidemiology for illicit drugs: A critical review on global data. *Water Research*, *207*, 117789. DOI: 10.1016/j.watres.2021.117789 PMID: 34731667

Humboldt-Dachroeden, S., & Mantovani, A. (2021). Assessing environmental factors within the one health approach. *Medicina (Kaunas, Lithuania)*, *57*(3), 240. DOI: 10.3390/medicina57030240 PMID: 33807528

Hyde, R., Hoflund, A. B., & Pautz, M. (2016). One HACCP, two approaches: Experiences with and perceptions of the hazard analysis and critical control point food safety management systems in the United States and the EU. *Administration & Society*, *48*(8), 962–987. DOI: 10.1177/0095399714548266

Idahor, C. O., Esomu, E. J. O., Ogbonna, N., Momoh, Z., Ogbeide, O. A., Ikhu-Omoregbe, O., Adigwe, A., Erhabor, O. M., Osaghae, O., Orons, N., & Ikhu-Omoregbe, O. A. (2025). Infectious disease surveillance in the era of big data and AI: Opportunities and pitfalls. *Cureus*, *17*(10). Advance online publication. DOI: 10.7759/cureus.93929 PMID: 41200593

Igwama, G. T., Nwankwo, E. I., Emeihe, E. V., & Ajegbile, M. D.Geneva Tamunobarafiri IgwamaEjike Innocent NwankwoEbube Victor EmeiheMojeed Dayo Ajegbile. (2024). AI and big data analytics for enhancing public health surveillance in rural communities. *International Journal of Applied Research in Social Sciences*, *6*(8), 1797–1823. DOI: 10.51594/ijarss.v6i8.1427

Inam, S. A. (2025). A review of artificial intelligence for predicting climate driven infectious disease outbreaks to enhance global health resilience. *Discover Public Health*, *22*(1), 738. Advance online publication. DOI: 10.1186/s12982-025-01167-4

Ioannou, A., Bataka, E., Kokosis, N., Billinis, C., & Laspidou, C. (2025). One Health in Coastal and Marine Contexts: A Critical Bibliometric Analysis Across Environmental, Animal, and Human Health Dimensions. *International Journal of Environmental Research and Public Health*, *22*(10), 1523. Advance online publication. DOI: 10.3390/ijerph22101523 PMID: 41154927

Isiaka, A. B., Anakwenze, V. N., Ilodinso, C. R., Anaukwu, C. G., Ezeokoli, C. M., Noi, S. M., Agboola, G. O., & Adonu, R. (2024). Harnessing artificial intelligence for early detection and management of infectious disease outbreaks. *International Journal of Innovative Research and Development*, *13*(2), 52–65.

Islam, S., Haque, C. E., Hossain, S., & Hanesiak, J. (2021). Climate Variability, Dengue Vector Abundance and Dengue Fever Cases in Dhaka, Bangladesh: A Time-Series Study. *Atmosphere (Basel)*, *12*(7), 905. DOI: 10.3390/atmos12070905

Iyamu, I., Gómez-Ramírez, O., Xu, A. X., Chang, H.-J., Watt, S., McKee, G., & Gilbert, M. (2022). Challenges in the development of digital public health interventions and mapped solutions: Findings from a scoping review. *Digital Health*, *8*, 20552076221102255. DOI: 10.1177/20552076221102255 PMID: 35656283

Iyamu, I., Xu, A. X., Gómez-Ramírez, O., Ablona, A., Chang, H.-J., McKee, G., & Gilbert, M. (2021). Defining digital public health and the role of digitization, digitalization, and digital transformation: Scoping review. *JMIR Public Health and Surveillance*, *7*(11), e30399. DOI: 10.2196/30399 PMID: 34842555

Izah, S. C. (2026). Risk assessment frameworks and predictive modeling for food safety decision-making. In Ogwu, M., & Izah, S. (Eds.), *Mitigation strategies and risk management for food contaminants* (pp. 85–110). IGI Global Scientific Publishing., DOI: 10.4018/979-8-3373-7052-1.ch003

Izah, S. C., Etebu, E. N., Aigberua, A. O., Odubo, T. C., & Iniamagha, I. (2022). A meta-analysis of microbial contaminants in selected ready-to-eat foods in Bayelsa State, Nigeria: Public health implications and risk-reduction strategies. *Hygiene and Environmental Health Advances*, *4*, 100017. DOI: 10.1016/j.heha.2022.100017

Izah, S. C., Nurmahanova, A., Ogwu, M. C., Toktarbay, Z., Umirbayeva, Z., Ussen, K., Koibasova, L., Nazarbekova, S., Tynybekov, B., & Guo, Z. (2025). Public health risks associated with antibiotic residues in poultry food products. *Journal of Agriculture and Food Research*, *21*, 101815. DOI: 10.1016/j.jafr.2025.101815

Izah, S. C., Odubo, T. C., Ogwu, M. C., & Hait, M. (2024). Aspects of microorganisms in the food industry. In Ogwu, M. C., Izah, S. C., & Ntuli, N. R. (Eds.), *Food safety and quality in the global south* (pp. 399–425). Springer., DOI: 10.1007/978-981-97-2428-4_13

Izah, S. C., & Ogwu, M. C. (2025). Modeling solutions for microbial water contamination in the Global South for public health protection. *Frontiers in Microbiology*, *16*, 1504829. DOI: 10.3389/fmicb.2025.1504829 PMID: 40241726

Izah, S. C., Ogwu, M. C., & Hamidifar, H. (Eds.). (2024). *Biomonitoring of pollutants in the Global South*. Springer Nature., DOI: 10.1007/978-981-97-1658-6

Izah, S. C., Ogwu, M. C., & Iyiola, A. O. (2026). *Artificial intelligence in fisheries: Transformative potentials and challenges* (1st ed.). CRC Press., DOI: 10.1201/9781003506287

Izah, S. C., Ogwu, M. C., & Shahsavani, A. (2024). *Air pollutants in the context of One Health: Fundamental, sources, and impacts*. Springer Cham., DOI: 10.1007/978-3-031-74165-4

Izah, S. C., Richard, G., Stanley, H. O., Sawyer, W. E., Ogwu, M. C., & Uwaeme, O. R. (2023). Integrating the One Health approach and statistical analysis for sustainable aquatic ecosystem management and trace contamination mitigation. *ES Food and Agroforestry*, *14*, 1012. DOI: 10.30919/esfaf1012

Jagadeesan, B., Gerner-Smidt, P., Allard, M. W., Leuillet, S., Winkler, A., Xiao, Y., Chaffron, S., Van Der Vossen, J., Tang, S., Katase, M., McClure, P., Kimura, B., Ching Chai, L., Chapman, J., & Grant, K. (2019). The use of next generation sequencing for improving food safety: Translation into practice. *Food Microbiology*, *79*, 96–115. DOI: 10.1016/j.fm.2018.11.005 PMID: 30621881

Jain, M., Olsen, H. E., Paten, B., & Akeson, M. (2016). The Oxford Nanopore MinION: Delivery of nanopore sequencing to the genomics community. *Genome Biology*, *17*(1), 239. Advance online publication. DOI: 10.1186/s13059-016-1103-0 PMID: 27887629

Jangra, A., & Tiwari, S. (2025). Empowering nutrition: Biofortification to combat hidden hunger. In Pathak, H., Lakra, W. S., Gopalakrishnan, A., & Bansal, K. C. (Eds.), *Advances in agri-food systems*. Springer., DOI: 10.1007/978-981-96-0763-1_17

Jansen-Vullers, M. H., van Dorp, C. A., & Beulens, A. J. (2003). Managing traceability information in manufacture. *International Journal of Information Management*, *23*(5), 395–413. DOI: 10.1016/S0268-4012(03)00066-5

Jha, S. B., & Babiceanu, R. F. (2023). Deep CNN-based visual defect detection: Survey of current literature. *Computers in Industry*, *148*, 103911. DOI: 10.1016/j.compind.2023.103911

Jiang, C., Zhang, X., Xie, W., Wang, R., Feng, C., Ma, L., Li, Q., Yang, Q., & Wang, H. (2022). Predicting the potential distribution of the fall armyworm Spodoptera frugiperda (J.E. Smith) under climate change in China. *Global Ecology and Conservation*, *33*, e01994. DOI: 10.1016/j.gecco.2021.e01994

Jiménez-Carvelo, A. M., Li, P., Erasmus, S. W., Wang, H., & van Ruth, S. M. (2023). Spatial-Temporal Event Analysis as a Prospective Approach for Signalling Emerging Food Fraud-Related Anomalies in Supply Chains. *Foods*, *12*(1), 61. DOI: 10.3390/foods12010061 PMID: 36613277

Jin, H., & Xic, Y. (2023). A rcvicw of rcscarch on the impact of digitalization on agricultural supply chain security. *Digitalization and Management Innovation*, 378–385.

Joel, O. S., Oyewole, A. T., Odunaiya, O. G., & Soyombo, O. T.Olorunyomi Stephen JoelAdedoyin Tolulope OyewoleOlusegun Gbenga OdunaiyaOluwatobi Timothy Soyombo. (2024). Leveraging artificial intelligence for enhanced supply chain optimization: A comprehensive review of current practices and future potentials. *International Journal of Management & Entrepreneurship Research*, *6*(3), 707–721. DOI: 10.51594/ijmer.v6i3.882

Johansson, M. A., Apfeldorf, K. M., Dobson, S., Devita, J., Buczak, A. L., Baugher, B., Moniz, L. J., Bagley, T., Babin, S. M., Guven, E., Yamana, T. K., Shaman, J., Moschou, T., Lothian, N., Lane, A., Osborne, G., Jiang, G., Brooks, L. C., Farrow, D. C., & Biggerstaff, M. (2019). An open challenge to advance probabilistic forecasting for dengue epidemics. *Proceedings of the National Academy of Sciences of the United States of America*, *116*(48), 24268–24274. DOI: 10.1073/pnas.1909865116 PMID: 31712420

Johnson, B. (2025). Artificial intelligence as an anti-corruption tool (AI-ACT): Potentials and pitfalls for top-down and bottom-up approaches.

Kahn, M. G., Raebel, M. A., Glanz, J. M., Riedlinger, K., & Steiner, J. F. (2012). A pragmatic framework for single-site and multisite data quality assessment in electronic health record-based clinical research. *Medical Care*, *50*(Suppl.), S21–S29. DOI: 10.1097/MLR.0b013e318257dd67 PMID: 22692254

Kakade, S. L., & Vitalkar, S. (2024). Inventory management in pharmaceutical industry. *International Journal of Scientific Research in Engineering and Management*, *8*(4), 1–5.

Kakarla, S. G., Caminade, C., Mutheneni, S. R., Morse, A. P., Upadhyayula, S. M., Kadiri, M. R., & Kumaraswamy, S. (2019). Lag effect of climatic variables on dengue burden in India. *Epidemiology and Infection*, *147*, e170. Advance online publication. DOI: 10.1017/S0950268819000608 PMID: 31063099

Kalli, M., Noutsopoulos, C., & Mamais, D. (2023). The Fate and Occurrence of Antibiotic-Resistant Bacteria and Antibiotic Resistance Genes during Advanced Wastewater Treatment and Disinfection: A Review. *Water (Basel)*, *15*(11), 2084. Advance online publication. DOI: 10.3390/w15112084

Kalluri, S., Gilruth, P., Rogers, D., & Szczur, M. (2007). Surveillance of Arthropod Vector-Borne Infectious Diseases Using Remote Sensing Techniques: A Review. *PLoS Pathogens*, *3*(10), e116. DOI: 10.1371/journal.ppat.0030116 PMID: 17967056

Kamana, E., Zhao, J., & Bai, D. (2022). Predicting the impact of climate change on the re-emergence of malaria cases in China using LSTMSeq2Seq deep learning model: A modelling and prediction analysis study. *BMJ Open*, *12*(3), e053922. DOI: 10.1136/bmjopen-2021-053922 PMID: 35361642

Kamara, R. D. (2025). Bridging the gap: Opportunities, challenges and strategies for AI deployment in public service delivery. *Public Administration and Regional Development*, (28), 334–356. DOI: 10.34132/pard2025.28.02

Kamble, S. S., Gunasekaran, A., & Sharma, R. (2020). Modeling the blockchain enabled traceability in agriculture supply chain. *International Journal of Information Management*, *52*, 101967. DOI: 10.1016/j.ijinfomgt.2019.05.023

Kangana, N., Kankanamge, N., De Silva, C., Goonetilleke, A., Mahamood, R., & Ranasinghe, D. (2024). Bridging community engagement and technological innovation for creating smart and resilient cities: A systematic literature review. *Smart Cities*, *7*(6), 3823–3852. DOI: 10.3390/smartcities7060147

Karlsen, K. M., Dreyer, B., Olsen, P., & Elvevoll, E. O. (2013). Literature review: Does a common theoretical framework to implement food traceability exist? *Food Control*, *32*(2), 409–417. DOI: 10.1016/j.foodcont.2012.12.011

Karnati, M., Seal, A., Bhattacharjee, D., Yazidi, A., & Krejcar, O. (2023). Understanding deep learning techniques for recognition of human emotions using facial expressions: A comprehensive survey. *IEEE Transactions on Instrumentation and Measurement*, *72*, 1–31. DOI: 10.1109/TIM.2023.3243661

Kassam, A., & Patten, S. B. (2006). Major depression, fibromyalgia and labour force participation: A population-based cross-sectional study. *BMC Musculoskeletal Disorders*, *7*(1), 4. DOI: 10.1186/1471-2474-7-4 PMID: 16423291

Kassem, A., Khattab, M. S., Ismael, E., Yassin, A. M., Hamza, D., & Osman, A. H. (2025). Climate change and the emergence of Rift Valley fever virus: A pathological and environmental study of sheep jaundice in Egyptian slaughterhouses. *BMC Veterinary Research*, *21*(1), 539. Advance online publication. DOI: 10.1186/s12917-025-05022-1 PMID: 40963143

Kathirvel, N., Gobinath, V. M., & Maheswaran, C. P. (2025). Effective leadership strategies for enhancing student success in higher education. In *New horizons in leadership: Inclusive explorations in health, technology, and education* (pp. 61–91). IGI Global., DOI: 10.4018/979-8-3693-6437-6.ch003

Kaur, P., Krishan, K., Sharma, S. K., & Kanchan, T. (2020). Facial-recognition algorithms: A literature review. *Medicine, Science, and the Law*, *60*(2), 131–139. DOI: 10.1177/0025802419893168 PMID: 31964224

Kayani, M., Huang, W., Feng, R., & Chen, L. (2021). Genome-resolved metagenomics using environmental and clinical samples. *Briefings in Bioinformatics*, *22*(5), bbab030. Advance online publication. DOI: 10.1093/bib/bbab030 PMID: 33758906

Keawdounglek, V., Chaiyakul, S., & Aunkham, A. (2025). The development of the "Laab Nuer Model" for food safety management in handling traditional Lanna cuisine in Thailand. *PLoS One*, *20*(9), e0331933. DOI: 10.1371/journal.pone.0331933 PMID: 41004446

Keenum, I., Liguori, K., Calarco, J., Davis, B., Milligan, E., Harwood, V., & Pruden, A. (2022). A framework for standardized qPCR-targets and protocols for quantifying antibiotic resistance in surface water, recycled water, and wastewater. *Critical Reviews in Environmental Science and Technology*, *52*(24), 4395–4419. DOI: 10.1080/10643389.2021.2024739

Keerthi, M., & Jayan, A. (2025). AI in global governance: Unleashing potential, overcoming challenges, and addressing ethical concerns. *MET SRUJAN*, *1*(1), 55–65. DOI: 10.34047/SJ/20251105

Keesara, S., Jonas, A., & Schulman, K. (2020). Covid-19 and health care's digital revolution. *The New England Journal of Medicine*, *382*(23), e82. DOI: 10.1056/NEJMp2005835 PMID: 32240581

Keller, M., Blench, M., Tolentino, H., Freifeld, C. C., Mandl, K. D., Mawudeku, A., Eysenbach, G., & Brownstein, J. S. (2009). Use of unstructured event-based reports for global infectious disease surveillance. *Emerging Infectious Diseases*, *15*(5), 689–695. DOI: 10.3201/eid1505.081114 PMID: 19402953

Kelley, K., Gozzi, N., Mazzoli, M., & Paolotti, D. (2025). Exploring influenza vaccination determinants through digital participatory surveillance. *BMC Public Health*, *25*(1), 1345. DOI: 10.1186/s12889-025-22496-8 PMID: 40211245

Kelly, C. J., Karthikesalingam, A., Suleyman, M., Corrado, G., & King, D. (2019). Key challenges for delivering clinical impact with artificial intelligence. *BMC Medicine*, *17*(1), 195. DOI: 10.1186/s12916-019-1426-2 PMID: 31665002

Khan, A., Zubair, M., & Ahmad, M. (2020). Data quality challenges in sensor networks: Issues, methods, and future directions. *Journal of Big Data*, *7*(1), 85. Advance online publication. DOI: 10.1186/s40537-026-01413-9 PMID: 32547903

Khan, I. A., & Ullah, A. (2024). The role of artificial intelligence in enhancing social governance: A framework for ethical implementation and policy development in Pakistan. *Journal of Management & Social Science*, *1*(4), 274–289.

Khan, M. R. I. (2026). AI-powered predictive analytics for enhancing supply chain resilience and delivery outcomes. *Discovery Analytics*, *4*(1), 13. DOI: 10.1007/s44257-026-00065-6

Khan, S., Haleem, A., Khan, M. I., Abidi, M. H., & Al-Ahmari, A. (2018). Implementing traceability systems in specific supply chain management through critical success factors. *Sustainability (Basel)*, *5*(1), 1515822. DOI: 10.3390/su10010204

Kher, S. V., Frewer, L. J., Jonge, J. D., Wentholt, M. T., Davies, O. H., Luijckx, N. B. L., & Cnossen, H. J. (2010). Experts' perspectives on the implementation of traceability in Europe. *British Food Journal*, *112*(3), 261–274. DOI: 10.1108/00070701011029138

Khurana, M. P., Raaschou-Pedersen, D. E., Kurtzhals, J., Bardram, J. E., Ostrowski, S. R., & Bundgaard, J. S. (2022). Digital health competencies in medical school education: A scoping review and Delphi method study. *BMC Medical Education*, *22*(1), 129. DOI: 10.1186/s12909-022-03163-7 PMID: 35216611

Kilaru, P., Hill, D., Anderson, K., Collins, M., Green, H., Kmush, B., & Larsen, D. (2021). Wastewater Surveillance for Infectious Disease: A Systematic Review. *American Journal of Epidemiology*. Advance online publication. DOI: 10.1101/2021.07.26.21261155 PMID: 36227259

Kim, A. J., & Tak, S. (2019). Implementation system of a biosurveillance system in the Republic of Korea and its legal ramifications. *Health Security*, *17*(6), 462–467. DOI: 10.1089/hs.2019.0071 PMID: 31800333

Kim, C., Pongpanich, M., & Porntaveetus, T. (2024). Unraveling metagenomics through long-read sequencing: A comprehensive review. *Journal of Translational Medicine*, *22*(1), 111. Advance online publication. DOI: 10.1186/s12967-024-04917-1 PMID: 38282030

Kim, H. M., & Laskowski, M. (2018). Toward an ontology-driven blockchain design for supply-chain provenance. *Intelligent Systems in Accounting, Finance & Management*, *25*(1), 18–27. DOI: 10.1002/isaf.1424

Kman, N. E., & Bachmann, D. J. (2012). Biosurveillance: A review and update. *Advances in Preventive Medicine*, *2012*, 301408. Advance online publication. DOI: 10.1155/2012/301408 PMID: 22242207

Kononowicz, A. A., Woodham, L. A., Edelbring, S., Stathakarou, N., Davies, D., Saxena, N., Tudor Car, L., Carlstedt-Duke, J., Car, J., & Zary, N. (2019). Virtual patient simulations in health professions education: Systematic review and meta-analysis by the Digital Health Education Collaboration. *Journal of Medical Internet Research, 21*(7), e14676. DOI: 10.2196/14676 PMID: 31267981

Konttila, J., Siira, H., Kyngäs, H., Lahtinen, M., Elo, S., Kääriäinen, M., Kaakinen, P., Oikarinen, A., Yamakawa, M., Fukui, S., Utsumi, M., Higami, Y., Higuchi, A., & Mikkonen, K. (2019). Healthcare professionals' competence in digitalisation: A systematic review. *Journal of Clinical Nursing*, *28*(5–6), 745–761. DOI: 10.1111/jocn.14710 PMID: 30376199

Koplan, J. P., Bond, T. C., Merson, M. H., Reddy, K. S., Rodriguez, M. H., Sewankambo, N. K., & Wasserheit, J. N. (2009). Towards a common definition of global health. *Lancet*, *373*(9679), 1993–1995. DOI: 10.1016/S0140-6736(09)60332-9 PMID: 19493564

Köster, J., & Rahmann, S. (2012). Snakemake: A scalable bioinformatics workflow engine. *Bioinformatics (Oxford, England)*, *28*(19), 2520–2522. DOI: 10.1093/bioinformatics/bts480 PMID: 22908215

Kostkova, P. (2013). A roadmap to integrated digital public health surveillance: The vision and the challenges. In *Proceedings of the 22nd International World Wide Web Conference (WWW '13)* (pp. 687–694). DOI: 10.1145/2487788.2488024

Kostkova, P. (2018). Disease surveillance data sharing for public health: The next ethical frontiers. *Life Sciences, Society and Policy*, *14*(1), 16. DOI: 10.1186/s40504-018-0078-x PMID: 29971516

Kostkova, P., Saigí-Rubió, F., Eguia, H., Borbolla, D., Verschuuren, M., Hamilton, C., Azzopardi-Muscat, N., & Novillo-Ortiz, D. (2021). Data and digital solutions to support surveillance strategies in the context of the COVID-19 pandemic. *Frontiers in Digital Health*, *3*, 707902. DOI: 10.3389/fdgth.2021.707902 PMID: 34713179

Kotchi, S., Bouchard, C., Ludwig, A., Rees, E., & Brazeau, S. (2019). Using Earth observation images to inform risk assessment and mapping of climate change-related infectious diseases. *Canada Communicable Disease Report = Relevé des Maladies Transmissibles au Canada*, *45*(5), 133–142. DOI: 10.14745/ccdr.v45i05a04 PMID: 31285704

Kruse, C. S., Karem, P., Shifflett, K., Vegi, L., Ravi, K., & Brooks, M. (2018). Evaluating barriers to adopting telemedicine worldwide: A systematic review. *Journal of Telemedicine and Telecare*, *24*(1), 4–12. DOI: 10.1177/1357633X16674087 PMID: 29320966

Kumar, M., Kumar, D. R., Khatti, J., Samui, P., & Grover, K. S. (2024). Prediction of bearing capacity of pile foundation using deep learning approaches. *Frontiers of Structural and Civil Engineering*, *18*(6), 870–886. DOI: 10.1007/s11709-024-1085-z

Kumar, N., & Khurana, S. P. (2025). An application of biopesticides in control of pest and crop protection: An eco-friendly management. In *Microbial biocontrol techniques: Importance in ensuring food security* (pp. 95–124). Springer Nature Singapore.

Kumar, S. M., Nath, S. N., Vani, H. C., Shrinivasa, B. M., Ahmed, N., Tomer, V., Bankoti, K., Negi, G., Ali, S. S. B., Jaiswal, M. K., & Warkade, V. (2025). The role of artificial intelligence (AI) in the elimination of malaria: A systematic literature review. *Malaria Journal*, *24*(1), 427. Advance online publication. DOI: 10.1186/s12936-025-05661-7 PMID: 41291780

Kumar, S., Rani, S., Jain, A., Verma, C., Raboaca, M. S., Illés, Z., & Neagu, B. C. (2022). Face spoofing, age, gender and facial expression recognition using advance neural network architecture-based biometric system. *Sensors (Basel)*, *22*(14), 5160. DOI: 10.3390/s22145160 PMID: 35890840

Kumblathan, T., Liu, Y., Uppal, G., Hrudey, S., & Li, X. (2021). Wastewater-Based Epidemiology for Community Monitoring of SARS-CoV-2: Progress and Challenges. *ACS Environmental Au*, *1*(1), 18–31. DOI: 10.1021/acsenvironau.1c00015 PMID: 37579255

Kurniawan, M., Suparno, S., & Vanany, I. (2023, December). Implementation blockchain technology and the internet of things in halal supply chain traceability and food safety: A systematic literature review. In *Proceeding International Conference of Technology on Community and Environmental Development* (Vol. 1, No. 2). DOI: 10.18196/ictced.v1i2.75

Kyaw, B. M., Saxena, N., Posadzki, P., Vseteckova, J., Nikolaou, C. K., George, P. P., Divakar, U., Masiello, I., Kononowicz, A. A., Zary, N., & Tudor Car, L. (2019). Virtual reality for health professions education: Systematic review and meta-analysis by the Digital Health Education Collaboration. *Journal of Medical Internet Research*, *21*(1), e12959. DOI: 10.2196/12959 PMID: 30668519

Kye, H., Kim, J., Ju, S., Lee, J., Lim, C., & Yoon, Y. (2023). Microplastics in water systems: A review of their impacts on the environment and their potential hazards. *Heliyon*, *9*(3), e14359. Advance online publication. DOI: 10.1016/j.heliyon.2023.e14359 PMID: 36950574

La Rosa, M., Maugeri, A., Favara, G., La Mastra, C., Lio, M., Barchitta, M., & Agodi, A. (2025). The Impact of Wastewater on Antimicrobial Resistance: A Scoping Review of Transmission Pathways and Contributing Factors. *Antibiotics (Basel, Switzerland)*, *14*(2), 131. Advance online publication. DOI: 10.3390/antibiotics14020131 PMID: 40001375

Labrique, A. B., Wadhwani, C., Williams, K. A., Lamptey, P., Hesp, C., Luk, R., & Aerts, A. (2018). Best practices in scaling digital health in low- and middle-income countries. *Globalization and Health*, *14*(1), 103. DOI: 10.1186/s12992-018-0424-z PMID: 30390686

Laicans, J., Dejus, B., Dejus, S., & Juhna, T. (2024). Precision and Accuracy Limits of Wastewater-Based Epidemiology—Lessons Learned from SARS-CoV-2: A Scoping Review. *Water (Basel)*, *16*(9), 1220. Advance online publication. DOI: 10.3390/w16091220

Laimou-Geraniou, M., Heath, D., & Heath, E. (2022). Analytical methods for the determination of antidepressants, antipsychotics, benzodiazepines and their metabolites through wastewater-based epidemiology. *Trends in Environmental Analytical Chemistry*. Advance online publication. DOI: 10.1016/j.teac.2022.e00192

Laine, J., Minkkinen, M., & Mäntymäki, M. (2024). Ethics-based AI auditing: A systematic literature review on conceptualizations of ethical principles and knowledge contributions to stakeholders. Information &amp. *Information & Management*, *61*(5), 103969. DOI: 10.1016/j.im.2024.103969

Lalor, J. P., Abbasi, A., Oketch, K., Yang, Y., & Forsgren, N. (2024). Should fairness be a metric or a model? A model-based framework for assessing bias in machine learning pipelines. *ACM Transactions on Information Systems*, *42*(4), 99. Advance online publication. DOI: 10.1145/3641276

Langmead, B., & Salzberg, S. L. (2012). Fast gapped-read alignment with Bowtie 2. *Nature Methods*, *9*(4), 357–359. DOI: 10.1038/nmeth.1923 PMID: 22388286

Lan, H. Y., Ubina, N. A., Cheng, S. C., Lin, S. S., & Huang, C. T. (2022). Digital twin architecture evaluation for intelligent fish farm management using modified analytic hierarchy process. *Applied Sciences (Basel, Switzerland)*, *13*(1), 141. DOI: 10.3390/app13010141

Lauer, S. A., Brown, A. C., & Reich, N. G. (2020). Infectious disease forecasting for public health. arXiv preprint arXiv:2006.00073. DOI: 10.1093/oso/9780198853244.003.0004

Lawal, O. P., Opara, I. J., Ayo-Ige, A., Eboh, N. A., Cos-Ibe, U., Forson, K. A. A. M., & Forson, K.. (2025). Artificial intelligence-integrated biosensors for antimicrobial resistance detection and surveillance: A review and future perspectives for global biosecurity. *Cureus*, *17*(11). Advance online publication. DOI: 10.7759/cureus.98098 PMID: 41477370

Laxminarayan, R., Duse, A., Wattal, C., Zaidi, A. K. M., Wertheim, H. F. L., Sumpradit, N., Vlieghe, E., Hara, G. L., Gould, I. M., Goossens, H., Greko, C., So, A. D., Bigdeli, M., Tomson, G., Woodhouse, W., Ombaka, E., Peralta, A. Q., Qamar, F. N., Mir, F., & Cars, O. (2013). Antibiotic resistance: The need for global solutions. *The Lancet. Infectious Diseases*, *13*(12), 1057–1098. DOI: 10.1016/S1473-3099(13)70318-9 PMID: 24252483

Lazer, D., Kennedy, R., King, G., & Vespignani, A. (2014). The parable of Google Flu: Traps in big data analysis. *Science*, *343*(6176), 1203–1205. DOI: 10.1126/science.1248506 PMID: 24626916

Lei, M., Liu, S., Luo, N., Yang, X., & Sun, C. (2022a). Trusted-auditing chain: A security blockchain prototype used in agriculture traceability. *Heliyon*, *8*(11), e11477. DOI: 10.1016/j.heliyon.2022.e11477 PMID: 36406715

Lei, M., Xu, L., Liu, T., Liu, S., & Sun, C. (2022b). Integration of privacy protection and blockchain-based food safety traceability: Potential and challenges. *Foods*, *11*(19), 2932. DOI: 10.3390/foods11152262 PMID: 35954029

Leinonen, R., Sugawara, H., & Shumway, M.International Nucleotide Sequence Database Collaboration. (2011). The Sequence Read Archive. *Nucleic Acids Research*, *39*(Database issue), D19–D21. DOI: 10.1093/nar/gkq1019 PMID: 21062823

Lekadir, K., Frangi, A. F., Porras, A. R., Glocker, B., Cintas, C., Langlotz, C. P., Weicken, E., Asselbergs, F. W., Prior, F., Collins, G. S., Kaissis, G., Tsakou, G., Buvat, I., Kalpathy-Cramer, J., Mongan, J., Schnabel, J. A., Kushibar, K., Riklund, K., Marias, K., & Starmans, M. P. A. (2025). FUTURE-AI: International consensus guideline for trustworthy and deployable artificial intelligence in healthcare. *BMJ (Clinical Research Ed.)*, *388*, e081554. Advance online publication. DOI: 10.1136/bmj-2024-081554 PMID: 39909534

Lemeilleur, S., Maître-d'Hôtel, É., Lepiller, O., & Hobeika, A. (2019). Uncertain impacts of new technologies: The case of digital agriculture and blockchains. FOOD SYSTEMS AT RISK, 91.

Leong, D., Alvarez-Ordóñez, A., & Jordan, K. (2014). Monitoring occurrence and persistence of Listeria monocytogenes in foods and food processing environments in the Republic of Ireland. *Frontiers in Microbiology*, *5*, 436. DOI: 10.3389/fmicb.2014.00436 PMID: 25191314

Li, C., Li, G., Wang, K., & Han, B. (2022). A multi-energy load forecasting method based on parallel architecture CNN-GRU and transfer learning for data deficient integrated energy systems. *Energy*, *259*, 124967. DOI: 10.1016/j.energy.2022.124967

Liguori, K., Keenum, I., Davis, B., Calarco, J., Milligan, E., Harwood, V., & Pruden, A. (2022). Antimicrobial Resistance Monitoring of Water Environments: A Framework for Standardized Methods and Quality Control. *Environmental Science & Technology*, *56*(13), 9149–9160. DOI: 10.1021/acs.est.1c08918 PMID: 35732277

Li, H. (2018). Minimap2: Pairwise alignment for nucleotide sequences. *Bioinformatics (Oxford, England)*, *34*(18), 3094–3100. DOI: 10.1093/bioinformatics/bty191 PMID: 29750242

Li, J., Zhu, Z., Liu, H., Su, Y., & Deng, L. (2023). Strawberry R-CNN: Recognition and counting model of strawberry based on improved faster R-CNN. *Ecological Informatics*, *77*, 102210. DOI: 10.1016/j.ecoinf.2023.102210

Lim, A., Shearer, F. M., Sewalk, K., Pigott, D. M., Clarke, J., Ghouse, A., Judge, C., Kang, H., Messina, J. P., Kraemer, M. U. G., Gaythorpe, K. A. M., de Souza, W. M., Nsoesie, E. O., Celone, M., Faria, N., Ryan, S. J., Rabe, I. B., Rojas, D. P., Hay, S. I., & Brady, O. J. (2025). The overlapping global distribution of dengue, chikungunya, Zika and yellow fever. *Nature Communications*, *16*(1), 3418. Advance online publication. DOI: 10.1038/s41467-025-58609-5 PMID: 40210848

Lindley, J. (2022). Food regulation and policing: Innovative technology to close the regulatory gap in Australia. *Frontiers in Sustainable Food Systems*, *6*(2), 836264. DOI: 10.1007/s00003-022-01372-2 PMID: 35282596

Linthicum, K. J., Anyamba, A., Tucker, C. J., Kelley, P. W., Myers, M. F., & Peters, C. J. (1999). Climate and Satellite Indicators to Forecast Rift Valley Fever Epidemics in Kenya. *Science*, *285*(5426), 397–400. DOI: 10.1126/science.285.5426.397 PMID: 10411500

Liu, D., Clemente, L., Poirier, C., Ding, X., Chinazzi, M., Davis, J. T., Vespignani, A., & Santillana, M. (2020). A machine learning methodology for real-time forecasting of the 2019-2020 COVID-19 outbreak using Internet searches, news alerts, and estimates from mechanistic models. arXiv preprint arXiv:2004.04019.

Liu, G., Peng, J., & El-Latif, A. A. A. (2023). SK-MobileNet: A lightweight adaptive network based on complex deep transfer learning for plant disease recognition. *Arabian Journal for Science and Engineering*, *48*(2), 1661–1675. DOI: 10.1007/s13369-022-06987-z

Liu, K., Xing, R., Sun, R., Ge, Y., & Chen, Y. (2022). An accurate and rapid way for identifying food geographical origin and authenticity: Editable DNA-traceable barcode. *Foods*, *12*(1), 82. DOI: 10.3390/foods12010017 PMID: 36613233

Liu, K., Zhang, W., Li, W., Wang, T., & Zheng, Y. (2023). Effectiveness of virtual reality in nursing education: A systematic review and meta-analysis. *BMC Medical Education*, *23*(1), 710. DOI: 10.1186/s12909-023-04662-x PMID: 37770884

Liu, P., Ren, S., Wang, J., Yuan, S., Nian, Y., & Li, Y. (2022). A blockchain consensus optimization-based algorithm for food traceability. *Mobile Information Systems*, *2022*(1), 1529983. DOI: 10.1155/2022/1529983

Liu, Q., Liu, M., Liang, W., Li, X., Jing, W., Chen, Z., & Liu, J. (2025). Global distribution and health impact of infectious disease outbreaks, 1996–2023: A worldwide retrospective analysis of World Health Organization emergency event reports. *Journal of Global Health*, *15*, 04151. Advance online publication. DOI: 10.7189/jogh.15.04151 PMID: 40375732

Liu, X., Faes, L., Kale, A. U., Wagner, S. K., Fu, D. J., Bruynseels, A., Mahendiran, T., Moraes, G., Shamdas, M., Kern, C., Ledsam, J. R., Schmid, M. K., Balaskas, K., Topol, E. J., Bachmann, L. M., Keane, P. A., & Denniston, A. K. (2019). A comparison of deep learning performance against health-care professionals in detecting diseases from medical imaging: A systematic review and meta-analysis. *The Lancet. Digital Health*, *1*(6), e271–e297. DOI: 10.1016/S2589-7500(19)30123-2 PMID: 33323251

Liu, Z., Deng, D., Lu, H., Sun, J., Lv, L., Li, S., Peng, G., Ma, X., Li, J., Li, Z., Rong, T., & Wang, G. (2020). Evaluation of machine learning models for predicting antimicrobial resistance of Actinobacillus pleuropneumoniae from whole genome sequences. *Frontiers in Microbiology*, *11*, 48. DOI: 10.3389/fmicb.2020.00048 PMID: 32117101

Li, X., Zhang, S., Shi, J., Luby, S., & Jiang, G. (2021). Uncertainties in estimating SARS-CoV-2 prevalence by wastewater-based epidemiology. *Chemical Engineering Journal*, *415*, 129039–129039. DOI: 10.1016/j.cej.2021.129039 PMID: 33642938

Li, Y., Zhang, X., Zhao, Z., Xu, J., Jiang, Z., Yu, J., & Cui, X. (2022). Research on grain food blockchain traceability information management model based on master–slave multichain. *Computational Intelligence and Neuroscience*, *2022*(1), 7498025. DOI: 10.1155/2022/7498025 PMID: 36605726

López-Ramos, L. M., Leiser, F., Rastogi, A., Hicks, S., Strümke, I., Madai, V. I., Budig, T., Sunyaev, A., & Hilbert, A. (2024). Interplay between federated learning and explainable artificial intelligence: A scoping review (arXiv:2411.05874v1). arXiv. https://arxiv.org/abs/2411.05874v1

Lo, S. W., & Jamrozy, D. (2020). Genomics and epidemiological surveillance. *Nature Reviews. Microbiology*, *18*(9), 478. DOI: 10.1038/s41579-020-0421-0 PMID: 32690878

Lowe, R., Lee, S. A., O'Reilly, K. M., Brady, O. J., Bastos, L., Carrasco-Escobar, G., de Castro Catão, R., Colón-González, F. J., Barcellos, C., Carvalho, M. S., Blangiardo, M., Rue, H., & Gasparrini, A. (2021). Combined effects of hydrometeorological hazards and urbanisation on dengue risk in Brazil: A spatiotemporal modelling study. *The Lancet. Planetary Health*, *5*(4), e209–e219. DOI: 10.1016/S2542-5196(20)30292-8 PMID: 33838736

Lu, H., Ziajahromi, S., Neale, P., & Leusch, F. (2021). A systematic review of freshwater microplastics in water and sediments: Recommendations for harmonisation to enhance future study comparisons. *The Science of the Total Environment*, *781*, 146693. DOI: 10.1016/j.scitotenv.2021.146693

Luo, J., Leng, S., & Bai, Y. (2022). Food supply chain safety research trends from 1997 to 2020: A bibliometric analysis. *Frontiers in Public Health*, *9*, 742980. DOI: 10.3389/fpubh.2021.742980 PMID: 35186862

Ma, C.-S., Wang, B.-X., Wang, X.-J., Lin, Q.-C., Zhang, W., Yang, X.-F., van Baaren, J., Bebber, D. P., Eigenbrode, S. D., Zalucki, M. P., Zeng, J., & Ma, G. (2025). Crop pest responses to global changes in climate and land management. Nature Reviews Earth &amp. *Nature Reviews. Earth & Environment*, *6*(4), 264–283. DOI: 10.1038/s43017-025-00652-3

Machleid, F., Kaczmarczyk, R., Johann, D., Balčiūnas, J., Atienza-Carbonell, B., von Maltzahn, F., & Mosch, L. (2020). Perceptions of digital health education among European medical students: Mixed methods survey. *Journal of Medical Internet Research*, *22*(8), e19827. DOI: 10.2196/19827 PMID: 32667899

MacIntyre, C. R., Chen, X., Kunasekaran, M., Quigley, A., Lim, S., Stone, H., Paik, H. Y., Yao, L., Heslop, D., Wei, W., & Gurdasani, D. (2023). Artificial intelligence in public health: The potential of epidemic early warning systems. *The Journal of International Medical Research*, *51*(3), 03000605231159335. Advance online publication. DOI: 10.1177/03000605231159335 PMID: 36967669

Madhavakumar, D., Nelson, A., Naren, A. R., & Prathap, S. (2025). An application of Pythagoras theorem: From Heron's formula to derive the foci of an ellipse. In *Educational AI humanoid computing devices for cyber nomads* (pp. 73–98). IGI Global., DOI: 10.4018/979-8-3693-8985-0.ch004

Mahi, R. (2024). Optimizing supply chain efficiency in the manufacturing sector through AI-powered analytics. *International Journal of Management Information Systems and Data Science*, *1*(1), 41–50.

Malcom, H., & Bowes, D. (2025). Use of Wastewater to Monitor Antimicrobial Resistance Trends in Communities and Implications for Wastewater-Based Epidemiology: A Review of the Recent Literature. *Microorganisms*, *13*(9), 2073. Advance online publication. DOI: 10.3390/microorganisms13092073 PMID: 41011405

Manning, L., & Soon, J. M. (2016). Building strategic resilience in the food supply chain. *British Food Journal*, *118*(6), 1477–1493. DOI: 10.1108/BFJ-10-2015-0350

Mao, J., Xu, R., Yin, X., Chang, Y., Nie, B., Huang, A., & Wang, Y. (2025). Poster++: A simpler and stronger facial expression recognition network. *Pattern Recognition*, *157*, 110951. DOI: 10.1016/j.patcog.2024.110951

Mao, K., Zhang, K., Du, W., Ali, W., Feng, X., & Zhang, H. (2020). The potential of wastewater-based epidemiology as surveillance and early warning of infectious disease outbreaks. *Current Opinion in Environmental Science & Health*, *17*, 1–7. DOI: 10.1016/j.coesh.2020.04.006 PMID: 32395676

Marcolino, M. S., Oliveira, J. A. Q., D'Agostino, M., Ribeiro, A. L., Alkmim, M. B. M., & Novillo-Ortiz, D. (2018). The impact of mHealth interventions: Systematic review of systematic reviews. *JMIR mHealth and uHealth*, *6*(1), e23. DOI: 10.2196/mhealth.8873 PMID: 29343463

Martínez, J. L. (2008). Antibiotics and antibiotic resistance genes in natural environments. *Science*, *321*(5887), 365–367. DOI: 10.1126/science.1159483 PMID: 18635792

Marvin, H. J. P., Bouzembrak, Y., Janssen, E. M., van der Fels-Klerx, H. J., van Asselt, E. D., & Kleter, G. A. (2016). A holistic approach to food safety risks: Food fraud as an example. *Food Research International*, *89*(1), 463–470. DOI: 10.1016/j.foodres.2016.08.028 PMID: 28460939

Mastrogiovanni, S. (2025). AI-driven data governance for smart cities: Balancing privacy, efficiency, and public trust. *South Florida Journal of Development, 6*(9), e5810–e5810. Mazimwe, A., Hammouda, I., & Gidudu, A. (2019). An empirical evaluation of data interoperability: A case of the disaster management sector in Uganda. *ISPRS International Journal of Geo-Information*, *8*(11), 484.

Ma, T., Tian, W., & Xie, Y. (2022). Multi-level knowledge distillation for low-resolution object detection and facial expression recognition. *Knowledge-Based Systems*, *240*, 108136. DOI: 10.1016/j.knosys.2022.108136

Mathews, S. C., McShea, M. J., Hanley, C. L., Ravitz, A., Labrique, A. B., & Cohen, A. B. (2019). Digital health: A path to validation. *npj. Digital Medicine*, *2*(1), 38. DOI: 10.1038/s41746-019-0111-3 PMID: 31304384

McArthur, A. G., Waglechner, N., Nizam, F., Yan, A., Azad, M. A., Baylay, A. J., Bhullar, K., Canova, M. J., De Pascale, G., Ejim, L., Kalan, L., King, A. M., Koteva, K., Morar, M., Mulvey, M. R., O'Brien, J. S., Pawlowski, A. C., Piddock, L. J. V., Spanogiannopoulos, P., & Wright, G. D. (2013). The Comprehensive Antibiotic Resistance Database. *Antimicrobial Agents and Chemotherapy*, *57*(7), 3348–3357. DOI: 10.1128/AAC.00419-13 PMID: 23650175

McCorquodale-Bauer, K., Grosshans, R., Zvomuya, F., & Cicek, N. (2023). Critical review of phytoremediation for the removal of antibiotics and antibiotic resistance genes in wastewater. *The Science of the Total Environment*, *161876*, 161876. Advance online publication. DOI: 10.1016/j.scitotenv.2023.161876 PMID: 36716878

McGough, S. F., Johansson, M. A., Lipsitch, M., & Menzies, N. A. (2020). Nowcasting by Bayesian smoothing: A flexible, generalizable model for real-time epidemic tracking. *PLoS Computational Biology*, *16*(4), e1007735. DOI: 10.1371/journal.pcbi.1007735 PMID: 32251464

McGowan, C. R., Takahashi, E., Romig, L., Bertram, K., Kadir, A., Cummings, R., & Cardinal, L. J. (2022). Community-based surveillance of infectious diseases: A systematic review of drivers of success. *BMJ Global Health*, *7*(8), e009934. DOI: 10.1136/bmjgh-2022-009934 PMID: 35985697

McMurry, A. J., Zipursky, A. R., Geva, A., Olson, K. L., Jones, J. R., Ignatov, V., Miller, T. A., & Mandl, K. D. (2024). AI-enabled diagnostic prediction within electronic health records to enhance biosurveillance and early outbreak detection. *PLoS Computational Biology*, *20*(6), e1012177. Advance online publication. DOI: 10.1371/journal.pcbi.1012177

McNamara, P. E., Miller, G. Y., Liu, X., & Barber, D. A. (2007). A farm-to-fork stochastic simulation model of pork-borne salmonellosis in humans: Lessons for risk ranking. *Agribusiness*, *23*(2), 157–172. DOI: 10.1002/agr.20115

McNeil, C., Verlander, S., Divi, N., & Smolinski, M. (2022). The Landscape of Participatory Surveillance Systems Across the One Health Spectrum: Systematic Review. *JMIR Public Health and Surveillance*, *8*(8), e38551. DOI: 10.2196/38551 PMID: 35930345

Meara, J. G., Leather, A. J., Hagander, L., Alkire, B. C., Alonso, N., Ameh, E. A., Bickler, S. W., Conteh, L., Dare, A. J., Davies, J., Mérisier, E. D., El-Halabi, S., Farmer, P. E., Gawande, A., Gillies, R., Greenberg, S. L. M., Grimes, C. E., Gruen, R. L., Ismail, E. A., & Yip, W. (2015). Global Surgery 2030: Evidence and solutions for achieving health, welfare, and economic development. *Lancet*, *386*(9993), 569–624. DOI: 10.1016/S0140-6736(15)60160-X PMID: 25924834

Meckawy, R., Stuckler, D., Mehta, A., Al-Ahdal, T., & Doebbeling, B. N. (2022). Effectiveness of early warning systems in the detection of infectious disease outbreaks: A systematic review. *BMC Public Health*, *22*(1), 2216. Advance online publication. DOI: 10.1186/s12889-022-14625-4 PMID: 36447171

Mei Soon, J., Krzyzaniak, S. A., Shuttlewood, Z., Smith, M., & Jack, L. (2019). Food fraud vulnerability assessment tools used in food industry. *Food Control*, *101*, 225–232. DOI: 10.1016/j.foodcont.2019.03.002

Meskó, B., Drobni, Z., Bényei, É., Gergely, B., & Győrffy, Z. (2017). Digital health is a cultural transformation of traditional healthcare. *mHealth*, *3*, 38. DOI: 10.21037/mhealth.2017.08.07 PMID: 29184890

Michelon, W., Santos, L. D., Santin, L. C., & Viancelli, A. (2025). Increased leptospirosis incidence following flooding in Rio Grande do Sul, Brazil. *International Journal of Environmental Health Research*, *35*(12), 3841–3846. DOI: 10.1080/09603123.2025.2498623 PMID: 40292553

Michener, L., Aguilar-Gaxiola, S., Alberti, P. M., Castaneda, M. J., Castrucci, B. C., Harrison, L. M., Hughes, L. S., Richmond, A., & Wallerstein, N. (2020). Engaging with communities — Lessons (re)learned from COVID-19. *Preventing Chronic Disease*, *17*, E125. Advance online publication. DOI: 10.5888/pcd17.200250 PMID: 32678059

Minaee, S., Kafieh, R., Sonka, M., Yazdani, S., & Soufi, G. J. (2020). Deep-COVID: Predicting COVID-19 from chest X-ray images using deep transfer learning. *Medical Image Analysis*, *65*, 101794. DOI: 10.1016/j.media.2020.101794 PMID: 32781377

Moen, E., Bannon, D., Kudo, T., Graf, W., Covert, M., & Van Valen, D. (2019). Deep learning for cellular image analysis. *Nature Methods*, *16*(12), 1233–1246. DOI: 10.1038/s41592-019-0403-1 PMID: 31133758

Mohsin Khan, M., Shah, N., Shaikh, N., Thabet, A., Alrabayah, T., & Belkhair, S. (2025). Towards secure and trusted AI in healthcare: A systematic review of emerging innovations and ethical challenges. *International Journal of Medical Informatics*, *195*, 105780. DOI: 10.1016/j.ijmedinf.2024.105780 PMID: 39753062

Monaghesh, E., & Hajizadeh, A. (2020). The role of telehealth during COVID-19 outbreak: A systematic review based on current evidence. *BMC Public Health*, *20*(1), 1193. DOI: 10.1186/s12889-020-09301-4 PMID: 32738884

Monaro, M., Maldera, S., Scarpazza, C., Sartori, G., & Navarin, N. (2022). Detecting deception through facial expressions in a dataset of videotaped interviews: A comparison between human judges and machine learning models. *Computers in Human Behavior*, *127*, 107063. DOI: 10.1016/j.chb.2021.107063

Mora-Cantallops, M., García-Barriocanal, E., & Sicilia, M.-Á. (2024). Trustworthy AI Guidelines in Biomedical Decision-Making Applications: A Scoping Review. *Big Data and Cognitive Computing*, *8*(7), 73. DOI: 10.3390/bdcc8070073

Moradigaravand, D., Palm, M., Farewell, A., Mustonen, V., Warringer, J., & Parts, L. (2018). Prediction of antibiotic resistance in Escherichia coli from large-scale pan-genome data. *PLoS Computational Biology*, *14*(12), e1006258. DOI: 10.1371/journal.pcbi.1006258 PMID: 30550564

Mordecai, E. A., Cohen, J. M., Evans, M. V., Gudapati, P., Johnson, L. R., Lippi, C. A., Miazgowicz, K., Murdock, C. C., Rohr, J. R., Ryan, S. J., Savage, V., Shocket, M. S., Stewart Ibarra, A., Thomas, M. B., & Weikel, D. P. (2017). Detecting the impact of temperature on transmission of Zika, dengue, and chikungunya using mechanistic models. *PLoS Neglected Tropical Diseases*, *11*(4), e0005568. DOI: 10.1371/journal.pntd.0005568 PMID: 28448507

Mordecai, E. A., Ryan, S. J., Caldwell, J. M., Shah, M. M., & LaBeaud, A. D. (2020). Climate change could shift disease burden from malaria to arboviruses in Africa. *The Lancet. Planetary Health*, *4*(9), e416–e423. DOI: 10.1016/S2542-5196(20)30178-9 PMID: 32918887

Morse, S. S. (2012). Public health surveillance and infectious disease detection. *Biosecurity and Bioterrorism*, *10*(1), 6–16. DOI: 10.1089/bsp.2011.0088 PMID: 22455675

Mo, Y., Ding, Y., Cao, Y., Hopkins, J., Ashley, E. A., Waithira, N., Wannapinij, P., Lee, S. J., Paterson, D. L., Doorn, H. R., & Turner, P. (2023). ACORN (A Clinically-Oriented Antimicrobial Resistance Surveillance Network) II: Protocol for case based antimicrobial resistance surveillance. *Wellcome Open Research*, *8*, 179. DOI: 10.12688/wellcomeopenres.19210.1 PMID: 37854055

Mukherjee, U., Dutta, S., & Bandyopadhyay, S. K. (2021). Assembling blockchain and IoT for smart food-supply chain. *Asian Journal of Advanced Agricultural Research*, *16*, 49–58. DOI: 10.9734/ajaar/2021/v16i330177

Mulwa, D., Kazuzuru, B., Misinzo, G., & Bett, B. (2024). An XGBoost Approach to Predictive Modelling of Rift Valley Fever Outbreaks in Kenya Using Climatic Factors. *Big Data and Cognitive Computing*, *8*(11), 148. DOI: 10.3390/bdcc8110148

Mulwa, D., Kazuzuru, B., Misinzo, G., & Bett, B. (2025). Machine Learning Approach to Predicting Rift Valley Fever Disease Outbreaks in Kenya. *Zoonotic Diseases*, *5*(3), 20. DOI: 10.3390/zoonoticdis5030020

Murray, C. J. L., Ikuta, K. S., Sharara, F., Swetschinski, L., Robles Aguilar, G., Gray, A., Han, C., Bisignano, C., Rao, P., Wool, E., Johnson, S. C., Browne, A. J., Chipeta, M. G., Fell, F., Hackett, S., Haines-Woodhouse, G., Hamadani, B. H. K., Kumaran, E. A. P., McManigal, B., & Naghavi, M. (2022). Global burden of bacterial antimicrobial resistance in 2019: A systematic analysis. *Lancet*, *399*(10325), 629–655. DOI: 10.1016/S0140-6736(21)02724-0 PMID: 35065702

Mushtaq, M. A., Ateeq, M., Ikram, M., Alam, S. M., Kaleem, M. M., Ashraf, M. A., Asim, M., Almutairi, K. F., Seleiman, M. F., & Shireen, F. (2025). Securing fruit trees future: AI-driven early warning and predictive systems for abiotic stress in changing climate. *Plant Stress*, *17*, 100953. DOI: 10.1016/j.stress.2025.100953

Nagendran, M., Chen, Y., Lovejoy, C. A., Gordon, A. C., Komorowski, M., Harvey, H., Topol, E. J., Ioannidis, J. P. A., Collins, G. S., & Maruthappu, M. (2020). Artificial intelligence versus clinicians: Systematic review of design, reporting standards, and claims of deep learning studies. *BMJ (Clinical Research Ed.)*, *368*, m689. DOI: 10.1136/bmj.m689 PMID: 32213531

Naing, C., Reid, S. A., Aye, S. N., Htet, N. H., & Ambu, S. (2019). Risk factors for human leptospirosis following flooding: A meta-analysis of observational studies. *PLoS One*, *14*(5), e0217643. DOI: 10.1371/journal.pone.0217643 PMID: 31141558

Nam, N., Khoa, H., Trinh, K., & Lee, N. (2023). Metagenomics: An Effective Approach for Exploring Microbial Diversity and Functions. *Foods*, *12*(11), 2140. Advance online publication. DOI: 10.3390/foods12112140 PMID: 37297385

Nangunoori, P. K. (2025). Explainable AI frameworks for transparent cloud database optimization. *Journal of International Crisis and Risk Communication Research*. https://doi.org/DOI: 10.63278/jicrcr.vi.3533

Naren, A., & Gunasekaran, K. (2025). Future cyber physical system security. In *Impact of digitalization on communication dynamics*. IGI Global., DOI: 10.4018/979-8-3693-3579-6.ch010

Naren, A., & Kaliappan, P. (2025). Mathematical approach: AI-based feature selection in machine learning. In *Serverless computing concepts in contemporary legal systems*. IGI Global., DOI: 10.4018/979-8-3693-7205-0.ch012

Naren, A., & Saravanan, N. (2025). Entrepreneurship and business culture in rural area. In *Digital disruption in hospitality: Sustainable hospitality: A global perspective on people, planet, and profit*. Emerald Publishing., DOI: 10.1108/978-1-83662-360-120251009

Nasci, R. S., & Mutebi, J.-P. (2019). Reducing West Nile Virus Risk Through Vector Management. *Journal of Medical Entomology*, *56*(6), 1516–1521. DOI: 10.1093/jme/tjz083 PMID: 31549724

Nasios, I. (2025). AI-driven multi-source data fusion for algal bloom severity classification in small inland water bodies: Leveraging Sentinel-2, DEM, and NOAA climate data. *Remote Sensing Applications: Society and Environment*, *40*, 101800. DOI: 10.1016/j.rsase.2025.101800

Nastasijevic, I., Kundacina, I., Jaric, S., Pavlovic, Z., Radovic, M., & Radonic, V. (2025). Recent advances in biosensor technologies for meat production chain. *Foods*, *14*(5), 744. DOI: 10.3390/foods14050744 PMID: 40077447

National Institute of Standards and Technology. (2023). *Artificial Intelligence Risk Management Framework (AI RMF 1.0)* (NIST AI 100-1). DOI: 10.6028/NIST.AI.100-1

Nazeha, N., Pavagadhi, D., Kyaw, B. M., Car, J., Jimenez, G., & Tudor Car, L. (2020). A digitally competent health workforce: Scoping review of educational frameworks. *Journal of Medical Internet Research*, *22*(11), e22706. DOI: 10.2196/22706 PMID: 33151152

Nazir, A., He, J., Zhu, N., Qureshi, S. S., Qureshi, S. U., Ullah, F., Wajahat, A., & Pathan, M. S. (2024). A deep learning-based novel hybrid CNN-LSTM architecture for efficient detection of threats in the IoT ecosystem. *Ain Shams Engineering Journal*, *15*(7), 102777. DOI: 10.1016/j.asej.2024.102777

Neher, R. A., & Bedford, T. (2018). Real-time analysis and visualization of pathogen sequence data. *Science, 359*(6378), 1537–1539. DOI: 10.1126/science.aaq1018 PMID: 30135232

Nguyen, A., Vu, H., Nguyen, L., Wang, Q., Djordjevic, S., Donner, E., Yin, H., & Nghiem, L. (2021). Monitoring antibiotic resistance genes in wastewater treatment: Current strategies and future challenges. *The Science of the Total Environment, 783*, 146964. DOI: 10.1016/j.scitotenv.2021.146964 PMID: 33866168

Nguyen, T. G., Dang, D. K., Ngo, C. T., Nguyen, H. D., Pham, V. T., Bui, Q.-T., Can, T.-V., & Van Luu, N. (2025). Prediction of riverbank erosion susceptibility areas of the Vietnam Mekong Delta using machine learning and remote sensing in the context of geographical studies. *Physical Geography*, •••, 1–25. DOI: 10.1080/02723646.2025.2579104

Nicholson, C., Jackson, C., & Marley, J. (2013). A governance model for integrated primary/secondary care for the health-reforming first world–results of a systematic review. *BMC Health Services Research, 13*(1), 528. DOI: 10.1186/1472-6963-13-528 PMID: 24359610

Nikita, M., Kashyap, L., Singh, B. P., & Tyagi, D. (2023). AI enabled advance surveillance system. *International Research Journal of Modernization in Engineering Technology and Science, 5*, 6056–6061.

Nisa'Habib Elias, S. K., Usman, S., & Chuprat, S. (2024). Discovering the global landscape of agri-food and blockchain: A bibliometric review. *International Journal of Advanced Computer Science and Applications, 15*(4).

Nissan, H., Ukawuba, I., & Thomson, M. (2021). Climate-proofing a malaria eradication strategy. *Malaria Journal, 20*(1), 190. Advance online publication. DOI: 10.1186/s12936-021-03718-x PMID: 33865383

Njotto, L. L., Senyoni, W., Cronie, O., & Stensgaard, A.-S. (2025). Bayesian spatio-temporal modeling and prediction of malaria cases in Tanzania mainland (2016-2023): Unveiling associations with climate and intervention factors. *International Journal of Health Geographics, 24*(1), 20. Advance online publication. DOI: 10.1186/s12942-025-00408-8 PMID: 40750887

Noman, S. M., Zeeshan, M., Arshad, J., Deressa Amentie, M., Shafiq, M., Yuan, Y., Zeng, M., Li, X., Xie, Q., & Jiao, X. (2023). Machine learning techniques for antimicrobial resistance prediction of Pseudomonas aeruginosa from whole genome sequence data. *Computational Intelligence and Neuroscience, 2023*(1), 5236168. DOI: 10.1155/2023/5236168 PMID: 36909968

Nouis, S. C., Uren, V., & Jariwala, S. (2025). Evaluating accountability, transparency, and bias in AI-assisted healthcare decision- making: A qualitative study of healthcare professionals' perspectives in the UK. *BMC Medical Ethics*, *26*(1), 89. DOI: 10.1186/s12910-025-01243-z PMID: 40629303

Nugroho, G., Tedjakusuma, F., Lo, D., Romulo, A., Pamungkas, D. H., & Kinardi, S. A. (2023). Review of the application of digital transformation in food industry. *Journal of Current Science and Technology*, *13*(3), 774–790. DOI: 10.59796/jcst.V13N3.2023.1285

Nuzzo, J. B. (2017). Improving biosurveillance systems to enable situational awareness during public health emergencies. *Health Security*, *15*(1), 17–19. DOI: 10.1089/hs.2016.0097 PMID: 28092456

Obermeyer, Z., Powers, B., Vogeli, C., & Mullainathan, S. (2019). Dissecting racial bias in an algorithm used to manage the health of populations. *Science*, *366*(6464), 447–453. DOI: 10.1126/science.aax2342 PMID: 31649194

Odimarha, A. C., Ayodeji, S. A., & Abaku, E. A.Agnes Clare OdimarhaSodrudeen Abolore AyodejiEmmanuel Adeyemi Abaku. (2024). Machine learning's influence on supply chain and logistics optimization in the oil and gas sector: A comprehensive analysis. *Computer Science & IT Research Journal*, *5*(3), 725–740. DOI: 10.51594/csitrj.v5i3.976

Ogidi, O. I., & Izah, S. C. (2024). Emerging technologies in the global south food industry: Prospects and challenges. In *Food safety and quality in the global south* (pp. 197–231).

Ogundokun, R. O., Misra, S., Akinrotimi, A. O., & Ogul, H. (2023). MobileNet-SVM: A lightweight deep transfer learning model to diagnose BCH scans for IoMT-based imaging sensors. *Sensors (Basel)*, *23*(2), 656. DOI: 10.3390/s23020656 PMID: 36679455

Ogwu, M. C., & Izah, S. C. (2023). *One Health implications of agrochemicals and their sustainable alternatives* (Sustainable Development and Biodiversity, Vol. 34). Springer Nature. https://doi.org/DOI: 10.1007/978-981-99-3439-3

Ogwu, M. C., & Izah, S. C. (2025). Honey as a natural antimicrobial. *Antibiotics (Basel, Switzerland)*, *14*(3), 255. DOI: 10.3390/antibiotics14030255 PMID: 40149066

Ogwu, M. C., & Izah, S. C. (2025). Nanotechnology for fungal pathogen control in crops: Innovations, public health impacts, and disease prevention. *Frontiers in Fungal Biology*, *6*, 1653214. DOI: 10.3389/ffunb.2025.1653214 PMID: 40959466

Ogwu, M. C., & Izah, S. C. (Eds.). (2026a). *Detection, identification, and monitoring of food contaminants*. IGI Global Scientific Publishing., DOI: 10.4018/979-8-3373-3982-5

Ogwu, M. C., Izah, S. C., Aigberua, A. O., & Ngun, C. T. (2024). Editorial: Detection, risk analysis, and monitoring of chemical contaminants from agro-aqua food production and processing: Implications on the One Health triad. *Frontiers in Sustainable Food Systems*, *8*, 1501930. DOI: 10.3389/fsufs.2024.1501930

Ogwu, M. C., Izah, S. C., & Ntuli, N. R. (2024). *Food safety and quality in the Global South*. Springer Nature., DOI: 10.1007/978-981-97-2428-4

Ogwu, M. C., Izah, S. C., Sawyer, W. E., & Amabie, T. (2025). Environmental risk assessment of trace metal pollution: A statistical perspective. *Environmental Geochemistry and Health*, *47*(4), 94. DOI: 10.1007/s10653-025-02405-z PMID: 40019615

Ojagh, S., Cauteruccio, F., Terracina, G., & Liang, S. H. (2021). Enhanced air quality prediction by edge-based spatiotemporal data preprocessing. *Computers & Electrical Engineering*, *96*, 107572. DOI: 10.1016/j.compeleceng.2021.107572

Olawade, D. B., Wada, O. Z., Ige, A. O., Egbewole, B. I., Olojo, A., & Oladapo, B. I. (2024). Artificial intelligence in environmental monitoring: Advancements, challenges, and future directions. *Hygiene and Environmental Health Advances*, *12*, 100114. Advance online publication. DOI: 10.1016/j.heha.2024.100114

Oloye, F., Xie, Y., Challis, J., Femi-Oloye, O., Brinkmann, M., McPhedran, K., Jones, P., Servos, M., & Giesy, J. (2023). Understanding common population markers for SARS-CoV-2 RNA normalization in wastewater – A review. *Chemosphere*, *333*, 138682–138682. DOI: 10.1016/j.chemosphere.2023.138682 PMID: 37201600

Olsen, P., & Borit, M. (2013). How to define traceability. *Trends in Food Science & Technology*, *29*(2), 142–150. DOI: 10.1016/j.tifs.2012.10.003

Opara, L. U. (2003). Traceability in agriculture and food supply chain: A review of basic concepts, technological implications, and future prospects. *Journal of Food Agriculture and Environment*, *1*(1), 101–106.

Oriekhoe, O. I., Ilugbusi, B. S., & Adisa, O. (2024). Ensuring global food safety: Integrating blockchain technology into food supply chains. *Engineering Science & Technology Journal*, *5*(3), 811–820. DOI: 10.51594/estj.v5i3.905

Ovuru, K. F., Izah, S. C., Ogidi, O. I., Imarhiagbe, I., & Ogwu, M. C. (2023). Slaughterhouse facilities in developing nations: Sanitation and hygiene practices, microbial contamination, and sustainable management system. *Food Science and Biotechnology*. Advance online publication. DOI: 10.1007/s10068-023-01406-x PMID: 38274182

Oyewole, A. T., Okoye, C. C., Ofodile, O. C., & Ejairu, E. (2024). Reviewing predictive analytics in supply chain management: Applications and benefits. *World Journal of Advanced Research and Reviews*, *21*(3), 568–574. DOI: 10.30574/wjarr.2024.21.3.0673

Paerl, H. W. (2017). Controlling harmful cyanobacterial blooms in a climatically more extreme world: Management options and research needs. *Journal of Plankton Research*, *39*(5), 763–771. DOI: 10.1093/plankt/fbx042

Palojoki, S., & Vuokko, R. (2023). Semantic interoperability of EHRs: Review of approaches for enhancing the goals of European Health Data Space.

Pan, Y., Wang, X., & Ye, Q. (2024). Enhancing supply chain management through artificial intelligence: A case study of JD Logistics. *Advances in Economics. Management and Political Sciences*, *109*(1), 116–121. DOI: 10.54254/2754-1169/109/2024BJ0127

Papyan, V., Romano, Y., Sulam, J., & Elad, M. (2018). Theoretical foundations of deep learning via sparse representations: A multilayer sparse model and its connection to convolutional neural networks. *IEEE Signal Processing Magazine*, *35*(4), 72–89. DOI: 10.1109/MSP.2018.2820224

Paquet, C., Coulombier, D., Kaiser, R., & Ciotti, M. (2006). Epidemic intelligence: A new framework for strengthening disease surveillance in Europe. *Eurosurveillance*, *11*(12), 212–214. DOI: 10.2807/esm.11.12.00665-en PMID: 17370970

Parker, M., Fraser, C., Abeler-Dörner, L., & Bonsall, D. (2020). Ethics of instantaneous contact tracing using mobile phone apps in the control of the COVID-19 pandemic. *Journal of Medical Ethics*, *46*(7), 427–431. DOI: 10.1136/medethics-2020-106314 PMID: 32366705

Parkins, M., Lee, B., Acosta, N., Bautista, M., Hubert, C., Hrudey, S., Frankowski, K., & Pang, X. (2023). Wastewater-based surveillance as a tool for public health action: SARS-CoV-2 and beyond. *Clinical Microbiology Reviews*, *37*(1), e00103-22. Advance online publication. DOI: 10.1128/cmr.00103-22 PMID: 38095438

Parks, D. H., Imelfort, M., Skennerton, C. T., Hugenholtz, P., & Tyson, G. W. (2015). CheckM: Assessing the quality of microbial genomes recovered from isolates, single cells, and metagenomes. *Genome Research, 25*(7), 1043–1055. DOI: 10.1101/gr.186072.114 PMID: 25977477

Patel, M., Emerich, D., & Singh, S. (2022). Review of antimicrobial resistance surveillance using wastewater-based approaches. *Environmental Monitoring and Assessment.*

Patin, N., & Goodwin, K. (2023). Capturing marine microbiomes and environmental DNA: A field sampling guide. *Frontiers in Microbiology, 13*, 1026596. Advance online publication. DOI: 10.3389/fmicb.2022.1026596 PMID: 36713215

Pawlowski, J., Bruce, K., Panksep, K., Aguirre, F., Amafitano, S., Apothéloz-Perret-Gentil, L., Baussant, T., Bouchez, A., Carugati, L., Cermakova, K., Cordier, T., Corinaldesi, C., Costa, F., Danovaro, R., Dell'Anno, A., Duarte, S., Eisendle, U., Ferrari, B., Frontalini, F., & Fazi, S. (2021). Environmental DNA metabarcoding for benthic monitoring: A review of sediment sampling and DNA extraction methods. *The Science of the Total Environment, 151783*. Advance online publication. DOI: 10.1016/j.scitotenv.2021.151783 PMID: 34801504

Pearson, S., May, D., Leontidis, G., Swainson, M., Brewer, S., Bidaut, L., Freyd, J. G., Parr, G., Maull, R., & Zisman, A. (2019). Are distributed ledger technologies the panacea for food traceability? *Global Food Security, 20*, 145–149. DOI: 10.1016/j.gfs.2019.02.002

Peek, N., Sujan, M., & Scott, P. (2020). Digital health and care in pandemic times: Impact of COVID-19. *BMJ Health & Care Informatics, 27*(1), e100166. DOI: 10.1136/bmjhci-2020-100166 PMID: 32565418

Pei, L., & Wu, H. (2019). Does online learning work better than offline learning in undergraduate medical education? A systematic review and meta-analysis. *Medical Education Online, 24*(1), 1666538. DOI: 10.1080/10872981.2019.1666538 PMID: 31526248

Peng, J., Zhou, X., Rensing, C., Liesack, W., & Zhu, Y. (2023). Soil microbial ecology through the lens of metatranscriptomics. *Soil Ecology Letters, 6*(3), 230217. Advance online publication. DOI: 10.1007/s42832-023-0217-z

Pennisi, F., Pinto, A., Ricciardi, G. E., Signorelli, C., & Gianfredi, V. (2025). The role of artificial intelligence and machine learning models in antimicrobial stewardship in public health: A narrative review. *Antibiotics (Basel, Switzerland), 14*(2), 134. DOI: 10.3390/antibiotics14020134 PMID: 40001378

Pfenning-Butterworth, A., Buckley, L. B., Drake, J. M., Farner, J. E., Farrell, M. J., Gehman, A.-L. M., Mordecai, E. A., Stephens, P. R., Gittleman, J. L., & Davies, T. J. (2024). Interconnecting global threats: Climate change, biodiversity loss, and infectious diseases. *The Lancet. Planetary Health*, *8*(4), e270–e283. DOI: 10.1016/S2542-5196(24)00021-4 PMID: 38580428

Pham, N.-L., Ta, Q.-B., & Kim, J.-T. (2025). Pseudo-damage simulation and CNN deep learning for damage identification of submerged structure-foundation systems. *Structural Health Monitoring*, *24*(4), 2149–2176. DOI: 10.1177/14759217241288773

Phillips, C., Chapman, B., Agunos, A., Carson, C. A., Parmley, E. J., Reid-Smith, R. J., Smith, B. A., & Murphy, C. P. (2022). A scoping review of factors potentially linked with antimicrobial-resistant bacteria from turkeys (iAM. AMR Project). *Epidemiology and Infection*, *150*, e153. DOI: 10.1017/S0950268822001224 PMID: 35843720

Pillai, S. V., & Kumar, R. S. (2021). The role of data-driven artificial intelligence on COVID-19 disease management in public sphere: A review. *Decision (Washington, D.C.)*, *48*(4), 375–389. DOI: 10.1007/s40622-021-00286-8

Prata, J., Reis, V., Da Costa, J., Mouneyrac, C., Duarte, A., & Rocha-Santos, T. (2021). Contamination issues as a challenge in quality control and quality assurance in microplastics analytics. *Journal of Hazardous Materials*, *403*, 123660. DOI: 10.1016/j.jhazmat.2020.123660 PMID: 33264868

Primavilla, S., Roila, R., Zicavo, A., Ortenzi, R., Branciari, R., Shtylla Kika, T., Valiani, A., & Ranucci, D. (2021). *Salmonella* spp. in pigs slaughtered in small and medium-sized abattoirs in central Italy: Preliminary results on occurrence and control strategies. *Applied Sciences (Basel, Switzerland)*, *11*(16), 7600. DOI: 10.3390/app11167600

Pubellier, C. (2005). Lithuanian geographic information infrastructure@ spatial data sharing with Lithuania. *Geodezija ir Kartografija, 31*(2), 47–53.

Pūraitė, A., Zuzevičiūtė, V., Bereikienė, D., Skrypko, T., & Shmorgun, L. (2020). Algorithmic governance in public sector: Is digitization a key to effective management. *Independent Journal of Management & Production*, *11*(9), 2149–2170. DOI: 10.14807/ijmp.v11i9.1400

Putro, M. D., Nguyen, D.-L., & Jo, K.-H. (2022). A fast CPU real-time facial expression detector using sequential attention network for human–robot interaction. *IEEE Transactions on Industrial Informatics*, *18*(11), 7665–7674. DOI: 10.1109/TII.2022.3145862

Pykes, K. (2024). *10 signs of bad data: How to spot poor quality data*. DataCamp. https://www.datacamp.com/blog/10-signs-bad-data-quality

Qiu, J. (2020). Covert coronavirus infections could be seeding new outbreaks. *Nature*. Advance online publication. DOI: 10.1038/d41586-020-00822-x PMID: 32203376

Quainoo, S., Coolen, J. P. M., van Hijum, S. A. F. T., Huynen, M. A., Melchers, W. J. G., van Schaik, W., & Wertheim, H. F. L. (2017). Whole-genome sequencing of bacterial pathogens: The future of nosocomial outbreak analysis. *Clinical Microbiology Reviews*, *30*(4), 1015–1063. DOI: 10.1128/CMR.00016-17 PMID: 28855266

Quick, J., Grubaugh, N. D., Pullan, S. T., Claro, I. M., Smith, A. D., Gangavarapu, K., Oliveira, G., Robles-Sikisaka, R., Rogers, T. F., Beutler, N. A., Burton, D. R., Lewis-Ximenez, L. L., de Jesus, J. G., Giovanetti, M., Hill, S. C., Black, A., Bedford, T., Carroll, M. W., Nunes, M., & Loman, N. J. (2017). Multiplex PCR method for MinION and Illumina sequencing of Zika and other virus genomes directly from clinical samples. *Nature Protocols*, *12*(6), 1261–1276. DOI: 10.1038/nprot.2017.066 PMID: 28538739

Quick, J., Loman, N. J., Duraffour, S., Simpson, J. T., Severi, E., Cowley, L., Bore, J. A., Koundouno, R., Dudas, G., Mikhail, A., Ouédraogo, N., Afrough, B., Bah, A., Baum, J. H. J., Becker-Ziaja, B., Boettcher, J. P., Cabeza-Cabrerizo, M., Camino-Sánchez, Á., Carter, L. L., & Carroll, M. W. (2016). Real-time, portable genome sequencing for Ebola surveillance. *Nature*, *530*(7589), 228–232. DOI: 10.1038/nature16996 PMID: 26840485

Quince, C., Walker, A. W., Simpson, J. T., Loman, N. J., & Segata, N. (2017). Shotgun metagenomics, from sampling to analysis. *Nature Biotechnology*, *35*(9), 833–844. DOI: 10.1038/nbt.3935 PMID: 28898207

R., B., N, K., Lp, T., Jt, C., Wp, H., Mt, B., & Tl, C. (2025). Advancing health equity in wastewater-based epidemiology: A global critical review and conceptual framework. *SSM - Population Health*, 30. https://doi.org/DOI: 10.1016/j.ssmph.2025.101786

Radanliev, P. (2025). Privacy, ethics, transparency, and accountability in AI systems for wearable devices. *Frontiers in Digital Health*, *7*, 1431246. Advance online publication. DOI: 10.3389/fdgth.2025.1431246 PMID: 40599875

Raghavan, K., B, S., & v, K. (2024). Attention guided grad-CAM: An improved explainable artificial intelligence model for infrared breast cancer detection. *Multimedia Tools and Applications*, *83*(19), 57551–57578. DOI: 10.1007/s11042-023-17776-7

Rahman, A., Hossain, M. S., Muhammad, G., Kundu, D., Debnath, T., Rahman, M., Khan, M. S. I., Tiwari, P., & Band, S. S. (2022). Federated learning-based AI approaches in smart healthcare: Concepts, taxonomies, challenges and open issues. *Cluster Computing*, •••, 1–41. Advance online publication. DOI: 10.1007/s10586-022-03658-4 PMID: 35996680

Rahman, M. M., & Szabó, G. (2023). Assessing the status of national spatial data infrastructure (NSDI) of Bangladesh. *ISPRS International Journal of Geo-Information*, *12*(6), 236. DOI: 10.3390/ijgi12060236

Rahman, M. U., Shah, J. A., Khan, M. N., Bilal, H., Zhu, D., Du, Z., & Mu, D. S. (2026). Innovative approaches to combat antimicrobial resistance: A review of emerging therapies and technologies. *Probiotics and Antimicrobial Proteins*, *18*(3), 3597–3614. DOI: 10.1007/s12602-025-10676-2 PMID: 41026406

Rajkomar, A., Hardt, M., Howell, M. D., Corrado, G., & Chin, M. H. (2018). Ensuring fairness in machine learning to advance health equity. *Annals of Internal Medicine*, *169*(12), 866–872. DOI: 10.7326/M18-1990 PMID: 30508424

Rajpurkar, P., Chen, E., Banerjee, O., & Topol, E. J. (2022). AI in health and medicine. *Nature Medicine*, *28*(1), 31–38. DOI: 10.1038/s41591-021-01614-0 PMID: 35058619

Ramanathan, A., Pullum, L. L., Hobson, T. C., Steed, C. A., Quinn, S., Chennubhotla, C., & Valkova, S. (2015). ORBiT: Oak Ridge biosurveillance toolkit for public health dynamics. *BMC Bioinformatics*, *16*(S17, Suppl 17), S4. Advance online publication. DOI: 10.1186/1471-2105-16-S17-S4 PMID: 26679008

Rambaut, A., Holmes, E. C., O'Toole, Á., Hill, V., McCrone, J. T., Ruis, C., du Plessis, L., & Pybus, O. G. (2020). A dynamic nomenclature proposal for SARS-CoV-2 lineages to assist genomic epidemiology. *Nature Microbiology*, *5*(11), 1403–1407. DOI: 10.1038/s41564-020-0770-5 PMID: 32669681

Rane, N., Paramesha, M., Choudhary, S., & Rane, J. (2024). Machine Learning and Deep Learning for Big Data Analytics: a Review of Methods and Applications. SSRN Electronic Journal. https://doi.org/DOI: 10.2139/ssrn.4835655

Rani, M., & Kumar, M. (2025). MobileNet for human activity recognition in smart surveillance using transfer learning. *Neural Computing & Applications*, *37*(5), 3907–3924. DOI: 10.1007/s00521-024-10882-z

Rashid, S., Rajendiran, S., Nazakat, R., Sham, N., Hasni, N., Anasir, M., Kamel, K., & Robat, R. (2024). A scoping review of global SARS-CoV-2 wastewater-based epidemiology in light of COVID-19 pandemic. *Heliyon*, *10*(9), e30600. Advance online publication. DOI: 10.1016/j.heliyon.2024.e30600 PMID: 38765075

Regattieri, A., Gamberi, M., & Manzini, R. (2007). Traceability of food products: General framework and experimental evidence. *Journal of Food Engineering*, *81*(2), 347–356. DOI: 10.1016/j.jfoodeng.2006.10.032

Regmi, K., & Jones, L. (2020). A systematic review of the factors—enablers and barriers—affecting e-learning in health sciences education. *BMC Medical Education*, *20*(1), 91. DOI: 10.1186/s12909-020-02007-6 PMID: 32228560

Regueira-Iglesias, A., Balsa-Castro, C., Blanco-Pintos, T., & Tomás, I. (2023). Critical review of 16S rRNA gene sequencing workflow in microbiome studies: From primer selection to advanced data analysis. *Molecular Oral Microbiology*, *38*(5), 347–399. Advance online publication. DOI: 10.1111/omi.12434 PMID: 37804481

Rehan, M. W., & Rehan, M. M. (2025). Survey, taxonomy, and emerging paradigms of societal digital twins for public health preparedness. *NPJ Digital Medicine*, *8*(1), 520. DOI: 10.1038/s41746-025-01737-5 PMID: 40804297

Reich, N. G., Brooks, L. C., Fox, S. J., Kandula, S., McGowan, C. J., Moore, E., Osthus, D., Ray, E. L., Tushar, A., Yamana, T. K., Biggerstaff, M., Johansson, M. A., Rosenfeld, R., Shaman, J., & Viboud, C. (2019). A collaborative multiyear, multimodel assessment of seasonal influenza forecasting in the United States. *Proceedings of the National Academy of Sciences of the United States of America*, *116*(8), 3146–3154. DOI: 10.1073/pnas.1812594116 PMID: 30647115

Ren, H., & Xu, N. (2024). Forecasting and mapping dengue fever epidemics in China: A spatiotemporal analysis. *Infectious Diseases of Poverty*, *13*(1), 50. Advance online publication. DOI: 10.1186/s40249-024-01219-y PMID: 38956632

Ren, J., Song, K., Deng, C., Ahlgren, N. A., Fuhrman, J. A., Li, Y., Xie, X., Poplin, R., & Sun, F. (2020). Identifying viruses from metagenomic data using deep learning. *Quantitative Biology*, *8*(1), 64–77. DOI: 10.1007/s40484-019-0187-4 PMID: 34084563

Ren, Y., Chakraborty, T., Doijad, S., Falgenhauer, L., Falgenhauer, J., Goesmann, A., Hauschild, A.-C., Schwengers, O., & Heider, D. (2022). Prediction of antimicrobial resistance based on whole-genome sequencing and machine learning. *Bioinformatics (Oxford, England)*, *38*(2), 325–334. DOI: 10.1093/bioinformatics/btab681 PMID: 34613360

Rice, B. T., Rasmus, S., Onders, R., Thomas, T., Day, G., Wood, J., Britton, C., Hernandez-Boussard, T., & Hiratsuka, V. (2025). Community-engaged artificial intelligence: An upstream, participatory design, development, testing, validation, use and monitoring framework for artificial intelligence and machine learning models in the Alaska Tribal Health System. *Frontiers in Artificial Intelligence*, *8*, 1568886. DOI: 10.3389/frai.2025.1568886 PMID: 40260415

Rieke, N., Hancox, J., Li, W., Milletarì, F., Roth, H. R., Albarqouni, S., Bakas, S., Galtier, M. N., Landman, B. A., Maier-Hein, K., Ourselin, S., Sheller, M., Summers, R. M., Trask, A., Xu, D., Baust, M., & Cardoso, M. J. (2020). The future of digital health with federated learning. *NPJ Digital Medicine*, *3*(1), 119. DOI: 10.1038/s41746-020-00323-1 PMID: 33015372

Ringsberg, H. (2014). Perspectives on food traceability: A systematic literature review. *Supply Chain Management*, *19*(5/6), 558–576. DOI: 10.1108/SCM-01-2014-0026

Rithik, K., & Naren, A. K. (2024). The IoT of robotics: The frontier of automation. In *Artificial intelligence-enabled digital twin for smart manufacturing*. Scrivener Publishing/Wiley., DOI: 10.1002/9781394303601.ch13

Roberts, M., Dobson, A., Restif, O., & Wells, K. (2021). Challenges in modelling the dynamics of infectious diseases at the wildlife–human interface. *Epidemics*, *37*, 100523. DOI: 10.1016/j.epidem.2021.100523 PMID: 34856500

Rodriguez, J. A., Clark, C. R., & Bates, D. W. (2020). Digital health equity as a necessity in the 21st Century Cures Act era. *Journal of the American Medical Association*, *323*(23), 2381–2382. DOI: 10.1001/jama.2020.7858 PMID: 32463421

Rossion, B., & Lochy, A. (2022). Is human face recognition lateralized to the right hemisphere due to neural competition with left-lateralized visual word recognition? A critical review. *Brain Structure & Function*, *227*(2), 599–629. DOI: 10.1007/s00429-021-02370-0 PMID: 34731327

Rout, A., Rout, S., Panda, A., Tripathy, P., Kumar, N., Parida, S., Dey, S., Dash, S., Behera, B., & Pandey, P. (2025). Potential applications and future prospects of metagenomics in aquatic ecosystems. *Gene*, *149720*, 149720. Advance online publication. DOI: 10.1016/j.gene.2025.149720 PMID: 40789383

Rouvière, E., & Caswell, J. A. (2012). From punishment to prevention: A French case study of the introduction of co-regulation in enforcing food safety. *Food Policy*, *37*(3), 246–254. DOI: 10.1016/j.foodpol.2012.02.009

Ruan, C., Dong, Y., Huang, W., Huang, L., Ye, H., Ma, H., Guo, A., & Ren, Y. (2021). Prediction of Wheat Stripe Rust Occurrence with Time Series Sentinel-2 Images. *Agriculture, 11*(11), 1079. DOI: 10.3390/agriculture11111079

Rudnick, W., Mukhi, S. N., Reid-Smith, R. J., German, G. J., Nichani, A., & Mulvey, M. R. (2022). Overview of Canada's Antimicrobial Resistance Network (AMRNet): A data-driven One Health approach to antimicrobial resistance surveillance. *Canada Communicable Disease Report = Relevé des Maladies Transmissibles au Canada, 48*(11-12), 522. PMID: 38173468

Russell, S. J. (2019). *Human compatible: Artificial intelligence and the problem of control.* Viking.

Saadatzadeh, R. (2024). The future of public health: Integrating artificial intelligence in disease surveillance and prevention. *Journal of Foresight and Health Governance, 1*(2), 1–15.

Saheb, T. (2022). Ethically contentious aspects of artificial intelligence surveillance: A social science perspective. *AI and Ethics, 3*(2), 369–379. DOI: 10.1007/s43681-022-00196-y PMID: 35874304

Sahoo, M., Panigrahi, C., & Aradwad, P. (2022). Management strategies emphasizing advanced food processing approaches to mitigate food borne zoonotic pathogens in food system. *Food Frontiers, 3*(4), 641–665. DOI: 10.1002/fft2.153

Salah, K., Rehman, M. H. U., Nizamuddin, N., & Al-Fuqaha, A. (2019). Blockchain for AI: Review and open research challenges. *IEEE Access : Practical Innovations, Open Solutions, 7*, 10127–10149. DOI: 10.1109/ACCESS.2018.2890507

Salardi, S. (2025). What is the role of institutional bioethics in addressing new technological challenges? The case of the ecological and carbon footprint of the healthcare sector. In *EACME Conference 2025 Book of Abstracts*.

Salathé, M., Althaus, C. L., Anderegg, N., Antonioli, D., Ballouz, T., Bugnon, E., Čapkun, S., Jackson, D., Kim, S. I., Larus, J. R., Low, N., Lueks, W., Menges, D., Moullet, C., Payer, M., Riou, J., Stadler, T., Troncoso, C., Vayena, E., & von Wyl, V. (2022). Participatory disease surveillance systems: Engaging communities in public health monitoring. *JMIR Public Health and Surveillance, 8*(4), e28159. Advance online publication. DOI: 10.2196/28159

Salathé, M., Althaus, C. L., Neher, R., Stringhini, S., Hodcroft, E., Fellay, J., Zwahlen, M., & Low, N. (2020). COVID-19 epidemic in Switzerland: On the importance of testing, contact tracing, and isolation. *Swiss Medical Weekly, 150*(1112), w20225. DOI: 10.4414/smw.2020.20225 PMID: 32191813

Salathé, M., Bengtsson, L., Bodnar, T. J., Brewer, D. D., Brownstein, J. S., Buckee, C., Campbell, E. M., Cattuto, C., Khandelwal, S., Mabry, P. L., & Vespignani, A. (2012). Digital epidemiology. *PLoS Computational Biology*, *8*(7), e1002616. DOI: 10.1371/journal.pcbi.1002616 PMID: 22844241

Salem, K. F. M., & Eid, N. A. (2025). Modern breeding strategies and tools in bread wheat (*Triticum aestivum* L.). In K. F. M. Salem, J. M. Al-Khayri, & S. M. Jain (Eds.), *Breeding and biotechnology of grass and bast fiber crops* (Advances in Plant Breeding Strategies, Vol. 12). Springer. https://doi.org/DOI: 10.1007/978-3-032-00403-1_8

Santillana, M., Nguyen, A. T., Dredze, M., Paul, M. J., Nsoesie, E. O., & Brownstein, J. S. (2015). Combining search, social media, and traditional data sources to improve influenza surveillance. *PLoS Computational Biology*, *11*(10), e1004513. DOI: 10.1371/journal.pcbi.1004513 PMID: 26513245

Saqib, S. M., Iqbal, M., Asghar, M. Z., Mazhar, T., Almogren, A., Rehman, A. U., & Hamam, H. (2024). Cataract and glaucoma detection based on Transfer Learning using MobileNet. *Heliyon*, *10*(17), e36759. DOI: 10.1016/j.heliyon.2024.e36759 PMID: 39281545

Sassi, A., Basher, N., Kirat, H., Meradji, S., Ibrahim, N., Idres, T., & Touati, A. (2025). The Role of the Environment (Water, Air, Soil) in the Emergence and Dissemination of Antimicrobial Resistance: A One Health Perspective. *Antibiotics (Basel, Switzerland)*, *14*(8), 764. Advance online publication. DOI: 10.3390/antibiotics14080764 PMID: 40867959

Schaeffer, B. A., Reynolds, N., Ferriby, H., Salls, W., Smith, D., Johnston, J. M., & Myer, M. (2024). Forecasting freshwater cyanobacterial harmful algal blooms for Sentinel-3 satellite resolved U.S. lakes and reservoirs. *Journal of Environmental Management*, *349*, 119518. DOI: 10.1016/j.jenvman.2023.119518 PMID: 37944321

Schoch, C. L., Ciufo, S., Domrachev, M., Hotton, C. L., Kannan, S., Khovanskaya, R., Leipe, D., McVeigh, R., O'Neill, K., Robbertse, B., Sharma, S., Soussov, V., Sullivan, J. P., Sun, L., Turner, S., & Karsch-Mizrachi, I. (2020). NCBI taxonomy: A comprehensive update on curation, resources, and tools. *Database : The Journal of Biological Databases and Curation*, *2020*, baaa062. Advance online publication. DOI: 10.1093/database/baaa062 PMID: 32761142

Schulte, T., & Bohnet-Joschko, S. (2022). How can big data analytics support people-centred and integrated health services: A scoping review. *International Journal of Integrated Care*, *22*(2), 23. Advance online publication. DOI: 10.5334/ijic.5543 PMID: 35756337

Seemann, T. (2014). Prokka: Rapid prokaryotic genome annotation. *Bioinformatics (Oxford, England)*, *30*(14), 2068–2069. DOI: 10.1093/bioinformatics/btu153 PMID: 24642063

Sengupta, A., Seal, A., Panigrahy, C., Krejcar, O., & Yazidi, A. (2020). Edge information based image fusion metrics using fractional order differentiation and sigmoidal functions. *IEEE Access : Practical Innovations, Open Solutions*, *8*, 88385–88398. DOI: 10.1109/ACCESS.2020.2993607

Shachar, C., Gerke, S., & Adashi, E. Y. (2020). AI Surveillance during Pandemics: Ethical Implementation Imperatives. *The Hastings Center Report*, *50*(3), 18–21. DOI: 10.1002/hast.1125 PMID: 32596887

Shahi, T. B., Xu, C.-Y., Neupane, A., & Guo, W. (2023). Recent Advances in Crop Disease Detection Using UAV and Deep Learning Techniques. *Remote Sensing (Basel)*, *15*(9), 2450. DOI: 10.3390/rs15092450

Shah, R. Y., Prajapati, P. N., & Agrawal, Y. K. (2010). Anticounterfeit packaging technologies. *Journal of Advanced Pharmaceutical Technology & Research*, *1*(4), 368–373. DOI: 10.4103/0110-5558.76434 PMID: 22247875

Shah, S., Gwee, S., Ng, J., Lau, N., Koh, J., & Pang, J. (2021). Wastewater surveillance to infer COVID-19 transmission: A systematic review. *The Science of the Total Environment*, *804*, 150060–150060. DOI: 10.1016/j.scitotenv.2021.150060 PMID: 34798721

Shakallis, A., Fallowfield, H., Ross, K., & Whiley, H. (2022). The Application of Passive Sampling Devices in Wastewater Surveillance. *Water (Basel)*, *14*(21), 3478. Advance online publication. DOI: 10.3390/w14213478

Shanahan, C., Kernan, B., Ayalew, G., McDonnell, K., Butler, F., & Ward, S. (2009). A framework for beef traceability from farm to slaughter using global standards: An Irish perspective. *Computers and Electronics in Agriculture*, *66*(1), 62–69. DOI: 10.1016/j.compag.2008.12.002

Sharma, K., & Maharshi, N. (2024). Social impacts of AI-powered online reviews: An ethical consideration. *Journal of Informatics Education and Research*, *4*(3).

Sharma, R., Kamble, S. S., Gunasekaran, A., Kumar, V., & Kumar, A. (2020). A systematic literature review on machine learning applications for sustainable agriculture supply chain performance. *Computers & Operations Research*, *119*, 104926. DOI: 10.1016/j.cor.2020.104926

Sharma, S., & Singh, S. (2023). ISL recognition system using integrated mobile-net and transfer learning method. *Expert Systems with Applications*, *221*, 119772. DOI: 10.1016/j.eswa.2023.119772

Sharuddin, S., Ramli, N., Yusoff, M., Muhammad, N., Ho, L., & Maeda, T. (2022). Advancement of Metatranscriptomics towards Productive Agriculture and Sustainable Environment: A Review. *International Journal of Molecular Sciences*, *23*(7), 3737. Advance online publication. DOI: 10.3390/ijms23073737 PMID: 35409097

Sheffield, J., Wood, E. F., Pan, M., Beck, H., Coccia, G., Serrat-Capdevila, A., & Verbist, K. (2018). Satellite Remote Sensing for Water Resources Management: Potential for Supporting Sustainable Development in Data-Poor Regions. *Water Resources Research*, *54*(12), 9724–9758. DOI: 10.1029/2017WR022437

Shenoy, D., Bhat, R., & Krishna Prakasha, K. (2025). Exploring privacy mechanisms and metrics in federated learning. *Artificial Intelligence Review*, *58*(8), 223. DOI: 10.1007/s10462-025-11170-5

Shen, Y., Ren, Z., Fan, J., Xiao, J., Zhang, Y., & Liu, X. (2025). Fine-Scale Risk Mapping for Dengue Vector Using Spatial Downscaling in Intra-Urban Areas of Guangzhou, China. *Insects*, *16*(7), 661. DOI: 10.3390/insects16070661 PMID: 40725294

Shevtsova, D., Ahmed, A., Boot, I. W. A., Sanges, C., Hudecek, M., Jacobs, J. J. L., Hort, S., & Vrijhoef, H. J. M. (2024). Trust in and acceptance of artificial intelligence applications in medicine: Mixed methods study. *JMIR Human Factors*, *11*, e47031. Advance online publication. DOI: 10.2196/47031 PMID: 38231544

Shi, Y., Liu, X., Kok, S.-Y., Rajarethinam, J., Liang, S., Yap, G., Chong, C.-S., Lee, K.-S., Tan, S. S. Y., Chin, C. K. Y., Lo, A., Kong, W., Ng, L. C., & Cook, A. R. (2016). Three-Month Real-Time Dengue Forecast Models: An Early Warning System for Outbreak Alerts and Policy Decision Support in Singapore. *Environmental Health Perspectives*, *124*(9), 1369–1375. DOI: 10.1289/ehp.1509981 PMID: 26662617

Shmueli, G., & Burkom, H. (2010). Statistical challenges facing early outbreak detection in biosurveillance. *Technometrics*, *52*(1), 39–51. DOI: 10.1198/TECH.2010.06134

Shome, N., Kashyap, R., & Laskar, R. H. (2024). Detection of tuberculosis using customized MobileNet and transfer learning from chest X-ray image. *Image and Vision Computing*, *147*, 105063. DOI: 10.1016/j.imavis.2024.105063

Shrestha, S., Yoshinaga, E., Chapagain, S., Mohan, G., Gasparatos, A., & Fukushi, K. (2021). Wastewater-Based Epidemiology for Cost-Effective Mass Surveillance of COVID-19 in Low- and Middle-Income Countries: Challenges and Opportunities. *Water (Basel)*, *13*(20), 2897. Advance online publication. DOI: 10.3390/w13202897

Shu, Y., & McCauley, J. (2017). GISAID: Global initiative on sharing all influenza data - from vision to reality. *Eurosurveillance*, *22*(13), 30494. DOI: 10.2807/1560-7917.ES.2017.22.13.30494 PMID: 28382917

Siddiqui, S. A., Bahmid, N. A., Taha, A., Khalifa, I., Khan, S., Rostamabadi, H., & Jafari, S. M. (2023). Recent advances in food applications of phenolic-loaded micro/nanodelivery systems. *Critical Reviews in Food Science and Nutrition*, *63*(27), 8939–8959. DOI: 10.1080/10408398.2022.2056870 PMID: 35426751

Sieck, C. J., Sheon, A., Ancker, J. S., Castek, J., Callahan, B., & Siefer, A. (2021). Digital inclusion as a social determinant of health. *npj. Digital Medicine*, *4*(1), 52. DOI: 10.1038/s41746-021-00413-8 PMID: 33731887

Sims, N., & Kasprzyk-Hordern, B. (2020). Future perspectives of wastewater-based epidemiology: Monitoring infectious disease spread and resistance to the community level. *Environment International*, *139*, 105689. DOI: 10.1016/j.envint.2020.105689 PMID: 32283358

Singh, A., Bhardwaj, P., Laux, P., Pradeep, P., Busse, M., Luch, A., Hirose, A., Osgood, C., & Stacey, M. (2024). AI and ML-based risk assessment of chemicals: Predicting carcinogenic risk from chemical-induced genomic instability. *Frontiers in Toxicology*, *6*, 1461587. Advance online publication. DOI: 10.3389/ftox.2024.1461587 PMID: 39659701

Singh, A., Pratap, S., & Raj, A. (2024). Occurrence and dissemination of antibiotics and antibiotic resistance in aquatic environment and its ecological implications: A review. *Environmental Science and Pollution Research International*, *31*(35), 47505–47529. DOI: 10.1007/s11356-024-34355-x PMID: 39028459

Singh, K. S., Anand, S., Dholpuria, S., Sharma, J. K., Blankenfeldt, W., & Shouche, Y. S. (2021). Antimicrobial resistance dynamics and the One Health strategy: A review. *Environmental Chemistry Letters*, *19*(4), 2995–3007. DOI: 10.1007/s10311-021-01238-3

Singh, S., Ahmed, A., Almansoori, S., Alameri, S., Adlan, A., Odivilas, G., Chattaway, M., Salem, S., Brudecki, G., & Elamin, W. (2024). A narrative review of wastewater surveillance: Pathogens of concern, applications, detection methods, and challenges. *Frontiers in Public Health*, *12*, 1445961. Advance online publication. DOI: 10.3389/fpubh.2024.1445961 PMID: 39139672

Sinha, R. K.Ritesh Kumar Sinha. (2025). Data engineering as social infrastructure: Building platforms for equitable global development. *World Journal of Advanced Research and Reviews*, *26*(3), 1128–1135. DOI: 10.30574/wjarr.2025.26.3.2172

Skendžić, S., Zovko, M., Živković, I. P., Lešić, V., & Lemić, D. (2021). The Impact of Climate Change on Agricultural Insect Pests. *Insects*, *12*(5), 440. DOI: 10.3390/insects12050440 PMID: 34066138

Smith, G. C., Tatum, J. D., Belk, K. E., Scanga, J. A., Grandin, T., & Sofos, J. N. (2005). Traceability from a US perspective. *Meat Science*, *71*(1), 174–193. DOI: 10.1016/j.meatsci.2005.04.002 PMID: 22064063

Smolinski, M. S., Crawley, A. W., Olsen, J. M., Jayaraman, T., & Libel, M. (2017). Participatory disease surveillance: Engaging communities directly in reporting, monitoring, and responding to health threats. *JMIR Public Health and Surveillance*, *3*(4), e62. DOI: 10.2196/publichealth.7540 PMID: 29021131

Smolinski, M. S., Crawley, A. W., Olsen, J. M., Jayaraman, T., Libel, M., & Cummings, D. A. T. (2015). Flu Near You: Crowdsourced symptom reporting spanning 2 influenza seasons. *American Journal of Public Health*, *105*(10), 2124–2130. DOI: 10.2105/AJPH.2015.302696 PMID: 26270299

Song, D., Yao, J., Jiang, Y., Shi, S., Cui, C., Wang, L., Wang, L., Wu, H., Tian, H., Ye, X., Ou, D., Li, W., Feng, N., Pan, W., Song, M., Xu, J., Xu, D., Wu, L., & Dong, F. (2023). A new xAI framework with feature explainability for tumors decision-making in Ultrasound data: Comparing with Grad-CAM. *Computer Methods and Programs in Biomedicine*, *235*, 107527. DOI: 10.1016/j.cmpb.2023.107527 PMID: 37086704

Sophia, Y., Roxy, M. K., Murtugudde, R., Karipot, A., Sapkota, A., Dasgupta, P., Baliwant, K., Saunik, S., Tiwari, A., Chattopadhyay, R., & Phalkey, R. K. (2025). Dengue dynamics, predictions, and future increase under changing monsoon climate in India. *Scientific Reports*, *15*(1), 1637. Advance online publication. DOI: 10.1038/s41598-025-85437-w PMID: 39837878

Southall, E., Brett, T. S., Tildesley, M. J., & Dyson, L. (2021). Early warning signals of infectious disease transitions: A review. *Journal of the Royal Society, Interface*, *18*(182), 20210555. Advance online publication. DOI: 10.1098/rsif.2021.0555 PMID: 34583561

Spink, J., & Moyer, D. C. (2011). Defining the public health threat of food fraud. *Journal of Food Science*, *76*(9), R157–R163. DOI: 10.1111/j.1750-3841.2011.02417.x PMID: 22416717

Sri Vigna Hema, V., & Manickavasagan, A. (2024). Blockchain implementation for food safety in supply chain: A review. *Comprehensive Reviews in Food Science and Food Safety*, *23*(5), e70002. DOI: 10.1111/1541-4337.70002 PMID: 39217509

Starling, M., Neto, R., Pires, G., Vilela, P., & Amorim, C. (2021). Combat of antimicrobial resistance in municipal wastewater treatment plant effluent via solar advanced oxidation processes: Achievements and perspectives. *The Science of the Total Environment*, *786*, 147448. DOI: 10.1016/j.scitotenv.2021.147448 PMID: 33965817

Storøy, J., Thakur, M., & Olsen, P. (2013). The TraceFood framework - Principles and guidelines for implementing traceability in food value chains. *Journal of Food Engineering*, *115*(1), 41–48. DOI: 10.1016/j.jfoodeng.2012.09.018

Suartana, M., & Putra, R. E. (2022, December). Software-defined networking (SDN) traffic analysis using big data analytic approach. In *International Joint Conference on Science and Engineering 2022 (IJCSE 2022)* (pp. 243–250). Atlantis Press. DOI: 10.2991/978-94-6463-100-5_25

Subedi, B., Poudel, A., & Aryal, S. (2023). The impact of climate change on insect pest biology and ecology: Implications for pest management strategies, crop production, and food security. *Journal of Agriculture and Food Research*, *14*, 100733. DOI: 10.1016/j.jafr.2023.100733

Subramanian, C. R., Yang, D. A., & Khanna, R. (2024). Enhancing health care communication with large language models—The role, challenges, and future directions. *JAMA Network Open*, *7*(3), e240347. Advance online publication. DOI: 10.1001/jamanetworkopen.2024.0347 PMID: 38466311

Sugg, M. M., Ryan, S., Spurlock, T., Duncan, S., Hege, A., Ogwu, M. C., Tyson, J., Shay, E., Renwick, K., Hendren, C., Roy, M., & Runkle, J. D. (2023). Climate change and health in rural mountain environments: Summary of a workshop on knowledge gaps, barriers, and opportunities for action. *GeoJournal*, *88*(5), 5397–5409. Advance online publication. DOI: 10.1007/s10708-023-10916-4 PMID: 40291811

Sun, J., Xu, L., Huang, C., & Ng, E. Y. K. (2025). Climate change and health: The role of artificial intelligence in predictive surgical treatment. *Innovation and Emerging Technologies*, *12*, 2550004. Advance online publication. DOI: 10.1142/S2737599425500045

Sutherst, R. W. (2004). Global Change and Human Vulnerability to Vector-Borne Diseases. *Clinical Microbiology Reviews*, *17*(1), 136–173. DOI: 10.1128/CMR.17.1.136-173.2004 PMID: 14726459

Sutton, R. T., Pincock, D., Baumgart, D. C., Sadowski, D. C., Fedorak, R. N., & Kroeker, K. I. (2020). An overview of clinical decision support systems: Benefits, risks, and strategies for success. *npj. Digital Medicine*, *3*(1), 17. DOI: 10.1038/s41746-020-0221-y PMID: 32047862

Tabassum, A., Ghaznavi, I., Abd-Alrazaq, A., & Qadir, J. (2025). Exploring the application of AI and extended reality technologies in metaverse-driven mental health solutions: Scoping review. *Journal of Medical Internet Research*, *27*, e72400. DOI: 10.2196/72400 PMID: 40829151

Tawfik, M., Abu-Ein, A. A., Noaman, H. M., Abdelhaliem, A. H., & Fathi, I. S. (2025). FedMedSecure: Federated few-shot learning with cross-attention mechanisms and explainable AI for collaborative healthcare cybersecurity. *Scientific Reports*, *15*(1), 40050. DOI: 10.1038/s41598-025-25107-z PMID: 41238815

Tengnah, M. A. J., Sooklall, R., & Nagowah, S. D. (2019). A predictive model for hypertension diagnosis using machine learning techniques. In *Telemedicine technologies* (pp. 139–152). Academic Press. DOI: 10.1016/B978-0-12-816948-3.00009-X

Tesla, B., Demakovsky, L. R., Mordecai, E. A., Ryan, S. J., Bonds, M. H., Ngonghala, C. N., Brindley, M. A., & Murdock, C. C. (2018). Temperature drives Zika virus transmission: evidence from empirical and mathematical models. Proceedings of the Royal Society B: Biological Sciences, 285(1884), 20180795. https://doi.org/ DOI: 10.1098/rspb.2018.0795

Tesson, V., Federighi, M., Cummins, E., de Oliveira Mota, J., Guillou, S., & Boué, G. (2020). A systematic review of beef meat quantitative microbial risk assessment models. *International Journal of Environmental Research and Public Health*, *17*(3), 688. DOI: 10.3390/ijerph17030688 PMID: 31973083

Thakur, S., & Gray, G. C. (2019). The mandate for a global One Health approach to antimicrobial resistance surveillance. *The American Journal of Tropical Medicine and Hygiene*, *100*(2), 227–228. DOI: 10.4269/ajtmh.18-0973 PMID: 30608047

Thapen, N., Simmie, D., Hankin, C., & Gillard, J. (2016). DEFENDER: Detecting and forecasting epidemics using novel data-analytics for enhanced response. *PLoS One*, *11*(5), e0155417. Advance online publication. DOI: 10.1371/journal.pone.0155417 PMID: 27192059

Tian, F. (2016). An agri-food supply chain traceability system for China based on RFID & blockchain technology. In 2016 13th International Conference on Service Systems and Service Management (pp. 1-6). IEEE. DOI: 10.1109/ICSSSM.2016.7538424

Tian, F. (2017). A supply chain traceability system for food safety based on HACCP, blockchain & Internet of things. In *2017 International Conference on Service Systems and Service Management* (pp. 1-6). IEEE. DOI: 10.1109/ICSSSM.2017.7996119

Tiggeloven, T., Pfeiffer, S., Matanó, A., van den Homberg, M., Thalheimer, L., Reichstein, M., & Torresan, S. (2025). The role of artificial intelligence for early warning systems: Status, applicability, guardrails, and ways forward. *iScience*, *28*(11), 113689. DOI: 10.1016/j.isci.2025.113689 PMID: 41280665

Tiwari, A., Adhikari, S., Zhang, S., Solomon, T., Lipponen, A., Islam, M., Thakali, O., Sangkham, S., Shaheen, M., Jiang, G., Haramoto, E., Mazumder, P., Malla, B., Kumar, M., Pitkänen, T., & Sherchan, S. (2023). Tracing COVID-19 Trails in Wastewater: A Systematic Review of SARS-CoV-2 Surveillance with Viral Variants. *Water (Basel)*, *15*(6), 1018. Advance online publication. DOI: 10.3390/w15061018

Tiwari, A., Krolicka, A., Tran, T. T., Räisänen, K., Ásmundsdóttir, Á. M., Wikmark, O. G., Lood, R., & Pitkänen, T. (2024). Antibiotic resistance monitoring in wastewater in the Nordic countries: A systematic review. *Environmental Research*, *246*, 118052. DOI: 10.1016/j.envres.2023.118052 PMID: 38163547

Tiwari, A., Kurittu, P., Al-Mustapha, A., Heljanko, V., Johansson, V., Thakali, O., Mishra, S., Lehto, K., Lipponen, A., Oikarinen, S., Pitkänen, T., & Heikinheimo, A. (2022). Wastewater surveillance of antibiotic-resistant bacterial pathogens: A systematic review. *Frontiers in Microbiology*, *13*, 977106. Advance online publication. DOI: 10.3389/fmicb.2022.977106 PMID: 36590429

Topol, E. J. (2019). High-performance medicine: The convergence of human and artificial intelligence. *Nature Medicine*, *25*(1), 44–56. DOI: 10.1038/s41591-018-0300-7 PMID: 30617339

Torab-Miandoab, A., Samad-Soltani, T., Jodati, A., & Rezaei-Hachesu, P. (2023). Interoperability of heterogeneous health information systems: A systematic literature review. *BMC Medical Informatics and Decision Making*, *23*(1), 18. DOI: 10.1186/s12911-023-02115-5 PMID: 36694161

Trang, S., Trenz, M., Weiger, W. H., Tarafdar, M., & Cheung, C. M. K. (2020). One app to trace them all? Examining app specifications for mass acceptance of contact-tracing apps. *European Journal of Information Systems*, *29*(4), 415–428. DOI: 10.1080/0960085X.2020.1784046

Trienekens, J., & Zuurbier, P. (2008). Quality and safety standards in the food industry, developments, and challenges. *International Journal of Production Economics*, *113*(1), 107–122. DOI: 10.1016/j.ijpe.2007.02.050

Tsang, Y. P., Choy, K. L., Wu, C. H., Ho, G. T. S., & Lam, H. Y. (2019). Blockchain-driven IoT for food traceability with an integrated consensus mechanism. *IEEE Access : Practical Innovations, Open Solutions*, *7*, 129000–129017. DOI: 10.1109/ACCESS.2019.2940227

Tsiouris, V., Mantzios, T., Kiskinis, K., Papadopoulos, G. A., Giannenas, I., Marín, C., Carrasquer, F., Karvelis, P., Filippitzi, M. E., & Fortomaris, P. (2024). Backyard flock sampling and artificial intelligence: A dual strategy for early detection of avian influenza and Newcastle disease. *German Journal of Veterinary Research*, *4*(4), 73–80. DOI: 10.51585/gjvr.2024.4.0111

Tudor Car, L., Kyaw, B. M., Nannan Panday, R. S., van der Kleij, R., Chavannes, N., Majeed, A., & Car, J. (2021). Digital health training programs for medical students: Scoping review. *JMIR Medical Education*, *7*(3), e28275. DOI: 10.2196/28275 PMID: 34287206

Tyagi, S., & Katara, P. (2024). Metatranscriptomics: A Tool for Clinical Metagenomics. *Omics : A Journal of Integrative Biology*, *28*(8), 394–407. DOI: 10.1089/omi.2024.0130 PMID: 39029911

U.S. Food and Drug Administration, Health Canada, & Medicines and Healthcare products Regulatory Agency. (2021). *Good machine learning practice for medical device development: Guiding principles.*

Ullah, F., Guru-Pirasanna-Pandi, G., Murtaza, G., Sarangi, S., Gul, H., Li, X., Chavarín-Gómez, L. E., Ramírez-Romero, R., Guedes, R. N. C., Desneux, N., & Lu, Y. (2025). Evolving strategies in agroecosystem pest control: Transitioning from chemical to green management. *Journal of Pest Science*, *98*(4), 2307–2324. DOI: 10.1007/s10340-025-01939-6

Ullah, N., Javed, A., Ghazanfar, M. A., Alsufyani, A., & Bourouis, S. (2022). A novel DeepMaskNet model for face mask detection and masked facial recognition. *Journal of King Saud University. Computer and Information Sciences*, *34*(10), 9905–9914. DOI: 10.1016/j.jksuci.2021.12.017 PMID: 37521179

Uluseker, C., Kaster, K., Thorsen, K., Basiry, D., Shobana, S., Jain, M., Kumar, G., Kommedal, R., & Pala-Ozkok, I. (2021). A Review on Occurrence and Spread of Antibiotic Resistance in Wastewaters and in Wastewater Treatment Plants: Mechanisms and Perspectives. *Frontiers in Microbiology*, *12*, 717809. Advance online publication. DOI: 10.3389/fmicb.2021.717809 PMID: 34707579

Unertl, K. M., Schaefbauer, C. L., Campbell, T. R., Senteio, C., Siek, K. A., Bakken, S., & Veinot, T. C. (2016). Integrating community-based participatory research and informatics approaches to improve the engagement and health of underserved populations. *Journal of the American Medical Informatics Association : JAMIA*, *23*(1), 60–73. DOI: 10.1093/jamia/ocv094 PMID: 26228766

United Nations Educational, Scientific and Cultural Organization. (2023). *Guidance for generative AI in education and research*. UNESCO.

Unkel, S., Farrington, C. P., Garthwaite, P. H., Robertson, C., & Andrews, N. (2012). Statistical methods for the prospective detection of infectious disease outbreaks: A review. *Journal of the Royal Statistical Society. Series A, (Statistics in Society)*, *175*(1), 49–82. DOI: 10.1111/j.1467-985X.2011.00714.x

Urbano, O., Perles, A., Pedraza, C., Rubio-Arraez, S., Castelló, M. L., Ortolá, M. D., & Mercado, R. (2020). Cost-effective implementation of a temperature traceability system based on smart RFID tags and IoT services. *Sensors (Basel)*, *20*(4), 1163. DOI: 10.3390/s20041163 PMID: 32093218

Uren, K. R., Van Schoor, G., Du Rand, C. P., & Botha, A. (2016). An integrated approach to sensor FDI and signal reconstruction in HTGRs–Part II: Case studies. *Annals of Nuclear Energy*, *87*, 739–749. DOI: 10.1016/j.anucene.2015.06.013

Uysal, G., & Ozturk, M. (2020). Alzheimer's diagnosis using ML. *Journal of Neuroscience Methods*, *33*, 108669. DOI: 10.1016/j.jneumeth.2020.108669 PMID: 32126274

Van Boxstael, S., Habib, I., Jacxsens, L., De Vocht, M., Baert, L., Van de Perre, E., Rajkovic, A., Lopez-Galvez, F., Sampers, I., Spanoghe, P., De Meulenaer, B., & Uyttendaele, M. (2013). Food safety issues in fresh produce: Bacterial pathogens, viruses and pesticide residues indicated as major concerns by stakeholders in the fresh produce chain. *Food Control*, *32*(1), 190–197. DOI: 10.1016/j.foodcont.2012.11.038

Van Der Drift, A., Welling, A., Arntzen, V., Nagelkerke, E., Van Der Beek, R., & De Roda Husman, A. (2025). Wastewater surveillance studies on pathogens and their use in public health decision-making: A scoping review. *The Science of the Total Environment*, *993*, 179982. DOI: 10.1016/j.scitotenv.2025.179982 PMID: 40592216

Van der Weyde, L. K., Tobler, M. W., Gielen, M. C., Cozzi, G., Weise, F. J., Adams, T., Bauer, D., Bennitt, E., Bowles, M., Brassine, A., Broekhuis, F., Chase, M., Collins, K., Finerty, G. E., Golabek, K., Hartley, R., Henley, S., Isden, J., Keeping, D., & Flyman, M. V. (2021). Collaboration for conservation: Assessing countrywide carnivore occupancy dynamics from sparse data. *Diversity & Distributions*, *28*(5), 917–929. DOI: 10.1111/ddi.13386 van Kessel, R., Wong, B. L. H., Clemens, T., & Brand, H. (2022). Digital health literacy as a super determinant of health: More than simply the sum of its parts. *Internet Interventions : The Application of Information Technology in Mental and Behavioural Health, 27*, 100500. DOI: 10.1016/j.invent.2022.100500 PMID: 35242586

van Rijswijk, W., & Frewer, L. J. (2008). Consumer perceptions of food quality and safety and their relation to traceability. *British Food Journal, 110*(10), 1034–1046. DOI: 10.1108/00070700810906642

Vanidhasri, R., & Kathirvel, A. (2025, August). Improving EEG Signal Classification in Brain-Computer Interfaces via Grey Wolf Optimizer-based Deep Learning. In *2025 3rd International Conference on Sustainable Computing and Data Communication Systems (ICSCDS)* (pp. 1312-1318). IEEE.

Vayena, E., Blasimme, A., & Cohen, I. G. (2018). Machine learning in medicine: Addressing ethical challenges. *PLoS Medicine, 15*(11), e1002689. DOI: 10.1371/journal.pmed.1002689 PMID: 30399149

Vayena, E., Haeusermann, T., Adjekum, A., & Blasimme, A. (2018). Digital health: Meeting the ethical and policy challenges. *Swiss Medical Weekly, 148*(304), w14571. DOI: 10.4414/smw.2018.14571 PMID: 29376547

Veernapu, K. K., Patil, B. K., Tharayil, A. S., Sindhura, C., Andy, A., Budhewar, A., Somwanshi, S. K., Kavya, K., Bharadwaj, M., & Raj, P. (2026). Real-time AI in Clinical Decision Support: Bridging Research and Practice. In Andy, A., Tharayil, A., & Budhewar, A. (Eds.), *Revolutionizing Drug Research and Personalized Medicine Through AI and Machine Learning* (pp. 431–452). IGI Global Scientific Publishing., DOI: 10.4018/979-8-3373-6400-1.ch015

Velasco, E., Agheneza, T., Denecke, K., Kirchner, G., & Eckmanns, T. (2014). Social media and internet-based data in global systems for public health surveillance: A systematic review. *The Milbank Quarterly, 92*(1), 7–33. DOI: 10.1111/1468-0009.12038 PMID: 24597553

Venkatraman, S., Parvin, S., Mansoor, W., & Gawanmeh, A. (2023). Big data analytics and internet of things for personalised healthcare: Opportunities and challenges. *International Journal of Electrical and Computer Engineering, 13*(4).

Verbeke, W., & Ward, R. W. (2006). Consumer interest in information cues denoting quality, traceability and origin: An application of ordered probit models to beef labels. *Food Quality and Preference, 17*(6), 453–467. DOI: 10.1016/j.foodqual.2005.05.010

Vianny, M., Ayyaswamy, K., Rithik, G., Naren, K., & Anthony Raj, A. (2025). Private AI: An overview. In *AI-driven breakthroughs in antimicrobial resistance*. IGI Global., DOI: 10.4018/979-8-3693-7550-1

Viboud, C., Sun, K., Gaffey, R., Ajelli, M., Fumanelli, L., Merler, S., Zhang, Q., Chowell, G., Simonsen, L., & Vespignani, A.RAPIDD Ebola Forecasting Challenge Group. (2018). The RAPIDD Ebola forecasting challenge: Synthesis and lessons learnt. *Epidemics*, *22*, 13–21. DOI: 10.1016/j.epidem.2017.08.002 PMID: 28958414

Villanueva-Miranda, I., Xiao, G., & Xie, Y. (2025). Artificial intelligence in early warning systems for infectious disease surveillance: A systematic review. *Frontiers in Public Health*, *13*, 1609615. Advance online publication. DOI: 10.3389/fpubh.2025.1609615 PMID: 40626156

Visave, J. (2025). Transparency in AI for emergency management: Building trust and accountability. *AI and Ethics*, *5*(4), 3967–3980. DOI: 10.1007/s43681-025-00692-x

Vitale, D., Suárez-Varela, M. M., & Picó, Y. (2021). Wastewater-based epidemiology, a tool to bridge biomarkers of exposure, contaminants, and human health. [doi.org in Bing]. *Current Opinion in Environmental Science & Health*, *20*, 100229. DOI: 10.1016/j.coesh.2021.100229

Waller, M. A., & Fawcett, S. E. (2013). Click here for a data scientist: Big data, predictive analytics, and theory development in the era of a maker movement supply chain. *Journal of Business Logistics*, *34*(4), 249–252. DOI: 10.1111/jbl.12024

Wan, Y., Peng, Z., Wu, F., Gao, J., & Li, M. (2024). Understanding public concerned ethic issues and ethical acceptance of AI surveillance technology: Analysis from social media data. In *Electronics, Communications and Networks* (pp. 717–722). IOS Press.

Wang, Y., Yang, A., Li, Z., Wang, P., & Yang, H. (2015, April). Blind drift calibration of sensor networks using signal space projection and Kalman filter. In *2015 IEEE tenth international conference on intelligent sensors, sensor networks and information processing (ISSNIP)* (pp. 1-6). IEEE.

Wang, A., Ramaswamy, V. V., & Russakovsky, O. (2022). Towards intersectionality in machine learning: Including more identities, handling underrepresentation, and performing evaluation. In *Proceedings of the 2022 ACM Conference on Fairness, Accountability, and Transparency* (pp. 336–349). Association for Computing Machinery. https://doi.org/DOI: 10.1145/3531146.3533101

Wang, L., Kwok, S. K., & Ip, W. H. (2010). A radio frequency identification and sensor-based system for the transportation of food. *Journal of Food Engineering*, *101*(1), 120–129. DOI: 10.1016/j.jfoodeng.2010.06.020

Wang, M. (2023). Design and implementation of the food supply chain traceability system based on the blockchain. *Highlights in Science. Engineering and Technology*, *32*, 14–23. DOI: 10.54097/hset.v32i.4931

Wang, X., Xu, Y., & Xi, X. (2025). Spatiotemporal Risk Assessment of H5 Avian Influenza in China: An Interpretable Machine Learning Approach to Uncover Multi-Scale Drivers. *Animals (Basel)*, *15*(16), 2447. DOI: 10.3390/ani15162447 PMID: 40867775

Wang, Y., Chen, C. H., & Zghari-Sales, A. (2021). Designing a blockchain enabled supply chain. *International Journal of Production Research*, *54*(1), 6–14. DOI: 10.1080/00207543.2020.1824086

Waqas, M., & Humphries, U. W. (2024). A critical review of RNN and LSTM variants in hydrological time series predictions. *MethodsX*, *13*, 102946. DOI: 10.1016/j.mex.2024.102946 PMID: 39324077

Waseh, S., & Dicker, A. P. (2019). Telemedicine training in undergraduate medical education: Mixed-methods review. *JMIR Medical Education*, *5*(1), e12515. DOI: 10.2196/12515 PMID: 30958269

Werkneh, A., & Islam, M. (2023). Post-treatment disinfection technologies for sustainable removal of antibiotic residues and antimicrobial resistance bacteria from hospital wastewater. *Heliyon*, *9*(4), e15360. Advance online publication. DOI: 10.1016/j.heliyon.2023.e15360 PMID: 37123966

Wheatley, M. C. (2024). *Ethics of surveillance technologies: Balancing privacy and security in a digital age*. Premier Journal of Data Science., DOI: 10.70389/PJDS.100001

Wiens, J., Saria, S., Sendak, M., Ghassemi, M., Liu, V. X., Doshi-Velez, F., Jung, K., Heller, K., Kale, D., Saeed, M., Ossorio, P. N., Thadaney-Israni, S., & Goldenberg, A. (2019). Do no harm: A roadmap for responsible machine learning for health care. *Nature Medicine*, *25*(9), 1337–1340. DOI: 10.1038/s41591-019-0548-6 PMID: 31427808

Wilkinson, M. D., Dumontier, M., Aalbersberg, I. J., Appleton, G., Axton, M., Baak, A., Blomberg, N., Boiten, J.-W., da Silva Santos, L. B., Bourne, P. E., Bouwman, J., Brookes, A. J., Clark, T., Crosas, M., Dillo, I., Dumon, O., Edmunds, S., Evelo, C. T., Finkers, R., & Mons, B. (2016). The FAIR Guiding Principles for scientific data management and stewardship. *Scientific Data*, *3*(1), 160018. Advance online publication. DOI: 10.1038/sdata.2016.18 PMID: 26978244

Willis, H. H., & Moore, M. (2014). Improving the value of analysis for biosurveillance. *Decision Analysis*, *11*(1), 63–81. DOI: 10.1287/deca.2013.0283

Wirata, G., Gunawan, M. S., Judijanto, L., Sarmento, A. D. A., & Permono, A. R. (2025). The role of e-government innovation in enhancing public service performance and strengthening transparent governance. *International Journal of Accounting and Economics Studies*, *12*(6), 710–717. DOI: 10.14419/zxpzp606

Wittmer, J., Prouse, C., & Arefin, M. R. (2025). Digitalizing sewage: The politics of producing, sharing, and operationalizing data from wastewater-based surveillance. *Environment and Planning C: Politics and Space*, 23996544241313454.

Wójcik, O. P., Brownstein, J. S., Chunara, R., & Johansson, M. A. (2014). Public health for the people: Participatory infectious disease surveillance in the digital age. *Emerging Themes in Epidemiology*, *11*(7), 1–7. DOI: 10.1186/1742-7622-11-7 PMID: 24991229

Woldaregay, A. Z., Årsand, E., Walderhaug, S., Albers, D., Mamykina, L., Botsis, T., & Hartvigsen, G. (2019). Data-driven modeling and prediction of blood glucose dynamics: Machine learning applications in type 1 diabetes. *Artificial Intelligence in Medicine*, *98*, 109–134. DOI: 10.1016/j.artmed.2019.07.007 PMID: 31383477

Womersley, F. C., Sousa, L. L., Humphries, N. E., Abrantes, K., Araujo, G., Bach, S. S., Barnett, A., Berumen, M. L., Lion, S. B., Braun, C. D., Clingham, E., Cochran, J. E. M., de la Parra, R., Diamant, S., Dove, A. D. M., Duarte, C. M., Dudgeon, C. L., Erdmann, M. V., Espinoza, E., & Sims, D. W. (2024). Climate-driven global redistribution of an ocean giant predicts increased threat from shipping. *Nature Climate Change*, *14*(12), 1282–1291. DOI: 10.1038/s41558-024-02129-5 PMID: 39650805

Wong, B. L. H., Khurana, M. P., Smith, R. D., El-Omrani, O., Pold, A., Lotfi, A., O'Leary, C. A., & Saminarsih, D. S. (2021). Harnessing the digital potential of the next generation of health professionals. *Human Resources for Health*, *19*(1), 50. DOI: 10.1186/s12960-021-00591-2 PMID: 33853625

Wood, D. E., Lu, J., & Langmead, B. (2019). Improved metagenomic analysis with Kraken 2. *Genome Biology*, *20*(1), 257. Advance online publication. DOI: 10.1186/s13059-019-1891-0 PMID: 31779668

World Health Organization. (2015). *Global action plan on antimicrobial resistance*. World Health Organization.

World Health Organization. (2019). *WHO guideline: Recommendations on digital interventions for health system strengthening*. World Health Organization.

World Health Organization. (2021a). *Global strategy on digital health 2020–2025*. World Health Organization.

World Health Organization. (2021b). *Ethics and governance of artificial intelligence for health*. World Health Organization.

World Health Organization. (2022). *Global antimicrobial resistance and use surveillance system (GLASS) report: 2022*. World Health Organization.

World Health Organization. (2022). Global genomic surveillance strategy for pathogens with pandemic and epidemic potential, 2022-2032. World Health Organization. https://www.who.int/publications/i/item/9789240046979

World Health Organization. (2023). *Classification of digital interventions, services and applications in health: A shared language to describe the uses of digital technology for health* (2nd ed.). World Health Organization.

World Health Organization. (2024). *Ethics and governance of artificial intelligence for health: Guidance on large multi-modal models*. World Health Organization.

World Health Organization. Disease Outbreaks. 2016. Available online: https://www.emro.who.int/health-topics/disease-outbreaks/index.html (accessed on 12 July 2016).

Wu, X., Liu, X., & Zhou, Y. (2021, October). Review of unsupervised learning techniques. In *Proceedings of 2021 Chinese Intelligent Systems Conference:* Volume II (pp. 576-590). Singapore: Springer Singapore.

Xagoraraki, I., & O'Brien, E. (2020). Wastewater-based epidemiology for early detection of viral outbreaks. In *Women in water quality: Investigations by prominent female engineers* (pp. 75–97). Springer International Publishing.

Xing, Y., Liu, X., & Wang, X. (2026). Integrating UAVs, satellite remote sensing, and machine learning in precision agriculture: Pathways to sustainable food production, resource efficiency, and scalable innovation. *Frontiers in Agronomy*, *7*, 1670380. Advance online publication. DOI: 10.3389/fagro.2025.1670380

Xi, X., Meng, X., Qin, Z., Nie, X., Yin, Y., & Chen, X. (2020). IA-net: Informative attention convolutional neural network for choroidal neovascularization segmentation in OCT images. *Biomedical Optics Express*, *11*(11), 6122–6136. DOI: 10.1364/BOE.400816 PMID: 33282479

Xueting, H. (2023). Blockchain-enabled smart packaging: Enhancing food traceability and consumer confidence in the Chinese food industry. *The Frontiers of Society. Science and Technology*, *5*(12), 1–7.

Xu, L., Chen, L., Gao, Z., Chang, Y., Iakovou, E., & Shi, W. (2018). Binding the physical and cyber worlds: A blockchain approach for cargo supply chain security enhancement. In *2018 IEEE International Symposium on Technologies for Homeland Security* (pp. 1-5). IEEE. DOI: 10.1109/THS.2018.8574184

Yadav, A., Yadav, S., Shahi, N., Yadav, P. K., & Kumar, A. (2023). Uses of artificial intelligence in agricultural pest management. In *Smart Agricultural Technologies for Sustainable Crop Production*. Springer., DOI: 10.1007/978-981-19-XXXX-X_13

Yadav, B., Sharma, G., Sankhla, M. S., Sehgal, S., Tomar, J. S., & Dubey, A. (2026). Integrative Forensic Methods in Food Contaminant Detection and Identification. In Ogwu, M., & Izah, S. (Eds.), *Detection, Identification, and Monitoring of Food Contaminants* (pp. 59–82). IGI Global Scientific Publishing., DOI: 10.4018/979-8-3373-3982-5.ch003

Yadav, S. K., Shrestha, L., Acharya, J., Gompo, T. R., Chapagain, S., & Jha, R. (2023). Integrative digital tools to strengthen data management for antimicrobial resistance surveillance in the "One Health" domain in Nepal. *Tropical Medicine and Infectious Disease*, *8*(6), 291. DOI: 10.3390/tropicalmed8060291 PMID: 37368709

Yang, C. (2024). Innovation in cross-border supply chain inventory management driven by big data. *Advances in Economics. Management and Political Sciences*, *76*(1), 66–73.

Yang, C., Chowdhury, D., Zhang, Z., Cheung, W., Lu, A., Bian, Z., & Zhang, L. (2021). A review of computational tools for generating metagenome-assembled genomes from metagenomic sequencing data. *Computational and Structural Biotechnology Journal*, *19*, 6301–6314. DOI: 10.1016/j.csbj.2021.11.028 PMID: 34900140

Yang, F., & Wang, M. (2020). Big data evaluation review. *Frontiers of Engineering Management*, *7*, 27–46. DOI: 10.1007/s42524-020-0092-6

Yang, H., Xu, J., & Li, S. (2025). Mapping research trends and hotspots in sports ethics: A visual bibliometric analysis using CiteSpace. *SAGE Open*, *15*(4), 21582440251404196. DOI: 10.1177/21582440251404196

Yang, S., Santillana, M., & Kou, S. C. (2015). Accurate estimation of influenza epidemics using Google search data via ARGO. *Proceedings of the National Academy of Sciences of the United States of America*, *112*(47), 14473–14478. DOI: 10.1073/pnas.1515373112 PMID: 26553980

Yang, Y. (2024). Applications and challenges of big data in market analytics. Transactions on Economics. *Business and Management Research*, *9*, 450–458.

Yatsyna, Y., & Kudinov, I. (2023). Directions for the implementation of innovative analytical and statistical technologies as a tool for corruption counteraction. *Laisvalaikio Tyrimai*, *2*(22), 17–29. DOI: 10.33607/elt.v2i22.1440

Ye, Y., Pandey, A., Bawden, C., Sumsuzzman, D., Rajput, R., Shoukat, A., Singer, B., Moghadas, S., & Galvani, A. (2025). Integrating artificial intelligence with mechanistic epidemiological modeling: A scoping review of opportunities and challenges. *Nature Communications*, *16*(1), 581. Advance online publication. DOI: 10.1038/s41467-024-55461-x PMID: 39794317

Yıldırım, Ö., Pławiak, P., Tan, R. S., & Acharya, U. R. (2018). Arrhythmia detection using deep convolutional neural network with long duration ECG signals. *Computers in Biology and Medicine*, *102*, 411–420. DOI: 10.1016/j.compbiomed.2018.09.009 PMID: 30245122

Yin, B., Tan, G., Muhammad, R., Liu, J., & Bi, J. (2025). AI-powered innovations in food safety from farm to fork. *Foods*, *14*(11), 1973. DOI: 10.3390/foods14111973 PMID: 40509500

Yoon, H., Seo, J., Shin, S., Abdelhamid, M., & Pack, S. (2025). Bioinformation and Monitoring Technology for Environmental DNA Analysis: A Review. *Biosensors (Basel)*, *15*(8), 494. Advance online publication. DOI: 10.3390/bios15080494 PMID: 40862956

Yudkowsky, E. (2011). Complex value systems in friendly AI. In International Conference on Artificial General Intelligence (pp. 388–393). Berlin, Heidelberg: Springer Berlin Heidelberg.

Yu, H. (2025). Climate change unveils hidden microbial dangers. *Environmental Science and Ecotechnology*, *24*, 100544. DOI: 10.1016/j.ese.2025.100544 PMID: 40115027

Yu, J., Jia, Y., Yu, Q., Lin, L., Li, C., Chen, B., Zhong, P., Lin, X., Li, H., Sun, Y., Zhong, X., He, Y., Huang, X., Lin, S., & Pan, Y. (2024). Deciphering complex antibiotic resistance patterns in Helicobacter pylori through whole genome sequencing and machine learning. *Frontiers in Cellular and Infection Microbiology*, *13*, 1306368. DOI: 10.3389/fcimb.2023.1306368 PMID: 38379956

Yu, Y., Wheeler, N. E., & Barquist, L. (2025). Biased sampling driven by bacterial population structure confounds machine learning prediction of antimicrobial resistance. *PLoS Biology*, *23*(12), e3003539. DOI: 10.1371/journal.pbio.3003539 PMID: 41401143

Zeghina, A., Leborgne, A., Le Ber, F., & Vacavant, A. (2024). Deep learning on spatiotemporal graphs: A systematic review, methodological landscape, and research opportunities. *Neurocomputing*, *594*, 127861. DOI: 10.1016/j.neucom.2024.127861

Zeng, D., Cao, Z., & Neill, D. B. (2021). Artificial intelligence–enabled public health surveillance—from local detection to global epidemic monitoring and control. In *Artificial Intelligence in Medicine* (pp. 437–453). Elsevier., DOI: 10.1016/B978-0-12-821259-2.00022-3

Zhang, D. (2024). AI integration in supply chain and operations management: Enhancing efficiency and resilience. *Applied and Computational Engineering*, *90*(1), 8–13. DOI: 10.54254/2755-2721/90/2024MELB0060

Zhang, J., & Bhatt, T. (2014). A Guidance Document on the Best Practices in Food Traceability. *Comprehensive Reviews in Food Science and Food Safety*, *13*(5), 1074–1103. DOI: 10.1111/1541-4337.12103

Zhang, S., Li, X., Wu, J., Coin, L., O'Brien, J., Hai, F., & Jiang, G. (2021). Molecular Methods for Pathogenic Bacteria Detection and Recent Advances in Wastewater Analysis. *Water (Basel)*, *13*(24), 3551. Advance online publication. DOI: 10.3390/w13243551

Zhang, T., Jiang, M., Li, H., Gao, Y., Yousuf, S., Yu, K., Yi, X., Wang, J., Yang, L., & Liu, Y. (2025). Computational Tools and Resources for Long-read Metagenomic Sequencing Using Nanopore and PacBio. *Genomics, Proteomics & Bioinformatics*, *23*(4), qzaf075. Advance online publication. DOI: 10.1093/gpbjnl/qzaf075 PMID: 40845163

Zhang, X., Li, Y., Peng, X., Zhao, Z., Han, J., & Xu, J. (2022). Information traceability model for the grain and oil food supply chain based on trusted identification and trusted blockchain. *International Journal of Environmental Research and Public Health*, *19*(11), 6594. DOI: 10.3390/ijerph19116594 PMID: 35682180

Zhang, Y., Wu, X., Ge, H., Jiang, Y., Sun, Z., Ji, X., Jia, Z., & Cui, G. (2023). A blockchain-based traceability model for grain and oil food supply chain. *Foods*, *12*(17), 3235. DOI: 10.3390/foods12173235 PMID: 37685168

Zhao, G., Liu, S., Lopez, C., Lu, H., Elgueta, S., Chen, H., & Boshkoska, B. M. (2019). Blockchain technology in agri-food value chain management: A synthesis of applications, challenges and future research directions. *Computers in Industry*, *109*, 83–99. DOI: 10.1016/j.compind.2019.04.002

Zhao, J., Xu, X., Jiang, H., & Ding, Y. (2020). The effectiveness of virtual reality-based technology on anatomy teaching: A meta-analysis of randomized controlled studies. *BMC Medical Education*, *20*(1), 127. DOI: 10.1186/s12909-020-1994-z PMID: 32334594

Zhong, X. (2025). Adaptive Multi-Scale Deep Learning Framework for Intelligent Environmental Anomaly Detection and Predictive Analytics in Smart Cities. *Journal of Computer Science and Frontier Technologies*, *1*(3), 65. DOI: 10.63313/JCSFT.9024

Zhorzholiani, T. (2025). The ethical and societal implications of AI in economic policy and public administration. In *AI use in social sciences* (pp. 125–156). IGI Global Scientific Publishing. DOI: 10.4018/979-8-3373-2612-2.ch005

Zhou, J. R., You, Z. H., Cheng, L., & Ji, B. Y. (2021). Prediction of lncRNA-disease associations via an embedding learning HOPE in heterogeneous information networks. *Molecular Therapy. Nucleic Acids*, *23*, 277–285. DOI: 10.1016/j.omtn.2020.10.040 PMID: 33425486

Zhou, Z., Jin, J., Yong, B., Huang, W., Yu, L., Yang, P., & Sun, D. (2025). Enhanced NDVI prediction accuracy in complex geographic regions by integrating machine learning and climate data—A case study of Southwest basin. *International Journal of Applied Earth Observation and Geoinformation*, *139*, 104498. DOI: 10.1016/j.jag.2025.104498

Zhu, W., Wang, D., Li, P., Deng, H., & Deng, Z. (2025). Advances in Wastewater-Based Epidemiology for Pandemic Surveillance: Methodological Frameworks and Future Perspectives. *Microorganisms*, *13*(5), 1169. Advance online publication. DOI: 10.3390/microorganisms13051169 PMID: 40431340

Zitnik, M., Nguyen, F., Wang, B., Leskovec, J., Goldenberg, A., & Hoffman, M. M. (2019). Machine learning for integrating data in biology and medicine: Principles, practice, and opportunities. *Information Fusion*, *50*, 71–91. DOI: 10.1016/j.inffus.2018.09.012 PMID: 30467459

About the Contributors

Matthew Chidozie Ogwu (Ph.D.) is a scholar of sustainability science and interdisciplinary STEM education at Appalachian State University, where he teaches and researches at the nexus of sustainable development, environmental systems, and data-driven decision-making. His work spans microbial and environmental sciences, ecological modeling, sustainability transitions, and policy-relevant research in both the Global South and the Global North. He has collaborated on multi-sector projects with universities, government agencies, and community partners, supervising students and early-career researchers and advancing applied methods for environmental health, food systems, and waste management. Dr. Ogwu's publications include peer-reviewed articles, technical reports, and several edited volumes and book projects, both ongoing and completed, across environmental management, predictive microbiology, and rural sustainability. His teaching portfolio covers sustainability science, ecosystem ecology, and resource management. He has coordinated cross-continental collaborations and secured support for fieldwork, sequencing, and dashboard/tool development, and he actively mentors students and practitioners on open science, reproducibility, and community-engaged research.

***

Esther Alum, a Biochemist and Senior Researcher at Kampala International University Uganda, holds a Ph.D. in Toxicological Biochemistry. She is a member of several professional societies, serves as an editorial member for journals, and is part of research teams that have successfully obtained and implemented research funds. Dr. Esther, a renowned writer with over 170 papers in prestigious journals, is a role model for contemporary researchers, known for her profound, meticulous, and clear research potentially accelerating the global search for alternative therapies with minimal toxicities.

Nimish Anand is currently pursuing a Bachelor of Technology in Information Technology at Velalar College of Engineering and Technology, Erode.

Adinarayana Andy is a licensed pharmacist and research scientist with over sixteen years of experience bridging clinical pharmacy practice, pharmaceutical innovation, and operations management. A multidisciplinary expert, Dr. Andy holds a unique combination of credentials: a PhD in Biotechnology, an MBA in Project Management, and active licensure as a pharmacist. Served as Pharmacist-in-Charge at Weatherwax Family Pharmacy in Michigan, Dr. Andy directs clinical operations with a focus on medication therapy management (MTM), immunization initiatives, and public health outreach. His leadership is grounded in extensive industry experience with global pharmaceutical organizations, including Bristol-Myers Squibb-Biocon, Cadila Pharmaceuticals, and Dr. Reddy's Laboratories. In these roles, he contributed significantly to drug development, regulatory affairs, and quality assurance frameworks. Andy earned his PhD in Biotechnology from Acharya Nagarjuna University and an MBA from the University of the Cumberlands, USA. He also holds a Master of Pharmacy in Pharmaceutical Analysis and Quality Assurance from Biju Patnaik University of Technology and a Bachelors of Pharmacy from Andhra University. An active contributor to the scientific community, Dr. Andy serves as a peer reviewer for international journals, championing evidence-based strategies to improve patient outcomes and healthcare delivery.

Eiman Ashiq is a Pharm-D student in the Department of Pharmacy, Faculty of Health and Pharmaceutical Sciences, University of Agriculture, Faisalabad, Pakistan.

Nimra Aslam is a distinguished Pharm.D. scholar in the Department of Pharmacy, Faculty of Health and Pharmaceutical Sciences, University of Agriculture, Faisalabad, Pakistan. She has authored a pioneering book chapter on artificial intelligence and machine learning applications in pharmaceutical research, focusing on computational drug design, predictive modeling, and data-driven clinical decision support. Her scholarly endeavors aim to advance pharmacy education and interdisciplinary innovation in healthcare through rigorous research and translational contributions.

Garima Awasthi earned her Ph.D. in Biological Sciences from the CSIR-National Botanical Research Institute (CSIR-NBRI), one of India's premier research institutions focused on plant biology and environmental science. Following her doctoral studies, she was awarded a SERB-National Postdoctoral Fellowship at the University of Lucknow, enabling advanced research experience early in her career. Her research expertise lies primarily in plant and environmental biology, with an emphasis on: Plant stress physiology and responses to toxic elements, especially heavy metals like arsenic. Biochemical and molecular mechanisms of plant tolerance to environmental stressors. Phytoremediation and nanotechnology-based environmental solutions. Dr. Awasthi's work spans molecular, biochemical, and proteomic investigations, contributing to a deeper understanding of how plants cope with environmental toxicity and how these mechanisms can be leveraged for sustainable environmental management. She has published more than 60 peer-reviewed scientific articles in internationally indexed journals, reflecting a strong research profile and influence in environmental and biological sciences. Her Google Scholar profile shows significant citations, indicating the impact of her work within the scientific community. She has held faculty positions in other academic institutions and actively participates in national and international scientific forums. Her research excellence has been recognized through awards such as Best Paper Award (CSIR) and Best Presentation Awards at national conferences. While Dr. Awasthi's core expertise is in plant and environmental sciences, she has also co-authored interdisciplinary research that intersects with environmental toxicology and forensic contexts (such as studies involving heavy metal contamination and biosorption mechanisms). These contributions illustrate a broader engagement with applied science fields that inform environmental and forensic investigations.

Jeeviga B is currently pursuing a Bachelor of Technology in Information Technology at Velalar College of Engineering and Technology, Erode.

Anmol Budhewar serves as an Assistant Professor in Computer Engineering, with research interests spanning artificial intelligence, machine learning, blockchain-enabled systems, and secure computing architectures. In the previous academic year, he published one Scopus-indexed research article and one IEEE international conference paper, demonstrating his commitment to high-quality scholarly dissemination. He contributes extensively to the academic peer-review ecosystem, serving as a reviewer for multiple international journals and edited books, and earning several Certificates of Excellence for review quality. Notably, he reviewed four research articles for SN Computer Science (Springer Nature) on topics including blockchain security, medical image analysis, fraud detection, and authentication systems. His academic work emphasizes integrating intelligent computational models with real-world applications in healthcare, cybersecurity, and data-driven decision systems. Through research, reviewing, and teaching, he strengthens methodological rigor and interdisciplinary innovation in computer engineering.

Fatima Umer Draz is a Pharm.D. student in the Department of Pharmacy, Faculty of Health and Pharmaceutical Sciences, University of Agriculture Faisalabad, Pakistan.

Baby K is an Assistant Professor in the Department of Information Technology at Velalar College of Engineering and Technology, Erode.

Ayibapreye Kari holds an MSc in Environmental Microbiology with distinction and a BSc in Microbiology. He has research experience in microbial systems and their applications in environmental sustainability, focusing on microbial interactions, energy microbiology, and innovative approaches to carbon conversion

Rachan Karmakar is an environmental scientist and academician currently working as an Assistant Professor in the Department of Environmental Science at Graphic Era (Deemed to be University), Dehradun, India. He holds a Ph.D. in Environmental Science and Engineering from Thapar Institute of Engineering and Technology, Patiala, and completed his M.Sc. and B.Sc. in Environmental Science from the University of Calcutta. His academic training has given him a strong base in environmental chemistry, bioenergy systems, and applied sustainability research. Dr. Karmakar has more than eight years of combined teaching and research experience. His main research interests include bioenergy and biofuel production, characterization of alternative fuels, emission and combustion analysis, bioremediation, waste valorization, and circular economy based approaches for sustainable development. He has worked extensively on experimental design, process optimization, and advanced analytical tools to address real environmental problems linked with energy and waste. He has published several research papers in reputed peer-reviewed international journals and conference proceedings, and has also contributed to patents related to renewable energy and environmental technologies. Along with research, he is actively involved in academic planning, student mentoring, field based learning, and organization of national and international conferences, workshops, and outreach programs. Dr. Karmakar is a member of professional bodies such as the Royal Society of Chemistry and the International Society for Development and Sustainability. His work reflects a strong commitment towards practical, science-based solutions for environmental protection and sustainable energy transitions.

Atin Kumar is an Assistant Professor and Soil Scientist at the School of Agriculture, Uttaranchal University, Dehradun, with expertise in soil fertility, soil remediation, water quality, and heavy metal contamination. He is the author and editor of 3 academic books, a patent holder, and an active editor and reviewer for several recognized international journals. With over 50 publications and multiple high-quality book chapters, he has been honored with the Young Scientist Awards. His research integrates innovative approaches such as nanotechnology to improve soil–water–environment interactions, with a focus on sustainable agriculture and environmental resilience.

Saman Mahmood is Pharm D student at Department of Pharmacy, Faculty of Health and Pharmaceutical Sciences, University of Agriculture, Faisalabad, Pakistan.

Kazi Mizan is an undergraduate student pursuing a Bachelor of Science in Forensic Science at K.R. Mangalam University, Gurugram, India. His academic interests include digital forensics, cybercrime investigation, and the application of artificial intelligence in forensic analysis, along with emerging intersections between biological and digital systems. He has gained practical exposure to laboratory procedures and investigative techniques, focusing on analytical problem-solving and evidence-based approaches. He is keen to explore advancements in interdisciplinary forensic research and contribute to innovation in the field.

Pratibha Naithani is working as Professor and Head of Environmental Science in Graphic Era Deemed to be University, Dehradun. She has received her Masters degree from Indian Institute of Technology, Roorkee and awarded PhD in Environmental Engineering from University of Illinois at Chicago, Chicago, USA. Dr. Pratibha Naithani has teaching experience of more than 14 years. She has also published research papers in famous national and international journals. Her research interest includes heavy metals in industrial waste water, fly ash, soil and its remediation and waste water treatment.

Adeyemi Olatunbosun is a plant scientist and academic with expertise in plant physiology, crop production, and soil–plant interactions under abiotic stress conditions. His research focuses on improving the growth, nutrient uptake, and yield performance of lowland rice and other staple crops under heavy metal toxicity, nutrient deficiency, and environmental stress, with particular emphasis on the use of arbuscular mycorrhizal fungi, plant growth regulators, and sustainable production techniques. He has conducted field and pot experiments across different agro-ecological zones in Nigeria and has contributed to studies on phytoremediation, nutrient cycling, and microbial–soil dynamics. Dr. Adeyemi is actively involved in teaching, research supervision, and academic publishing, with a strong interest in translating scientific findings into practical solutions for sustainable agriculture and food security.

Dhanalakshmi Perumal is a Pharm.D graduate currently pursuing her Ph.D. She has over three years of experience in teaching and research. Her academic interests include clinical pharmacy, pharmacovigilance, and community pharmacy practice.

She is actively involved in research activities and is committed to contributing to pharmaceutical education and patient care through evidence-based practice.

Sabyasachi Pramanik is a professional IEEE member. He obtained a PhD in Computer Science and Engineering from Sri Satya Sai University of Technology and Medical Sciences, Bhopal, India. Presently, he is an Associate Professor, Department of Computer Science and Engineering, Haldia Institute of Technology, India. He has many publications in various reputed international conferences, journals, and book chapters (Indexed by SCIE, Scopus, ESCI, etc). He is doing research in the fields of Artificial Intelligence, Data Privacy, Cybersecurity, Network Security, and Machine Learning. He also serves on the editorial boards of several international journals. He is a reviewer of journal articles from IEEE, Springer, Elsevier, Inderscience, IET and IGI Global. He has reviewed many conference papers, has been a keynote speaker, session chair, and technical program committee member at many international conferences. He has authored a book on Wireless Sensor Network. He has edited 8 books from IGI Global, CRC Press, Springer and Wiley Publications.

Bindhuja R is currently pursuing a Bachelor of Technology in Information Technology at Velalar College of Engineering and Technology, Erode.

Morufu Raimi (MNES, REHO, LEHO, FAIWMES) is a lecturer at Federal University Otuoke, Bayelsa State, Nigeria, specializing in Environmental Epidemiology, Pollution Control Management, and Toxicology. He is actively engaged in academia, research, and environmental management, holding multiple leadership roles. Dr. Raimi serves as the Deputy Director of the Niger-Delta Institute for Emerging and Re-Emerging Infectious Diseases (NDIERID) and Program Manager at the Centre for Niger-Delta Studies and Sustainability (CNDSS), Federal University Otuoke. He is also a consultant to the National Agency for Food and Drug Administration and Control (NAFDAC) on Mpox (Pharmacovigilance) in Bayelsa State. Internationally, Dr. Raimi is an Assessor for the 2025/26 eLife Community Ambassadors Programme, a Guest Editor for iScience (Cell Press Journal), and provides scientific and technical support to the European Food Safety Authority (EFSA) – EOI/EFSA/2022/01. He serves as the Country Coordinator for the Global Research Institute and Training Centre and the South-South Zonal Team Lead for the Civil Alliance for Climate Protection and Education (CACPE). He is a Fellow and Bayelsa State.

Sayali Somwanshi is a Quality Analyst who has a professional experience of more than three years working in the field of software testing in Telecom and Retail spheres. She is a specialist in Manual Testing, Functional and Non-Functional Testing, API Testing and in Automation Testing with Selenium WebDriver and

Java. Her knowledge of SDLC, STLC, defect life cycle management, regression testing, and end to end validation of web applications is high. Sayali is also well-acquainted with practical experience of MySQL, Adobe campaign, Cosmos DB and Azure storage platforms. She received SRTMUN University, Nanded Bachelor of Engineering (Computer) first class. Her work is based on a high level of quality assurance, constant learning, and provision of best software solutions to organizational objectives.

Amareswrapu V Surendra is Assistant Professor in the Department of Pharmaceutics at K L College of Pharmacy, KLEF, Guntur, with nearly one decade of academic, research, and administrative experience. He earned his Ph.D. in Pharmaceutics from Acharya Nagarjuna University for his work on Novel Local Drug Delivery systems for the treatment of Periodontal diseases and holds an M. Pharm in Pharmaceutics. His research interests include Novel Local drug delivery systems, gastro-retentive formulations, and nanotechnology-based approaches. He has published over 15 research articles, presented at more than 7 national and international conferences. He is an active member of professional bodies such as IPGA and IARA. Dr. Amareswrapu V Surendra has delivered several guest lectures and expert talks on Quality by Design (QbD), nanotechnology, and targeted drug delivery, and has actively participated in FDPs, STTPs, and international webinars. He brings a well-rounded perspective to pharmaceutical education through his academic leadership and practical insights into formulation science.

Aman Suresh Tharayil is an Assistant Professor in Pharmacy Practice, currently serving at the Faculty of Pharmacy, Dr. M.G.R. Educational and Research Institute, Chennai. With a Pharm.D and ongoing PhD in Pharmacy Practice from SIMATS, he brings academic depth and clinical insight to his teaching and research. His teaching repertoire includes subjects such as pharmacotherapeutics, biostatistics, and pharmacoepidemiology for undergraduate and doctoral pharmacy programs. Dr. Aman is an active contributor to scholarly literature with multiple publications in Scopus and Web of Science-indexed journals, including research on diabetes, nephrology, and pharmacovigilance. His work has earned recognition such as subject-wise ranks at RGUHS and a prestigious ASCPT LMIC Travel Grant. Beyond academia, Dr. Aman has reviewed over 200 manuscripts and serves on editorial boards of reputed journals like Acta Scientific Pharmaceutical Sciences, PLOS ONE, and Cureus. He is also a frequent workshop resource person and guest lecturer on clinical pharmacy and regulatory topics. With strong skills in academic mentorship, interdisciplinary collaboration, and public health education, he actively bridges classroom instruction with real-world clinical practice.

Indhiresh V is currently pursuing a Bachelor of Technology in Information Technology at Velalar College of Engineering and Technology, Erode.

Fizza Waheed is a Pharm D student in the Department of Pharmacy, Faculty of Health and Pharmaceutical Sciences, University of Agriculture, Faisalabad, Pakistan.

Bekri Xhemali is a plant pathologist at the Kosovo Institute of Agriculture and an external teaching associate at the University Haxhi Zeka. He has more than 13 years of professional experience, primarily focused on diagnostics, epidemiology, and the management of plant diseases. He obtained his PhD in Food and Agriculture, Technology and Biotechnology from the University of Modena, and holds an MSc in Integrated Pest Management from the Mediterranean Agronomic Institute of Bari, Italy.

Baljeet Yadav is a distinguished academic and researcher in the field of forensic science, with recognized contributions to forensic education, research, and professional training in India. He is known for his work in areas such as forensic investigation techniques, forensic chemistry, crime scene management, and emerging applications of science and technology in forensics. Dr. Yadav has actively contributed to teaching, curriculum development, academic leadership, and capacity building in forensic science programs. He has served as a resource person, session chair, and expert speaker at national conferences, workshops, and training programs, reflecting his engagement with both academic and applied forensic communities. His professional involvement supports the advancement of scientific rigor, ethical practices, and innovation in forensic investigations and forensic education.

Index

A

B

C

D

E

F

G

I

M

O

P

R

S

W

www.ingramcontent.com/pod-product-compliance
Lightning Source LLC
LaVergne TN
LVHW082051011026
849162LV00019B/736